THE NEW BOOK OF
POPULAR SCIENCE

THE NEW BOOK OF POPULAR SCIENCE

VOLUME 1

Astronomy & Space Science
Mathematics

COPYRIGHT © 1992 BY

Copyright © 1990, 1988, 1987, 1984, 1982, 1981, 1980, 1979, 1978 by GROLIER INCORPORATED

Copyright © Philippines 1979 by GROLIER INTERNATIONAL, INC.

Copyright © Republic of China 1978 by GROLIER INTERNATIONAL, INC.

Library of Congress Cataloging in Publication Data

Main entry under title:
The New book of popular science.
      p.   cm.
   Includes bibliographical references.
   Summary: Discusses the major sciences and their applications in today's world.
      Contents: v. 1.   Astronomy & space science, mathematics.
     ISBN 0-7172-1218-1 (6 v. set)
    1. Science—Popular works. 2. Technology—Popular works. 3. Natural history—Popular works. I. Grolier Incorporated.
Q162.N437 1992   91-38365
500—dc20   CIP

Cover photos: (front, clockwise from upper left) © 1968 Metro-Goldwyn-Mayer, Inc.; Eileen Tanson/Photo Researchers, Inc.; NASA; Nancy Sefton/Photo Researchers, Inc.; Solarfilma, Iceland; Barasch/Photo Researchers, Inc.; O.R.T.F.; Hewlett Packard; (front, center) Peg Esty/Photo Researchers, Inc.; (back, clockwise from upper left) Bell Laboratories; NASA; Irving Dolin; Art Resource; Harry Rogers, Audubon/Photo Researchers, Inc.; Mayo Clinic; Richard Weymouth Brooks/Photo Researchers, Inc.; T. Cook, RCA Laboratories; Bell Laboratories; John Lewis Stage/Photo Researchers, Inc.

All rights reserved. No part of this book may be reproduced or transmitted in any form by any means electronic, mechanical, or otherwise, whether now or hereafter devised, including photocopying, recording, or by any information storage and retrieval system without express written prior permission from the publisher.

THE NEW BOOK OF POPULAR SCIENCE

EDITORIAL

Editorial Director Lawrence T. Lorimer
Executive Editor Joseph M. Castagno

Editor Lisa Holland   **Managing Editor** Doris E. Lechner
Associate Editor Jessica Snyder   **Art Director** Eric Akerman

Editors

Robert S. Anderson   Edwin Rosenblum
Judith C. Cuddihy   Ronald B. Roth
Stephen P. Elliott   Barbara Tchabovsky
Sylvia Helm   Jenny Tesar
Peter Margolin   Linda Triegel
Steven Moll   Jo Ann White

Copy Editors David M. Buskus, Meghan O'Reilly LeBlanc

Indexer Pauline M. Sholtys
Production Editor Sheila Rourk   **Production Assistant** Wendy J. McDougall
Art Assistant Elizabeth Farrington
Editorial Assistant Karen A. Fairchild   **Proofreader** Stephan Romanoff
Editorial Financial Manager S. Jean Gianazza   **Librarian** Charles Chang

Picture Research

Head, Photo Research Ann Eriksen   **Manager, Picture Library** Jane H. Carruth
Photo Researchers Paula K. Wehde, Lisa J. Grize
Photo Research Assistant Linda R. Kubinski

MANUFACTURING

Director of Manufacturing Joseph J. Corlett
Senior Production Manager Christine L. Matta
Assistant Production Manager Pamela J. Murphy
Production Assistant A. Rhianon Michaud

CONTRIBUTORS

ELISABETH ACHELIS, *Former President, World Calendar Association, Inc.*

JOHN G. ALBRIGHT, Ph.D., *Former Chairman, Department of Physics, University of Rhode Island*

FIKRI ALICAN, M.D., *Assistant Professor of Surgery, The University of Mississippi Medical Center, Jackson, Mississippi*

LAWRENCE K. ALTMAN, M.D., *Medical Reporter,* New York Times

KENNETH ANDERSON, *Science writer*

ROBERT S. ANDERSON, *Senior Science Editor,* The Encyclopedia Americana

STANLEY W. ANGRIST, Ph.D., *Department of Mechanical Engineering, Carnegie-Mellon University*

ELIAS M. AWAD, M.B.A., M.A., *Associate Professor, Graduate School of Business, DePaul University*

S. HOWARD BARTLEY, Ph.D., *Professor of Psychology, Michigan State University*

J. FREMONT BATEMAN, M.D., *Former Superintendent, Columbus State Hospital, Columbus, Ohio*

W. W. BAUER, M.D., *Former consultant in health education; former Director, Department of Health Education, American Medical Association*

J. KELLY BEATTY, *Staff Member,* Sky and Telescope

HANS J. BEHM, *Science writer*

CHARLES BEICHMAN, Ph.D., *Astronomer, Department of Astrophysics, California Institute of Technology*

LYMAN BENSON, *Professor Emeritus of Botany, Pomona College; author,* Cacti of the United States and Canada

M. H. BERRY, M.S., *Chairman, Division of Science and Mathematics, and Professor of Botany, West Liberty State College*

CHRISTOPHER J. BISE, Ph.D., *Assistant Professor of Mining Engineering, College of Earth & Mineral Sciences, Pennsylvania State University*

LOUIS FAUGERES BISHOP, M.D., *Cardiologist; visiting physician, Bellevue Hospital, New York City*

WILLIAM BLAIR, Ph.D., *Astronomer and research scientist, Department of Physics and Astronomy, Johns Hopkins University, Baltimore*

NICHOLAS T. BOBROVNIKOV, Ph.D., *Professor Emeritus, Astronomy, Ohio State University; former Director, Perkins Observatory*

BART J. BOK, Ph.D., *Chairman, Astronomy Department, University of Arizona*

YVONNE BONNAFOUS, *Science writer*

BARBARA BRANCA, *Science writer*

FRANKLYN M. BRANLEY, Ph.D., *Astronomer, American Museum of Natural History—Hayden Planetarium*

FRANK A. BROWN, Jr., Ph.D., *Morrison Professor of Biology, Northwestern University*

JEFF BRUNE, M.S., *Science writer and former editor,* Discover *magazine*

KEITH E. BULLEN, *Science writer*

BRYAN H. BUNCH, *President, Scientific Publishing, Inc.*

ALAN C. BURTON, Ph.D., *Professor of Biophysics, The University of Western Ontario Faculty of Medicine*

MARY CASTELLION, *Science writer*

H. A. CATES, M.D., *Former Professor of Anatomy, University of Toronto*

MORRIS CHAFETZ, M.D., *Principal Research Scientist, The Johns Hopkins University; former Director, National Institute of Alcohol Abuse and Alcoholism*

JANE CHESNUTT, *Editor, Environmental Information Center*

CLYDE M. CHRISTENSEN, Ph.D., *Professor of Plant Pathology, University of Minnesota*

DANIEL M. COHEN, Ph.D., *Laboratory Director, Ichthyological Laboratory, Bureau of Commercial Fisheries, U.S. Fish and Wildlife Service*

JAMES S. COLES, Ph.D., *President, Bowdoin College*

DEBORAH M. COLLIER, *Technical Information Specialist, Centers for Disease Control*

IAN McT. COWAN, Ph.D., *Dean, Faculty of Graduate Studies; Prof. of Zoology, University of British Columbia*

JOSEPH G. COWLEY, *Research Institute of America*

WILLIAM J. CROMIE, *Executive Director, Council for the Advancement of Science Writing*

F. JOE CROSSWHITE, Ph.D., *Associate Professor of Mathematics Education, Ohio State University*

G. EDWARD DAMON, *Former Consumer Writer, Bureau of Foods, U.S. Food and Drug Administration*

BRUCE DAVIES, *Account Executive, Michael Bobrin Public Relations/Advertising*

GÉRARD De VAUCOULEURS, Ph.D., *Professor, Department of Astronomy, University of Texas*

THEODOSIUS DOBZHANSKY, D.Sc., *Former Professor, Rockefeller University*

DAVID DOOLING, *research associate, Essex Corporation; former science editor,* The Huntsville (Alabama) Times

ROY DUBISCH, Ph.D., *Professor, Department of Mathematics, University of Washington*

BETH DWORETZKY, M.S., *Research Analyst, Woods Hole, Massachusetts*

MAC V. EDDS, Jr., Ph.D., *Chairman, Division of Medical Science, Brown University*

HOWARD F. FEHR, Ph.D., *Professor of Mathematics, Head of the Department of the Teaching of Mathematics, Teachers College, Columbia University*

GERALD FEINBERG, Ph.D., *Professor, Department of Physics, Columbia University*

IRWIN K. FEINSTEIN, Ph.D., *Instructor, Mathematics, University of Illinois at Chicago Circle*

DAVID FISHMAN, M.S., *Science writer and former editor,* Discover *magazine*

F. L. FITZPATRICK, Ph.D., *Former Chairman, Department of Natural Sciences, Teachers College, Columbia University*

JOHN A. FLEMING, B.S., D.Sc., *Former Director of the Department of Terrestrial Magnetism, Carnegie Institute*

WILLIAM S. FOSTER, B.S. in Ch.E., *Editor,* The American City

EUGENE E. FOWLER, M.S., *Director, Division of Isotopes Development, U.S. Atomic Energy Commission*

PAUL J. FOX, Ph.D., *Lamont-Doherty Geological Observatory, Columbia University*

SMITH FREEMAN, M.D., Ph.D., *Chairman, Department of Biochemistry, Northwestern University*

M. A. FREIBERG, Ph.D., *Argentina Museum of Natural History*

HARRY J. FULLER, Ph.D., *Former Professor of Botany, University of Illinois*

ELDON J. GARDNER, *Science writer*

A. B. GARRETT, Ph.D., *Vice President for Research, Ohio State University*

CHALMERS L. GEMMILL, M.D., *Chairman, Department of Pharmacology, University of Virginia School of Medicine*

WILLIS J. GERTSCH, Ph.D., *Curator of the Department of Entomology, American Museum of Natural History*

BENTLEY GLASS, Ph.D., *Vice President for Academic Affairs and Distinguished Professor of Biological Sciences, State University of New York at Stony Brook*

PHILLIP GOLDSTEIN, M.S., *Chairman, Department of Biology, Abraham Lincoln High School, Brooklyn, New York*

RICHARD P. GOLDTHWAIT, Ph.D., *Chairman, Department of Geology, Ohio State University*

MARVIN R. GORE, *Dean, School of Business Administration, Mount San Antonio College*

DONALD C. GREGG, Ph.D., *Pomeroy Professor of Chemistry, University of Vermont*

DIANA HADLEY, *Science writer*

JAMES D. HARDY, M.D., *Professor and Chairman, Department of Surgery, The University of Mississippi Medical Center, Jackson, Mississippi*

E. NEWTON HARVEY, Ph.D., *Former Professor Emeritus of Biology, Princeton University*

JAMES HASSETT, Ph.D., *Department of Psychology, Boston University*

MARK A. HAWES, *Engineer, Applied Technology Associates, Inc.*

JOEL W. HEDGPETH, Ph.D., *Professor of Zoology and Director of Pacific Marine Station, College of the Pacific*

LOUIS M. HEIL, Ph.D., *Director, Office of Testing and Research, Brooklyn College*

ALEXANDER HELLEMANS, *Commissioning Editor for Physical Sciences & Chemistry, Academic Press, London*

ANTONY HEWISH, Ph.D., F.R.S., *Nobel Prize winner in Physics (1974); Professor of Radio Astronomy, University of Cambridge*

NORMAN E. A. HINDS, Ph.D., *Former Professor of Geology, University of California*

LILLIAN HODDESON, Ph.D., *Department of Physics, Rutgers University*

GEORGE HOEFER, B.S. in E.E., *former Associate Editor,* Electronic Equipment Engineering *magazine*

ROBERT W. HOWE, Ed.D., *Assistant Professor of Science Education, Ohio State University*

ALEXANDER JOSEPH, D.Ed., *Professor, Department of Physics, Bronx Community College, New York*

CORLISS G. KARASOV, *Science writer*

MORLEY KARE, Ph.D., *Professor of Physiology, University of Pennsylvania*

JOHN KENTON, *Science writer*

HAROLD P. KNAUSS, Ph.D., *Former Head, Department of Physics, University of Connecticut*

C. ARTHUR KNIGHT, Ph.D., *Professor, Molecular Biology; Research Biochemist, University of California, Berkeley*

SERGE A. KORFF, Ph.D., *Professor of Physics, New York University*

LILLIAN KOZLOSKI, *Archivist, Smithsonian National Air & Space Museum*

STEPHEN N. KREITZMAN, Ph.D., *Assistant Professor, Dental Research, Emory University*

LEONARD C. LABOWITZ, Ph.D., *Physical chemist and science writer*

MORT LA BREQUE, *Editor,* The Sciences

DONALD A. LAIRD, Ph.D., *Author; former Director, Colgate University Psychological Laboratory*

LYNN LAMOREAUX, Ph.D., *MRC Radiobiology Unit, Great Britain*

T. K. LANDAUER, Ph.D., *Bell Telephone Laboratories*

HENRY LANSFORD, M.A., *National Center for Atmospheric Research*

PETER A. LARKIN, D. Phil., *Institute of Animal Research Ecology, University of British Columbia*

THOMAS GORDON LAWRENCE, A.M., *Former Chairman, Biology Department, Erasmus Hall High School, Brooklyn, New York*

BENEDICT A. LEERBURGER, *Consulting editor, contributor, science writer*

RICHARD G. LEFKON, M.S., *Assistant Vice President, Citibank*

AARON A. LERNER, M.D., *School of Medicine, Yale University*

W. THOMAS LIPPINCOTT, Ph.D., *Professor and Academic Vice Chairman, Department of Chemistry, Ohio State University*

ARMIN K. LOBECK, Ph.D., *Former Professor of Geology, Columbia University*

MARIAN LOCKWOOD, *Science writer*

SUZANNE LOEBL, *Science editor, The Arthritis Foundation*

DONALD B. LOURIA, M.D., *Professor and Chairman, Department of Medicine and Community Health, New Jersey Medical School*

CHARLES C. MACKLIN, M.D., *Former Professor of Histology and Embryology, University of Western Ontario*

MAURA MACKOWSKI, *Science writer*

DENNIS MAMMANA, Ph.D., *Astronomer, Reuben H. Fleet Space Sciences Center, San Diego*

OSCAR E. MEINZER, Ph.D., *Former Chief, Division of Ground Water, United States Geological Survey*

DENNIS MEREDITH, *Public information officer, Duke University, Durham, North Carolina (formerly astronomy and physical science writer for Cornell University)*

HELEN HYNSON MERRICK, *Former Senior Editor, Grolier Incorporated*

LON W. MORREY, D.D.S., *Editor Emeritus, Journal of the American Dental Association*

WALTER MUNK, Ph.D., *Director, La Jolla Unit, and Associate Director, Institute of Geophysics, UCLA*

CHARLES MERRICK NEVIN, *Science writer*

IAN NISBET, *Environmental consultant*

LINDSAY S. OLIVE, Ph.D., *Professor of Botany, Columbia University*

JAMES A. OLIVER, Ph.D., *Director, New York Zoological Society*

DAMON R. OLSZOWY, M.S., Ph.D., *Biology Department, State University of New York at Farmingdale*

JOHN C. PALLISTER, *Research Associate, Department of Entomology, American Museum of Natural History*

CECILIA PAYNE-GAPOSCHKIN, D.Sc., *Smithsonian Astrophysical Observatory, Harvard University*

RICHARD F. POST, Ph.D., *Director, Magnetic Mirror Program, Lawrence Radiation Laboratory, University of California*

FREDERICK H. POUGH, Ph.D., *Consulting mineralogist; former Curator of Minerals and Gems, American Museum of Natural History*

TERENCE T. QUIRKE, Ph.D., *Former Professor of Geology, University of Illinois*

HOWARD G. RAPAPORT, M.D., *Associate Professor of Clinical Pediatrics (Allergy), Albert Einstein College of Medicine, Yeshiva University*

KARIN RHINES, *Science writer*

ERIC RODGERS, Ph.D., *Dean of the Graduate School, University of Alabama*

ALBERT J. RUF, *Science writer*

BETTE RUNCK, *Science writer*

J. H. RUSH, Ph.D., *Physicist, National Center for Atmospheric Research, Boulder, Colorado*

FRANCIS J. RYAN, Ph.D., *Former Professor of Biology, Columbia University*

JOHN D. RYDER, Ph.D., *Dean, College of Engineering, Michigan State University*

MARC SACERDOTE, *Teacher, New York Public School system*

BETH SCHULTZ, Ed.D., *Department of Biology, Western Michigan University*

GARY E. SCHWARTZ, Ph.D., *Department of Psychology, Harvard Medical School*

FRANCIS P. SHEPARD, Ph.D., *Professor Emeritus of Oceanography, Scripps Institution of Oceanography*

FERDINAND L. SINGER, *Science writer*

BERNHARDT G. A. SKROTZKI, M.E., *Former Engineering Editor and Managing Editor, Power*

W. H. SLABAUGH, Ph.D., *Professor of Chemistry and Associate Dean, Graduate School, Oregon State University*

HAROLD T. U. SMITH, Ph.D., *Professor of Geology, University of Massachusetts*

HILTON A. SMITH, Ph.D., *Dean of the Graduate School and Coordinator of Research, University of Tennessee*

JESSICA SNYDER, Science writer, former editor, Science Digest

MURRAY SPIEGEL, Ph.D., Professor of Mathematics and Chairman, Department of Mathematics, Rensselaer Polytechnic Institute

R. CLAY SPROWLS, Ph.D., Professor of Information Systems, UCLA

ANTHONY STANDEN, M.S., Executive Editor, Kirk-Othmer Encyclopedia of Chemical Technology

HARLAN T. STETSON, Ph.D., Former Director of Laboratory for Cosmic Research, Needham, Massachusetts

JAMES STOKLEY, D.Sc., Associate Professor of Journalism, College of Communication Arts, Michigan State University

MARK STRAUSS, M.S., Science writer and former editor, Discover magazine

IVAN RAY TANNEHILL, D.Sc., Former Chief of the Division of Synoptic Reports of the United States Weather Bureau

SANFORD S. TEPFER, Ph.D., Associate Professor, Department of Biology, University of Oregon

JENNY TESAR, Science writer

JAMES TREFIL, Ph.D., Professor of Physics, University of Virginia

RICHARD G. VAN GELDER, Ph.D., Chairman and Curator, Department of Mammalogy, The American Museum of Natural History

RAY VILLARD, M.S., Public information manager, Space Telescope Science Institute, Johns Hopkins University, Baltimore

TOM WATERS, M.S., Science writer and former editor, Discover magazine

BRUCE WETTERAU, Science writer

FRED L. WHIPPLE, Ph.D., Former Professor of Astronomy, Harvard University; Former Director, Astrophysical Observatory, Smithsonian Institution

ORAN R. WHITE, Ph.D., High Altitude Observatory, National Center for Atmospheric Research

GRAY WILLIAMS, Science writer

J. TUZO WILSON, Sc.D., F.R.S., Director, Ontario Science Centre

VOLNEY C. WILSON, Ph.D., Former physicist, Research Laboratory, General Electric Company

LYNN YARRIS, Public Relations writer, Lawrence Berkeley Laboratory

GEORG ZAPPLER, Publications Editor, Texas Memorial Museum, Austin, Texas

RICHARD G. ZWEIFEL, Ph.D., Curator, Department of Amphibians and Reptiles, American Museum of Natural History

GUIDE TO THE NEW BOOK OF POPULAR SCIENCE

Astronauts blast to the moon and bring back samples from another world; one person's heart beats in another's body; newspapers report work on creating new forms of life in a test tube...These are just some of the most startling of the advances in science that have excited people in the last few decades. Is nature now under human control? The first rumblings of an earthquake, the appearance of a new island where only ocean was visible, the discovery of a hitherto unknown form of life deep in the dark and unfavorable ocean abysses answer that question, once again revealing the power and mystery of nature.

The 12 sections of which THE NEW BOOK OF POPULAR SCIENCE consists present the major fields of science and discuss their applications in the world of today. The section Astronomy & Space Science, for example, explores people's long fascination with the heavens, explains what we have learned through the centuries, and discusses how this knowledge and the quest for more has led to the exciting age of manned space exploration. The section on Earth Science and the closely related Energy and Environmental Sciences sections take us on a tour of our home base, probe how its vast energy is being and can be used, and describe how nature and people both work to change the earth.

The arrangement of articles in each section provides, as far as possible, a logical, step-by-step presentation of the subject matter in a particular field. Each article, however, constitutes a unit in itself and can be read as such. The articles are essay length, providing a well-rounded introduction to a particular topic. Subheads and sideheads within the text of each article help give the reader a quick and concise overall view of the article's contents. The metric system of measurement, long used by scientists throughout the world, is used throughout THE NEW BOOK OF POPULAR SCIENCE.

Illustrations are a very important part of THE NEW BOOK OF POPULAR SCIENCE. Plentiful and beautiful color photographs and artwork as well as black-and-white illustrations make THE NEW BOOK OF POPULAR SCIENCE a very attractive set that invites you to wander through its pages. A description of a solar flare is made real when you see the striking yellow and red outbursts from the solar disk. The entire story of the exploration of the moon comes alive as you see the astronauts on the moon, gathering and examining lunar rocks. Complex concepts in chemistry and physics become simple and the human body reveals its marvelous organization through clear diagrams. Electron micrographs of viruses and bacteria, underwater photographs of deep ocean life, views of animals in their natural habitats provide a beautiful panorama of life's diverse forms. Throughout, the illustrations complement the text, explaining, expanding and beautifying it.

For readers who wish to have additional information on a given subject, the editors have provided a bibliography, called Selected Readings, at the end of each volume. It contains not only a list of informative books but also a brief evaluation of each book that is listed so that the reader may have some idea of what it contains.

An alphabetical index for the six volumes of THE NEW BOOK OF POPULAR SCIENCE is given at the end of Volume 6. It enables the reader to find easily and quickly specific items of factual information in the articles. Instructions for the use of the index are given on its first page. The index makes it possible for the reader to obtain the fullest possible benefit from the set.

In the pages that follow, we list the 12 sections of THE NEW BOOK OF POPULAR SCIENCE, and give brief accounts of each.

ASTRONOMY & SPACE SCIENCE — Vol. 1

No areas of science excite the imagination in the ways that astronomy and space science do. We've all dreamed of floating weightlessly through intergalactic space, perhaps visiting the surface of Mars, seeing the rings of Saturn, or meeting some unusual extraterrestrial creatures. At the same time, we all enjoy the Earth-bound pleasures of astronomy, whether it be a brilliant full moon, a shooting star, a total eclipse, or just a beautiful sunset.

The Astronomy & Space Science section opens with a look at our universe and what we know about it. We see how ancient and modern people have used astronomy to perfect the calendar and track the constellations. Next, we examine the equipment and methods used to study the heavens—everything from the Hubble space telescope to ultraviolet light. Turning to our own solar system, we learn about the Sun, then the planets and moons, comets, asteroids, and meteors. After that, we leave our solar system to explore the stars and galaxies—the world of supernovas and black holes, quasars and brown dwarfs. Finally, we return to Earth to study the space program, manned and unmanned—where we've been and where we're going. The section concludes with a look at the search for intelligent alien life and at the always-fascinating topic of UFOs.

MATHEMATICS — Vol. 1

Each time we pay a bill, count change, weigh ourselves, draw a circle or square, or do anything that involves numbers or shapes, we use mathematics. Mathematics, then, can be defined as the study of numbers and shapes, and their many interrelationships.

The Mathematics section first looks at numerals—the symbols we use to represent numbers—and the many systems people used until the Arabic notation came into general use centuries ago. Next we turn to the basic mathematical operations of numbers, called arithmetic, and the substitution of those numbers by letters, known as algebra. We then explore the study of shapes: flat shapes, or plane geometry; three-dimensional shapes, or solid geometry; and triangles, or trigonometry. By expressing geometric figures in terms of algebra, we come to analytical geometry. Finally, we explore the realm of non-Euclidian geometry. Mathematics, of course, finds numerous applications, and the rest of this section concentrates on mathematically based topics of statistics, probability, game theory, calculus, set theory, and binary numbers.

EARTH SCIENCES — Vol. 2

The earth is our home and we are, of course, interested in its foundations, its landscape, its features, its crust. In the section on Earth Science, we examine the earth's crust and some of the ways it has formed and is still changing through the action of earthquakes, volcanoes, and other agents of the tremendous power within the earth's interior. We look at some crustal features—mountains and caves, for example—and piece together theories on how they have formed and how they are now working to affect the landscape. Then we turn our attention to the protective envelope that surrounds the earth—the atmosphere. We study lightning, wind, rain, and other atmospheric occurrences before we begin our exploration of the earth's third major part: water. Hidden in the ground, coursing through rivers, locked in huge oceans with their own life cycles, water covers three fourths of the earth's surface. We explore its characteristics, its actions, its life. We close this section with a brief look at the early history of our home in space.

ENERGY   Vol. 2

Perhaps the greatest challenge facing us today is the quest for sources of energy to power civilization and at the same time preserve the earth for future generations. We open the section on Energy by taking an overall look at how much and in what ways we use energy, at how we now obtain it, and at what our future needs are likely to be. Then we explore the major conventional sources of energy: oil, coal, and natural gas. Then we turn our attention to some of the more exotic sources: geothermal energy and solar energy—heat from the earth's interior and directly from the sun. We also discuss nuclear energy—both fission and fusion. Is it safe? Will it answer future needs for energy? We end the section with a discussion of the way in which most of our energy is used—in producing electricity.

ENVIRONMENTAL SCIENCES   Vol. 2

Will earth be a suitable home for future generations? In the section on Environmental Sciences we examine our total environment, or surroundings, and see how the forces of nature and the hand of man have changed and are continuing to change the earth. First, we examine earth's natural resources—water, wood, minerals—and discuss how they can be both used and preserved. Then we see how nature's forces—running water and wind, for example—are carving the land, changing its face. We next see how human beings—through industry, agriculture, transportation, and other activities—are changing the earth. We wonder what will be left, and in what condition, for future generations. We take some plant and animal species as examples illustrating the role of a changing environment and discuss how the existence of some species is threatened and the steps that can be taken to safeguard them and other endangered species. We also explore some of the areas now set aside as specially protected regions where nature wins and people are allowed little or no influence—so-called wilderness areas.

PHYSICAL SCIENCES   Vol. 3

What do all things—living and nonliving alike—have in common? They are all forms of matter—that which occupies space. The different forms of matter possess and can be made to possess energy. Matter and energy are the twin bases of all things. In the Physical Sciences section we study chemistry, "the matter science," and physics, "the energy science." We explore matter's different forms and its basic component—the atom. We see how various elements combine to form acids, bases, salts, solutions, and colloids.

As we turn more particularly to the study of energy in the physics subsection, we see how energy is the basis of work and movement. We investigate the nature of electricity and its relationship to magnetism. We explore how energy is involved in such phenomena as heat, sound, optics, and color. Then we turn to what is probably the scientific theory most closely associated with modern science—relativity—and try to understand its basic premises and applications. We end the section exploring two of matter and energy's most intriguing phenomena: the world of plasma, or superhot, energized gases; and the world of supercooled, superconducting elements.

BIOLOGY Vol. 3

What is life? How did it arise and diversify into the countless forms it now takes on land, in water, and even in the air? In the General Biology section we explore the nature of life. We study its simplest and most basic unit—the cell—learning its structure, tracing the steps of its activities, and marveling at its complexity. We consider how living things—small and large alike—develop from a single cell and how the many varied organisms have attained their present organization through a long series of changes to which we give the name evolution. We see how organisms react to their environments and to one another in a particular community and habitat, forming a complex and completely interdependent web of life. Finally, we see how life forms are classified, or grouped.

PLANT LIFE Vol. 4

The plant kingdom includes forms from the tiny diatom visible only through a microscope to the mighty redwood trees towering over one hundred meters high. In the section on Plant Life we see how these two seemingly diverse forms are related—how they are both plants—and go on to study the characteristics of all plant life. We start with the soil, the anchor for all land-dwelling plants, and proceed from the simplest plants to the highest forms. We see how algae and fungi form an essential part of the cycle of life on earth and how ferns, which beautify the earth in dense and luxurious growths, begin to show some of the characteristics of higher plants. We devote particular attention to the seed plants, the dominant form of vegetation on earth. We study their roots, stems, leaves, flowers, and fruits. We also consider how various plants adapt to their environments. We end this section with a discussion of some plants of particular importance and interest to humans—vegetables, cacti, houseplants, and trees.

ANIMAL LIFE Vol. 4

The animal kingdom is vast and includes strikingly diverse forms—from tiny one-celled protozoans to highly complex mammals—including people. The section on Animal Life surveys the vast panorama of animal life. The story begins with the invertebrates, or animals without a backbone: protozoans, sponges, starfish, worms, mollusks, lobsters and other crustaceans, spiders, and many types of insects. Then comes the backboned animals, or vertebrates: fishes, amphibians, snakes, and birds—similar in some ways, yet each exhibiting a unique adaptation to a particular way of life.

MAMMALS Vol. 5

In the section on Mammals we explore in somewhat greater detail many of the most important mammal groups. We start with two highly unusual types of mammals—egg-layers and marsupials. Then we discuss the highly varied placentals—rabbits, rodents, aquatic mammals, dogs, bears, weasels, cats, hoofed animals—to name just a few. We end the Mammals section with a discussion of the primates, the mammal group to which we belong.

HUMAN SCIENCES Vol. 5

People are similar in many ways to animals and are in fact a part of the primate mammal group. Yet people are different from even their nearest mammal relatives. In the section on Human Sciences we deal with peoples' similarities to other animals and also with their differences. We start with a survey of the human body, studying its structure and how the various organ systems work. We then consider what personality is, what emotions are, and how memory works. We go on to see the human life cycle—heredity, development, growth, decline, and old age. We find that many factors—diet, sleep, and exercise—play important roles in our health. We see how a person's mind and body interact and how psychological stress can affect physical health. We discuss particular problems such as alcoholism, smoking, and drug abuse. We also consider a few of the particular diseases we commonly encounter—allergies, influenza, cancer, and arthritis—and end with a brief discussion of organ transplants and antibiotics.

TECHNOLOGY Vol. 6

Technology brings science to our everyday lives. Each time we turn on a light, use the telephone, watch television, or perform virtually any act, we actually witness thousands of years of technological advances in a split second. Technology and society have grown in parallel and, as the opening pages of this volume note, the Stone Age, Space Age, and other great eras of human history take their names from the level of technology achieved at the time.

For centuries, discoveries and advancements in chemistry, physics, and biology have been transformed into practical means of improving the quality of our existence. The Technology volume first explores how some of the fundamental needs of humans—building shelters, storing water, and growing food—served as an early creative impulse for ancient technologists, and set the stage for the soaring skyscrapers, massive dams, and genetically engineered farming of today. From there, the volume explores the various means by which people get from place to place, from primitive canoes and simple bicycles to magnetically levitated trains and hypersonic aircraft. The volume examines the modern factory and the manufacturing techniques and automation that make these advances possible, as well as the state-of-the-art plastics and ceramics on which they depend. Next, the volume turns to communication and the many devices that people use to stay in touch with one another. In health care, technology has ushered in entirely new ways of diagnosing, treating, and coping with medical disorders. Finally, the volume explores the tools we are using to tap the technology of tomorrow: lasers, particle accelerators, and computers.

Volume 1

Contents

Astronomy & Space Science	1 - 329
Mathematics	330 - 460

ASTRONOMY & SPACE SCIENCE

both photos, NASA

Mysterious Mars began to reveal its secrets as the U.S. Viking space probes approached and landed on the planet. The photomosaic at left reveals Vallis Marineris, the "Grand Canyon" of Mars (upper left of darkened area), and the large impact basin Argyre (lower left of dark area). The photo above shows a Martian sunset.

Pages	Topic
2–8	The Study of the Universe
9–13	The Early Sky Watchers
14–19	The Origin of the Universe
20–31	The Night Sky & Constellations
32–36	The Calendar
37–44	The Big Scopes
45–52	Eyes in the Sky
53–54	Invisible Astronomy
55–57	Radio Astronomy
58–59	X-Ray Astronomy
60–64	Infrared and Ultraviolet Astronomy
65–72	Planetariums
73–79	The Solar System
80–93	The Sun
94–98	Mercury
99–102	Venus
103–112	Earth
113–126	Moon
127–132	Mars
133–138	Jupiter
139–146	Saturn
147–150	Uranus
151–155	Neptune
156–157	Pluto
158–166	Comets
167–169	Asteroids
170–174	Meteors and Meteorites
175–180	Eclipses
181–189	Milky Way
190–196	The Stars
197–203	Collapsing and Failed Stars
204–214	Black Holes
215–217	Quasars and Energetic Galaxies
218–223	Interstellar Space
224–231	Cosmic Rays
232–251	The Space Program
252–257	The Future of the Space Program
258–263	The Space Station
264–269	The Mission to Mars
270–279	Dressing for Space
280–289	Astronauts in Training
290–301	Space Probes
302–315	Space Satellites
316–325	The Search for Intelligent Life
326–329	Unidentified Flying Objects

The secrets locked behind a starlit night have long intrigued the human race. Perhaps Einstein was contemplating the heavens when he wrote, "The most beautiful thing we can experience is the mysterious."

STUDY OF THE UNIVERSE

by Ray Villard

If you've ever observed constellation patterns, peered at the Moon through a small telescope, or simply lain back and counted stars, then you've practiced astronomy. The word "astronomy" comes from the two Greek words *astron*—"star" —and *nomos*—"law." Early peoples attempted to determine the "laws" that govern the motions of the Sun, Moon, and planets. Modern astronomy seeks to understand the underlying physical laws behind the origin, evolution, and composition of the universe. Dating back to the beginning of civilization, astronomy is the oldest of the natural sciences largely because it addresses some of the most fundamental and profound questions humans have ever asked: How did the universe begin? How were the Earth and Sun created? What process brought about the origin of life? What is the ultimate fate of the universe?

A UNIVERSAL VIEW

Earth was once thought to be the center of the universe. We now know that Earth and its neighboring planets are merely tiny specks of agglomerated debris, left behind from the formation of the Sun 5 billion years ago. Except for being our parent star, the Sun is a relatively unremarkable star. It lies in the hinterlands of the Milky Way galaxy, a vast, pancake-shaped city of 300 billion stars. Like islands scattered across an archipelago, many billions of other galaxies lie beyond the Milky Way, out to the horizon of the visible universe. On the largest scale, the structure in the universe resembles a great sponge, where long filamentary clusters of galaxies are the fabric in the sponge, and mysterious, as yet unidentified dark voids in space are the "holes."

Cosmic Evolution

This picture of a vast and complex universe is made even more mind-boggling by the realization that the universe is evolving. This is a starkly different view from that of less than a century ago, when the universe was portrayed as static and eternal.

Today we know that the universe is active, dynamic, and ever changing. Stars explode; galaxies collide; black holes devour matter. The very fabric of space and time was apparently forged in an incredible explosion, called the Big Bang, some 15 billion years ago.

Our past and destiny are intertwined with the violent cosmic events that shaped the cosmos. Many of the atoms in our bodies—calcium, nitrogen, and iron, for instance—were forged in the hearts of ancient stars that exploded long ago, spewing material back into space. Other cosmic catastrophes—an asteroid's collision with Earth, radiation from a nearby supernova, or changes in the Sun's brightness—may have affected the evolution of life in the past and may do so again in the future.

MODERN TOOLS AND TECHNIQUES

Astronomers have learned a great deal about the universe in only the past century. As with other sciences, progress in astronomy is driven by advances in technology. The past few decades have seen a revolution in the development of instruments for looking deep into the cosmos. Astronomy is the only science in which the scientists cannot directly collect samples of their subjects for analysis in a laboratory. Virtually everything we know about the universe is gleaned by collecting and analyzing starlight.

Telescopes are essentially "light buckets" for harvesting the extremely faint light from distant stars. The larger the mirror or lens used for collecting starlight, the greater the telescope's sensitivity to faint objects. The 20th century has seen telescope size rapidly increase. The largest telescope today, the 39-foot (10-meter) Keck Telescope in Mauna Kea, Hawaii, consists of segmented mirrors that are "tiled" together to form one gigantic mirror. This huge "eye" can detect the light from an object 1 billion times fainter than the human eye can see. Such optics, coupled with extremely sensitive solid-state electronic detectors (akin to those found in television cameras), enable astronomers to collect light from some of the most distant objects in the universe.

Most of astronomy involves simply the process of looking. Astronomers take pictures to study an object's shape, to determine its position in space, and to note any changes or movement. But pictures yield

Optical telescopes, which capture the light wavelengths reflected from celestial objects, are just one "window" we have to view the Milky Way.

© Lund Observatory

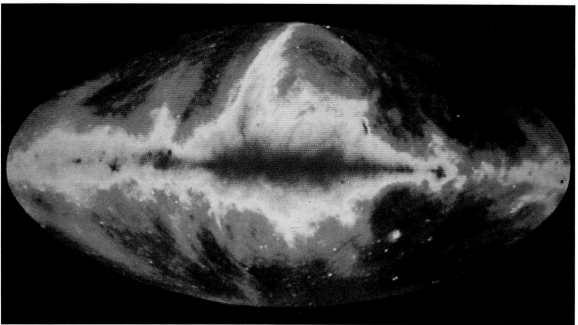

Stellar bodies emit all types of radiation, from X rays to radio waves (pictured above), each of which provides a different view of the universe.

only two-dimensional information. *Spectroscopy* complements "picture-taking." A spectrogram dissects starlight into its component colors, in much the same way that water droplets break up sunlight to create a rainbow. The atoms that make up a star emit light in very specific patterns of color when they are heated. Every element has its own unique spectral "fingerprint."

When analyzed and properly "decoded," a star's spectrum tells us about its temperature, chemical composition, rate of rotation, magnetic fields, and other physical characteristics. Through painstaking laboratory work, astronomers have learned to decode starlight to reveal that the universe is made out of the same elements that are found on Earth. Even more remarkable, spectroscopy proves that the same physical laws apply throughout the universe. Spectroscopy can also tell how fast an object is moving toward or away from Earth. This helps astronomers to assemble a three-dimensional view of how stars and galaxies are distributed across space.

The Invisible Universe

Ironically, astronomers have been working almost in the blind until only recently. That's because the universe is flooded with a symphony of radiation, from radio waves to X rays. Only certain wavelengths, light and radio energy, pass through the Earth's atmosphere to reach ground-based telescopes. Studying the universe from Earth's surface is like trying to listen to an orchestra and hearing only the clarinets. The advent of rocketry in the 1950s allowed simple telescopes and other instruments to be placed above Earth's obscuring atmosphere. The first detectors were placed aboard sounding rockets and then Earth-circling satellites.

These space-based instruments have revealed a remarkably different universe in infrared, ultraviolet, X-ray, and gamma-ray "light." When put together, this radiation forms the electromagnetic spectrum. By studying the whole spectrum, astronomers are assembling a complete picture of the universe.

Telescopes in space provide a much clearer view of the universe. That's because the message of starlight is scrambled when it hits the thin veil of our atmosphere, which is why stars appear to twinkle. The atmosphere bends starlight and blurs pinpoint stellar images, turning them into shimmering, fuzzy blobs of light in the eyepiece of a telescope. Major observatories are located on mountaintops to get above some of the atmosphere. Astronomers have developed clever ways to take the "twinkle" out of stars by distorting a telescope's mirror to "reassemble" the star's image. Ideally, major optical telescopes should be located in space. NASA's Hubble Space Telescope, the first of such "great observatories," reveals celestial objects with 10 times greater detail than can be achieved by any ground-based behemoths. Future space telescopes will likely be located on the Moon, where there are a number of advantages, including two-week-long nights!

Telescopes as Time Machines

Telescopes are the closest instruments we have to true time machines. They allow us to look into the past because light obeys a "speed limit" as it travels across the universe. This "posted speed" is 186,400 miles (300,000 kilometers) per second. Though this seems unimaginably fast, the universe is so vast that, for example, light from the Sun takes eight minutes to reach Earth, and light from nearby stars takes years to reach us. Light from distant galaxies can take billions of years to reach Earth. Astronomy really is a sort of grand "cosmic archaeology," where astronomers study the record of events that actually occurred long, long ago. It's a bit like watching an old silent movie. The actors have long since disappeared, but their behavior is recorded on film. In a similar way, light is a recording of the past 15 billion years of cosmic evolution.

THE ASTRONOMICAL ZOO

Planets

Our solar system consists of nine known planets, several dozen moons, thousands of rocky and metallic asteroids, and trillions of icy bodies called comets. All of these objects are gravitationally bound to the Sun. The solar system has two types of planets: the tiny rocky planets of Mercury, Venus, Earth, and Mars, which all lie close to the Sun; and the immense liquid and gaseous outer planets of Jupiter, Saturn, Uranus, and Neptune, which lie in the colder reaches of the solar system. The farthest planet from the Sun, icy Pluto, may actually be a "double planet," because it is orbited by a moon nearly one-quarter its size. Thousands of icy dwarf bodies similar to Pluto may have inhabited the early solar system.

Stars

Stars are massive, self-luminous objects that form the basic building blocks of the cosmos. They coalesce under gravity to form great clusters and galaxies. Stars are also fundamental engines that generate energy by smashing lighter elements together to form heavier elements through a process called nuclear fusion. Our Sun is the closest star to Earth, and very typical of stars in general. Every second, the Sun converts 540 million tons (490 million metric tons) of hydrogen into 595 million tons (540 million metric tons) of helium. In the process, 49 million tons (45 million metric tons) of matter is converted to pure energy, which eventually reaches Earth as light.

The basic recipe for making a star is fairly simple: compress a huge cloud of in-

Orbiting space "junk" poses a danger to all spacecraft. This computer-generated image shows some of the 7,000 objects larger than a baseball that are currently being tracked in Earth's orbit.

© New York Times Company

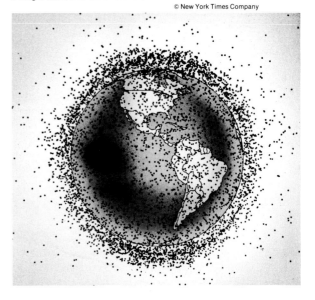

terstellar dust and gas into a relatively small, dense globule of hydrogen. Then collapse this cloud, perhaps with a shock wave from a nearby stellar explosion. The cloud continues to collapse under gravity until nuclear fusion begins, its outward force counterbalancing any further collapse. When this equilibrium is reached, a star is born.

Even a casual glance into the evening sky reveals that stars come in a wide range of brightness and colors. Some stars appear bright because they are close neighbors to the Sun. Others are intrinsically bright because they are much hotter than the Sun. Star colors also provide clues to their intrinsic nature. Bluish-colored stars are much hotter than the Sun; reddish stars are cooler. Extremely hot stars can be 10 or even 100 times more massive than our Sun.

More than half of the stars in the night sky travel in pairs. Such *binary* systems sometimes trigger spectacular fireworks as their stars evolve. If one star spills some of its atmosphere onto a more compact companion star, there may be a periodic explosion known as a *nova*. Such outbursts enrich interstellar space with heavier elements that are used in new star formation.

Stellar Death

Because they are basically "fusion engines," all stars eventually run out of fuel. As their nuclear furnaces readjust to the diminishing fuel supply, stars may expand, shrink, and oscillate like great cosmic yo-yos. Many release luminescent shells of gas that form eerie, doughnut-shaped planetary nebulas. When a star's nuclear furnace shuts down, the crushing force of gravity immediately collapses the star to a *white dwarf*—a superhot cinder no bigger than Earth.

An extremely massive star ends its life with a bang. When the core implodes, a powerful shock wave tears apart the upper layers of the star. In just a few hours, the doomed star unleashes more energy than our Sun would in 5 billion years. The star briefly grows millions of times brighter, a phenomenon called a *supernova*.

But the supernova's core almost instantaneously shrinks, crunching down to the size of, say, New York City. Such an ultra-compact relic is called a *neutron star*, since the electrons and protons in atoms have been forced together to make neutrons. Some neutron stars, called *pulsars*, behave like interstellar lighthouse bea-

Adaptive optics, in which mirrors are bent to compensate for atmospheric distortions that occur to light, is just one of the revolutionary techniques developed in recent decades to view the cosmos. Below left, a normal image of a binary star and at right, an image using adaptive optics.

United States Air Force

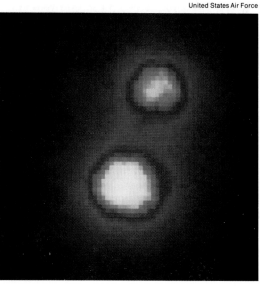

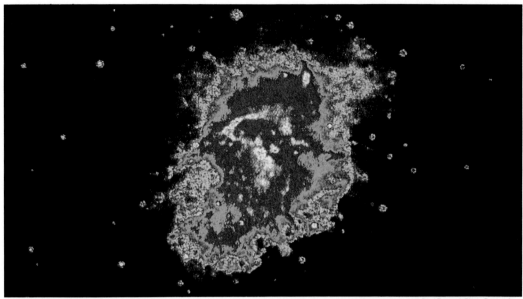

A supernova, or exploding star, was observed by Chinese astronomers in A.D. 1054. The remnant of this explosion forms the Crab Nebula in the constellation Taurus.

cons. As they spin hundreds or even thousands of times per second, they blast out two powerful beams of radiation.

The Ghosts of Stars

Imagine squeezing a fleet of battleships into an object the size of a baseball, or squeezing a mountain down to the size of a pinhead. No earthly trash compactor has that kind of muscle, but gravity does. The massive core of a burned-out star can implode beyond the neutron-star stage, to a point smaller than the dot at the end of this sentence (the mathematical term is a *singularity*, a point in space having no dimension). Though unimaginably small, such a collapsed core retains an incredibly powerful gravitational field that traps everything that passes close by. Even light itself cannot escape. Such stellar corpses are called *black holes*. You can't see directly into a black hole, but its gravitational field remains as a ghostly "footprint" that warps the space around it. Though there is increasing circumstantial evidence for black holes, their existence is still considered theoretical.

The Milky Way

The drama of star birth and death is continually played out across the Milky Way, our island city of 300 billion stars. Nearly 100,000 light-years across the Milky Way is a broad, flattened disk of stars, having the same relative dimensions as that of a phonograph record. A tennis ball glued at the center of the record would approximate the spherical nucleus of our galaxy, where the oldest stars dwell. The thin plane of our galaxy is so rich in raw material—mostly molecular hydrogen—that stellar birth is an ongoing process. Vast, glowing nebulas (clouds) mark the location where newborn stars have popped out of their stellar nurseries to light up the surrounding clouds of gas, much like light bulbs in a grotto. Bright young star clusters trace out the Milky Way's spiral shape.

Galaxies

Our Milky Way was once thought to contain all the stars in the universe. In the 1920s, however, American astronomer Edwin Hubble discovered that the universe is filled with other island cities of stars as well. Hubble classified galaxies according to shape. Many are spiral, or pinwheel-shaped, like our Milky Way. Others are elliptical, or football-shaped; and still others are irregular. Today astronomers are presented with a truly dizzying variety of galaxies—radio galaxies, infrared galaxies, X-ray galaxies, active galaxies—but there is as yet no coherent scheme of galactic evolution.

The cores of some galaxies are extraordinarily bright, shining at a level equal to compressing 1 million Suns into a piece of space no larger than our solar system. These *active galactic nuclei* may be powered by immense black holes that have grown perhaps from the merger of individual stellar black holes. Star dust and gas swirling down into the hole heats up to millions of degrees, allowing a prodigious amount of energy to be radiated.

Brighter than 1 million supernovas going off in unison, *quasars* are probably the most energetic type of active galactic nucleus. Quasar stands for quasi-stellar object, because it is brilliant, pointlike, and virtually indistinguishable from stars in photographic sky surveys. However, quasars are many billions of light-years away, far too distant to be individual stars.

HOW DID GALAXIES FORM?

If the universe started out as a dense "soup" of particles of matter and energy, then how did it get "lumpy" enough for galaxies to form? Did they start out as huge clouds of gas that contracted—or, instead, as small clumps of stars that merged? Even more puzzling, how did galaxies coalesce into vast clusters? The largest such structure, called the "Great Wall" of galaxies, is nearly 500 million light-years long. According to current theories, the universe simply hasn't been around long enough to build such unimaginably vast structures.

One idea is that the universe contains infinitely long but infinitesimally thin "cracks" called cosmic strings. Like crystal growing on a piece of thread, matter might have accumulated along these strings, early in the universe's history, to eventually form great filamentary clusters of galaxies. There is no direct evidence for cosmic strings as yet, however.

A more widely accepted possibility is that galaxies may just be "puddles" or "trace sediments" in a "haze" of invisible dark matter that may have been clumpy before galaxies formed. Since astronomers can't see dark matter, they don't know what it is. Possibilities range from primordial black holes to an as-yet-undetected new type of subatomic particle that pervades all of space. Though they can't see dark matter, astronomers can measure its ghostly influence on the motions of galaxies. If the universe has enough dark matter, it will eventually slow down and collapse back onto itself. When that happens, all matter will be engulfed in a fireball similar to the one from which it emerged.

THE SEARCH FOR OTHER SOLAR SYSTEMS

Astronomers are utilizing a number of strategies for determining if other solar systems exist. Such a discovery would be a watershed in astronomy as revolutionary as the concept of the Sun-centered solar system. In only the past decade, disks of gas and dust have been detected around young, nearby stars. The nearest such disk—around the star Beta Pictoris—appears to be lumpy, suggesting that a planet may be forming. Astronomers will survey stars to look for other circumstellar disks.

Extremely precise observations of stars within 500 light-years will be made to indirectly detect planets by measuring if a star "wobbles" due to the gravitational pull of an unseen planetary companion. Besides wobbling in their apparent motion across the sky, stars within a solar system should also "jiggle" due to a gravitational "tug-of-war" with planets. Ultimately astronomers may directly image a planet, using a large, space-based infrared telescope. This is an extraordinarily difficult task because a planet would be millions or even a billion times fainter than its parent star. The challenge is somewhat like trying to see a gnat flying around a distant streetlight.

Stonehenge is just one of many ancient monuments thought to be an early astronomical tool.

THE EARLY SKY WATCHERS

by Ray Villard

Simple astronomy played an important role in the cultures of many early peoples. They used the motions of the heavens largely for religious and ceremonial purposes. More practically, celestial motions allowed for the development of the first calendars for planning growing seasons. It was important to know the Sun's rising point on the first day of winter and summer (solstices) as well as the first day of spring and fall (equinoxes). More mathematically sophisticated peoples, such as the Babylonians and the Mayans, plotted the position of the bright planet Venus and even attempted to predict solar and lunar eclipses.

The predawn rising times of selected bright stars were also used to keep track of the seasons. For example, the Egyptians needed to predict when the Nile River would annually flood and thus fertilize the

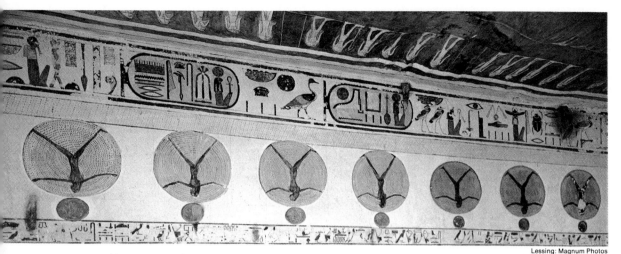

Many artifacts bear testimony to ancient Egypt's astronomical tradition. The painting above, found on the tomb of Ramses IX, illustrates the death and resurrection of the Sun.

A 16th-century drawing depicts Danish astronomer Tycho Brahe (pointing) and the wall quadrant he used to make the detailed planetary observations that supported a Sun-centered solar system theory.

land for crops to be planted. They discovered that when the bright star Sirius first appears in the early-morning sky (called helical rising), flooding soon would follow. This predawn appearance of Sirius happens annually, which allowed the Egyptians to accurately calculate the length of the year to 365 days.

There is abundant evidence that even primitive societies needed to keep track of celestial happenings. Stone patterns arranged on the ground mark the first astronomical "observatories" ever built. The most celebrated of these is Stonehenge, a mysterious 100-foot (30.48-meter)-diameter ring of massive, monument-sized (25 tons or 23 metric tons each) stones set on end. Located on the Salisbury Plain in southern England, Stonehenge was built in three distinct phases between 2700 and 1700 B.C. No one knows who built Stonehenge or for what purpose. Certain pairs of stones seem to align to points on the horizon where the Sun would rise and set on special dates. There are a number of other such prehistoric monuments in the British Isles and in Brittany in northwestern France.

At least 50 mysterious, wheel-shaped monuments built by the Native American Indians also have an eerie similarity to Stonehenge. The best example is the Bighorn Medicine Wheel in eastern Wyoming, built around A.D. 1500. Nearly 80 feet (24 meters) across, the wheel has 28 spokes radiating from the center. Besides aligning to the Sun's solstice rising point, the wheel aligns to the rising point of certain key bright stars on selected dates.

An even more fascinating structure, dubbed the "American Woodhenge," can be found in modern-day St. Louis, Missouri. This 410-foot (125-meter)-diameter circle, built around A.D. 1000, has remains of holes dug to anchor 48 large wooden posts. No one knows how high the posts were, but it is believed that one could sight the Sun's solstice from the center of the circle.

Remains of several villages of the Wichita Indians (A.D. 1500) have "council circles," each consisting of a central mound surrounded by a series of depressions arranged in an elliptical pattern. Pairs of ellipses are arranged so that their long axes align with the summer-solstice sunrise and winter-solstice sunset.

The Mayans (A.D. 1000), who inhabited Mexico's Yucatán Peninsula, aligned many features of their cities and buildings to astronomical phenomena. Mayan records show that they were extremely sophisticated at predicting eclipses and the position of Venus. A building called the Caracol, at the ancient city of Chichén Itzá, contains a complex series of windows, walls, and horizontal shafts that seem to point to the location of the Sun and Venus on significant days.

Similar structures are found in the southwestern United States. The Hohokam Indians, who occupied the Casa Grande area near Coolidge, Arizona, from 200 B.C. to A.D. 1475, had a unique building dubbed the "Big House." This three-story structure contains an array of windows and holes that give lines of sight to the horizon. At least half of these openings align to the Sun and Moon on significant dates.

Are many of these alignments coincidence, or did the builders of these monuments have a sophisticated knowledge of

The significance of the Medicine Wheel, in Wyoming's Bighorn Mountains, remains a mystery. One theory suggests that its 28 spokes represent the days of the lunar month; early Indians may have used the wheel as a calendar.

THE EARLY SKY WATCHERS

astronomy and the calendar? If some distant future society were to excavate a 20th-century airport runway, or even a baseball-park diamond, might they, too, find that it pointed to the Sun's equinox rising and setting points? Nobody knows, of course, but fascinating research continues in this young field of archeoastronomy.

More than any other people, the Chinese kept careful records of major astronomical events. The Chinese recorded sunspots as far back as 800 B.C., and were aware of the Sun's 11-year sunspot cycle fully 2,000 years before modern astronomers "rediscovered" it. At least 20 stellar explosions (supernovas), called "guest stars" by the Chinese, were recorded as far back as A.D. 185. Comets, which were called "broom stars," were duly noted, including every appearance of Halley's comet back to the year 239 B.C.

Driven largely by the superstitions of astrology, which attempted to link celestial alignment to human behavior, Babylonian astronomers kept detailed diaries of celestial events. Unlike the Chinese, the Babylonians were primarily interested in repeatable, cyclical events, so they did not keep records of unpredictable events such as the appearance of a comet or supernova. The Babylonians mapped the apparent path of the Sun across the heavens and established the 12 constellations of the zodiac, which lay along the Sun's path.

In the Chinese and Babylonian civilizations, astronomy and astrology were closely linked. An astronomer was kept in the court of the king or emperor to advise of celestial portents and predict future events. By contrast, in the Western world, the early Greek astronomers were private philosophers, and, in fact, often at odds with established religion. The Greeks were extraordinarily clever and intuitive at conducting purely scientific astronomical observations. They made extraordinary leaps in comprehending the clockwork of the heavens. By 500 B.C., they correctly explained the cause of solar and lunar eclipses. They realized that the Moon reflected sunlight (because of its phases), and that the Sun was much farther away than

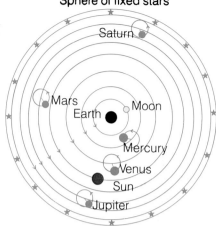

The Greek astronomer Ptolemy believed that the Earth was at the center of the universe, with the Sun and planets revolving around it.

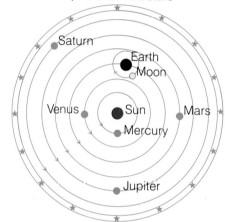

Copernicus challenged Ptolemy's geocentric theory by proposing that the Earth, Moon, and planets revolve around the Sun.

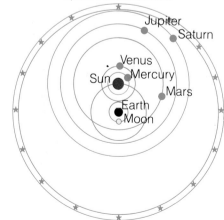

Tycho Brahe combined the theories of Ptolemy and Copernicus to propose that the Sun and Moon revolve around the Earth, but the other planets revolve around the Sun.

HIGHLIGHTS OF ASTRONOMY HISTORY

1610	Galileo uses telescope for astronomical observations
1665	Newton formulates law of universal gravitation
1695	Edmund Halley predicts the return of a periodic comet
1781	Uranus discovered
1815	Spectrum taken of Sun
1846	Neptune discovered
1905	Einstein develops theory of relativity
1923	Edwin Hubble measures distance to Andromeda galaxy
1926	Hubble discovers stars in other galaxies
1929	The Hubble constant formulated
1930	Pluto discovered
1931	First extraterrestrial radio waves detected
1939	Nuclear fusion proposed as the main source of stellar energy
1942	Radio emissions detected from the Sun
1951	Spiral structure of Milky Way mapped in radio waves
1958–67	Robot spacecraft survey of Moon
1960	First search for artificial interstellar signals
1962	Quasars discovered
1963	First interstellar X-ray sources detected
1963	Space probes survey Mercury and Venus
1963	Cosmic background radiation detected
1967	First pulsar detected in Crab nebula
1969–72	Manned expeditions to Moon
1970	First X-ray detection of a black-hole candidate
1971	First X-ray catalog of sky
1975	First radio message from Earth transmitted to the stars
1976	Experiments for microorganism on Mars (Viking mission)
1977–89	Initial survey of the outer planets via unmanned probes
1979	Pluto's moon Charon discovered
1987	Nearest supernova in 400 years
1990	Astronomers confirm cosmological nature of the microwave background

the Moon. Through numerous experiments and observations, the Greeks concluded that the Earth was a sphere, and they even calculated its diameter. Many philosophers concluded that the Sun, Moon, and planets revolved around Earth, which lay at the center of a huge "celestial sphere." Aristarchus of Samos correctly proposed that, instead, Earth orbited the Sun. However, this idea was far too revolutionary for even the Greek thinkers. The notion of an Earth-centered universe would persist until the 16th century.

The period from the mid-1500s to mid-1600s saw an unprecedented revolution in astronomy. Polish mathematician Nicolaus Copernicus reinvented the notion of a Sun-centered solar system. Detailed planetary observations by Danish astronomer Tycho Brahe provided an invaluable database for supporting Copernicus' theory. Tycho's heir to the astronomical data, Johannes Kepler, formulated the laws of planetary motion.

While Kepler was refining his theories, the Italian inventor and scientist Galileo Galilei used the telescope, a recent Dutch invention, to study the heavens. The telescope allowed Galileo to see objects 10 times more clearly than with the unaided human eye. This unveiled a universe that was more complex than sky watchers had dared to imagine, including literally millions of stars in the Milky Way, mountains and valleys on the Moon, shifting phases of Venus as it orbits the Sun, and a tiny moon orbiting the distant planet Jupiter.

Kepler and Galileo assembled evidence that proved the planets orbited the Sun. Neither scientist knew exactly what invisible, all-pervasive force kept the solar system "glued" together. Mathematician Isaac Newton combined Copernicus' theory, Kepler's laws, and Galileo's observations to formulate the law of universal gravitation. This law states that all objects are attracted to other objects, and that the strength of this bond depends on how much matter the objects contain and the distances between them. These laws describe the motion of all bodies in the universe, and remained unchanged until the 20th century.

Some scientists theorize that the creation of our universe began with a titanic explosion that spewed out matter into the far reaches of the universe.

THE ORIGIN OF THE UNIVERSE

by David Fishman

How and when did the universe begin? No other scientific question is more fundamental or provokes such spirited debate among researchers. After all, no one was around when the universe began, so who can say what happened? The best that scientists can do is work out the most foolproof theory, backed up by observations of the universe. The trouble is, so far, no one has come up with an absolutely foolproof theory.

THE BIG BANG

Since the early part of the 1900s, one explanation of the origin and fate of the universe, the Big Bang theory, has dominated the discussion. Proponents of the Big Bang maintain that, around 20 billion years ago, all the matter and energy in the known cosmos was crammed into a tiny, compact point. In fact, according to this theory, matter and energy back then were the same thing, and it was impossible to distinguish one from the other.

Adherents of the Big Bang believe that this small but incredibly dense point of primitive matter/energy exploded. Within seconds the fireball ejected matter/energy in all directions at velocities approaching the speed of light. At some later time— maybe seconds later, maybe years later— energy and matter became separate entities. All of the different elements in the universe today developed from what spewed out of this original explosion.

Big Bang theorists claim that all of the galaxies, stars, and planets still retain the explosive motion of the moment of creation and are moving away from each other at great speed. This supposition came from an unusual finding about our neighboring galaxies. In 1929 astronomer Edwin Hubble, working at the Mount Wilson Observatory in California, announced that all of the galaxies he had observed were receding from us, and from each other, at speeds of up to several thousand miles per second.

The Redshift

To clock the speeds of these galaxies, Hubble took advantage of the Doppler effect. This phenomenon occurs when a source of waves, such as light or sound, is moving with respect to an observer or listener. If the source of sound or light is moving toward you, you perceive the waves as rising in frequency: sound becomes higher in pitch, whereas light becomes shifted toward the blue end of the visible spectrum. If the source is moving away from you, the waves drop in frequency: sound becomes lower in pitch, and light tends to shift toward the red end of the spectrum. You may have noticed the Doppler effect when you listen to an ambulance siren: the sound rises in pitch as the vehicle approaches, and falls in pitch as the vehicle races away.

To examine the light from the galaxies, Hubble used a spectroscope, a device that analyzes the different frequencies present in light. He discovered that the light from galaxies far off in space was shifted down toward the red end of the spectrum. Where in the sky each galaxy lay didn't matter—all were redshifted. Hubble explained this shift by concluding that the galaxies were in motion, whizzing away from the Earth. The greater the redshift, Hubble assumed, the greater the galaxy's speed.

Some galaxies showed just a slight redshift. But light from others was shifted far past red into the infrared, even down into microwaves. Fainter, more distant galaxies seemed to have the greatest red shifts, meaning they were traveling fastest of all.

An Expanding Universe

So if all the galaxies are moving away from the Earth, does that mean Earth is at the center of the universe? The very vortex of the Big Bang? At first glance, it would seem so. But astrophysicists use a clever analogy to explain why it isn't. Imagine the universe as a cake full of raisins sitting in an oven. As the cake is baked and rises, it expands. The raisins inside begin to spread apart from each other. If you could select one raisin from which to look at the others, you'd notice that they were all moving away from your special raisin. It wouldn't matter which raisin you picked, because all the raisins are getting farther apart from each other as the cake expands. What's more, the raisins farthest away would be moving away the fastest, because there'd be more cake to expand between your raisin and these distant ones.

The American astronomer Edwin Hubble (below) was a pioneer in extragalactic astronomy. In 1925, he composed the classification scheme for the structure of the galaxies that is still used today.

The Observatories of the Carnegie Institution of Washington

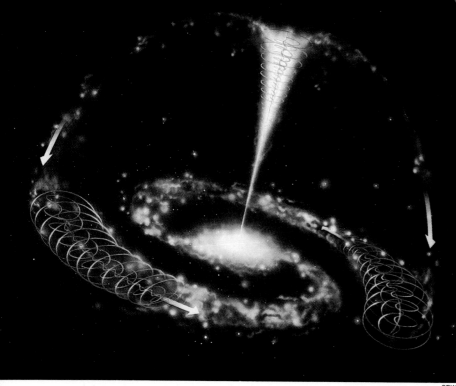

Plasma, the predominant form of matter in the universe, is made of electrically charged particles subject to both gravity and electromagnetism. Some theorists unhappy with the Big Bang theory believe that the many-stranded electromagnetic currents thought to permeate plasmas may be strong enough to supplant gravity as the primary force for sculpting the cosmos.

That's how it is with the universe, say Big Bang theorists. Since the Big Bang explosion, they reason, the universe has been expanding. Space itself is expanding, just as the cake expanded between the raisins in their analogy. No matter whether you're looking from Earth or from an alien planet billions of miles away, all other galaxies are moving away from you as space expands. Galaxies farther from you move faster away from you, because there's more space expanding between you and those galaxies. That's how Big Bang theorists explain why light from the more distant galaxies is shifted farther to the red end of the spectrum. In fact, most astronomers now use this rule, known as Hubble's law, to measure the distance of an object from Earth—the bigger the redshift, the more distant the object.

In 1965 two scientists made a blockbuster discovery that solidified the Big Bang theory. Arno Penzias and Robert Wilson of Bell Telephone Laboratories detected faint microwave radiation that came from all points of the sky. They and other physicists theorized that they were seeing the afterglow from the Big Bang's explosion. Since the Big Bang affected the entire universe at the same moment in time, the afterglow should permeate the entire universe and could be detected no matter what direction you looked. This afterglow is called the cosmic background radiation. Its wavelength and uniformity fit nicely with other astronomers' mathematical calculations about the Big Bang.

How Lumpy Do You Like Your Universe?

The Big Bang model is not uniformly accepted, however. One problem with the theory is that it predicts a smooth universe. That is, the distribution of matter, on a large scale, should be roughly the same wherever you look. No place in the universe should be unduly lumpy.

But recently astronomers have discovered huge clumps of matter populating the universe. Collections of galaxies and other matter line up across regions more than a billion light-years long. Instead of an even distribution of matter, the universe seems to contain great empty spaces punctuated by densely packed streaks of matter.

Big Bang proponents maintain that their theory is not flawed. Instead, they argue, gravity from huge, undetected objects in space (clouds of cold, dark matter we can't see with telescopes, or so-called "cosmic strings") attracts matter into clumps. Other astronomers, reluctant to believe in invisible objects just to solve a

Huge nebulas containing glowing gas and interstellar debris are thought to serve as the birthplace of the stars.

problem, continue to question fundamental aspects of the Big Bang theory.

In spite of its problems, the Big Bang is still considered by most astronomers to be the best theory we have. Like any scientific hypothesis, however, more observation and experimentation are needed to determine its credibility. Advances ranging from more-sensitive telescopes to experiments in physics should add more fuel to the cosmological debate during the coming decades.

THE STEADY STATE THEORY

But the Big Bang is not the only proposed theory concerning our universe's origin. In the 1940s a competing hypothesis arose, called the Steady State theory. Some astronomers turned to this idea simply because, at the time, there wasn't enough information to test the Big Bang. British astrophysicist Fred Hoyle and others argued that the universe was not only uniform in space—an idea called the cosmological principle—but also unchanging in time, a concept called the perfect cosmological principle. This theory didn't depend on a specific event like the Big Bang. Under the Steady State theory, stars and galaxies may change, but on the whole the universe has always looked the way it does now, and it always will.

The Big Bang predicts that as galaxies recede from one another, space becomes progressively emptier. The Steady State theorists admit that the universe is expanding, but predict that new matter continually comes to life in the spaces between the receding galaxies. Astronomers propose that this new material is made up of atoms of hydrogen, which slowly coalesced in open space to form new stars.

Naturally, continuous creation of matter from empty space has met with criticism. How can you get something from nothing? The idea violates a fundamental law of physics: the conservation of matter. According to this law, matter can neither be created nor destroyed, but only converted into other forms of matter, or into energy. But skeptical astronomers have found it hard to directly disprove the continuous creation of matter, because the amount of matter formed under the Steady State theory is so very tiny: about one atom every billion years for every several cubic feet of space.

The Steady State theory fails, however, in one important way. If matter is continuously created everywhere, then the average age of stars in any section of the

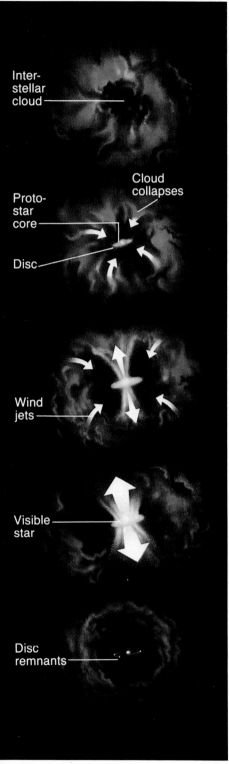

A STAR IS BORN

After the primordial explosion of the Big Bang, matter was dispersed in every direction.

1. A cloud forms that is 3 trillion miles wide.

2. The denser inner regions begin to collapse, a core is formed, and rotating gases create a disc.

3. As the collapse continues, the core grows, and wind jets carry mass outward.

4. Winds expand, blowing the cloud away to reveal an infant star inside the disc.

5. Remnants of the disc are shaped by gravity to create planets surrounding the star.

SRW

universe should be the same. But astronomers have found that not to be true.

Astronomers can figure out how old a galaxy or star is by measuring its distance from Earth. The farther away from Earth an object is, the longer it has taken light from the object to travel across space and reach Earth. That means that the most distant objects we can see are also the oldest.

For example, take quasars, the small points of light that give off enormous amounts of radio energy (see also page 215). Because the light from quasars is shifted so far to the red end of the spectrum, astronomers use Hubble's law to calculate that these powerhouses lie at a great distance from Earth, and hence are very old. But quasars exist only at these great distances—none are found nearer. If the Steady State theory were true, there ought to be both young and old quasars. Since astronomers haven't found quasars that formed recently, they conclude the universe must have changed over time. The discovery of quasars has put the Steady State theory on unsteady ground.

THE PLASMA UNIVERSE AND LITTLE BANGS

Not happy with either the Big Bang or the Steady State theory? A minority of astronomers are formulating other views of the creation of the universe. One model comes from the mind of Nobel laureate Hannes Alfvén, a Swedish plasma physicist. Called the Plasma Universe, his model starts by noting that 99 percent of the observable universe (including the stars) is made of plasma. Plasma, an ionized gas that conducts electricity, is sometimes called the fourth state of matter. This theory states that the Big Bang never happened, and that the universe is crisscrossed by gigantic electric currents and huge magnetic fields.

Under this view the universe has existed forever, chiefly under the influence of an electromagnetic force. Such a universe has no distinct beginning and no predictable end. In the Plasma Universe, galaxies come together slowly over a much greater

Quasars (above) only exist at very great distances, and are thus very old. If the Steady State theory were viable, then both young and old quasars would exist.

time span than in the Big Bang theory, perhaps taking as long as 100 billion years.

Little of the evidence for the Plasma Universe comes from direct observations of the sky. Instead, it comes from laboratory experiments. Computer simulations of plasmas subjected to high-energy fields reveal patterns that look like simulated galaxies. Using actual electromagnetic fields in the laboratory, researchers have also been able to replicate the plasma patterns seen in galaxies. While still a minority view, the Plasma Universe is gaining favor with younger, more laboratory-minded astronomers who value hard empirical evidence over mathematical proofs.

Meanwhile, another group of astronomers is developing a steady-state theory that actually conforms to astronomical observations. Like its predecessor, this steady-state theory proposes a universe with no beginning and no end. Rather, matter is continuously created via a succession of "Little Bangs," perhaps associated with mysterious quasars. In this new theory, galaxies would form at a rate determined by the pace at which the universe expands. These theorists can even account for the cosmic background radiation: they maintain that the microwaves are actually coming from a cloud of tiny iron particles—not some primordial explosion.

THE END OF THE UNIVERSE

Will the universe continue expanding? Will it just stop or even begin to contract? The answer depends on the amount of mass that the universe contains, according to work done in the 1920s by Russian mathematician Alexander Friedmann. If the universe's mass exceeds a certain crucial value, then gravity should eventually stop everything from flying away from everything else—just as, if you throw a ball straight up, you can depend on gravity to stop the ball and return it to you.

With enough mass the universe will eventually succumb to the attractive force of gravity and collapse once again into a single point—a fate often called the Big Crunch. But if there isn't enough mass, the universe will just keep on expanding. Scientists are still trying to find evidence for which of these fates our universe faces. Some say our universe alternates between Big Bangs and Big Crunches in a perpetual display of the power of gravity.

THE NIGHT SKY AND CONSTELLATIONS

Before the dawn of history, people peering up into the night sky began seeing designs in the grouping of stars. One saw a crab, its body and legs sketched in stars. Another saw a great hunter, his belt studded in stars, and his outstretched arm culminating in a starry club. Yet another saw a winged horse. And so, through the ages, the stars became divided into fanciful groups, or constellations, of which there are now 88. These constellations fill the entire sky, with no overlapping or empty spaces between them. Every visible star belongs in one of them.

Many of the constellations are located in what may be the most dramatic starry structure in the night sky: the rich, luminous band of stars called the Milky Way. We now realize that when we look at the Milky Way, we are in fact peering into the center of our own galaxy. Our Sun is a rather average-size star located on a spiral arm of this galaxy, about three-quarters of the way out from the galactic center.

Some stars—usually the brighter ones—have individual names in addition to their constellation name. Sirius and Aldebaran are just two well-known examples. Many more stars are designated simply by letters of the Greek alphabet followed by the Latin name of the constellation.

In most cases, though not all, the brightest star in a constellation is the alpha of that group, the next brightest is beta, the next gamma, and so on, down through the Greek alphabet. For example, Sirius, the brightest star in the constellation Canis Major (the Greater Dog), is also known as Alpha Canis Majoris.

We cannot, however, tell how intrinsically bright a star is just by looking at it. A brilliant star that is far away may seem much fainter than a less brilliant, but closer one. The astronomer refers to the brightness of a star as it appears to us as its "apparent magnitude" or, simply, its *magnitude*. A first-magnitude star looks about 2.5 times as bright as a second-magnitude star, which in turn is about 2.5 times as bright as one of third magnitude, and so on. The few stars brighter than first magnitude are given minus magnitudes.

With a simple telescope, we can not only see the stars—which shine with the energy of their own nuclear-powered furnaces—but also our own planet's moon, most of our solar system's planets, and several of their moons. Planets and moons shine with the reflected light of the Sun.

On a crystal-clear night, a person may be able to see some 4,000 shining heavenly bodies. But remember: at any one time, one sees only half the sky. The other half is visible from a point on the exact opposite side of the Earth.

READING A SKY MAP

The Northern Constellations

Map 1 shows the main northern constellations that are visible year-round and throughout the night north of 40 degrees north latitude (which passes through Philadelphia; Columbus, Ohio; and just south of the northern California border). To use *Map 1,* find the approximate date in the outer circle. If you are observing on November 15, for instance, and facing north, hold the map so that the middle of the section labeled November is at the top. You may find it helpful to hold the map over your head as you face north. Those constellations at the bottom of the map (such as Ursa Major) will now be close to the northern horizon, and those at the top of the map (such as Cassiopeia) will now be higher in the sky than Polaris, the star that almost exactly marks the north pole—the central point of the northern heavens.

Your meridian—the imaginary line running from the north point of your horizon, through a point directly overhead, to the south point of your horizon—will then run exactly from the top of the map to the bottom. The stars on that line will be on your meridian about 9:00 P.M. standard time.

The circle of numbers within the outer rim of the chart mark hours. These tell you how far east or west a star is in the sky, and will help you find the stars that will be on your meridian later or earlier in the evening. Since *Map 1* reflects the position of stars at 9:00 P.M., the hour marked 0 actually reflects 9:00 P.M. So if you want to locate stars at, say, 11:00 P.M., turn the map counterclockwise from the position for 9:00 P.M. (marked by the 0 hour), through two hours. The top of the map will then be about halfway between the hours 2 and 3. Again, the stars at the bottom of the map will be near the northern horizon; and those at the top, above the pole. If you want to observe at 7:00 P.M. on the same night, simply turn the map two hours in a clockwise direction.

Perhaps the most familiar constellation in the northern sky is the Big Dipper, which is part of the constellation Ursa Major, or the Greater Bear. You can easily find the Big Dipper at the bottom of *Map 1,* clearly outlined by its seven bright stars. If you imagine a line passing through the two brightest stars on the front edge of the Dipper's bowl, and then extend that line

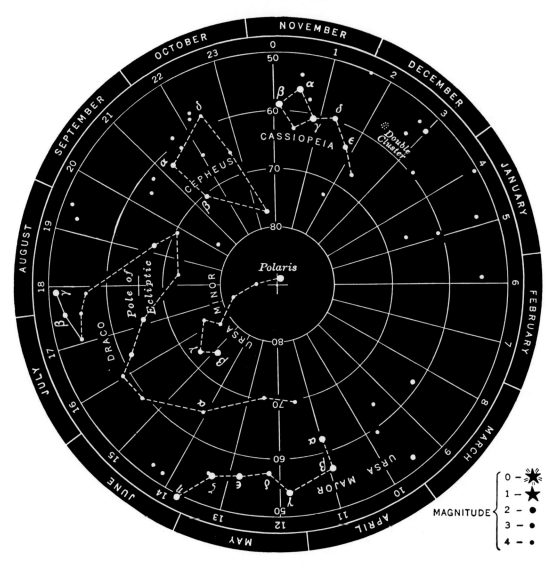

Map 1. The northern constellations.

straight up (above the bowl), you will find the bright polestar Polaris.

Polaris lies approximately over the Earth's North Pole and is the central point around which the stars of the Northern Hemisphere turn in a counterclockwise direction. You can always use Polaris like a compass, to find the north. Polaris is also the star forming the end of the handle of the Little Dipper. The Little Dipper, in turn, belongs to the constellation Ursa Minor, the Lesser Bear. Polaris also marks the midpoint in the sky between Ursa Major (the Big Dipper) and the constellation Cassiopeia, the Lady in the Chair. You will recognize Cassiopeia best as a huge *W* or *M*. Next to Cassiopeia is Cepheus, the King. This constellation is not as easy to see as the other groups, although, with care, you can make out its figure, almost like a tent or a building with a steeple. The long, winding constellation between Cepheus and the Big Dipper is Draco, the Dragon. His head is formed by a V-shaped group of stars about halfway around the sky between Cassiopeia and the Dipper's bowl. The end of his tail is marked by faint stars.

The Southern Constellations

Those viewing the sky from below the equator will see the constellations in *Map 2*. The south celestial pole is at the center of the cross in the middle of the diagram, but there is no bright star to mark it con-

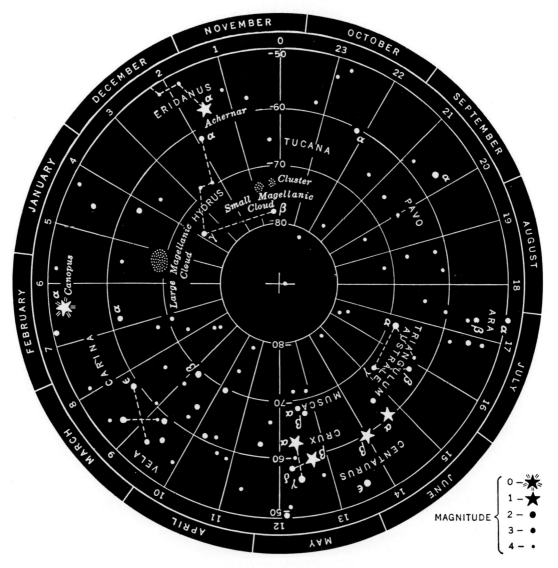

Map 2. The southern constellations.

veniently for navigators in southern seas or skies. There is a guide to the south celestial pole, however: the constellation Crux, the Southern Cross. The longer axis of the Cross points almost directly to the south pole of the heavens.

The two bright stars Alpha and Beta Centauri point to the top of the Southern Cross—the star Gamma Crucis. Alpha Centauri is a double star. It is generally held to be the star nearest to the Earth (except for the Sun); it is about 25 trillion miles (40 trillion kilometers) away from Earth. However, some astronomers think that its faint neighbor, Proxima Centauri, may be a bit closer. Nearby the Southern Cross is a famous dark nebula, the Coalsack. The Magellanic Clouds appear as hazy objects to the unaided eye. A telescope reveals, however, that they are actually masses of stars, nebulas, and star clusters.

Around the first of December, at the end of the constellation Eridanus, the star Achernar is seen above the South Pole. It is the one brilliant object in that long stream of stars. It forms, roughly, a right-angle triangle with the two Magellanic Clouds.

Canopus, second only to Sirius among the stars in brightness, is in the constellation Carina, the Keel. This constellation was once part of the big constellation Argo Navis, the Ship Argo, which in modern times has been broken up into several smaller constellations.

THE NIGHT SKY AND CONSTELLATIONS

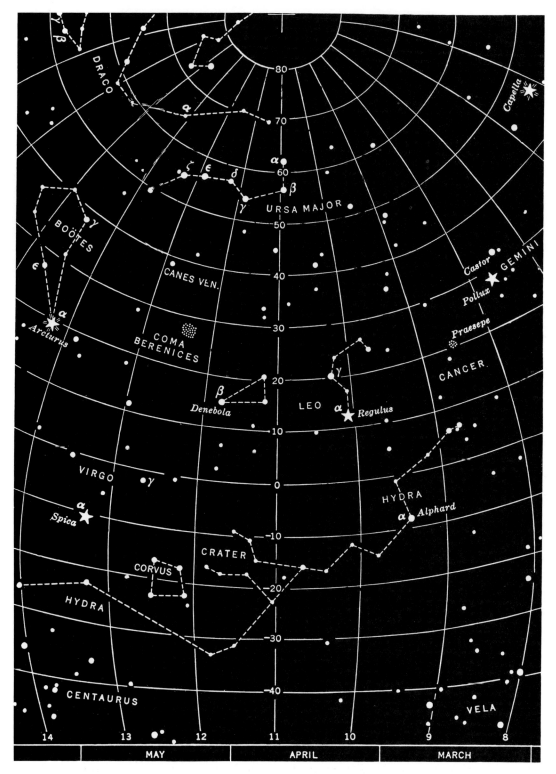

Map 3. The spring constellations, as seen from the middle northern latitudes.

Reading the Seasonal Maps

The four seasonal star maps reproduced in this book show the heavens as they appear from middle northern latitudes in spring, summer, fall, and winter, respectively. In using the maps, remember that each night, any given star passes by the same place in the heavens about four minutes earlier than it did the night before. This makes a difference of about two hours a month. For instance, a star that rises at 10:00 P.M. tonight will rise about 8:00 P.M. a month from now.

Choose the map for the season in which you are viewing the sky. If you are viewing in the middle of a season—say, April, using the spring map (*Map 3*)—the stars along the meridian of the map (the middle line that runs from top to bottom) will appear along the meridian of your night sky (the imaginary line running from south to north, directly over your head). If you are viewing later in the season—say, in May—look farther to the left of the meridian on your map to find the stars along the sky's meridian. If you are viewing earlier in the season, look farther to the right of the meridian on the map.

Remember that the star maps show constellations in their positions at 9:00 P.M. If you are viewing later in the evening, look farther to the left on your map. If you are viewing earlier in the night, look farther to the right. Also, don't forget to hold the map over your head, with the top of the map (where the plus sign indicating Polaris is located) pointed north. Or imagine that the maps in the book are mirror images of the sky. Otherwise, the discussions of east and west will be confusing.

SPRING SKIES

Use *Map 3* to find the Big Dipper, and continue the curve of its handle away from the bowl to locate the bright star Arcturus, in the constellation Boötes, the Herdsman. Arcturus is one of the first bright stars that appears over the eastern horizon in the bright spring evenings.

Going back to the Big Dipper, again continue the curve of the handle to Arcturus and on down to the next-brightest star, Spica, in the constellation Virgo, the Maiden. Virgo is one of the 12 constellations of the zodiac, the narrow belt in the sky through which the planets, the Sun, and the Moon appear to move.

Near the southern horizon is the small, four-sided constellation Corvus, the Crow, which precedes Spica into the sky; its top stars point to Spica. To the west of Arcturus is a splendid star cluster known as Coma Berenices—the Hair of Berenice.

Once again, go back to the Big Dipper as a guide group. Extend the line of the pointer stars (the last two stars that form the cup of the dipper) backward, away from the pole, and they will lead you to Leo, the Lion, another constellation of the zodiac. The part of the Lion to the west resembles a sickle. The other conspicuous group of stars in Leo is a right-angle triangle, to the east of the sickle. Denebola, or Beta, marks the tip of the Lion's tail; and Regulus, or Little King, marks the Lion's heart, at the end of the sickle's handle.

To the west of Leo is Cancer, the Crab, also in the zodiac. It is not a well-defined constellation, but if you look out of the corner of your eye, you will be able to make out, on a clear night, the faint star cluster Praesepe, sometimes called the Beehive. Stretching across the southern sky south of Virgo and Leo is the long, faint figure of Hydra, the Sea Serpent. The head is composed of five stars south of Praesepe.

SUMMER SKIES

On *Map 4*, find the bright star Arcturus within the constellation Boötes. Just to the east of Boötes is the beautiful crown of stars called Corona Borealis, the Northern Crown. Its brightest star is Alphecca, sometimes known as Gemma, the gem in the crown. All the other stars of the group are of fourth magnitude.

Look carefully, just south of Corona Borealis, for an *X* made up of five faint stars. It is the head of Serpens, the Serpent. This constellation is closely associated with the large, clearly defined figure of Ophiuchus, the Serpent Bearer of Greek legend.

Another group of stars, one to the east of Ophiuchus, is also called Serpens. The group containing the head is known as Serpens Caput (head). The other is Serpens Cauda (tail). By including the triangle of stars to the west of Ophiuchus, imagine this constellation forming one long serpent stretching across the figure of the Serpent Bearer.

One of the most striking constellations in the heavens is Scorpius, the Scorpion, visible in the summer just above the southern horizon. To some people, it looks a bit like a fishhook. Antares, the brightest star in the constellation, marks the heart of the Scorpion. To the west of Scorpius is Libra, the Scales, another zodiacal constellation.

To the east of Scorpius is Sagittarius, the Archer, yet another constellation of the zodiac. Some stargazers see in this constellation a teakettle, or a little dipper. Through it runs the brightest part of the Milky Way.

To the east of Corona Borealis is the H-shaped figure of Hercules. The Alpha of this constellation, Ras Algethi, is very close to Ras Alhague, the Alpha of Ophiuchus. Between the Zeta and Eta stars is one of the most beautiful star clusters in the whole sky. It can just barely be glimpsed with the unaided eye.

Following Hercules over the northeastern horizon comes Lyra, the Lyre, a small but extremely beautiful constellation, composed of a parallelogram and a triangle joined together. The outstanding star of this group is Vega, which is brighter than first magnitude. It is the brightest star that we can see in northern latitudes in the summertime, and the third-brightest star in the whole sky. The northernmost star of the triangle of which Vega forms part is Epsilon Lyrae; it is a famous double-double star, which may be possible to make out with the unaided eye.

To the east of Lyra is Cygnus, the Swan. It is often called the Northern Cross, though the stars that form the cross are only some of those that make up the Swan. The star at the Swan's tail is Deneb, which forms the top of the Cross.

In the Milky Way, not very far from the foot of the Northern Cross, is a first-magnitude star that is the central one of three. The bright star is Altair in the constellation Aquila, the Eagle; the two stars that flank it are much fainter.

AUTUMN SKIES

Use *Map 5* to find Altair and its companion stars in Aquila. Using them as pointers to the south, find the rather faint but large group of stars that comprises the zodiacal constellation Capricorn, the Sea Goat. Actually, the constellation looks like a tricorn hat upside down.

Between Capricorn and Cygnus are two small, inconspicuous groups called Delphinus, the Dolphin, or Job's Coffin, and Sagitta, the Arrow. East of these groups is Pegasus, the Winged Horse, best known for a conspicuous square of stars. The star Alpheratz, in the northeast corner of "The Square," forms part of both Pegasus and the constellation Andromeda, the Chained Lady.

Stretching from the Square's northeastern corner, you will see an almost-straight line of three fairly bright stars— Alpha, Beta, and Gamma Andromeda. Almost directly above Beta, which is called Mirach, it is possible to make out with the unaided eye a faint, elongated patch of light. This is the Great Nebula in Andromeda—a galaxy of thousands of millions of stars very much like our own Milky Way galaxy. By extending the imaginary line connecting the three stars of Andromeda, you can follow it to Algenib, the Alpha of the constellation Perseus, the Champion.

Perseus is in the part of the sky from which a famous meteor shower—the Perseids—appears around the 10th to the 12th of August. The Beta of Perseus—Algol—is one of the most fascinating stars that the amateur astronomer can watch. For almost exactly two days and 11 hours, Algol shines steadily as a star of 2.3 magnitude. Then, in a period of about five hours, it decreases to magnitude 3.5. In a second period of five hours, it regains its former brilliance and once more remains at that brightness for about 59 hours. This startling change is due to the fact that Algol consists of two stars

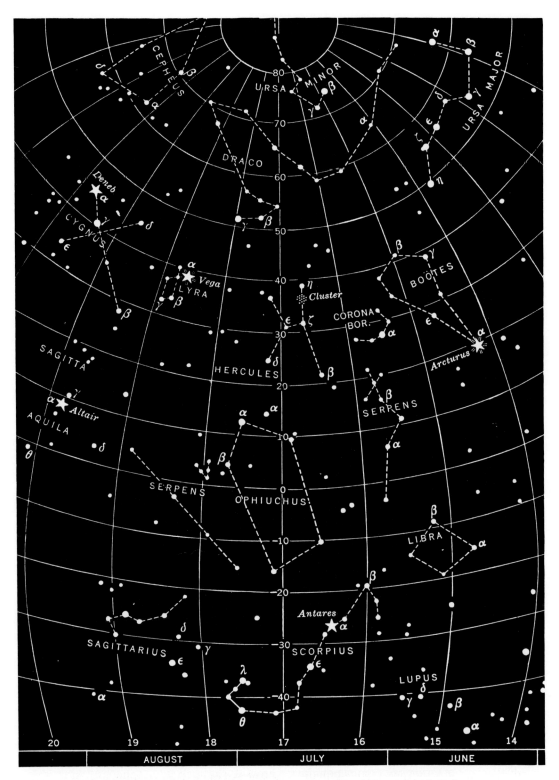

Map 4. The summer constellations, as seen from the middle northern latitudes.

THE NIGHT SKY AND CONSTELLATIONS

revolving around their common center of gravity, with the edge of the system turned toward the Earth. One star is much brighter than the other. When the fainter star comes between us and the brighter component, Algol seems to be dimmed.

Directly below the Great Square of Pegasus, in *Map 5*, is a circle of stars that marks the Western Fish, also called the Circlet, in the long, ribbonlike constellation Pisces, the Fishes. To the east of Pisces is the constellation Aries, the Ram. To the west, there is a veritable cascade of stars—the constellation Aquarius, the Water Carrier, in the zodiac. None of the stars of Aquarius is bright. Below Aquarius and close to the southern horizon is the bright star Fomalhaut in Piscis Austrinus, the Southern Fish.

WINTER SKIES

There is no more superb celestial sight than the sky of winter, illustrated in *Map 6*. One of the most splendid of all the constellations is Orion, the giant Hunter of the heavens. To find the rectangular figure of Orion, first look for his right shoulder (he is facing you). It is marked by the first-magnitude star Betelgeuse, a gigantic red sun hundreds of times as large as our own Sun. Rigel, a gorgeous blue-white star, marks Orion's left foot. Dividing Orion's rectangle in two are the three stars of the belt, from which hangs the Hunter's sword. The top star of the belt, Delta, is almost exactly on the celestial equator. One of the most interesting objects in this constellation is the great gaseous nebula—M42—that surrounds the middle star of the sword. You can barely glimpse it with the unaided eye.

South of Aries, we find the head of the constellation Cetus, the Whale. The head is made up of five stars. Cetus has a remarkable long-period variable star Mira, known as the Wonderful. When Mira is at its most brilliant, it is usually about magnitude 3.5. At its faintest, it is about ninth magnitude, far below the level of naked-eye vision. Mira cycles through these changing magnitudes over a period of about 330 days.

Northwest of Orion, we come upon the first-magnitude star Aldebaran, which represents the eye of Taurus, the Bull, a constellation of the zodiac. The face of the Bull is a V-shaped group of stars—an open star cluster called the Hyades. Taurus boasts another, even more beautiful, open cluster of stars, the Pleiades, to the northwest of the Bull's face. The Pleiades mark the shoulder of the Bull. (Only the forepart of the Bull is represented.) With the unaided eye, most people can make out six of the Pleiades. Beta Tauri, known as El Nath, "that which butts," marks the tip of the left horn. It also belongs to the five-sided constellation Auriga, the Charioteer. Its brightest star is Capella, the She-goat. Near Capella is a small triangle of three stars representing Capella's kids, a very good guide group to help you make sure you have actually found Capella.

By extending the line of Orion's belt to the southeast, we discover Sirius, the Dog Star, and the brightest star in the entire sky. It is in the constellation Canis Major, the Greater Dog. Forming an almost equilateral triangle with Betelgeuse and Sirius is Procyon, another first-magnitude star, in Canis Minor, the Lesser Dog. At Orion's left foot, Rigel, the long, meandering constellation Eridanus, the River, begins. Most of its stars are faint.

Gemini, the Twins, a zodiacal constellation, is easily recognized by its two parallel lines of stars, with Pollux (Beta) at the northern head of one line, and Castor (Alpha) at the northern end of the other. The constellation can be located by drawing a line from Rigel, in Orion, through Betelgeuse and as far again beyond.

THE CELESTIAL SPHERE

Just as geographers draw reference lines of latitude and longitude on models of the Earth, astronomers have developed a celestial sphere with reference lines that can be used to locate heavenly bodies. The Earth is the center of this imaginary celestial sphere (see diagram on page 31).

Astronomers generally use the "equatorial system" to establish reference lines around the celestial sphere. Using this reference system, you can find your

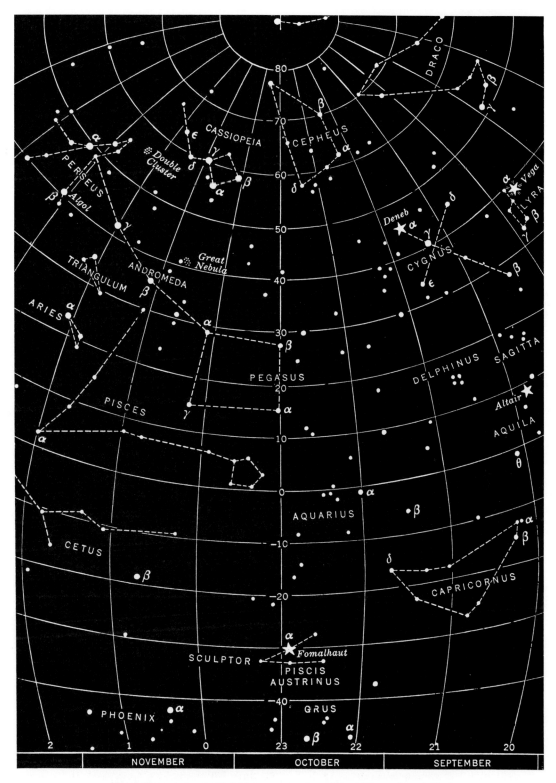

Map 5. The autumn constellations, as seen from the middle northern latitudes.

THE NIGHT SKY AND CONSTELLATIONS

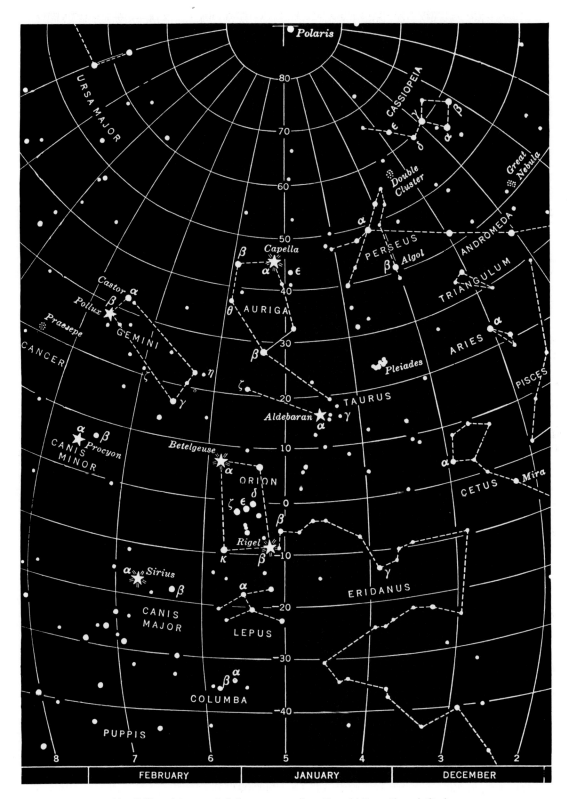

Map 6. The winter constellations, as seen from the middle northern latitudes.

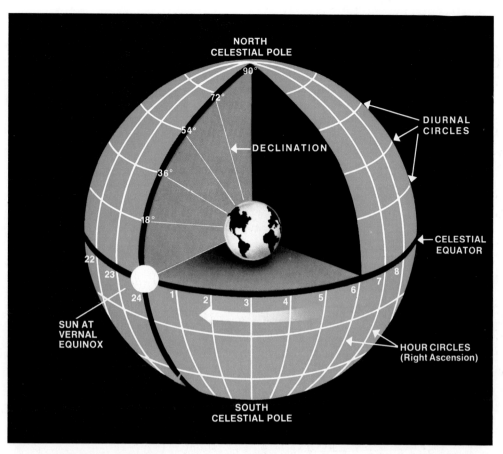

A celestial sphere uses reference lines similar to a geographer's lines of latitude and longitude to map the position of celestial bodies in relation to Earth. A star's east-west position, known as its right ascension, is similar to lines of longitude, while a star's north-south position, its declination, is akin to lines of latitude.

way through the most complicated of star maps.

In the equatorial system, it is helpful to imagine a pole running through the Earth's axis and extending through the North and South Poles—out into space till it touches our imaginary celestial sphere.

At one end of our imaginary pole is the north celestial pole, and at the other is the south celestial pole. Equidistant between the celestial poles is the imaginary line known as the celestial equator, dividing the heavens into the Northern and Southern Hemispheres.

Lines of declination—somewhat akin to the geographer's latitude lines—divide the sphere between the equator and the poles. Declination is given in degrees, minutes, and seconds, north or south of the celestial equator. A star that is 10 degrees north of the celestial equator is given a declination of +10 degrees; one 10 degrees to the south, −10 degrees.

Running through the celestial poles and at right angles to the celestial equator are 24 hour circles. Remembering that the circumference of any circle is 360 degrees, we know that each of our 24 "hours" contains 15 degrees. Each degree contains 60 minutes (60'), and each minute contains 60 seconds (60"). We use these lines to designate a star's east-west position, referred to by astronomers as its R.A., for hours of right ascension.

A star's right ascension is always measured eastward from the vernal equinox (one of the two places where the celestial equator crosses the annual path of the Sun across the sky) to the place where the star's hour circle crosses the celestial equator.

THE NIGHT SKY AND CONSTELLATIONS

THE CALENDAR

by Elisabeth Achelis

A calendar page from the medieval manuscript Très Riches Heures showing the pageantry of celebrations in the month of May.

The calendar we use today is the product of a great many centuries of patient study and of constant trial and error. When people first looked to heavenly bodies for a way to measure time, they observed that the Sun seemed to make a constantly repeated journey in the heavens, always returning to the same place after many days. (Actually, of course, it is the Earth that makes a yearly revolution around the Sun.) They observed that the Moon also followed a definite cycle.

Most of the earliest calendars were based on Moon cycles. These calendars were made to fit as best they could within the larger framework of the Sun cycle. The year, in these calendars, generally consisted of 12 Moon cycles, or months. Since 12 Moon cycles are not quite equal to a solar year, an extra month—called an *intercalary*, or inserted, month—was added from time to time. A number of ancient peoples—including the Babylonians, Hebrews, Greeks, and Romans—adopted this method of computation.

FIRST SOLAR CALENDAR

The Egyptians were the first to base their calendar on a year of 365 days, approximating the Sun cycle, or solar year. They also made the month a purely arbitrary unit, not corresponding to the actual lunar cycle. The year was divided into 12 months of 30 days each, totaling 360 days, to which were added five extra "feast days." The 365-day calendar is believed to have been adopted by the Egyptians in the year 4236 B.C.

In the course of the centuries that followed, it was discovered that the year really consisted of 365¼ days. This additional quarter of a day was causing a gradual shift of the seasons as recorded in the calendar. In 238 B.C., the pharaoh Ptolemy III, also known to history as Euergetes, tried to correct this error in calculation by adding another day to the calendar every four years. It was to be a religious holiday, but unfortunately, the priests were unwilling to accept the extra day. As a result the Egyptian calendar continued to be defective as a measure of the seasons.

THE MAYAN CALENDAR

Another seasonal Sun calendar that was used in antiquity was that of the Mayas of Mexico. Their invention probably goes back to the year 580 B.C., and it was the first seasonal and agricultural calendar produced in America.

The Mayan calendar was arranged differently from that of the Egyptians. The Mayan solar year, called a *tun*, had 18 months of 20 days each. It had a five-day unlucky period at its end to make 365 days. Each month had its own name, and the days were numbered from 0 to 19.

Dovetailed with the Mayan Sun calendar was a religious year, sometimes called a *tzolkin*. The tzolkin contained 13 months of 20 days each. Each day had a name that was combined with the numbers 1 to 13 to count out the 260 days of the tzolkin.

THE JULIAN CALENDAR

The ancient Romans had a moon calendar, which was complicated and most confusing. Originally only 10 months long (March to December), it was soon extended to 12 months by the addition of January and February. A 13th month, called Mercedonius, was occasionally inserted. The 12 months of the Roman year consisted of seven months of 29 days each; four months of 31 days each; and one month, Februarius (February), with 28 days—making a year of 355 days. The names of the 12 months of the Roman year were as follows:

Name of month	Origin of name
Martius	Month of Mars
Aprilis	"Opening" month, when the Earth opens to produce new fruits
Maius	Month of the great god Jupiter
Junius	Month of the Junii, a Roman clan
Quintilis	Fifth month
Sextilis	Sixth month
September	Seventh month
October	Eighth month
November	Ninth month
December	Tenth month
Januarius	Month of the god Janus
Februarius	Month of the Februa, a purification feast

In 153 B.C., January was designated as the first month of the year instead of Martius (our March).

The Romans used a complicated system of reckoning within the month. There were three more-or-less fixed dates—the *calends*, the *ides*, and the *nones*. The cal-

The ancient Maya used several calendars simultaneously, some that rivaled the Julian calendar for precision. This stone tablet is a calendar from Yaxchilan, Mexico, from the 8th century A.D.

© Michael Holford

This Aztec calendar stone depicts Huitzilopochtli, the Sun god, surrounded by symbols for historic eras, the days of the month, and the stars.

ends always fell on the first of the month. The ides came on the 15th in Martius, Maius, Sextilis, and October and on the 13th in other months. The nones always came on the eighth day before the ides. (The Romans, however, used inclusive numbering and counted the ides themselves, making the nones, meaning "nine," the ninth day before the ides.)

The calendar was entrusted to a council of priests—the College of Pontiffs, presided over by a *pontifex maximus*. The pontiffs were state officials charged with the regulation of certain religious matters, including the fixing of dates for ceremonies.

Julius Caesar was elected pontifex maximus in 63 B.C., but it was not until 47 B.C. that he took the first steps to reform the calendar. Following the suggestions of the famous Greek astronomer Sosigenes, Caesar adopted the solar year for the Roman calendar. He gave it 365 days, plus a quarter-day of six hours. Quarter-days were withheld until a full day had accumulated. The day was then added to the common year as a leap-year day. This happened once every four years.

The year 46 B.C. bridged the old and the new calendar. The following year, 45 B.C., was actually the first one using the reformed calendar. Caesar retained the complicated system of calends, nones, and ides within the months. January continued to be the first month of the year. The Roman Senate changed the name of the month Quintilis to Julius (our July) in honor of Caesar. The new calendar was known as the Julian calendar. Later the Roman Senate changed the name of the month Sextilis to Augustus (August) to honor the Emperor Augustus.

THE SEVEN-DAY WEEK

In A.D. 321, the Emperor Constantine issued an edict introducing the seven-day week in the calendar, doing away once and for all with the system of calends, ides, and nones. Constantine established Sunday as the first day of the week, and set it aside as the Christian day of worship.

Although the introduction of the week greatly simplified matters, it brought about a serious defect in the calendar, a defect that is still present today. Both the Egyptian and Julian calendars had been stabilized. That is, every year had been like every other year. Through Constantine's reform the Julian calendar became a shifting one. Now that there were 52 seven-day weeks, totaling 364 days, there was always one day left over in ordinary years, and two days in leap years. This meant that in successive years, the Julian calendar began on different days of the week.

THE GREGORIAN CALENDAR

The true length of the solar year is a trifle less than 365¼ days. It is 365.242199 days, or 365 days, five hours, 48 minutes, and 46 seconds, to be exact. Therefore, the Julian calendar was too long by about 11 minutes. After a number of centuries, the error amounted to several days.

In the year 1582, another momentous calendar reform took place. Pope Gregory XIII determined to adjust the calendar to the seasons. For this purpose, he called

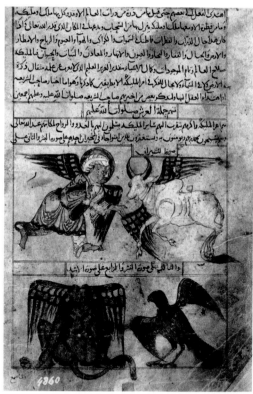

A manuscript page from an Islamic calendar from Damascus, Syria, circa 1366. The Islamic calendar begins with the first day of Muhammad's journey to Medina, which corresponds to July 15, 622, of the Christian Era.

upon the services of mathematician Christopher Clavius and astronomer-physician Aloysius Lilius. They found that the error caused by the excessive length of the Julian calendar amounted to 10 days. To set the year aright, they canceled 10 days from the Julian calendar, so that October 4, 1582, was followed by October 15.

Naturally, this loss of 10 days in the month of October created a certain amount of confusion. To avoid confusion, dates prior to October 15, 1582, were often given thereafter (and are still often given) as O.S. (old style), and dates after as N.S. (new style). If neither O.S. nor N.S. is given after a date, it's assumed to be N.S.

To avoid further error in the calendar, the leap-year rule was changed. In the case of centurial years (those ending in "00"), only the ones that were divisible by 400 were to be leap years. Noncenturial leap years continued to receive an extra day. No attempt was made to equalize the lengths of the months or to stabilize the calendar.

This Gregorian calendar is the one that we use today.

All Roman Catholic countries adopted the Gregorian reform, but other groups in Christendom were slow in accepting it. The English did not adopt the Gregorian calendar until 1752. France, like the other Catholic countries of Europe, had adopted the Gregorian calendar in 1582; however, for a period beginning in 1792, it was replaced by the "Revolutionary Calendar." In line with other antireligious developments of the time, the days and months of this calendar were given symbolic names, such as "Brumaire" ("month of fog"), denoting the "natural" order of things. In 1806 Napoleon restored the Gregorian calendar as a gesture of reconciliation toward the Church.

Japan adopted the Gregorian calendar in 1873, China in 1912, Greece in 1924, and Turkey in 1927. Russia began to use the

In 45 B.C., Julius Caesar was persuaded by the astronomer Sosigenes to adopt a solar year. The resultant Julian calendar was used extensively for centuries. Below, a Julian calendar from the 5th century.

calendar in 1918, replaced it with another calendar when the Bolsheviks took over the country, and returned to the Gregorian calendar in 1940.

OTHER CALENDARS IN USE

The Gregorian calendar is not the only one used at the present time. For religious purposes, Jews employ the Hebrew calendar, which begins with the year of creation, set at 3,760 years before the beginning of the Christian Era. This calendar is based on the cycles of the Moon. There are 12 months, which are alternately 29 and 30 days in length. An extra month of 29 days is intercalated seven times in every cycle of 19 years. Whenever this is done, one of the 29-day months receives an extra day. The year begins in the autumn.

Another important calendar is the Islamic, or Muslim, calendar. It also is based on the cycles of the Moon. There are 354 days and 12 months, half of which have 29 days, and the other half 30 days. Thirty years form a cycle; 11 times in every cycle, an extra day is added at the end of the year. The Muslim calendar begins with the first day of the year of the Hegira—that is, the journey of Muhammad to Medina. This date corresponds to July 15, 622, of the Christian Era.

Although the Gregorian calendar is China's official calendar, the Chinese New Year is still calculated by the ancient Chinese lunar calendar. The months of this lunar calendar are known by the names of the 12 animals of the Chinese zodiac: rat, ox, tiger, hare, dragon, serpent, horse, sheep, monkey, rooster, dog, and boar.

MODERN CALENDAR REFORM

The Gregorian calendar has served people well for almost four centuries. Yet some thoughtful people have tried to bring about reforms that would restore stability to the calendar within the framework of the seasonal year.

In 1834 Abbé Marco Mastrofini put forward a plan in which every year would be the same, and the lost stability of the calendar would be restored. In his calendar, there are 364 days in the year—a number easily divisible in various ways. The 365th day and the 366th in leap years were inserted as extra days within the year. Each year would begin on Sunday, January 1. The abbé's idea was so simple that most modern calendar reformers have made it the basis of their own proposals.

Calendar reform lagged until the League of Nations took up the question in 1923. One proposal, the World Calendar, based on the easily divisible number 12, once seemed particularly promising. In this calendar, each equal quarter-year of 91 days, or 13 weeks, or three months, corresponds to a season period. Every year in this calendar is like every other year. The first of every year, for example, falls on a Sunday; Christmas, December 25, falls on a Monday. To provide the necessary 365th day, a day—known as Worldsday—is inserted after December 30 and before January 1. The 366th day in leap years is inserted between June 30 and July 1. The Gregorian 400 centurial leap-year rule is retained. However, the idea was not adopted, and it has met with scant success in the years since.

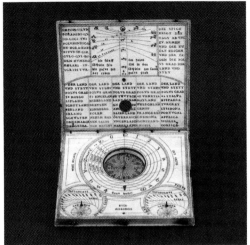

The beautifully crafted calendar below was created during the Italian Renaissance. It is a unique calendar in that it uses both Sun-based and Moon-based systems to measure time.

G. Tomsich/Photo Researchers

THE BIG SCOPES

by Dennis Meredith

Since Galileo first peered through his modest telescope to view the craters of the Moon in 1609, humans have strived for ever-larger telescopes to study the heavens. Galileo's telescope stunned scientists by showing that the Moon was no heavenly, perfect object made of an otherworldly "aether," but a craggy body as imperfect as the Earth.

Could Galileo have imagined the gargantuan instruments of today—telescopes weighing as much as whales, with mirrors dozens of feet across? Like his hand-held tube, today's big scopes have yielded startling new images of the exotic objects that populate the universe.

The telescopes with which we are most familiar are optical telescopes, which gather light. The world's largest optical telescope is at the Keck Observatory, located on Mauna Kea, Hawaii, which captures light on a mirrored surface 33 feet (10 meters) across. Powerful optical telescopes have to be big, because they are basically "light buckets." That is, they gather and focus as much light as possible from a distant planet, star, or galaxy. The larger the light-collecting surface—a curved mirror or lens—the more light can be gathered, and hence the more information scientists can extract from it.

But light isn't the only kind of radiation from which astronomers can gather information. The biggest telescopes on Earth pick up radio waves from distant objects. Others collect X rays, and ultraviolet and infrared light. Some float in orbit clear of Earth's atmosphere, which absorbs and distorts many radiation wavelengths (see also page 45). Each of these types of radiation describes a different aspect of the object that emits the radiation: how bright it is, what it's made of, how warm it is, how energetic it is, and so on.

© Roger Ressmeyer/Starlight

The powerful telescopes housed in today's observatories bear little resemblance to the primitive telescope first used by Galileo in 1609. Yet the telescopes from both eras have expanded our horizons immeasurably.

HOW OPTICAL TELESCOPES WORK

There are three general types of optical telescopes: refracting telescopes, or refractors; reflecting telescopes, or reflectors; and telescopes that combine features of both types.

A refracting telescope is a long tube. At one end sits a large convex lens. This lens, called the objective, is the part of the telescope that gathers light from a celestial object. The lens bends, or refracts, the light rays. This refraction brings the incoming light rays to a focus, to produce a sharp image. The image is then magnified and

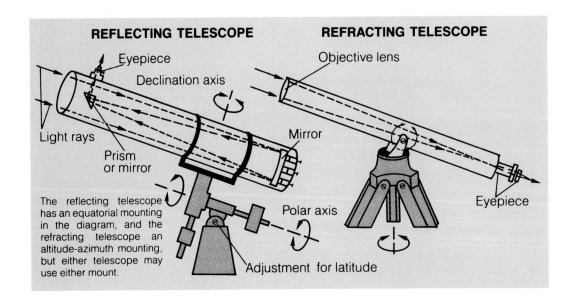

REFLECTING TELESCOPE / **REFRACTING TELESCOPE**

The reflecting telescope has an equatorial mounting in the diagram, and the refracting telescope an altitude-azimuth mounting, but either telescope may use either mount.

viewed through another lens device known as an eyepiece, which is at the opposite end of the telescope tube from the lens.

Reflecting telescopes have a more complex design. In these telescopes, the objective is a glass mirror with a curved surface shaped like the headlamp of an automobile. This is known as a parabolic mirror. The mirror gathers and focuses incoming light rays to a point in front of the mirror, called the prime focus. A second mirror, near the prime focus, relays the light beam elsewhere, to a place convenient for the observer. Several different mirror arrangements can accomplish this.

For example, a flat mirror, set at a 45-degree angle to the telescope tube, is used to move the focus to a spot just outside the tube. This arrangement is known as *Newtonian focus*, because 17th-century mathematician-astronomer Isaac Newton first developed it. Newtonian focus is popular in small reflecting telescopes used by amateur astronomers.

In another design a curved mirror intercepts the light beam from the main mirror. The curved mirror sends the beam back through a hole in the main mirror so that it comes to a focus behind that mirror. This arrangement, called *Cassegrain focus*, is what most professional astronomers use in their reflecting telescopes.

A similar arrangement is the *coudé focus*, whereby light is also bounced from a small mirror back toward the primary mirror. However, before it reaches the primary mirror, the light is reflected out one side of the lower end of the telescope. With a coudé-focus telescope, instruments at its focus need not shift as the telescope tracks objects.

Besides refractors and reflectors, there are the combination telescopes, which use both mirrors and lenses. Such telescopes can achieve a wide field of view or make a large-scale image in a short distance.

In a combination telescope, light passes through a weak refracting lens called a corrector plate. The light then strikes a spherical mirror and comes to a focus. Unlike parabolic mirrors, spherical mirrors do not bring all of the light to the same focus. That is why the corrector plate is used—to correct this unwanted condition.

The common caricature of an astronomer squinting into the eyepiece of a telescope is really not accurate. Most astronomers use television monitors to see what is being viewed through the telescope. The scopes are guided automatically, and the data they gather are likewise recorded automatically by sophisticated instruments.

Optical telescopes may contain arrays of electronic sensors, called charge-coupled devices, to capture the images of stars and galaxies. Computers immediately transform these images into data to be displayed and analyzed. Similarly, astronomers can train optical telescopes on the faint infrared, or heat, emanations from celestial objects such as warm dust clouds. They use infrared detectors supercooled for maximum sensitivity. However, photographic plates still work for such purposes as broad surveys of the sky, which produce catalogs of stars and galaxies. The combination telescopes, with their wide field of view, capture such images on photographic plates.

ALL DONE WITH MIRRORS

The very first telescopes, like Galileo's, were simple magnifying lenses. But astronomers quickly recognized the advantages of mirrors. Mirrors can be supported at the back as well as at the sides, making larger systems easier to design. Also, the mirror glass need not be perfect except at its reflecting surface, since light does not pass through it. In a lens, by contrast, the glass must be free of faults all the way through.

The earliest large reflectors had mirrors made, not of glass, but a metal called speculum, a nickel compound that polished up to a nice shine. Speculum tarnished easily, however, and it was a nuisance to take the telescope apart every few weeks for repolishing. In addition, the speculum expanded and contracted when the temperature changed, producing distorted images. Therefore, these telescopes fell into disfavor. Until the craft of making large mirrors developed further, very large reflectors could only be built in astronomers' dreams.

Meanwhile, a major problem in early refracting telescopes was that the lens broke up the image into different colors. A lens bends light of different colors at different angles, resulting in a *chromatic aberration*. In the late 1600s, astronomers learned how to combine two or more lenses to correct for chromatic aberration. Refractors became the most commonly built telescopes over the next two centuries. The largest refracting telescope is the 40-inch (102-centimeter) telescope at Yerkes Observatory in Wisconsin, in use since the 1890s. Larger lenses have never been built, because huge pieces of glass bend under their own weight.

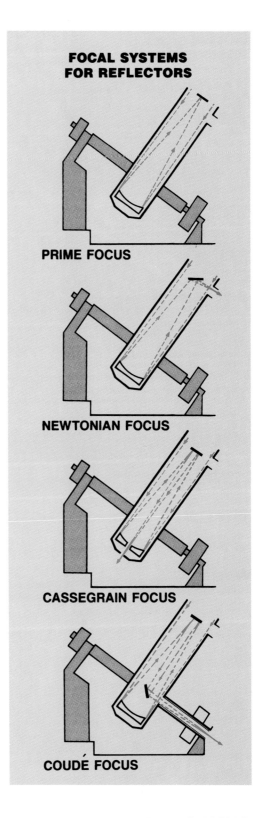

FOCAL SYSTEMS FOR REFLECTORS

PRIME FOCUS

NEWTONIAN FOCUS

CASSEGRAIN FOCUS

COUDÉ FOCUS

The 16-foot Hale refracting telescope was built in 1949. For years a telescope with a larger mirror was considered impossible to build due to its huge weight. Not until 1992 did scientists devise a more powerful telescope—an instrument in which many lightweight mirrors function as a single mirror.

But the discovery of more-suitable materials for giant mirrors permitted scientists to build larger and larger reflecting telescopes in the 20th century. For reflectors to perform well, their mirrors must expand and contract as little as possible in response to temperature changes. Instead of glass and metal, astronomers found that mirrors made of Pyrex and quartz are well suited. Today materials are available that scarcely change at all with temperature. None of the great modern reflecting telescopes has a mirror made of common glass.

THE BIGGEST AND BEST

George Ellery Hale, the mastermind behind some of the largest telescopes ever built, expressed astronomers' urge to build ever-larger telescopes when he wrote in 1928: "Starlight is falling on every square mile of the Earth's surface and the best we

A replica of the telescope used by Galileo to discover lunar mountains, the starry nature of the Milky Way, and previously unnoted Jovian moons.

can do is to gather up and concentrate the rays that strike an area 100 inches [254 centimeters] in diameter?" The giant of Hale's day was the Mount Wilson Telescope near Los Angeles, but Hale yearned for a telescope twice as big.

Hale's dream came true in 1949, in the form of the monster telescope at the California Institute of Technology's (Caltech's) Palomar Observatory near San Diego. The Hale Telescope's 16-foot (5-meter), 14.5-ton (13-metric-ton) mirror made it the world's premier optical telescope for decades.

Scientists were prevented from building larger telescopes due to the huge increase in the weight of mirrors as their diameter increases. The weight problem had prevented construction of a reflector any larger than the Hale Telescope. A mirror that massive would sag out of shape. What's more, the mirror would warp as the telescope shifted its angle to follow heavenly objects across the sky. But in 1992 the frontiers of telescope building expanded again with the completion of the 33-foot (10-meter) Keck Telescope, located on Mauna Kea in Hawaii, by Caltech and the University of California.

Instead of one mirror, the Keck Telescope consists of 36 lightweight hexagonal mirrors, each weighing about 880 pounds (400 kilograms). These fit together like the segments of a honeycomb to form the 33-foot (10-meter) main mirror. Each segment rests on a computer-controlled piston that constantly adjusts the segment, so the precise shape of the overall mirror stays the same as the telescope scans across the heavens.

At the University of Arizona, engineers have taken another approach to building larger, lighter telescope mirrors. They are casting giant mirrors inside a spinning furnace. The centrifugal force pressing evenly against a slab of molten glass will cause it to flow outward, naturally forming the concave surface of a telescope mirror without the need for extensive grinding and polishing. To stand up to grinding, most giant mirrors have to be very stiff and thick. In contrast, the furnace-spun mirrors

The Pic du Midi Observatory stands 8,600 feet above sea level in the French Pyrenees. The rarified air at that altitude allows for better imaging of stars and galaxies.

© Roger Ressmeyer/Starlight

can be light and hollowed out on the bottom. Astronomers plan to apply the Arizona technique to a number of telescopes with mirrors 26 feet (8 meters) in diameter.

Building a single giant collecting surface is only one way to increase a telescope's ability. For example, the European Very Large Telescope, to be built by the next century in Chile, will feature four 26-foot (8-meter) mirrors mounted together. They will combine their light to form a single image.

RADIO TELESCOPES

Ground-based optical telescopes can capture infrared, visible, and some wavelengths of ultraviolet light from outer space. Astronomers also build gigantic radio telescopes to gather radio waves from cosmic objects.

Just as reflecting telescopes use large parabolic mirrors to collect light from the heavens and concentrate it, so radio telescopes do the same with radio waves from celestial objects. But because radio wavelengths are far longer than those of light, the surfaces of radio dishes need not be as precise. Typically, the surfaces of radio-telescope dishes are made of perforated metal or wire mesh, which reflect radio waves just as glass reflects light. These metal reflectors focus the radio signals at a point above the dish, where other antennas capture the signals and send them to sensitive detectors housed nearby.

The world's largest radio telescope is Cornell University's Arecibo radio/radar dish in Puerto Rico. Its 105-foot (32-meter)-diameter collecting surface covers more than 18.5 acres (7.5 hectares). Nestled in the natural bowl of a small valley, the dish consists of some 39,000 aluminum panels. Such a gigantic telescope provides incredible sensitivity, which astronomers need because radio waves from space aren't very powerful. All the radio energy ever collected by radio telescopes on Earth is equal to the energy released when only a few raindrops hit the ground!

Cornell's Arecibo dish can also be used as a radar telescope, sending out powerful radio pulses to bounce off objects such as moons, asteroids, and planets in

© Roger Ressmeyer/Starlight

The Sun rises over the Very Large Array (VLA) radio telescope near Socorro, New Mexico. Consisting of 27 movable dishes arranged in a Y-shaped configuration, the VLA has a resolution equivalent to that of a single dish about 20 miles in diameter.

THE BIG SCOPES

The Keck Observatory on the 13,700-foot summit of Mauna Kea, Hawaii, houses the largest optical telescope. The telescope has a novel, segmented reflecting mirror.

our immediate celestial neighborhood. The radar-reflection data can be used to map these objects and determine their composition.

Because radio waves have such long wavelengths, radio astronomers rely on enormous collecting surfaces like Arecibo's dish for resolving power, the ability to focus fine detail. To match the resolving power of an optical telescope, however, a telescope working at a radio wavelength of 20 inches (50 centimeters) would need a collecting surface a million times as wide. This could amount to a width of 3,100 miles (5,000 kilometers), clearly an impractical feat.

But astronomers have found a way to work around this problem. Two radio telescopes placed far apart but connected by a cable can act in concert. The effect is a telescope as wide as the distance between the two receivers. Of course, the two smaller telescopes do not collect as much radio energy as would a single enormous radio telescope—they are unable to pick out weak radio signals from small or distant objects.

One such multiple-telescope system is the Very Large Array in New Mexico. This collection of 27 giant radio dishes, each 85 feet (26 meters) across, stands on a Y-shaped set of railroad tracks with arms about 13 miles (21 kilometers) long. The 213-ton (190-metric-ton) dishes roll into various arrangements along the tracks for different observations. When the radio dishes are moved farthest apart, they have the greatest resolving power. When they are bunched closest together, they sacrifice some resolving power, but can detect fainter radio signals.

Another array of radio telescopes, called the Very Large Baseline Array, is spread out even farther for increased resolving power. These 10 telescopes are scattered throughout the U.S., and their combined operation yields our most detailed maps of distant galaxies.

TELESCOPES AND RESOLVING POWER

A telescope's resolving power depends largely on the wavelength of the incoming radiation it is being used to view. Light, radio waves, and X rays are all part of the electromagnetic spectrum. They exist in the form of electromagnetic waves, and these waves have different lengths.

All other things being equal, a telescope working to collect visible light, whose wavelength is about 5,000 angstroms (one angstrom equals 0.00000001 centimeter), has twice as much resolving power as that same telescope does when it's collecting infrared waves, at the longer wavelength of 10,000 angstroms. In other words, telescopes can focus the shorter wavelengths twice as well as they can the longer ones.

Opticians have to polish a telescope's mirror, its "light-ray collector," to an accuracy of one-eighth the wavelength at which the telescope is to operate. So a telescope being operated at 200,000 angstroms —the far infrared—need be only 1/40th as precisely polished as a telescope that collects visible light at 5,000 angstroms.

The resolving power of a telescope also depends on the width of its collecting surface, or aperture. Thus the aperture of a telescope operating at 10,000 angstroms must be twice as wide as that of a telescope operating at 5,000 angstroms if it is to have the same resolving power.

For this reason, astronomers want to build optical telescopes that can work together with other telescopes in order to get greater resolution. For example, in 1992 construction began on a second Keck Telescope, a twin of the first located on Mauna Kea, Hawaii. Used together, these telescopes will be able to resolve detail comparable to seeing the two individual headlights on a car at a distance of 16,000 miles (26,000 kilometers)!

Of course, the resolving power of even the largest ground-based optical telescopes is limited by the Earth's atmosphere. The same air turbulence that causes stars to twinkle also bedevils astronomers trying to make the finest possible images of stars or galaxies.

Astronomers try to get above this turbulence by building their observatories on high mountaintops. Such high-altitude locations also reduce absorption of infrared radiation by water vapor in the air and cut down on "light pollution" from cities. Astronomers are even contemplating building optical and radio telescopes on the Moon, where atmospheric distortion, light pollution, and radio interference are nonexistent.

Although ground-based astronomers cannot completely eliminate atmospheric distortion, in the future they may be able to reduce it enormously by using "smart mirrors" in a technology called adaptive optics. These systems use computer links to automatically sense and correct distortions by subtly changing the mirrors' shapes. Such advanced optics may allow astronomers to come close to achieving on Earth the resolution of telescopes orbiting above the atmosphere.

TELESCOPES IN MOTION

Few telescopes stand rigidly in one position. They must turn to follow celestial objects as those objects travel across the sky. Most of the movement we observe is due to the fact that the Earth is rotating on its axis. As the Earth rotates from west to east, celestial objects appear to rise in the east and move westward.

The earliest telescopes were simply moved by hand. For a long while thereafter, telescopes were driven by weights, the same mechanism used by cuckoo clocks. Today almost all telescopes have motor drives controlled by electronic systems. Many have automatic devices, and some use computers. Today's telescopes are very carefully balanced on their mountings so that they can move smoothly and with little force.

The world's largest telescope, the 18.5-acre Arecibo radio/radar telescope, however, is rigidly fixed in place. To collect radio signals from objects that are not directly above the dish, radio astronomers adjust the angle of the receiver above the dish, so that it "sees" one side of the dish or another, thereby collecting signals coming in at an angle.

As the number of telescopes launched into orbit expands, so too will our ability to make in-space repairs.

EYES IN THE SKY

by Dennis Meredith

Peering into a velvet-black night sky teeming with crystalline stars, you may think that our atmosphere is a crystal-clear window to the universe. But to astronomers seeking to capture and analyze cosmic images, the atmosphere is a frustrating, ever-shifting veil.

The same atmospheric turbulence that makes stars twinkle on even the clearest night prevents scientists from capturing high-resolution pictures of cosmic bodies. Earthbound astronomers also have to contend with clouds, storms, atmospheric dust, and the glow of city lights. Their vision is spoiled by the interfering shine of the Moon at night and the Sun in the daytime. Even in the depths of a moonless night, the faint "airglow" of the atoms and molecules in the upper atmosphere adds an unwelcome glare to observations.

Even if all this reflected glow were to suddenly vanish, the molecules of the atmosphere would still pose a barrier, because they absorb many kinds of radiation from the heavens. To understand this, imagine if you could tune your eyes to individual wavelengths of light. The heavens would change drastically in appearance as you proceeded through the spectrum from our Earthbound vantage point.

If you began at the longest wavelengths—radio waves—the heavens would appear bright, because those waves easily penetrate the atmosphere. Then, as you adjusted your eyes to perceive shorter-wavelength "microwaves," the stars would

High-altitude balloons, while still unable to surmount our distorting atmosphere, are used for observing the heavens in regions where the atmosphere is relatively thin. This balloon, launched by Caltech, will be sent to float 100,000 feet over Australia.

© Roger Ressmeyer/Starlight

entirely disappear behind the atmosphere's veil. The sky would remain dark through most of the infrared wavelengths, as there are only a few wavelengths of infrared radiation that can penetrate the atmosphere to reach the ground.

As we tune our eyes through the visible wavelengths, the stars would suddenly burst forth again, only to disappear through most of the ultraviolet and all X-ray and gamma-ray regions.

So you see, from the surface of our planet, we can view only a tiny fraction of the immense panoply of radiation from the cosmos. Yet astronomers know that every wavelength carries an immense amount of information about the stars and other celestial objects.

Realizing their blindness, astronomers have long tried to overcome the atmospheric veil. In the 1920s and 1930s, they tried—largely unsuccessfully—to detect ultraviolet radiation from the ground by moving their instruments to the far northern latitudes, where they believed the ultraviolet-absorbing ozone layer was relatively thin. They also carried their instruments high into the atmosphere in manned balloons.

But it was not until 1946, with the use of captured German V-2 rockets, that the dreamed-of era of space-based astronomy finally began. The V-2s and other rockets allowed astronomers to throw instruments above the atmosphere, where they began mapping X rays and ultraviolet rays from the Sun.

High-altitude balloons and sounding rockets are still useful for space astronomy. While balloons cannot surmount the atmosphere entirely, they do offer long observing times in regions where the atmosphere is relatively thin. High-altitude aircraft are likewise useful for viewing the heavens through the thin upper atmosphere. Astronomers use them to study certain infrared wavelengths.

Sounding rockets, which can thrust well above the atmosphere, remain there only for a few minutes before plunging back to Earth. Astronomers knew they would have to surmount the atmosphere for considerable amounts of time to study the heavens thoroughly.

The most important breakthrough for space astronomy came on October 4, 1957, when the U.S.S.R launched the first Earth-orbiting satellite, Sputnik. Astronomers quickly seized on the idea of orbiting telescopes. From the late 1950s onward, the size and sophistication of astronomy satellites has steadily risen.

Astronomers plan for a full range of space-based telescopes that together can

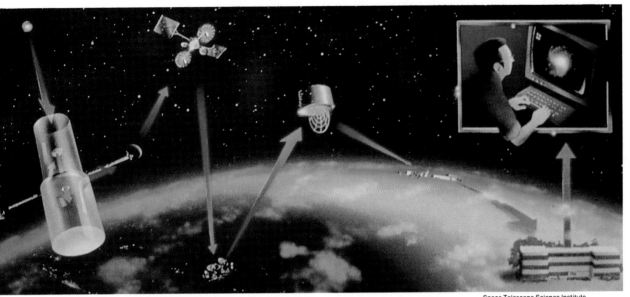

Space Telescope Science Institute

Space-based telescopes typically collect light from a distant star or planet and convert it into radio signals. These signals are relayed via satellite and ground stations until reaching the command center.

In April 1990, after years of precision design and engineering on the ground, the Hubble Space Telescope was launched into orbit. Its five light-detecting instruments make it the first orbiting telescope to operate in the visible part of the spectrum.

NASA

close the gaps in our vision of stellar radiation. The centerpiece of this effort is the National Aeronautics and Space Administration's (NASA's) "Great Observatory Program," a grandly ambitious plan that has already put the 11.5-ton (10.4-metric-ton) optical Hubble Space Telescope, launched in 1990, and the 17-ton (15.4-metric-ton) Gamma Ray Observatory, launched in 1991, into orbit. These giants will be joined in the later 1990s by the Space Infrared Telescope Facility (SIRTF) and the Advanced X-Ray Astrophysics Facility (AXAF).

Now let's survey the past, present, and future of space-based astronomy by proceeding through the electromagnetic spectrum, from the longest wavelengths to the shortest, from radio and microwaves through infrared, visible, ultraviolet, X rays, and gamma rays.

RADIO ASTRONOMY IN SPACE

While radio waves penetrate the atmosphere with little distortion, space-based radio telescopes would still have an edge on Earthbound instruments. One important advantage is that the gigantic radio dishes needed for radio astronomy could maintain their shape much more precisely in the microgravity of Earth's orbit. Radio

EYES IN THE SKY 47

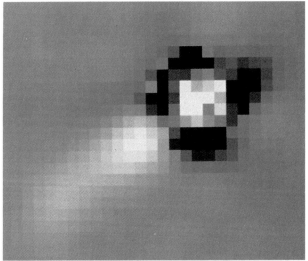

Space Telescope Science Institute

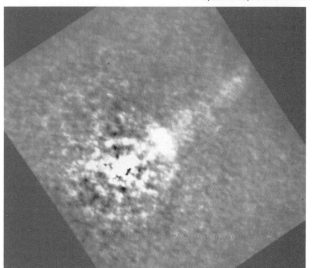

NASA/ESA

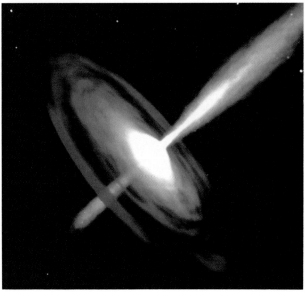

Berry/STSCI

astronomers using the technique of Very Long Baseline Interferometry (VLBI) would also find space useful—not for its lack of atmosphere, but for its immense roominess. VLBI is a technique of observing the same object with widely separated radio telescopes and then combining the data. The resulting power is virtually equivalent to what astronomers would get from one immense dish spanning the distance separating these individual telescopes. By launching a radio telescope into orbit, astronomers can combine its orbital observations with those of Earth-based telescopes thousands of miles away. This will triple the resolution, or sharpness, of images of radio-emitting objects such as quasars and certain energetic galaxies.

While no true radio telescope has yet been launched into space, the Cosmic Background Explorer (COBE), launched in November 1989, has enabled astronomers to study radiation nearly as long as radio waves: microwaves and long-infrared wavelengths. COBE is dedicated to studying the extremely faint radiation left over from the Big Bang, the theoretical explosion that many scientists believe gave birth to our universe. These echoes of the Big Bang may be 100 million times fainter than the electromagnetic radiation emitted by a birthday candle. This radiation would have begun as gamma rays that would have been stretched out into longer wavelengths over billions of years by the expansion of our universe.

INFRARED ASTRONOMY IN SPACE

Infrared radiation is emitted by such warm objects as distant galaxies undergoing extensive star birth, and interstellar clouds of gas and dust that likewise give birth to new stars. Astronomers have long

Space-based telescopes have expanded our knowledge of the heavenly horizons enormously. From a ground-based telescope (top), the phenomenon of plasma jets ejected from distant galaxies is hardly visible. The Hubble (center) produces the clearest image to date of these jets. From the Hubble images, artists are able to conceptualize what the jets would look like close up (bottom).

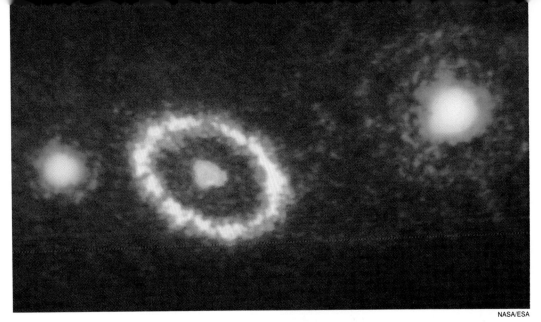

NASA/ESA

When aimed at the remnant of Supernova 1987A that exploded in February 1987, the Hubble discovered a ring of stellar debris that is about 1.3 light-years across and has a temperature of over 20,000° F.

used ground-based optical telescopes fitted with special infrared detectors to peek at certain wavelengths of infrared radiation that can slip through our atmosphere. Then, in 1983, the 10-month flight of the Infrared Astronomical Satellite (IRAS) added an immense wealth of infrared data to astronomy's store. With supercold sensors cooled by liquid helium, IRAS mapped the entire sky at infrared wavelengths. It discovered 250,000 new sources of infrared radiation, including some 10,000 new galaxies that cannot be seen at visible wavelengths.

The next major orbiting infrared telescope will be the Space Infrared Telescope Facility (SIRTF), scheduled for launch before the end of the 1990s. According to plans, SIRTF may be about 1,000 times more sensitive to infrared radiation than are the most powerful Earth-based telescopes, and 1 million times more sensitive than IRAS. Also, unlike IRAS, which could only scan the sky, SIRTF will focus on specific objects in space.

VISIBLE ASTRONOMY IN SPACE

The most famous—some would say infamous—of all orbiting telescopes is undoubtedly the 11.5-ton (10.4-metric-ton) Hubble Space Telescope (HST), carried aloft by the space shuttle on April 12, 1990.

The HST is the first orbiting telescope to operate in the visible part of the spectrum. Its telescope actually consists of five light-detecting instruments that can capture optical and ultraviolet light reflected from its 94.5-inch (242-centimeter) primary mirror. They include the Wide-Field/Planetary Camera (WF/PC), HST's chief instrument. In a wide-field mode, this camera captures images of large, faint objects such as galaxies, galactic clusters, and quasars. In its small-aperture mode, it can focus on individual bright objects such as planets.

Among HST's other instruments is the Faint-Object Camera (FOC), with a relatively small field of view. It is quite sensitive to ultraviolet light. The third instrument is the Faint-Object Spectrograph, which measures the wavelengths of dim objects at both ultraviolet and optical wavelengths. The fourth, the Goddard High-Resolution Spectrograph, operates only at ultraviolet wavelengths. HST's fifth and final "eye in the sky" consists of a high-speed photometer that measures the brightness of objects and gauges variations in brightness.

Shortly after HST's launch, astronomers discovered that the manufacturers of its main mirror had polished its shape too flatly. Ideally, the mirror's concave surface should be shaped such that all incoming light is focused at a single point, producing

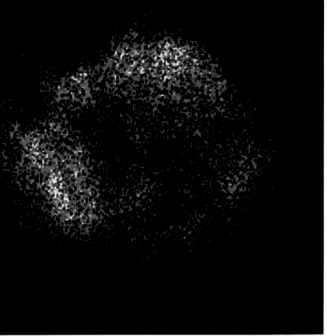

An X-ray image of the supernova remnant Cas A, which exploded in A.D. 1670. This image was taken by ROSAT, a joint satellite project of the Federal Republic of Germany, the United Kingdom, and the United States.

a sharp image. But the mistake in polishing had produced a flaw, called "spherical aberration," that caused light striking the edge of the primary mirror to focus at a different point than light striking near the center of the image. The result was fuzzy, out-of-focus pictures.

The size of the error—the edges of the mirror seem to be about 2 micrometers (millionths of a meter) lower than they should be—is surprisingly large given that the mirror was ground to an overall precision of better than one-hundredth of a micrometer.

Investigators determined that the problem occurred when a measuring rod was inserted backward in an optical device used to guide grinding of the mirror. This rod was vital to properly spacing the mirrors and a lens in a device known as a reflective null corrector. By shining a laser light through this null corrector and onto HST's mirror, technicians would test the mirror's curvature. But with the measuring rod inserted backward, the null corrector's components were misspaced by about the thickness of a dime. And as a result, the shape of HST's mirror was off by less than the thickness of a sheet of paper.

The investigators also determined that budget restrictions and managerial problems led to incomplete testing by the manufacturer and inadequate supervision by NASA. The space telescope also experiences slight shaking due to expansion and contraction of its solar panels as they pass in and out of the Earth's shadow. This shaking further reduces the sharpness of HST's images.

NASA has scheduled a shuttle flight for late 1993 to repair HST's optical flaws. One of the planned correction devices, the Corrective Optics Space Telescope Axial Replacement (COSTAR), is a set of postage-stamp-sized mirrors, some of which are overly concave to compensate for the overflatness of the main mirror. Astronauts may also replace the scope's Wide-Field/Planetary Camera with a new and improved version, and its shaky solar panels with a more stable set.

Despite some serious flaws, the $1.5 billion Hubble Space Telescope has yielded important scientific information on stars, galaxies, planets, and nebulas. Astronomers have been able to compensate for some of the telescope's fuzzy images with computer processing. Fortunately, HST's spectrographic studies have not been significantly affected.

ULTRAVIOLET ASTRONOMY IN SPACE

Perhaps the most widely used space telescope—at any wavelength—has been the International Ultraviolet Explorer (IUE), first launched in 1978 as a joint project of NASA, the European Space Agency (ESA), and Great Britain. Its 18-inch (45-centimeter) telescope remains in a stationary orbit over the eastern U.S. Astronomers around the world keep it in use nearly 24 hours a day in their ultraviolet studies of individual stars, planets, nebulas, galaxies, and quasars.

To date, the IUE has captured more than 90,000 images, including those of virtually every class of astronomical object. Though it was originally designed for a three-year mission, IUE is still returning high- and low-resolution images of a wide

Before the Gamma Ray Observatory was deployed from the Atlantis shuttle in April 1991, an unexpected space walk was required to manually extend one of the satellite's antennas.

NASA

range of astronomical sources emanating in the far-ultraviolet wavelengths.

But the first thorough and detailed map of the extreme-ultraviolet and the X-ray regions of the sky was provided by the Roentgen Satellite (ROSAT), launched in June 1990. ROSAT is also an international project. It was built in Germany, contains instruments provided by Britain and the United States, and was launched by a NASA Delta-II rocket. ROSAT has already been used for extensive ultraviolet studies as well as an all-sky X-ray survey.

Another important space mission for ultraviolet exploration was Astro-1, carried aboard the space shuttle in December 1990. With its three ultraviolet instruments and one X-ray telescope, Astro-1 could study an object at both wavelengths simultaneously. From the cargo bay of the space shuttle, Astro-1 analyzed radiation from a wide variety of objects, including black holes, stars in the process of birth and death, distant galaxies, planets, and comets. Scientists hope to fly the Astro-1 telescope on another shuttle mission in 1993.

Yet another ultraviolet telescope, the Extreme Ultraviolet Explorer (EUE), is scheduled for launch by NASA sometime in 1992. Its two-and-one-half-year mission is to survey the entire sky for radiation that falls between ultraviolet light and X rays. According to plans, the EUE will then be refurbished—while still in orbit—to become the X-Ray Timing Explorer. In this second incarnation, it will study neutron stars and black holes for fluctuations in their brightness and spectra.

X-RAY ASTRONOMY IN SPACE

While scientists got their first glimpse of X rays from space long ago, using balloons and sounding rockets, it was not until 1970 that they launched the first X-ray satellite. The Small Astronomy Satellite-1, nicknamed Uhuru (Swahili for "freedom"), was not a telescope in the traditional sense. That is, it did not actually capture images of celestial bodies. It just pointed astronomers in the right direction, working in a manner much like a radio antenna, enabling scientists to find the direction of X-ray "transmitters." Uhuru enabled astronomers to revise the number of known X-ray sources from the 30-odd they had detected from the ground to several hundred.

The next major X-ray space telescope was the High Energy Astrophysical Observatory-2 (HERO-2), nicknamed the Einstein Observatory, launched in 1978. HERO-2 could actually form pictures of the celestial objects that emit X rays. During its two-and-one-half-year lifetime, it produced more than 7,000 such images.

As mentioned earlier, ROSAT, in addition to its ultraviolet studies, has produced a complete X-ray survey of the heavens, which revealed some 6,000 images of previously unknown X-ray sources.

ROSAT's X-ray mirror is considered to be the finest one ever made. Its instruments are upgraded versions of those that performed so well in the Einstein space observatory.

Over the next decade, the most important orbiting X-ray observatory will be the Advanced X-Ray Astrophysics Facility, or AXAF. It is scheduled for launch in the late 1990s. It will allow astronomers to study in detail for the first time thousands of known sources of X rays, such as suspected black holes and exploding stars called supernovas. AXAF's X-ray measurements will also act as a "tracer" to reveal where large amounts of dark matter in the universe may lie. This dark matter, which produces no radiation of its own, is believed to make up the largest portion of the universe. Yet astronomers have yet to identify even a fraction of it. By measuring X rays from large, visible clumps of matter floating in space, scientists can determine whether large amounts of nearby, unseen dark matter are holding it in place with their gravity.

GAMMA-RAY ASTRONOMY IN SPACE

Gamma rays, the highest-energy emanations from the heavens, are produced by the universe's most violent objects: black holes, neutron stars, supernovas, and mysterious "gamma-ray bursters," which suddenly blast forth huge amounts of gamma rays for wholly unknown reasons.

Gamma rays are also the most difficult type of radiation for astronomers to study. At present, they need huge instruments to capture these energetic rays. Like X rays, gamma rays were studied in the 1960s by a series of small satellites and high-altitude balloons. With these instruments, astronomers were able only to glimpse gamma rays in certain wavelengths or to study only small sections of the sky.

Astronomers' gamma-ray eyes were fully opened with the April 1991 launch of the Gamma Ray Observatory (GRO). GRO carries four instruments, three of which are the size of small cars. To detect gamma rays, these instruments use "scintillators" —crystals, liquids, and other materials that emit brief flashes of light when they are struck by gamma rays. GRO will provide the first detailed map of the sky at gamma-ray wavelengths, the last remaining unmapped region of the spectrum.

THE LASTING NEED FOR GROUND-BASED TELESCOPES

Astronomers do not expect space telescopes to take the place of ground-based observatories. For example, ground-based telescopes can be far larger than those that can be launched into space. As a result, these larger telescopes can gather more light for such critical studies as spectrographic analysis. In spectrographic analysis, instead of imaging the star or other object, the light from it is separated into its different wavelengths to obtain important "fingerprints" that tell astronomers about the object's composition.

Also, ground-based telescopes can be more easily repaired and refurbished than can space telescopes, and they do not require expensive backup systems or devices for remote control. In fact, some say that some of the newest ground-based telescopes can approach, and may someday even match, the resolution of space telescopes. Such advancements in ground-level astronomy are due to the continuing development of "adaptive optics," which can automatically adjust the shape of a telescope's mirrors to compensate for atmospheric distortion.

Images from ground-based telescopes can also be improved through optical interferometry. In this process, images of the same object taken through separate telescopes are combined to give the effect of one huge telescope, virtually the size of the distance separating the two instruments.

But most important, space- and ground-based telescopes complement each other. For example, when a space telescope discovers a new object by using its high-resolution imaging capability, that information can be relayed to astronomers on the ground, who can use their Earthbound telescopes to better analyze the spectrum of the new object's light qualities.

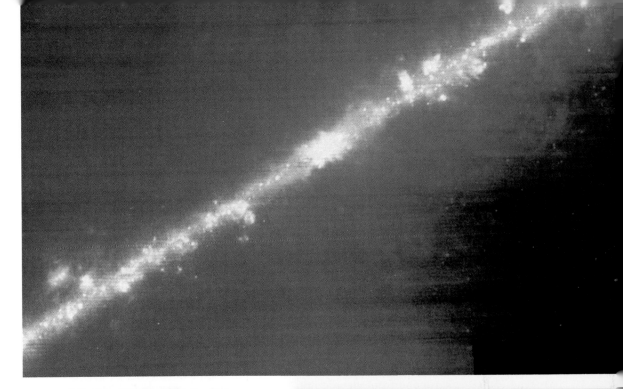

INVISIBLE ASTRONOMY

by William P. Blair

Early astronomers used only the naked eye to view the heavens. Later, observations became more refined with the aid of optical telescopes. But in the last 30 years, astronomers have exploited other wavelengths, such as the infrared radiation used to image the Milky Way (above) and the Andromeda Galaxy (right).

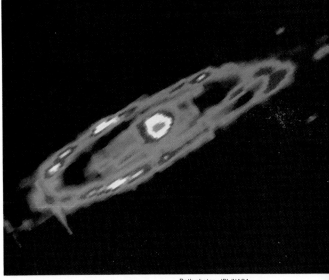

Both photos: JPL/NASA

The field of astronomy has changed dramatically over the past 30 years, and astronomers have changed along with it. Gone is the image of the grizzled astronomer, hunched over with an eye to the telescope, recording notes and measurements in a notebook. Astronomy today is high tech, exploiting every kind of electronic wizardry imaginable.

The tremendous advances in our understanding of the universe are due largely to discoveries made through invisible astronomy, that is, astronomy using wavelengths of light outside the visible spectrum. And with the prospect of new, more powerful and precise instruments and computer technology, these discoveries will continue for years to come.

When optical light is passed through a prism, it breaks up into a rainbow of colors —red, orange, yellow, green, blue, indigo, and violet. Each color represents a different wavelength of light. This optical spectrum is a tiny portion of a much larger range of wavelengths called the "electromagnetic spectrum," which extends outward from both the blue and red portions of the optical spectrum. Hence, the ends of the optical spectrum are not really "ends," but rather limits to what our eyes can detect. Beyond the red limit of the optical spectrum (toward longer wavelengths), one encounters first infrared and then radio light. Beyond the violet limit, one finds ultraviolet, X-ray, and gamma-ray light. There are no "borders" between parts of the spectrum—they are all part of the same continuous spectrum of electromagnetic radiation.

Although these extreme wavelengths of light are invisible to the naked eye, we can build instruments to sense and measure these wavelengths, and computers to transform their signals into information we can see and analyze.

Observations at invisible wavelengths tell us much about our universe that could not be learned from visible astronomy alone. For instance, the center of our Milky Way galaxy is shrouded from view at optical wavelengths by clouds of interstellar gas and dust. Infrared and radio light can penetrate this dust, however, and allow us to see such regions directly. Also, many objects in the universe emit primarily at wavelengths other than optical. Consequently, hot stars, active galaxies, and the exploded remnants of dead stars can be most effectively studied at ultraviolet and X-ray wavelengths. Some of the galaxies discovered using infrared instruments are 1,000 times more luminous than our own galaxy, yet they are undetectable at optical wavelengths!

It is no coincidence that our eyes are sensitive to visible light—visible wavelengths pass through our atmosphere to reach the Earth's surface. But among the invisible wavelengths, only radio waves and a few narrow regions of infrared penetrate the atmosphere. The rest of the electromagnetic spectrum gets blocked by our atmosphere. While this blockage is fortunate for life on Earth, since ultraviolet and X-ray emissions are deadly, the atmosphere's veil does cause some obvious problems for invisible astronomy. Not surprisingly, the vast majority of advances in invisible astronomy have occurred since the advent of the space age and the concurrent technical ability to place instruments above the Earth's atmosphere.

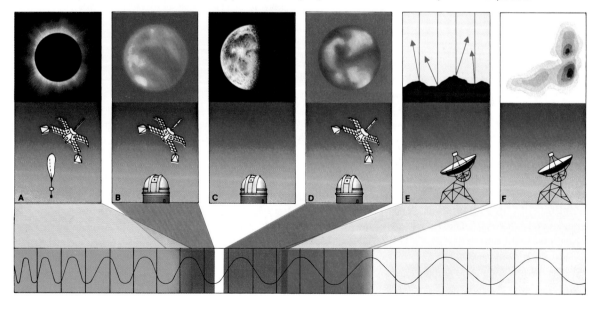

Astronomers can now study the wide variety of celestial bodies using all the wavelengths of the electromagnetic spectrum. (A) Balloons and orbiting satellites have collected X-ray and gamma-ray radiation emitted from the Sun's corona, providing invaluable information about the Sun's structure, density, and temperature. (B) Ultraviolet studies of Mars and other planets have revealed atmospheric compositions and pressures. (C) Observations of all objects using the visible portion of the spectrum remain important to astronomers. (D) Infrared studies of planets such as Mars, as well as the stars, furnish details about their origin. (E) Radar units yield data about distances and rotation rates of celestial bodies. (F) Radio telescopes provide information about pulsars and quasars.

The radio-radar telescope facility at Arecibo Observatory in Puerto Rico is used to monitor radio emissions from inside and outside our galaxy.

RADIO ASTRONOMY

Radio astronomy was the first part of invisible astronomy to be exploited—largely because a "window" in the Earth's atmosphere allows radio waves from less than an inch to about 98 feet (30 meters) in length to reach the ground. The first cosmic-radio sources were discovered in December 1931 by Karl G. Jansky, an engineer at Bell Telephone Laboratories who was working on the problem of radio static. After eliminating Earthly sources of static, he found that he was still left with a persistent hiss, whose source moved across the sky from east to west. With careful tests, he showed that this source of cosmic-radio noise came from our own galaxy, the Milky Way. In 1936 an American amateur radio operator named Grote Reber built a radio receiver in his backyard in Wheaton, Illinois, and produced the first radio map of the sky at 24-inch (61-centimeter) wavelengths. Yet such pioneering work received little attention for years, obscured as it was by the focus on military science during World War II.

In the long run, however, the war greatly advanced radio astronomy. Once peace was achieved, it was a relatively straightforward task to turn military radio receivers upward to investigate the heavens. Early discoveries included radio emissions from the Sun, and radar studies of the Moon and meteors as they entered our atmosphere. Although rather mundane by today's standards, these observations opened the door to much greater discoveries. Ever-larger radio telescopes were built, and a wider range of radio wavelengths was investigated, through the 1950s and early 1960s. Radio emissions at the 8-inch (21-centimeter) wavelength, emanating from interstellar hydrogen, were used to map the great clouds of gas in our galaxy. Many discrete sources of radio emissions were also found, including the enigmatic quasars—their name an acronym for "quasi-stellar radio sources." Radio astronomy was also responsible for the detection of pulsars—rapidly spinning neutron stars formed in some supernova explosions.

The resolution of a telescope (that is, its ability to see fine spatial detail) is a function of both the wavelength of light being sampled and the diameter of the telescope. At good sites, optical telescopes (which are also limited by turbulence in the atmosphere) achieve angular resolutions of about one second of arc (or 1/3600 of a degree). Because wavelengths of radio light are so much longer than those of optical light, radio observations were intrinsically of lower resolution than optical observations. Hence, astronomers wanting to see faint, distant objects and to improve resolution of closer ones were highly motivated to build very large radio telescopes.

The largest fully steerable radio telescope is the 328-foot (100-meter) dish in Bonn, Germany. Its curved, metallic dish is the size of a football field—yet readily movable. In Arecibo, Puerto Rico, astronomers use a 984-foot (300-meter) radio dish built right into the landscape. Instead of moving the telescope, a radio receiver at the focus of the dish is moved.

INTERFEROMETRY

One of the revolutions in radio astronomy has been the development of a technique called interferometry, which permits astronomers to combine radio signals from separate telescopes and, by doing so, see much finer detail. At least two radio telescopes are needed for interferometry, and the wider their separation, the higher the resolution achieved. Some information about the objects being observed is lost with this technique, but the more radio telescopes you add to such a network, the

The Very Large Array radio telescope in New Mexico combines radio signals from 27 separate radio dishes arranged in a Y configuration that extends as far as 17 miles.

NRAO

closer it comes to simulating a single, enormous instrument. The practical use of interferometry is by and large dependent on advanced computer technology, since it requires complex processing to combine and coordinate the signals from separate telescopes.

One of the most powerful interferometers today is the Very Large Array of radio telescopes (or VLA) near Socorro, New Mexico. Twenty-seven separate radio dishes, each 82 feet (25 meters) in diameter, are arrayed in one of four alternate Y-shaped configurations for observations. The most extended of these configurations has a diameter of 17 miles (27 kilometers). The VLA's radio antennas can be moved when necessary by placing them on railroad tracks. Socorro's VLA radio telescope produces radio images with a resolution comparable to ground-based optical images. And it has been used by hundreds of astronomers to observe everything from planets to quasars. The VLA has observed some of the largest single structures in the universe—mysterious double plumes of radio emissions that extend outward from the cores of some very active galaxies.

The best resolutions can be obtained by combining signals from radio telescopes separated by thousands of miles, even continents. Such Very Long Baseline Interferometry (VLBI) lacks the full imaging capability of the VLA, but has nonetheless achieved impressive resolutions of certain celestial objects. VLBI measurements are so sensitive that they can be used to detect movements of less than an inch in the Earth's crust between the individual telescopes!

VLBI measurements are accomplished largely by special arrangements made between the researchers who control radio telescopes around the world. Today an array of radio telescopes dedicated solely to VLBI measurements is being built by the U.S. National Science Foundation (NSF). Called the Very Long Baseline Array, or VLBA, it will include 10 radio telescopes located from Hawaii to the Virgin Islands. It is scheduled to go into operation in 1992.

Mount Wilson and Palomar Observatories

Radio and optical astronomy complement each other in the study of the heavens. After radio emissions have been detected from a particular part of the sky, the optical astronomer then knows where to focus his telescope to discern the object and discover more about it. The radio source emitted by Cygnus A led optical astronomers to discover that it was actually a double nebula (above), possibly a collision of two galaxies.

THE FUTURE OF RADIO ASTRONOMY

Although astronomers don't have to go into space to detect radio light, an orbiting radio telescope could be used in conjunction with ground-based telescopes to obtain even longer baselines and higher resolutions than ever before. An international coalition will be needed to fund and execute such a project, however.

Another reason to pursue radio astronomy from space would be to get away from Earth-based "radio pollution," man-made signals that interfere with radio astronomy in much the same way that "light pollution" plagues optical astronomers. Although specific parts of the radio spectrum are set aside for the sole use of radio astronomy, it is becoming increasingly difficult to find sites where radio telescopes are not affected by television- and radio-station signals. Hence, the idea of putting radio telescopes on the back side of the Moon, away from Earthly interference, is a hot topic among future-minded astronomers.

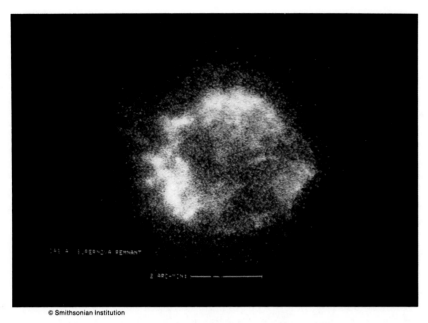

X-ray radiation glows in this spectacular image of Cassiopeia A, a supernova remnant of gas and debris that was left when a star exploded. Supernova explosions can continue to emit X rays for hundreds or thousands of years.

X-RAY ASTRONOMY

X-ray light and its high-energy cousin, gamma-ray light, are found at extremely short wavelengths. X rays typically have wavelengths about 4,000 times shorter than those of visible light, or less than a few angstroms (one angstrom is equal to one ten-billionth of a meter). Gamma-ray wavelengths are still smaller by a factor of 100 to 1,000. To put it another way, X-ray wavelengths are roughly the size of atoms, while gamma rays have wavelengths about the size of atomic nuclei. Since wavelengths are inversely proportional to the energy needed to create the light, these emissions come from the most energetic phenomena.

X-ray astronomy provides a very different perspective on our universe. This new way of looking at the skies has brought into focus hot young stars, exploding stars, and neutron stars, as well as active and exploding galaxies, and perhaps even black holes.

EARLY X-RAY ASTRONOMY

As early as the late 1940s, scientists fitted suborbital rockets with crude detectors to monitor X rays coming from the Sun. For nearly 15 years, the Sun was the only object bright enough to be detected in X-ray light. But in 1962, with better X-ray detectors and improved rockets, astronomers discovered an X-ray-emitting object in the constellation Scorpius. This object, dubbed Sco X-1, turned out to be a member of one of the primary classes of X-ray sources: a close binary star where solar material is actually jumping from one star to the other. With the discovery of Sco X-1, X-ray astronomy was born.

In 1970 astronomers launched the first satellite designed to study X-ray emissions. Named Uhuru, this satellite carried a simple scanning device that could tell bright from dark X-ray areas. The early instruments used in X-ray astronomy were like the light meters used by photographers. They could tell whether a patch of the sky was dim or bright in X rays, but they could not take pictures. While Uhuru could map the sky at X-ray wavelengths, it could not really "see." As X-ray-detector technology developed, scientists learned how to convert X rays to visible light. With this great advance, they were able to make X-ray pictures.

THE EINSTEIN AND ROSAT MISSIONS

The first satellite capable of taking X-ray pictures of the sky was launched in 1978. It was named the Einstein X-ray Observatory. The Einstein telescope consisted of a special X-ray grazing-incidence mirror, two X-ray cameras, and two X-ray spectrographs. For nearly two and a half years of operation, this telescope and spacecraft were controlled by onboard computers and radio commands from a ground-based control center. The telescope detected many thousands of X-ray sources, some 1,000 times fainter than any previously observed.

Scientists learned that X-ray emitters include normal stars, young and old; neutron stars and black holes; remnants of supernova explosions; galaxies; quasars; and even clusters of galaxies. Researchers today continue to learn from the data returned by the Einstein satellite.

On June 1, 1990, NASA launched the ROSAT X-ray mission into orbit on a Delta II rocket. ROSAT was built by the Germans, includes instruments built by Great Britain and the U.S., and was launched by the U.S. The ROSAT X-ray mirror is the finest one ever made, and its instruments are upgraded versions of the ones that worked so well on the Einstein satellite. After an initial checkout period, ROSAT performed an all-sky X-ray survey, and has now moved into a phase of "pointed observations" at specific objects of interest. With three times more sensitivity and higher spatial resolution than Einstein, ROSAT promises many new discoveries.

FUTURE OF X-RAY ASTRONOMY

The U.S., Japan, and Europe all have active X-ray-astronomy research programs. Toward the end of the 1990s, the National Aeronautics and Space Administration (NASA) plans to fly a mission called the Advanced X-ray Astronomy Facility (AXAF). It will have an enlarged and improved version of the Einstein telescope, and will provide pictures 10 times sharper of distant X-ray objects. It will also cover a wider range of X-ray wavelengths and provide higher-quality X-ray spectra. AXAF represents the X-ray portion of NASA's Great Observatories program.

In the realm of gamma rays, in April 1991, NASA launched the Gamma Ray Observatory (GRO) to investigate the highest energies of electromagnetic radiation known. Although gamma-ray light had previously been detected by balloon-borne detectors, relatively little is yet known about this extreme end of the spectrum.

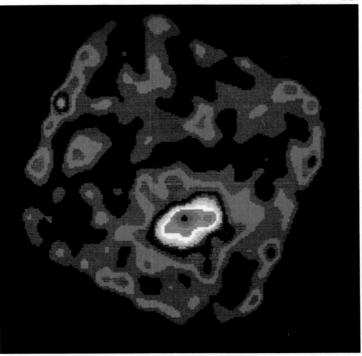

The ROSAT satellite captured this X-ray image of the cluster galaxy Abell 2256. Cluster galaxies are traditionally strong X-ray emitters, due to the tremendously hot gases which are swept out of colliding galaxies to accumulate in intergalactic space.

An edge-on view of the galaxy NGC 4565. The reddish area represents infrared light, which passes more easily through obscuring interstellar dust than does visible light.

INFRARED AND ULTRAVIOLET ASTRONOMY

Infrared Astronomy

The infrared wavelengths lie between the optical part of the light spectrum and the radio region. A few small portions of the near-infrared spectrum (those wavelengths closest to optical light) are able to slip through the Earth's atmosphere to where they can be observed directly from the ground. The longer far-infrared wavelengths, however, are completely absorbed by the slightest amount of atmospheric water vapor and gas.

Infrared light, also known as heat radiation, was first measured in an astronomical context by the English astronomer Sir William Herschel in the early 19th century. After breaking the Sun's light into a spectrum, Herschel used a thermometer to measure the temperature of various colors of sunlight, and noticed that he continued to get readings beyond the red end of the visible spectrum.

In the 20th century, astronomers attached electrical devices called thermocouples to optical telescopes to discover that stars appearing red radiate relatively more infrared energy than stars of other colors. In this way, they have also discovered that the planet Jupiter gives off twice as much energy as it receives from the Sun. (This discovery led to the conclusion that the massive planet is producing its own energy by gravitational contraction.)

Since infrared light readily passes through interstellar dust, while optical light gets absorbed, infrared observations have been used to learn much about the dusty regions where stars are being born.

Water vapor, which is the main block to infrared wavelengths, is concentrated in the lower atmosphere. Hence, one need not go all the way out of the atmosphere to observe some parts of the infrared spectrum. Infrared telescopes have been built in very dry and high places and sent aloft aboard special balloons and suborbital sounding rockets. Such technology enabled astronomers to obtain detailed information on individual sources of infrared and regions of star formation, but it could not provide a complete infrared survey and map of the heavens.

Refrigerators in Space

In actuality, everything in the universe emits some energy at infrared wavelengths —that is, everything gives off some heat. One of the ramifications of this is that even an Earth-based telescope "glows" at infrared wavelengths, and so can obscure faint infrared signals from objects in space. Astronomers recognized that one solution to this problem would be to place an infrared telescope in space (see also page 45). Besides providing a viewing platform far above the atmosphere, space provides an insulating vacuum. Using liquid helium, an orbiting infrared telescope could be chilled to within a few degrees of absolute zero ($-459°$ F, or $-273°$ C), reducing the instrument's own infrared background radiation and making the telescope much more sensitive. Also, an orbiting infrared satellite would at last make it possible to survey the entire sky for radio objects.

Thus, in cooperation with Great Britain and the Netherlands, the United States launched the InfraRed Astronomical Satellite (IRAS) aboard a Delta rocket on January 25, 1983. IRAS was the largest cryogenic (operating at very low temperatures) telescope ever put into Earth orbit. Its detectors operated at four far-infrared wavelengths, providing a wealth of temperature information on the objects being observed. IRAS was placed in a polar orbit at an altitude of 560 miles (900 kilometers), and was programmed to observe a 2-degree-wide band perpendicular to the line joining the Earth and Sun. Thus, as the Earth slowly moved around the Sun, IRAS slowly built up an infrared map encompassing nearly the entire celestial sphere.

During its 10 months of operation, IRAS provided astronomers with a new view of the universe. This 22-inch (57-centimeter) telescope was 1,000 times more sensitive than any infrared instrument on Earth. It peered through the interstellar dust that obscures our visible view of the center of the Milky Way galaxy, providing for the first time a clear, panoramic view of the center of our own galaxy. Within our solar system, IRAS found five new comets. One of these, named IRAS-Araki-Alcock after the satellite and two discoverers, was IRAS' first major find. This comet passed within 3 million miles (4.8 million kilometers) of Earth in May 1983.

The comet IRAS-Araki-Alcock, long invisible to optical telescopes, was discovered using the infrared-sensitive IRAS telescope.

JPL/NASA

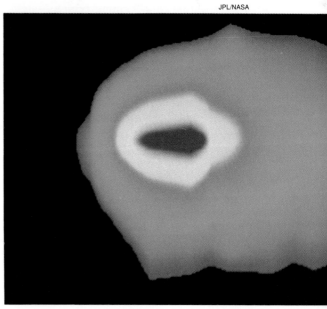

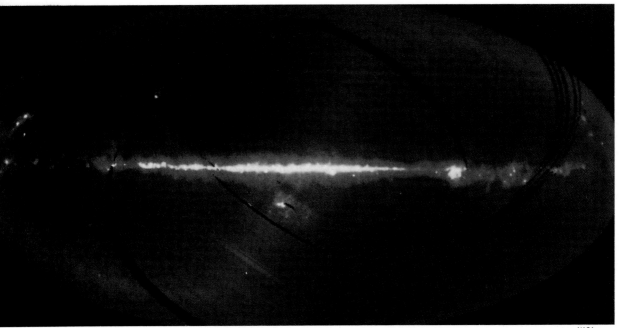

Most of our knowledge of the infrared-emitting objects in our universe is due to the IRAS satellite.

NASA

Outside our solar system, IRAS found that Vega, the brightest star in the constellation Lyra, was surrounded by a ring of solid particles 15 billion miles (24 billion kilometers) across—the kind of space debris that may eventually coalesce to form planets around this star. IRAS also peered deep into interstellar clouds of gas and dust to show that star-forming regions—stellar nurseries—are much more prevalent than previously thought. In all, IRAS cataloged more than 250,000 individual infrared sources and covered more than 98 percent of the sky.

NASA recently launched another refrigerated telescope. The Cosmic Background Explorer (COBE) was designed to survey infrared radiation thought to be the remains of the titanic explosion called the "Big Bang," which is believed to have started the universe expanding some 15 billion years ago. Hence, COBE looks past the stars, past the galaxies, and past the quasars to the very earliest times in the universe. COBE has found that this cosmic background is exceedingly uniform in all directions, so much so that it is difficult for scientists to understand how our present "lumpy" universe could have evolved.

The Future of Infrared Astronomy

Scientists will spend many more years studying the infrared data produced by IRAS and COBE. In the meantime, NASA is already planning an infrared telescope 1,000 times more sensitive than IRAS. This telescope, the Space InfraRed Telescope Facility (SIRTF), is to be launched from the space shuttle sometime during the next decade as part of NASA's Great Observatories program. NASA also has plans to build a larger flying infrared observatory.

ULTRAVIOLET ASTRONOMY

Ultraviolet (UV) astronomy encompasses the range of wavelengths from about 50 angstroms up to about 4,000 angstroms. Astronomers find it convenient to subdivide this range into the near UV (3,000 to 4,000 angstroms), the far UV (912 to 3,000 angstroms), and the extreme UV (below 912 angstroms). Near-UV light can penetrate our atmosphere and be observed from the ground. Far- and extreme-UV light can be observed only from above the atmosphere.

The wavelength at 912 angstroms is particularly significant because, at this wavelength, interstellar hydrogen gas cuts off our "sight" of nearly all objects outside our solar system. Hydrogen, the most abundant element in the universe, effectively absorbs wavelengths for several hundred angstroms short of 912 angstroms, making observations in this range impossible. However, as one moves to shorter wavelengths, hydrogen becomes less effective at absorbing light, and our observations improve. Given this difficulty, relatively little work has been done in the extreme UV.

Early Discoveries in Ultraviolet Astronomy

As with infrared astronomy, astronomers have lofted UV telescopes above the atmosphere by balloon and sounding rockets. But the majority of discoveries in UV astronomy have come by way of true, space-based telescopes. In 1972 the Copernicus satellite was the first to obtain high-resolution images at the short-wavelength limit of the far-UV range. This satellite was largely restricted to observing the hottest bright stars, and returned a wealth of information about these stars, as well as about interstellar gas.

In the mid-1970s, a satellite called TD-1 performed a survey of the entire sky at four far-UV wavelengths. The satellite cataloged over 31,000 objects with bright UV emission.

The biggest workhorse in UV astronomy, however, has been the International Ultraviolet Explorer (IUE) satellite, a joint venture of NASA, the European Space Agency (ESA), and Great Britain. Launched in 1978 for a planned three-year mission, IUE is still operating today. The IUE telescope is a mere 16 inches (41 centimeters) in size. Yet its location above the atmosphere and its geosynchronous orbit (above the equator and moving at the same speed as the Earth rotates) have enabled it to obtain long exposures and observe fairly faint objects. To date, the IUE has captured more than 90,000 individual spectral

The data provided by the International Ultraviolet Explorer satellite, launched in 1978, has provided information on virtually every type of astronomical object.

© Roger Ressmeyer/Starlight

Satellites far above the Earth's obscuring atmosphere, containing telescopes sensitive to the invisible portion of the spectrum, are revealing a universe that, until now, has been inaccessible to astronomers.

NASA

exposures. Virtually every class of astronomical object has been examined with IUE. And it is still returning high- and low-resolution images of a wide range of astronomical sources emanating in the far UV.

Of course, IUE is a relatively small telescope. Many years of planning and effort were culminated in April 1990 with the launch of the Hubble Space Telescope, which has both far-UV and optical telescopes. Hubble is pushing UV investigations to fainter objects and greater distances than ever before possible.

The Future of UV Astronomy

Because of absorption by interstellar hydrogen, relatively little exploration has occurred in the extreme UV region beyond 400 angstroms. Recently the X-ray satellite ROSAT carried aloft a Wide Field Camera developed in Great Britain. This camera is producing the first all-sky survey at extreme-UV wavelengths. A much more extensive instrument, called the Extreme Ultraviolet Explorer (EUE) mission, is scheduled for launch by NASA in 1992. This satellite will observe the entire sky at four extreme UV-wavelengths, and then follow up with detailed analysis of many important sources. EUE will thus open one of the last "windows" on the universe for detailed scrutiny.

Another mission, called the Lyman Far Ultraviolet Spectroscopic Explorer, or Lyman/FUSE, is under development by NASA for a launch in the late 1990s. This satellite is expected to produce very-high-resolution spectra of objects in the far- and extreme-UV part of the spectrum. Lyman/FUSE will be about 50,000 times more sensitive than the Copernicus satellite that flew in the early 1970s, and will extend high-resolution UV spectroscopy to many more exciting objects.

Courtesy of Audio-Visual Imagineering, Inc.

Planetariums, long renowned for showcasing the heavens, are now used for a variety of entertainment purposes, including spectacular laser light shows.

PLANETARIUMS

by Dennis L. Mammana

The doors open, and a large, circular theater looms before us. As we take our seats and look around, we can feel the excitement build as powerful music rises and darkness falls. The floor slowly drops away. We are now surrounded by stars, more than we've ever seen at one time in our lives. Some are familiar: There's the Big Dipper. Over there is Orion, the Hunter. And there is. . . . Wait. The sky is beginning to change. No, we're moving, drifting up and away from . . . wherever we are in the universe. Planets and asteroids rush by. A strange orange planet rises in front of us, growing ever larger. Suddenly we are immersed in the planet's clouds, a swirling reddish brown haze. Above, we see two suns—one blue and one yellow—engaged in a graceful cosmic dance. We soon forget them, however, as a huge lightning bolt tears through the clouds, literally shaking us to the core with its tremendous blast. Now what?! We reach for something to cover us as rain begins to fall. Is it water sprinkling onto our skin, or some strange, alien chemical?

Obviously, this is no ordinary theater. It is a remarkable multimedia coliseum of illusion, an engulfing experience that can capture our five senses and our imagination and carry us places we thought impossible. This is a modern planetarium.

EARLY PLANETARY MODELS

The fantastic space theaters of today have evolved from a long line of planetariums with a rich history extending back many centuries. The earliest planetariums began as scale models of the sky—though we might hardly recognize them as such.

The ancient Greek astronomer Eratosthenes may have been the first to build a sky model, around 250 B.C. His device, a metal sphere and surrounding rings, moved on two axes. The metal "celestial sphere" moved on one axis, while the rings, which represented the paths of the Sun and other planetary bodies, moved on the other. The device and those like it became known as armillary spheres.

The Greek mathematician and inventor Archimedes built an impressive armillary sphere powered by water. It is said to have been so accurate that it could reproduce eclipses of the Sun and Moon. Another ancestor of the modern planetarium is the celestial globe. This was a sphere on which the positions of the stars and constellations were marked so they could be viewed from the outside.

One other popular model, built in the 1700s, was the *orrery*, named after the British earl of Orrery, for whom it was built. It was notable for showing that the Earth was just one of the solar system's children planets, all of which were mounted on rotating arms surrounding a central sun. Orreries are still popular in science classrooms today, and are sometimes incorporated into a clock that coordinates their movements.

The first planetarium theater was built in 1657 by Andreas Busch. He constructed a large globe into which twelve people could climb. The stars were fixed on the inside of the globe, and the planets moved along a set of internal rings. Visitors could sit and gaze at the artificial stars and planets as if sitting under a perfectly clear night sky.

Astronomers continued to team with engineers to build such globes right up to the 20th century. Dr. Wallace Atwood of the Chicago Academy of Science built the last of these globes. Fifteen feet (4.5 meters) wide and powered by electricity, it is still on display at the academy.

Scala

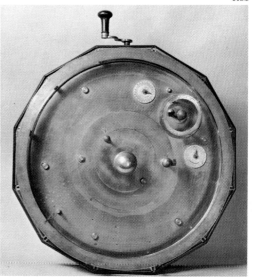

Photo Tomsich Rome

These early models of the sky are actually precursors to the modern planetarium. An 18th-century Copernican model (top) shows the Earth, Moon, Venus, and Mercury revolving around the Sun. An armillary sphere (right) is surrounded by metal rings, each mounted on a different axis to represent the paths of the Sun and other celestial bodies across the sky.

THE MODERN PLANETARIUM

For most of us, the word "planetarium" conjures up the image of a domed room with a large, dumbbell-shaped projector, or "bug," in its center. Here is a place where stars, the Moon, and planets can be realistically viewed in the middle of the day.

This modern form of the planetarium was developed in the early 1900s by Drs. Max Wolf of the Heidelberg Observatory and Walter Bauersfeld of the Carl Zeiss Company in Jena, Germany. These men were the first to envision and design a planetarium projector. Their idea, as Bauersfeld described it in 1919, revolutionized our ability to re-create the cosmos: "The great sphere shall be fixed; its inner white surface shall serve as the projection surface for many small projectors which shall be placed in the center of the sphere. The reciprocal positions and motions of the little projectors shall be interconnected by suitable driving gears in such a manner that the little images of the heavenly bodies, thrown upon the fixed hemisphere, shall represent the stars visible to the naked eye, in position and in motion, just as we are accustomed to see them in the natural clear sky."

In August 1923, Bauersfeld demonstrated the first "Zeiss Model I" planetarium projector in a makeshift plaster dome on the roof of the Zeiss factory. It showed in stunning detail the stars, Sun, Moon, and planets as they would normally be visible from Munich. Two months later the projector was installed in the Deutsches Museum, opening to the public on October 21, 1923. "The Wonder of Jena," as it soon became known, captured the public's imagination as no astronomical novelty ever had.

The Zeiss Company soon began constructing more planetarium projectors and developed the dumbbell-shaped projecting instrument. Descendants of these projectors are still seen in many planetariums today. At the center of each end of the dumbbell is a high-powered lamp. Arranged around each lamp are numerous lens systems, each containing a metal slide

Cie generale de physique

The Zeiss projector (above), the mainstay of many modern planetariums, projects images of the stars and planets as they appear at different times during the day or year, in both the northern and southern hemisphere.

of an area of the sky. Tiny holes in the slides represent individual stars and constellations. As light passes through each hole, a point of light—or star—is formed on the dome ceiling. Bigger holes are needed to project brighter stars. Re-creating the brightest stars requires separate projectors equipped with filters to produce each star's correct color. Similar projectors are used to show the images of planets, while still larger projectors create images of the Sun and Moon.

The first such planetarium in the United States was built in 1930 in Chicago. It became known as the Adler Planetarium, and was one of only half a dozen planetariums in the world. The equipment, backup facilities, and staff needed to run these planetariums were tremendously expensive. Only the largest of major-city museums could afford them.

In 1947 Armand Spitz, an amateur astronomer from Philadelphia, designed and built a smaller, simpler, and less expensive planetarium projector. His invention en-

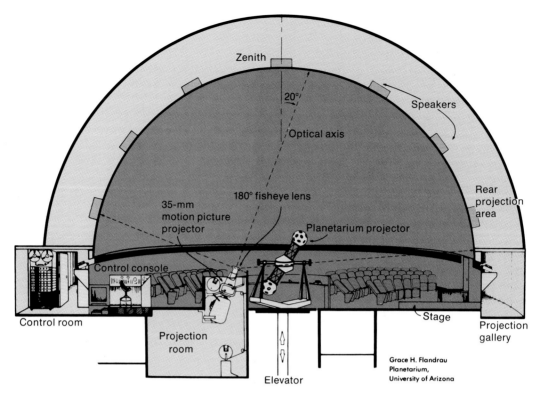

A typical planetarium has a characteristic dome ceiling, a Zeiss projector, a control console for the lecturer, and seats arranged for easy viewing of the ceiling.

In addition to the motions of celestial objects, planetariums often are equipped to project images of outer space, such as this lunar landscape exhibit at New York City's Hayden Planetarium.

© Mark Romanelli/The Image Bank

A show at the Virginia Living Museum captures the vibrant intensity of the Orion nebula, which is composed of gas and dust that reflect the light of nearby stars.

abled many small museums, libraries, and schools to install their own, more economical planetariums.

Since that time, planetariums have been springing up at an average rate of about 30 a year. Today there are nearly 1,400 operating around the world. In addition to the Zeiss and Spitz models, some 165 different types can be found in 64 countries on six continents.

HOW DOES IT ALL WORK?

A planetarium instrument is more than a static projector. It is a complicated machine of optical and mechanical parts that moves the stars, Sun, Moon, and planets across the "heavens." The gear systems that perform the work have been developed through detailed mathematical analysis of the paths of the heavenly bodies. A planetarium projector can show four basic motions: daily motion, annual motion, latitude motion, and precession, the motion created by the subtle wobbling of our planet on its axis.

The sky's daily motion—the sunrise in the east and the sunset in the west, for example—is really a manifestation of the Earth's daily rotation. While it takes 24 hours for the Earth to turn once, most planetarium projectors can perform the same feat in less than a minute. Using this daily motion, the planetarium operator can select any time of day to be shown on the dome.

Annual motion is the movement of the Earth and other planets as they move around the Sun. The planetarium operator can take the audience through a year in less than a minute, and re-create the sky on any day of any given year.

Latitude motion allows the planetarium operator to simulate the changing sky as seen from different latitudes on Earth. This enables the audience to view the heavens as they would appear above the North

The "Space Transporters" program at the Adler Planetarium in Chicago allows visitors to explore the atmosphere and environment of the Sun, Moon, and planets.

Pole, from New York City, Rio de Janeiro, or from virtually any point on the Earth's surface. One trip around our planet can be accomplished in a matter of seconds.

The subtle planetary motion called precession, which changes the celestial north pole in the sky, brings a slightly different array of stars into view over a given point on Earth. This subtle motion takes some 26,000 years to complete one cycle. Yet in the planetarium, audiences can experience the effects of a complete precession cycle in less than a minute.

By moving the projector along its many separate axes, the operator can also turn a planetarium into a space-and-time machine. Not only can it show you what the stars, Sun, Moon, and planets look like from any point on Earth, it can let you view the sky as it was on the night you were born, or how it will appear 50 or 100 years from now. Some planetariums now have the capability of re-creating solar and lunar eclipses, variable stars such as supernovas, meteors, weather phenomena like racing clouds, and beautiful sunrises and sunsets. All of this is done with remarkable scientific accuracy.

Today all this wizardry can be controlled by a solitary operator, who creates the desired effects with the knobs and buttons on the planetarium console. These professionals usually have training in both astronomy and the operation of electromechanical devices. Thanks to computer technology, operators can even preprogram a show, then just sit back and enjoy it.

And just as the hardware behind the shows has changed to match the times, our concept of a planetarium "show" has changed as well. Live lectures, with an astronomer describing the sky as audiences would normally see it at night, are still an important part of a planetarium's schedule. But the big crowd pleasers are spectacular multimedia extravaganzas.

LASER SHOWS AND SPACE THEATERS

Modern planetariums can create remarkable three-dimensional illusions of space travel. Shows now combine the projection of stars with flashy special effects that whisk the audience into space. Virtually anything real or imagined can be produced in the planetarium. Planetarium devices can take you on a journey across our galaxy, race you across the surface of Mars, or cast you into a black hole.

The shows literally surround an audience with realistic otherworldly landscapes and distant skyscapes. Colorful lasers, rotating planets, swirling clouds, and whizzing spacecraft fill the planetarium dome. Several planetariums are even equipped with special plumbing systems that actually sprinkle audiences with purified water to simulate a rainstorm! All this wizardry is controlled by computers and accompanied by powerful sound tracks. Often the prerecorded voices of movie stars and other well-known narrators guide the audience in their journeys across the universe.

Behind such spectacular illusions lies the hardware that forms the planetarium itself. The dome is often made of sheet aluminum, perforated with tiny holes. These holes, invisible to the casual visitor, make the dome virtually translucent. They also keep air ventilating through the planetarium theater, make the dome lighter, and allow the placement of immense audio speakers, which provide the spectacular sound tracks. Modern planetariums also hide elaborate special effects behind this inner dome. Exploding stars and space-walking astronauts stand invisible until the operator throws a switch.

Still, the planetariums continue to produce new magic tricks that amaze audi-

This unique space exhibit includes three-dimensional scale models of the Sun, planets, and the largest moons of the solar system.

Courtesy of the Adler Planetarium

ences. In 1983 a totally new type of planetarium instrument burst onto the scene. Introduced by Evans and Sutherland Computer Corporation of Salt Lake City, Utah, the "Digistar," a computer-graphics-imaging device, projects first-generation computer animation onto the planetarium dome. With this device, it has become possible to move at will within a simulated three-dimensional scene built up out of computerized data fields. In other words, audiences can now experience scientifically accurate simulated spaceflights to stars within a few hundred light-years of Earth. With amazing realism, we can watch how our solar neighborhood changes over a given year or over the next million years.

Digistar technology has even been used to create animated three-dimensional models of atoms and molecules—projected from the same device that re-creates entire galaxies. These instruments may one day entirely replace the electromechanical projectors used in most of today's modern planetariums.

Through the advance of technology, the planetarium, which has long been used to educate the public about astronomy, now serves many functions. It has truly become an awe-inspiring vehicle—not only for re-creating space travel or the view of a star-filled night as seen from another world —but also for journeying through history and learning about the evolution of Earth.

OTHER USES FOR PLANETARIUMS

Many modern planetariums also use their domed theaters for spectacular laser light shows, accompanied by rock, jazz, and classical music. They are likewise used for live plays, science fiction programs, and live concerts performed under the stars and planets and accompanied by special light, sound, and motion effects.

In addition, some planetarium theaters are used to show motion pictures with immense film formats and gigantic, yet crisp, images. "Omnimax" projectors, for example, use 70-mm-wide film projected across an entire dome. Smaller theaters may use 35-mm projectors such as the "Cine 360" to create similar wraparound illusions. Such "space theaters" are designed to fill the audience's entire field of vision with images, and so create a three-dimensional illusion of being transported through space and time.

Planetarium theaters have come a long way since the idea was born just a century ago. Today they are visited by hundreds of millions of people around the world, who leave inspired and awed by the wonder and mystery of the universe.

Courtesy of the Adler Planetarium

The Adler Planetarium in Chicago opened its doors to the public in May 1930; it was the first modern planetarium in the Western Hemisphere.

The inner planets of our solar system, Mercury, Venus, Earth, and Mars (pictured above), travel in nearly circular orbits around the Sun. The outer planets, Jupiter, Saturn, Uranus, Neptune, and Pluto, have more eccentric orbits.

THE SOLAR SYSTEM

by Ray Villard

Imagine you were on a starship deep within interstellar space, heading toward the solar system where Earth resides. As you approached from tens of billions of miles away, the Sun would appear to grow ever brighter. Eventually you would detect Earth as a faint pinpoint of light. If you observed for long enough, you would notice that Earth follows a wide path around the Sun. You would also see, at various distances from the Sun, eight other objects of various sizes. You might detect that many of these planets are circled by still smaller objects—their moons. In the space between the orbits of two of the planets, Mars and Jupiter, you would see thousands of very small "planets," or asteroids, also revolving around the Sun. You might even spot a few comets, their long, streaming tails slicing across the planetary orbits.

THE SUN

The Sun lies at the very heart of our solar system. It is a typical star, one of the 300 billion in the Milky Way galaxy. Because the Sun is much closer to us than is any other star, it appears many, many times larger than the more distant bodies. Its disk is about the size of a full moon.

Compared with the other stars of our galaxy, the Sun is an average-size star. But

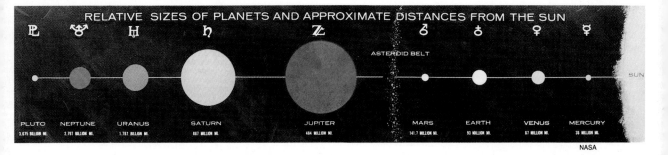

it is giant compared with even the largest planets. Its diameter of 869,960 miles (1.4 million kilometers) is 109 times greater than that of Earth. Even though it is of gaseous composition, the Sun weighs more than 300,000 times as much as Earth. Its surface temperature is 9,932° F (5,500° C). At its center, the temperature may reach as high as 27,000,000° F (15,000,000° C)—hot enough to smash atoms and generate energy through a process called nuclear fusion. Each second, the Sun converts 650 million tons of hydrogen into 600 million tons of helium. In the process, 50 million tons of matter are converted to pure energy. This energy initially takes the form of deadly gamma rays, but by the time it bubbles to the surface of the Sun, the energy has been transformed into a torrent of light, illuminating the planets and nurturing life on Earth.

CHILDREN OF THE SUN

The nine planets, in order of their distance from the Sun, are Mercury, Venus, Earth, Mars, Jupiter, Saturn, Uranus, Neptune, and Pluto. The planets all lie in about the same orbital plane, and they orbit the Sun in the same direction. This suggests that the solar system is a relic of a vast disk of dust and gas that surrounded the Sun as it formed 4.6 billion years ago. In the first few million years after the Sun ignited, major planets, ranging from several thousand to tens of thousands of miles across, formed within this gaseous disk. The largest chunks of leftover debris became trapped in the gravitational fields of the newly formed planets, and began orbiting them as moons. Gravity pulled the smaller chunks to the surfaces of the planets and moons. Many of the craters that pepper the surfaces of these bodies are relics of this early period of intensive bombardment by interplanetary debris.

The solar system has two types of planets. The tiny rock, or terrestrial, planets all lie close to the Sun, like campers huddled around a bonfire. The immense outer planets—Jupiter, Saturn, Uranus, and Neptune—lie in the colder reaches of the solar system. They consist mostly of liquid and gas. The farthest planet from the Sun, Pluto, is really a "double planet," because its moon is nearly one-quarter Pluto's size. Pluto may be the last remaining "fossil" of a population of thousands of "icy dwarf" bodies that once inhabited the solar system. These icy dwarfs were either absorbed into the major planets or tossed out of the solar system altogether.

In the space age, we have used robot spacecraft to fly by, or even land on, all of the planets in our solar system. Probes have returned spectacular close-up pictures of all the planets (except Pluto), and in the process revolutionized our understanding of how these celestial objects evolved. The manned Apollo expeditions to the Moon, in the late 1960s and early 1970s, returned with the first samples of rock from another world. The Moon rocks were found to be 4.5 billion years old, providing additional evidence that the formation of the planets and moons accompanied that of the Sun.

Moons

A moon is any natural body that orbits a planet. There are at least 35 known moons in our solar system. The majority of them orbit the giant planets Jupiter and Sat-

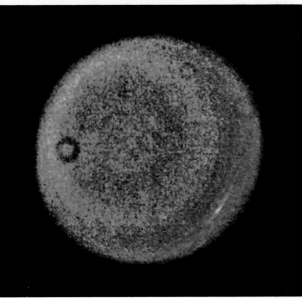

En route to Neptune, Voyager 2 collected data that greatly expanded our knowledge of other outer planets. Passing by Jupiter (above left), the probe identified volcanic eruptions on the Jovian moon Io. At Saturn (above), Voyager 2 determined the composition of Saturn's rings, information that may give clues to their origin. Whizzing by Uranus (left), Voyager 2 made the startling discovery of 10 new moons.

Top left: NASA; others: NASA/JPL

urn, and are little more than huge, airless balls of ice, ranging from hundreds to more than a thousand miles across. One of the largest moons, Saturn's Titan, is so big (3,169 miles, or 5,100 kilometers, in diameter) that it retains its own atmosphere of nitrogen. Mars has some of the smallest moons, a pair called Deimos and Phobos, each no bigger than an asteroid, which indeed they may have been at one time.

Asteroids

In the 18th century, astronomers calculated astrophysical laws that predicted they would find an as-yet-unseen planet between Mars and Jupiter. And they eagerly searched the skies for it. On the night of January 1, 1801, the Italian astronomer Giuseppe Piazzi discovered a small celestial body, which he took to be a planet, in the space between the orbits of Mars and Jupiter. This body, which was later called Ceres, was found to have a diameter of only 478 miles (770 kilometers). Over the years, many more small, planetlike bodies were found in the gap between Mars and Jupiter. Today more than 1,000 of these small bodies have been discovered, leading astronomers to estimate there may be more than 50,000 in all.

These small bodies are now known as minor planets, or asteroids. The orbits of some extend beyond the Mars-Jupiter gap. But their combined mass is only a fraction of the Earth's.

Astronomers once thought that asteroids were the fragments of a big planet that once orbited between Mars and Jupiter and then broke apart for unknown reason. But in recent years, scientists have come to believe that asteroids are probably debris left over from the solar system's formation, debris that simply never coalesced to form a planet.

THE SOLAR SYSTEM 75

Comets

Comets are among the strangest members of the solar system. Instead of moving as the planets do, in nearly circular orbits in the same direction, comets revolve around the Sun in very elongated ellipses, and from every conceivable direction. Much of the time they are so far away from the Sun that they are invisible even to our largest telescopes.

It was once thought that some comets approached the Sun from far beyond the solar system, and that once they withdrew from the Sun, they would never return. Today it is generally agreed that comets are members of the Sun's family. They make up a vast shell of icy debris called the Oort Cloud. Though this region lies 50,000 times farther from the Sun than does Earth, the trillions of icy comet bodies that inhabit it are all gravitationally bound to the Sun.

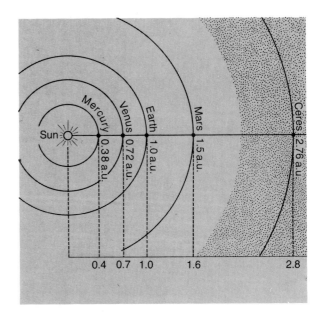

In this photomosaic of the solar system as seen from the Moon, the Earth is backlit by the Sun. The planet just above the Moon is Venus, and at top, from left to right, are the planets Jupiter, Mercury, Mars, and Saturn, replete with rings.

NASA

BODE'S LAW OF PLANETARY DISTANCES

The 18th-century astronomer Johann Bode devised a simple rule for determining the distances of the planets from the Sun. Begin by using the numbers 0, 3, 6, 12, 24, 48, 96, 192, 384, and 768. After the number 3, each of these is double its predecessor. Add four to each number, and then divide each by 10. The resulting series of numbers gives the approximate distance from the Sun of all the planets.

Jupiter
5.2 a.u.

Bode's Law is remarkably accurate for the first 6 planets from the Sun when compared to the measured astronomical units (a.u.), the mean distance between the Earth and the Sun. The rule fails to provide the correct distances for Uranus, Neptune, and Pluto, however.

Saturn
9.5 a.u.

5.2   10.0

← BODE'S SEQUENCE →

When astronomers first discover a comet, it usually appears as a faint, diffused, fuzzy star, with a dense, starlike center and a veil-like region, known as its *coma*. As the comet approaches the Sun, its coma becomes brighter, as more and more material vaporizes off the surface of the comet's solid, icy nucleus. When they are some 100 million miles (160 million kilometers) from the Sun, some comets begin to show a tail streaming behind them, pointing directly away from the Sun. Comet tails appear to consist of very thin gases that fluoresce, or glow, under sunlight, as well as a fine stream of dust particles. This material is forced away from the Sun by the pressure of the solar wind.

Meteors

Comets eventually break up into particles, which are sometimes seen entering the Earth's atmosphere as meteors. Meteors—some of which originate in comets, and others as chunks from asteroids, moons, or other planets—range in size from specks the size of a pinhead to huge stones weighing many tons. We become aware of meteors only through the bright light produced when they collide with air molecules in our atmosphere. Most meteors disintegrate once they strike the atmosphere. Those that reach the ground are called meteorites. Most meteorites are fragments of asteroids but a small number may have come from the Moon or Mars.

EARLY IDEAS OF THE SOLAR SYSTEM

The ancient Greek philosophers did not realize that the Earth itself is a planet, or wanderer in the heavens (which, incidentally, is what *planet* means). The Earth, they thought, hung motionless at the very center of the universe. They believed that each of the five planets they had seen (Mercury, Venus, Mars, Jupiter, and Saturn) were attached to concentric, invisible crystal spheres. The Moon and the Sun were attached to other spheres. These crystalline spheres, set one within the other, revolved around the Earth, carrying with them the heavenly bodies. This theory could not explain certain phenomena, however.

For one thing, the planets do not move at an even rate across the sky. At times they move more rapidly than at others. An even greater mystery was the observation that a planet such as Mars occasionally ceases its apparent eastward motion among the stars and reverses itself to head westward for a time. To explain this "retrograde motion" of the planets, early astronomers invented a complicated system of "epicycles." They held that each planet traveled along the circumference of a small circle, the center of which traveled along the circumference of a larger circle. The Earth, it was maintained, was at the center of the larger circle.

This model of the universe prevailed for over 1,000 years. In the first half of the

16th century, however, Polish astronomer Nicolaus Copernicus revived an idea that had been first proposed by the Greek philosopher Aristarchus of Samos—that the Earth and other planets move around the Sun. This system was called the heliocentric theory, since it placed the Sun (*helios*, in Greek) at the center of the universe.

Motions of the Planets

It required a lifetime effort on the part of several great astronomers to prove the Copernican heliocentric system. The 16th-century Danish nobleman Tycho Brahe made a long and accurate series of observations of the planets. Johannes Kepler, a German disciple of Brahe, drew up three laws of planetary motion that still hold true today. Kepler also improved on the Copernican model, which maintained that the planets move in circular orbits around the Sun. This belief led to inaccuracies in predicting planetary positions. Kepler was able to show, instead, that orbits are ellipses, rather than true circles.

While Kepler was refining his theories, Italian inventor and scientist Galileo Galilei used the telescope, a recent Dutch invention, to gather additional evidence supporting Copernicus' theory. The telescope allowed Galileo to see the phases of Venus, which proved that it orbited the Sun, not Earth. Galileo also saw four tiny moons orbiting the distant planet Jupiter, in perfect accordance with Kepler's laws of motion.

The research of Kepler and Galileo clearly explained the nature of the planets' movements around the Sun, but neither scientist understood the force that governed these movements. This force was first revealed in 1687, when the great English scientist Isaac Newton presented his law of universal gravitation. This law states that every particle of matter in the universe attracts every other particle. This force of gravitation increases with the mass of an object, and depends on the distance between two objects. Newton showed mathematically that this is truly a universal law, since it applies not only to objects upon the Earth, but to heavenly bodies as well. The law of universal gravitation explains why planets, asteroids, and meteors keep orbiting the Sun, which is by far the most massive object in the solar system.

Using the law of universal gravitation, we can now analyze the motions of the planets with great accuracy. We can ac-

Charles Eames, IBM

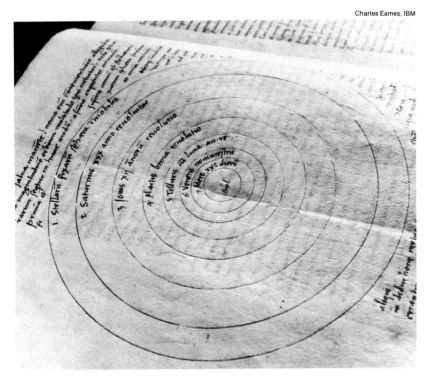

The Polish astronomer Nicolaus Copernicus published this diagram of a Sun-centered, or heliocentric, solar system in 1543. The theory wasn't fully accepted until the late 17th century.

count for the small deviations that arise as one planet affects the orbit of another.

It was the study of such deviations that led directly to the discovery of the planet Neptune. After Uranus had been discovered by Sir William Herschel in 1781, careful studies showed that it did not follow the orbit predicted by the law of universal gravitation. This led young Englishman John Couch Adams and French astronomer Urbain-Jean-Joseph Leverrier to conclude that Uranus was being attracted by another planet even more distant from the Sun. Both men calculated the position in the sky of the unknown planet without ever having seen it. On September 23, 1846, on the basis of Leverrier's calculations, German astronomer Johann Gottfried Galle located Neptune.

Astronomers suspected the existence of Pluto because of the disturbances of motion they'd seen in the orbits of Uranus and Neptune. Such deviations suggested that the two planets were being gravitationally tugged by yet another unseen body. Pluto was discovered in 1930 after a yearlong detailed search by astronomers at Flagstaff Observatory in Arizona.

In 1978 astronomers discovered that Pluto has at least one moon, which they called Charon. By plotting the moon's six-day orbit, astronomers were able to calculate the mass of Pluto, which turns out to be only 1/500 that of Earth. Pluto's orbit is rather unusual, taking a path some 17 degrees inclined to the plane taken by the other planets. Pluto's orbit is also highly elliptical, so much so that Pluto moves inside Neptune's path for about 20 years out of its 248-year orbit.

Because of Pluto's puny size, astrophysicists say its gravity cannot entirely account for the changes seen in the orbits of Neptune and Uranus. This suggests that there is a 10th planet, dubbed Planet X, which may be twice as far from the Sun as Pluto. Elaborate computer simulations of planetary orbits suggest that Planet X may be five times the mass of Earth and take 1,000 years to orbit the Sun. Astronomers continue to search for it with a detailed survey of the sky.

Yerkes Observatory

Galileo used the telescope, recently invented in Denmark, to gather information about the motion of the planets. In his records of Jupiter and its moons (above), a sphere represents the planet and an asterisk represents moons.

THE SOLAR SYSTEM

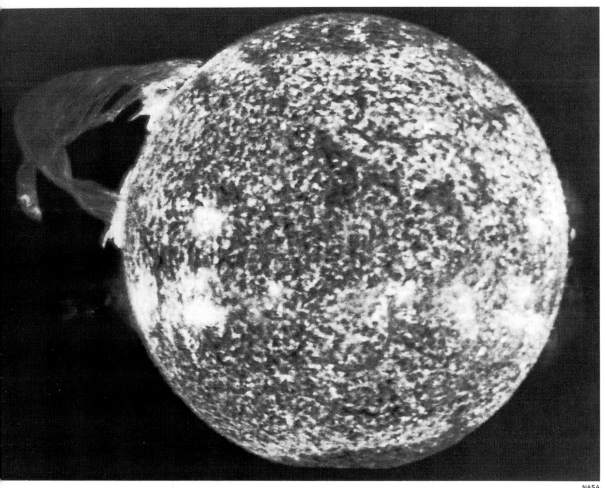

The Sun is the center of our solar system and the star upon which the Earth depends for the existence of life.

THE SUN

by Oran R. White

"All the lines were busy, all the screens lit up. Bells sounded, and dashed lines of every color traced across a dozen monitors. An enormous explosion appeared on one console, while measurements of its effects were displayed on others, overprinted with alert notices in bold red letters. A U.S. Air Force duty officer paced back and forth behind civilian technical specialists, and next door a teletype swiftly transmitted an urgent warning to the Soviet Union. From Alaska to Omaha, bases and field posts were reporting to the National Oceanic and Atmospheric Administration (NOAA). I had never seen anything like this, except in space-war movies," recalls astronomer Steve Maran of the American Astronomical Society. "But this was no cinematic attraction or civil-defense drill. This was a storm on the Sun, an electromagnetic maelstrom so fierce that Earth, 93 million miles (150 million kilometers) across space, was being bathed in high-energy radiation."

It was March 13, 1989, and astronomers were in the midst of recording the greatest peak of sunspot activity in history. A solar flare thousands of miles across was producing magnetic storms across the Earth, knocking out the power of entire cities, scrambling radio and telephone communications, and producing spectacular aurora borealis displays as far south as Arizona. The Solar Maximum of 1989 would stretch into 1992, wreaking electromagnetic havoc, not only on Earth, but also in the skies. Increased radiation from the Sun heated and expanded the Earth's atmosphere, dragging vital satellites to their premature and fiery deaths. Even the Solar Maximum Mission (Solar Max), a satellite lofted specifically to gather data on the Sun, was pulled into the heated atmosphere, where it disintegrated much earlier than originally planned.

During such solar peaks, occurring every 11 years or so, the Sun—taken for granted perhaps even more than the air we breathe—becomes front-page news. Yet even in the most mundane of times, no other heavenly body comes close to being as important to life on Earth. Not surprisingly, then, people have been observing the Sun for centuries. Today spacecraft are providing new data about the Sun and about how it affects the Earth.

PHYSICAL DATA

The Sun is the center around which the Earth and the other planets of our solar system revolve. It is a rather ordinary star of average size. Of course, the Sun appears much bigger and brighter to us because it is much closer to the Earth than is any other star. It is about 93 million miles (150 million kilometers) away. The next-nearest star, Alpha Centauri, is more than 25 trillion miles (40 trillion kilometers) away.

Our Sun is only one of about 100 billion stars in our galaxy, the Milky Way. It is located in one of the outer, spiral arms of the Milky Way, about three-quarters of the way from the galactic center.

The Sun is a vast ball of hot, glowing gas, some 870,000 miles (1.4 million kilometers) across—more than 100 times the diameter of the Earth. The Sun's mass, however, equals that of 333,420 Earths. This tremendous weight produces a pressure at the center of the Sun of more than 1 million metric tons per square centimeter.

The Sun's gravity is 28 times stronger than that of the Earth. So a man weighing 150 pounds (70 kilograms) on Earth would weigh 28 x 150 pounds (28 x 70 kilograms), or 4,200 pounds (1,960 kilograms), *if* he could stand on the surface of the Sun.

In spite of the great mass of the Sun, its average density—the weight of a standard volume of its matter—is only 1.4 times the weight of an equal volume of water. The Earth, on the other hand, is 5.5 times denser than water. This low solar density is easy to explain. The center of the Sun, because of the enormous pressure, is more than 100 times denser than water. But much of the Sun beyond the center is composed of gas that is often thinner than the Earth's atmosphere. When these densities are averaged together, the general density of the Sun is quite low.

The Sun is like a huge furnace, fired by nuclear, or atomic, energy at its core. Temperatures at the center may be 25,000,000° F (14,000,000° C) or more. At the surface, temperatures are much cooler—between 9,000° and 11,000° F (5,000° and 6,000° C)—

STRUCTURE OF THE SUN

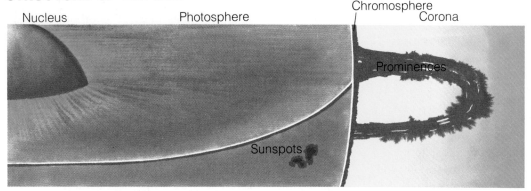

Nucleus   Photosphere   Chromosphere   Corona   Prominences   Sunspots

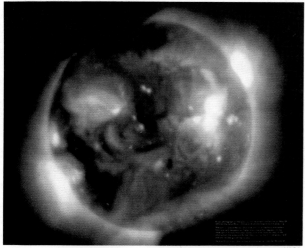

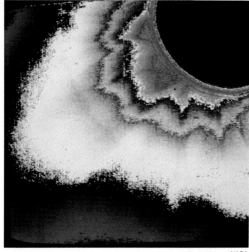

This Skylab photo captured the Sun's hot outer layer, or corona (top), which has a temperature of 4,000,000° F. The image above represents the densities of the corona, with the inner blue area being the most dense and the outer yellow area having the least density.

still hot enough to vaporize nearly all substances that exist as solids or liquids on the Earth.

ATMOSPHERE OF THE SUN

The Sun is composed of several distinct regions. It has a two-layer atmosphere. The solar atmosphere extends far upward from the *photosphere*, or solar surface. Consisting mostly of hydrogen gas, the atmosphere is also much less dense than the photosphere.

The lower atmospheric layer is the *chromosphere*, or "sphere of color." It extends as high as 7,500 miles (12,000 kilometers) above the surface. The upper layer is the *corona*, or "crown." The corona forms a beautiful white halo around the entire Sun, sending long streamers millions of miles out into space.

Normally, the effects of our own atmosphere and the bright glare of the photosphere blot out our view of the solar atmospheres. They become visible during a total eclipse, when the Moon covers the photosphere. Astronomers also study the corona through a special telescope called a coronagraph, which produces an artificial solar eclipse. The corona is also visible to astronauts flying above the Earth's atmosphere.

Temperatures in the corona and the chromosphere vary in a very unexpected manner. The lower part of the chromosphere may be less than 9,000° F (5,000° C), while the temperature rises in the outer reaches of the chromosphere, from 18,000° to 180,000° F (10,000° to 100,700° C). The corona, in turn, is much hotter than the chromosphere. Astronomers have estimated an astounding temperature of 3,600,000° F (2,000,000° C) for the corona's outer reaches.

Why should the corona, which is so far from the source of the Sun's energy, be so much hotter than the photosphere? One explanatory theory holds that strong shock waves, caused by turbulent movements of the photosphere, intensely heat the very thin gases of the corona.

A spectacular activity occurring in the chromosphere is the presence of *prominences*. These huge streamers of glowing gas can reach heights of hundreds of thousands of miles into the overlying corona. They take a great variety of shapes. Prominences are best observed during a solar eclipse or with a coronagraph.

Some prominences are eruptions or explosions, rising quickly and soon fading away. Other types last much longer. Still others seem to originate high in the chromosphere and then rain gas downward toward the Sun. The occurrence and lifetimes of prominences are influenced by solar magnetic fields.

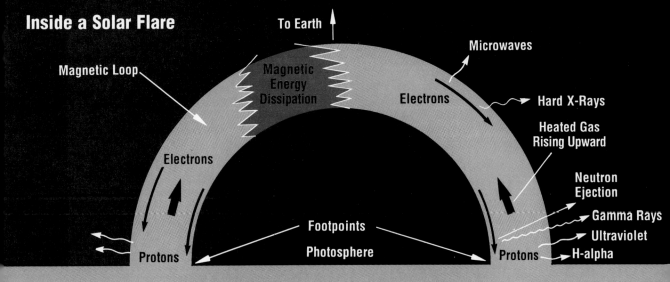

Astronomers believe solar flares occur when magnetic loops projecting from the Sun's surface "snap" from the buildup of stresses, accelerating large numbers of electrons and protons. These charged particles spiral along magnetic field lines, emitting electromagnetic radiation and heating the atmosphere at the base of the loop. This heated material then surges up the loop while gas at the "footpoints" emits H-alpha light.

Astronomy magazine

The chromosphere also displays much smaller jets or filaments of gas, called *spicules*. These may result from strong movements of hot chromospheric gas. These gas movements appear in the chromosphere as coarse cells, called supergranulation.

From time to time, the chromosphere is active in other ways. Astronomers often notice hot, bright markings: *plages*, areas of hot brightness, and *flares*, high-energy outbursts of radiation and subatomic particles. These particles sometimes reach the Earth's atmosphere.

SURFACE OF THE SUN

It may seem strange to talk of the "surface" of a wholly gaseous globe. Yet the Sun's surface has a definite border. This layer, called the solar disk, or photosphere, is the deepest visible part of the Sun.

The photosphere is a relatively thin region, some 200 miles (320 kilometers) deep. This is less than 1/2,000 the solar radius.

The photosphere was once thought to be a uniform orb of light. But even in ancient times, observers occasionally saw spots on it. In the early 1600s, Italian scientist Galileo Galilei became the first person to study the Sun and its spots through a telescope. These so-called sunspots are dark, irregular patches. In addition to sunspots, two other main features have been found on the solar surface: bright, irregular areas called *faculae*, or "little torches," and a network of fine cells, the *photospheric granulation*.

Faculae are hot, glowing regions, ranging from tiny, bright marks to huge splotches that have a coarse-grained structure. They resemble the plages in the chromosphere. They often surround sunspot groups, but they may occur alone. Faculae often arise where sunspots later appear, and linger after the sunspots have vanished. Many astronomers consider them to be huge masses of gas that are hotter than the rest of the gaseous solar surface.

Through the telescope, photospheric granulation looks like bright grains of rice. The grains are separated from one another by dark boundaries. A typical grain, or cell, measures about 1,000 miles (1,600 kilometers) across. Astronomers consider the granulation to be photospheric gas in continuous and violent motion because of heat. On film the cells look like boiling fluid bringing up gas from the depths of the Sun.

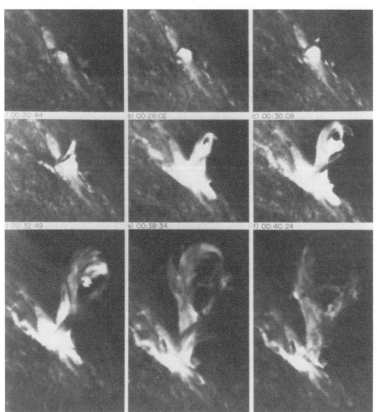

Big Bear Solar Observatory, photo courtesy R. L. Moore

To the naked eye the Sun appears constant. But a telescope reveals that the Sun's atmosphere is wracked by the tremendous explosions of solar flares. These solar flares grow rapidly, with the sequence at left occurring over only 34 minutes. During this time billions of tons of material are ejected from the Sun at speeds approaching 1,000,000 miles per hour.

Sunspots

Sunspots appear dark because, at a mere 7,000° F (4,000° C), they are cooler and less bright than the rest of the photosphere. A typical sunspot has two distinct parts: a dark central region, called the *umbra* ("shadow"), and a lighter surrounding area, the *penumbra* ("almost shadow"). Like the photosphere in general, the umbra shows a granular structure, which suggests the circulation of hot gas. At certain positions near the edge of the Sun, the spots look like depressions, or hollows, in the photosphere.

A single spot may be tens of thousands of miles wide. Its duration may last anywhere from a few days to a few months. Sunspots often develop in pairs, which tend to drift apart slowly. At times, they may form by the hundreds, while at other times, there may be practically no sunspots at all.

Many theories have been advanced to at least partially explain the nature of sunspots. They have been compared to low-pressure areas, tornadoes, or huge whirlwinds in the solar gas. And, in fact, complex movements of gas both into and away from a sunspot have been observed.

More-recent theories hold that sunspots are cool areas produced by reactions between the charged gases of the Sun and solar magnetic fields. That is, a local magnetic field breaks through the surface of the photosphere, producing a spot at that point. This disturbance also affects the solar atmosphere overlying the spot.

Solar Cycle

Sunspots appear and then disappear in a definite 11-year cycle, called the sunspot cycle, or the solar cycle. A sunspot cycle begins with a few small spots at solar latitudes of 30 to 40 degrees north and south of the Sun's equator. With the passage of time, more sunspots of larger size appear. These new spots arise closer and closer to the solar equator, until much of the Sun is covered by dark patches.

The cycle nears its end when spots in the higher solar latitudes begin to vanish. Finally only a few spots are left around the equator. Then the spots of the next cycle start to appear, at 30 to 40 degrees north and south of the solar equator.

Some astronomers have linked the sunspot cycle to a complex circulation of solar gas from the surface of the Sun down and then up again, and from the solar poles to the equator and back. The heat and rotation of the Sun are supposed to produce this effect. This circulation has been compared to that of the Earth's winds and ocean currents.

More-modern sunspot theories make use of new discoveries about magnetic fields. Scientists have found that a hot, electrically charged fluid, such as the solar gas, produces magnetism. As the gas moves, the lines of magnetic force follow. Regular movements of solar gas and its accompanying magnetic fields may cause the sunspot cycle.

Astronomers have discovered that the north and south magnetic poles in the Sun switch polarity in a regular, repeated fashion. These changes take place at the beginning of each sunspot cycle. The result is a magnetic cycle of 22 years, or double the duration of the sunspot cycle. This seems to indicate that the spot cycle is definitely connected with magnetic forces in the Sun, but exactly how is a mystery.

Sunspots are not the only features affected by the solar cycle. During a sunspot maximum, when many spots exist, the entire Sun becomes more active. Prominences are larger and more common. Huge solar flares, with temperatures in the millions of degrees, burst forth.

The corona undergoes changes in shape and brightness during different phases of the solar cycle. Man-made satellites have also photographed a strange feature, known as solar polar caps, in the corona. These coronal caps, centered over the north and south poles of the Sun, are

The solar wind is a continuous stream of high-speed protons and electrons emitted from the Sun's corona. This wind distorts the Earth's magnetic field, producing such observable effects as the auroras in high latitudes.

NASA

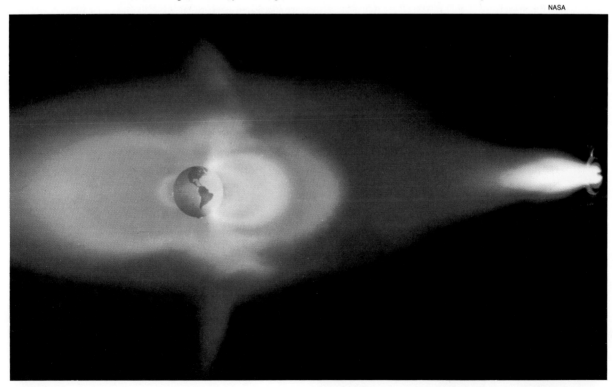

relatively cool masses of gas. At about 1,800,000° F (1,000,000° C), the caps are only about half as hot as the rest of the corona. During a sunspot maximum, the polar caps are very small or absent. During a sunspot minimum, when only a few spots exist, the caps become much larger.

During sunspot maximums, other parts of the corona become about twice as hot as usual—7,200,000° F (4,000,000° C), with the highest temperatures in the regions of solar flares.

The activities associated with sunspot cycles often affect the Earth and its atmosphere. These effects are due in large part to the many electrically charged, high-speed particles, and also to ultraviolet and X rays emitted from the Sun during a solar maximum. Many solar particles are trapped in the Earth's magnetic field and may create magnetic storms on Earth. They can also interfere with compasses, communications, and electrical-power transmission.

To observe an eclipse, the eyes must be protected from the direct rays of the Sun at all times. Since even an instant's glance can cause damage, indirect methods should be used.

© Davis Sams/Sipa Press

INTERIOR OF THE SUN

There is little direct evidence available to scientists concerning the inside of the Sun. Astronomers studying the Sun know mostly about the heat, light, magnetism, and particles that come from the upper layers of the photosphere and solar atmosphere.

There are, however, some clues to the nature of the Sun's interior. Scientists believe that certain atomic particles, called *neutrinos*, reach the Earth directly and very quickly from the core of the Sun. These minute particles have no charge and virtually no mass, but move at the speed of light. Because of these properties, neutrinos easily pass through great thicknesses of matter, making them hard to detect. Nevertheless, scientists have managed to detect and count some of the solar neutrinos by using elaborate laboratory chambers filled with tracer gases. From neutrino research and from what is known about the outer layers of the Sun, scientists have constructed the following theoretical model of the solar interior.

A "nuclear-energy furnace" probably forms the Sun's hot and dense core. It fills a relatively small volume. The intense conditions strip the atoms of their electrons, so they consist only of nuclei.

Circulating currents bring hydrogen—atomic fuel—to the "furnace" and carry away the resulting product: helium. The tremendous energy produced in this way is transferred outward, away from the core. As the energy reaches the Sun's surface, circulating currents of solar gas carry it away.

Like boiling water or hot air shimmering over a fire, the gas of the photosphere rises and falls from the energy. These photospheric movements produce the many phenomena seen at the Sun's surface: granulation, faculae, and spots.

Scientists estimate that it takes several million years for energy at the core of the Sun (with the exception of the rapid neutrinos) to reach the surface. From the solar surface, this energy then radiates in all directions.

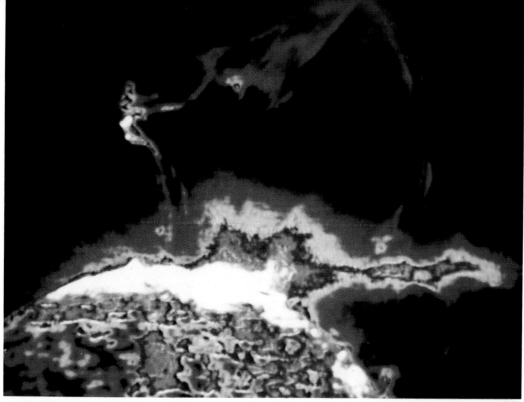

The huge streamers of glowing gas, called the solar prominences, show amazing diversity. Some rise to great heights above the chromosphere, while others originate above the chromosphere and rain down toward the Sun.

ROTATION OF THE SUN

Like the Earth, the Sun spins on its axis, rotating from west to east. But unlike a solid planet, the gaseous Sun spins faster in some places than in others.

Astronomers have several ways to calculate the rate of the Sun's rotation. Sunspots and faculae that last for several weeks or months are carried along as the Sun rotates. So astronomers can track them to measure the speed of solar rotation.

In latitudes where spots and faculae are absent, other methods must be used to measure the Sun's period—the time it takes the Sun to complete one turn on its axis. The spectroscope breaks up ordinary white sunlight into a spectrum. The Sun's spectrum is crossed by many dark lines, which provide clues to solar rotation. When the Sun moves away from an observer, the dark lines shift their positions toward the red end of the spectrum. If the Sun approaches the observer, the lines shift in the opposite direction, toward the violet end of the spectrum. These shifts of the spectrum lines are called the Doppler effect.

As the Sun spins on its axis, one side of the Sun approaches the observer, while, at the same time, the other side moves away. From studying the resulting Doppler effect, astronomers can determine the period of the Sun at any latitude.

Investigators have discovered that the period is shortest at the equator—26.9 days. Rotation becomes slower farther away from the equator. At the north and south poles, the period is 34 days. This difference of nearly 10 days in period between the solar equator and the poles is due to the fact that the Sun is not solid.

CHEMISTRY OF THE SUN

Our knowledge of the chemical elements in the Sun is based mostly on study of the solar spectrum. The Sun's spectrum is covered by many dark lines. These are called *Fraunhofer lines,* after the German physicist Joseph von Fraunhofer. Fraunhofer lines are also known as *absorption lines.* They represent certain colors, or wavelengths, of light absorbed by different elements in the Sun's atmosphere.

The atoms of an element, when hot enough, emit light of certain colors. The atoms also absorb light of these colors, thus producing dark absorption lines. The combination of colors and lines forms a spectrum characteristic of the element. Spectrums of earthly elements have been produced in laboratories and compared with the solar spectrum. In this way, chemists have learned what elements exist in the Sun.

A number of the absorption lines seen in the solar spectrum, however, are caused by certain atoms in the Earth's atmosphere. These atoms absorb some of the sunlight passing through the atmosphere. There are about 6,000 of these so-called *telluric* ("earth") lines. But scientists can easily distinguish telluric from solar lines. For example, telluric lines show no Doppler effect from the Sun's rotation.

Astronomers have learned that most, if not all, of the chemical elements present on Earth also exist in the Sun. Hydrogen is the most common solar element, making up more than 80 percent of the Sun's mass.

Helium is second, at 19 percent. The remaining 1 percent of the solar mass consists mostly of the following important elements, in descending order of concentration: oxygen, magnesium, nitrogen, silicon, carbon, sulfur, iron, sodium, calcium, nickel, and some trace elements.

The Sun is a mixture of gas atoms, atomic nuclei, and still smaller atomic particles. These atomic particles are electrons (negatively charged), protons (positively charged), neutrons (no charge), positrons (positively charged), and neutrinos (no charge).

This entire mass of hot, gaseous solar material is called a *plasma*. The high temperatures make it almost impossible for most chemical compounds to exist in the Sun.

The computerized color display of the Sun's corona at right, in which the north and south poles are black and the hottest regions are white, was obtained by the Orbiting Solar Observatory (far right). The satellite contains a coronagraph, an instrument that creates an artificial eclipse for the express purpose of studying the Sun's corona.

SOLAR RADIATION

The Sun radiates energy at practically all wavelengths. This electromagnetic energy ranges from long radio waves to the shorter microwaves and the infrared, light, ultraviolet, and X rays. We can see only the light waves; the infrared we can sense as heat; the other forms of radiation can be detected only by means of special instruments and films.

There is some question whether the rate of solar radiation is always exactly the same. The amount of light leaving the Sun seems to be steady, or constant. But the quantity of other kinds of radiation emitted by the Sun may depend on the number of sunspots present.

The Sun also sends subatomic particles into space. Particle emission increases sharply during a sunspot maximum, when solar flares are exceptionally strong. Flares release vast numbers of protons, electrons, and atomic nuclei.

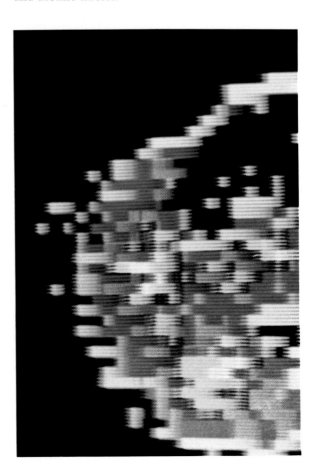

Even during a sunspot minimum, the Sun is always emitting particles. The fine "rain" of particles that passes from the corona toward and around the Earth is called the *solar wind*.

Many solar particles do not reach the Earth or its atmosphere. They are trapped by the Earth's magnetic field and become part of a belt of radiation, called the Van Allen belt, surrounding the Earth.

About 30 percent of the solar radiation is screened away from the ground by our atmosphere. It is well that this is so, for some of this radiation is deadly.

Strong solar radiation ionizes many of the Earth's higher atmospheric gases, producing electrically charged layers. Many scientists call these layers the ionosphere. The ionosphere shields the Earth below from harmful solar radiation. It also makes long-distance radio communication possible on Earth. This is because certain radio waves bounce off the ionosphere back toward the ground instead of going off straight into space. Radio communication on Earth may be disrupted from time to time during a sunspot maximum, because intense solar radiation and particles disturb the upper atmospheric layer.

ATOMIC ENERGY OF THE SUN

In 1939 German-American physicist Hans Bethe proposed an atomic, or nuclear, explanation of the Sun's energy. What goes on deep in the Sun's interior is probably what happens in a fusion reaction. Four nuclei of hydrogen atoms fuse, or join, to form the nucleus of a helium atom.

Bethe's fusion cycle takes place in six steps, which chiefly involve the element carbon, as well as hydrogen. This complex cycle is thus also called the carbon cycle. The carbon in the Sun is both used in the fusion process and produced during the reactions. The net result is that the number of carbon atoms is not changed. On the other hand, hydrogen is used up.

We now know that the carbon cycle occurs in very hot stars. The Sun is very cool compared to other stars, however, so another reaction plays a more important role: the proton-proton reaction. In this process, solar energy comes from the direct fusion of hydrogen nuclei, or protons, without going through the six-step carbon

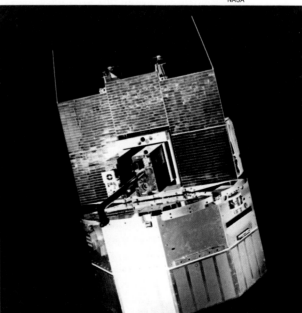

cycle. The proton-proton process also consumes hydrogen.

Scientists have calculated that 110 tons (100 metric tons) of hydrogen changing into helium yields more energy than all humans on Earth now use in a single year. At the present rate of solar energy production—about 4.4 million tons (4 million metric tons) of mass or matter turning into pure energy per second—the Sun contains enough hydrogen to remain as bright and hot as it is today for the next 30 billion years, and perhaps longer.

THE SUN'S INFLUENCE ON EARTH

Most of the Earth exists as we know it only because of the Sun's light and heat. The Sun makes all life possible. There would even be no weather without the Sun. The Sun directly or indirectly provides us with energy to light and heat our homes and to power our machines.

Because it is so small and so distant from the Sun, the Earth gets only about ½ billionth of the total solar-energy output. The quantity of solar energy reaching the edge of the Earth's atmosphere equals about 2 calories per square centimeter per minute. A calorie is the quantity of heat needed to raise the temperature of 1 gram of water 1° C. This rate of 2 calories per square centimeter per minute is called the solar constant. About 70 percent of this energy, on the average, reaches the ground. The Earth's atmosphere cuts off the rest.

The Sun provides light and warmth. It provides us with food and oxygen by way of green plants, which use solar energy to create themselves out of carbon dioxide and water. As the plants do this, they release oxygen to the environment.

SOLAR ENERGY

Most fuels are a chemical form of solar energy. Coal, for example, is actually the transformed residue of ancient green plants that lived and died many ages ago and were buried in the rocks. When we burn coal, we release solar energy once stored as chemical energy by green plants in their tissues. Petroleum, or oil, is similar to coal in this respect.

Hydroelectric plants, which generate electricity from running water, also depend on the Sun. Without the weather cycle, whose energizing force is the Sun's heat, no rain would fall. There would be no running water to move the turbines of hydroelectric-power stations.

Solar energy is used more directly in various ways by humans, but on a very limited scale as yet. Sunlight heats our homes in summertime or year-round in the tropics. Solar heat evaporates seawater in one rather primitive technology used in the salt-making industry.

Solar energy is strong, but it is so scattered that complex systems of mirrors and lenses are needed to collect it for uses demanding more power than in the examples just given. Once concentrated, however, the power of the Sun's heat is fearsome. In a solar furnace, for example, it can melt iron or steel at temperatures of several thousand degrees. Sunlight generates electricity in light-sensitive cells known as solar batteries. They are used most often in space vehicles and satellites to power their equipment. Solar engines usually produce steam for power from water heated by the Sun. The use of these engines is rather limited at present, however.

As the Earth's fuel reserves are used up, we may have to turn more and more to the Sun as our primary source of energy. Efforts are being made to harness the Sun's power for human use. For a discussion of some of these efforts, see the articles in *The New Book of Popular Science* section on Energy.

SOLAR TELESCOPES

Astronomers studying the Sun gain much information on the basic nature of other stars. Remember, however, that astronomers today use elaborate instruments to protect their eyes. One should never look directly at the Sun, either with the naked eye or through an ordinary telescope or binoculars, because the Sun's rays may damage the eyes, causing blindness.

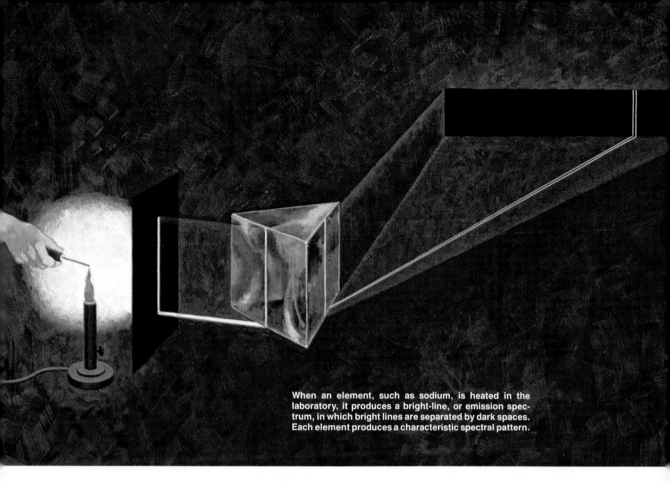

When an element, such as sodium, is heated in the laboratory, it produces a bright-line, or emission spectrum, in which bright lines are separated by dark spaces. Each element produces a characteristic spectral pattern.

Scientists use spectroscopes to analyze the atmosphere of the Sun and other stars. The spectroscope works by dispersing light or radiation to form a spectrum. Since each chemical element has its own particular spectrum, scientists can compare the spectra produced by a star with known spectral patterns produced in the laboratory.

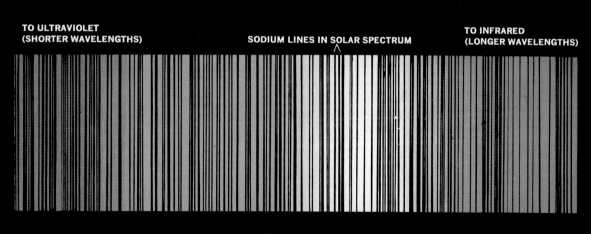

TO ULTRAVIOLET (SHORTER WAVELENGTHS)   **SODIUM LINES IN SOLAR SPECTRUM**   **TO INFRARED (LONGER WAVELENGTHS)**

The radiation emitted by the Sun is blocked by atmospheric gases. The resulting dark-line, or absorption, spectrum is the exact reverse of a bright-line spectrum. For example, the dark lines in the solar spectrum (above) match the bright-line spectrum of sodium.

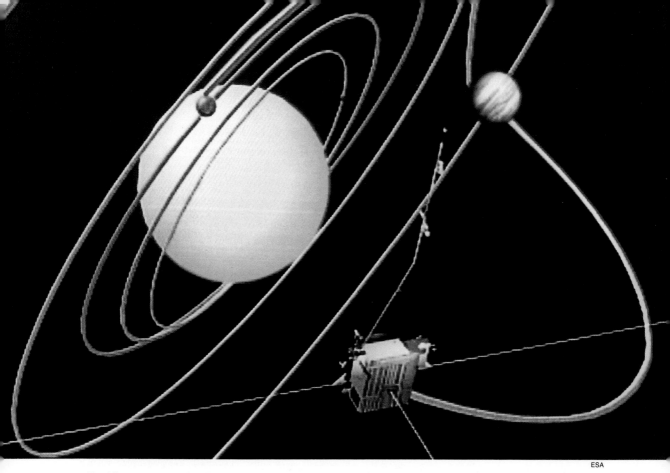

The Ulysses spacecraft will use Jupiter's gravity to jump into the orbit around the Sun's north and south poles.

Instruments for observing the Sun have existed for many hundreds of years. Before the invention of the telescope and other modern astronomical devices, people used simple mechanical solar instruments to measure the positions and paths of the Sun.

Most modern solar telescopes are very different from ordinary astronomical telescopes. One kind of solar telescope is called a *tower telescope* because the mirrors and lenses for concentrating sunlight are located atop a tower. A tower telescope projects an enlarged image of the Sun down onto a screen at ground level. Below the tower are instruments such as spectrographs and spectroheliographs that analyze the image.

Another type of solar telescope is the *coronagraph*. It reduces the glare emitted from the solar surface or photosphere, by means of a polished metal disk and nearly perfect lenses. In this way, astronomers can study the corona, the outermost atmosphere of the Sun, even when there is no total solar eclipse. The coronagraph is usually located on a mountaintop, where the air is thin and dust-free. Special types of radio telescopes, called radioheliographs, are used to detect radio waves, which are normally broadcast by the chromosphere (the lower part of the Sun's atmosphere) and the corona.

PAST AND FUTURE

How long has the Sun been radiating energy? How long can it continue to do so? Scientists are attempting to answer these questions. The Sun certainly has been shining for at least 5 billion years. At its present rate, as stated earlier, it could go on shining for another 30 billion years at least, if not longer.

But has the Sun always been radiating energy at the same rate? Will it continue at the same rate? The answer to the first question is maybe; to the second, probably not.

At present the amount of visible solar radiation, or light, varies only slightly, if at all. Many astronomers think that this has always been the case. Others, however, disagree. They point to the great ice sheets that from time to time have engulfed much of the Earth's surface. These, they say, are evidence that the Sun's total radiation may drop off. Even slight decreases would freeze vast areas on Earth.

Some astronomers believe that, as the Sun grows older, it will use up hydrogen at an increasingly faster rate. This would cut its future life span to about 10 billion years. As radiation increases, the Sun will become so hot that our oceans will boil away, and most life on Earth will be killed.

When its hydrogen supply finally gives out, the Sun will shrink into a very small star—a so-called white dwarf. Later it will die out completely.

SATELLITES AND SUN PROBES

Solar instruments are being sent into space aboard small, unmanned astronomical observatories. The craft are placed into orbit far above the Earth's distorting atmosphere. The U.S. manned space observatory, Skylab, and its Soviet counterpart, Soyuz, have both already conducted solar investigations.

The latest and most ambitious solar probe is the 815-pound (370-kilogram) Ulysses spacecraft, launched in October 1990 from the cargo bay of the space shuttle *Discovery*. Ulysses (named for the Greek mythical hero who set out to explore the uninhabited world beyond the Sun) is a joint effort of the National Aeronautics and Space Administration (NASA) and the European Space Agency (ESA).

Shortly after escaping Earth's gravitational influence, Ulysses became the fastest human-made object in history, reaching a velocity relative to the Earth of 112,000 miles (181,000 kilometers) per hour.

It is the first spacecraft aimed at orbiting the Sun perpendicular to the ecliptic plane, the plane where all the planets circle the Sun. Enormous energy is required to boost the spacecraft out of the ecliptic plane, so Ulysses is using a gravity-assist maneuver. It first traveled outward from the Sun toward Jupiter, and it will use Jupiter's gravity to swing back toward the Sun at an angle that takes it under the south solar pole.

Though the spacecraft will travel around the Sun at about twice the distance that the Earth does, its unique orbit will put it in a prime position to observe the Sun's magnetic field, or magnetosphere. Ulysses reaches the most southern point of its orbit in the spring of 1994. Then it will circle up to cross over the northern solar pole a year later.

Ulysses carries nine instruments to study the Sun. Its primary mission is to observe the solar wind emitted from the high latitudes of the Sun's surface. The solar wind is comprised of relatively low-energy protons and electrons forming a hot magnetized gas that continuously streams out from the Sun. Scientists want to learn how the solar wind interacts with the Sun's magnetic field near the poles. In the equatorial regions of the Sun, the magnetic-force lines spiral outward, creating a very complex picture of activity. Near the poles, however, the lines are more radial, suggesting that the magnetic field of the Sun may be less complicated at its poles.

As Ulysses traverses the solar poles, other onboard instruments will gather data on cosmic rays, the high-energy particles that cross space at incredible speeds and energies. With its less convoluted magnetic patterns, the solar polar regions will also enable Ulysses to study cosmic rays in a more pristine or unaltered energy state.

Another experiment will measure the speed and direction of the tiny motes of dust in the solar system. These dust particles are either drawn inward to the Sun by gravity, or are pushed outward by solar radiation, and acquire an electrical charge as they interact with the solar wind.

Ulysses also carries radio instruments that will enable scientists to study the basic physics of the ionized clouds of solar gases called *plasma*.

Mercury, the closest planet to the Sun, reflects light so poorly that its sunlit side is only dimly illuminated.

MERCURY

by Jeffrey Brune

Mercury lies closer to the Sun than do any of the other planets. The smallest planet save Pluto, Mercury is, nonetheless, one of the brightest at certain intervals. Unfortunately, even under the best of conditions, Mercury is difficult to see with the naked eye. Mercury appears in the heavens only at dawn and twilight, a time when even the brightest stars seem dim. Its position near the horizon often causes the planet to be obscured by haze. In fact, through all his years of heavenly observation, the great Polish astronomer Nicolaus Copernicus never once saw Mercury, a strange twist of fate perhaps due to the low, misty region where he lived.

MORNING AND EVENING STAR

Mercury makes a small circuit around the Sun. As a result, it never rises in the morning or sets in the evening much before or after the Sun does. Because it appears sometimes in the east and sometimes in the west, some ancient peoples—including the Egyptians, Hindus, and Greeks—thought of it as two separate heavenly bodies—a morning star and an evening star. The Greeks called the morning star "Apollon," after the god of the sun. They called their evening star "Hermes," the name of the swift messenger of the gods, because the planet's apparent motion among the stars was so swift. The Greek philosopher Pythagoras, who lived in the 6th century B.C., is thought to be the first to recognize that the morning star and evening star were one and the same heavenly body. That fact was well known to Roman astronomers, who worshiped Hermes under the name of Mercury (or Mercurius, in Latin).

In the Northern Hemisphere, Mercury can be seen with the naked eye for only a few days at dawn in late summer or early autumn, and also at twilight early in the spring. Fortunately, using telescopes, astronomers do not have to confine their observations of the planet to these periods.

When viewed through the telescope, Mercury looks a good deal like the Moon as seen with the naked eye. The telescope reveals that the disk of Mercury has phases like those of our Moon. It increases from a thin crescent to a full disk, and decreases again to a crescent. Then it disappears altogether when the planet is almost directly between the Earth and the Sun. These phases are not visible to the naked eye.

HEAVY, BUT FAST

Mercury is almost perfectly spherical. It has an equatorial diameter of 3,031 miles (4,878 kilometers), almost two-fifths that of the Earth. Mercury is only one-eighteenth as massive as the Earth. Yet its density—the average amount of mass it contains per unit volume—is very similar to that of our planet: 5.43 grams per cubic centimeter for Mercury versus 5.52 grams per cubic centimeter for Earth. In fact, even though Mercury is smaller than Venus or Mars, its density is greater than that of either planet.

If Mercury is so dense for its small size, it must be composed mostly of heavy materials. Scientists think that Mercury is like a Ping-Pong ball filled with heavy cement: it has a very thin, light covering, or crust, and a much thicker and heavier core. Mercury's crust is probably made of light silicate rocks, while most of the planet below the surface is composed of iron, a very heavy metallic element.

At one time in its early history, Mercury probably had a much thicker outer crust. That would mean it had more light rock than it does today, and thus a lower overall density. Astronomers think that, soon after the formation of the solar system, Mercury may have been bombarded by a shower of rocky asteroids. One of these asteroids hit Mercury so hard that it blasted most of the light crust away, leaving the heavy iron core behind.

A mosaic of photos taken by Mariner 10 shows how much the heavily cratered surface of Mercury resembles the surface of the Moon. The largest crater seen here measures 125 miles in diameter.

NASA

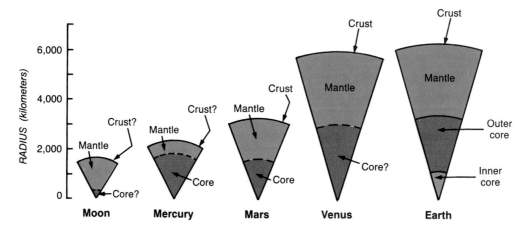

Mercury's size and density suggest that the planet's iron core is unusually large compared to the cores of the Moon and the three other terrestrial planets—Mars, Venus, and the Earth (above). Indeed, Mercury's core may account for 40 percent of the planet's entire volume. Astronomers debate whether or not Mercury has a liquid outer core and, if not, what phenomenon gives rise to the planet's weak magnetic field.

Above: Courtesy of *Sky & Telescope* magazine; below: NASA/JPL

Mercury has unusual systems of very long cliffs, or scarps, on its surface. The scarp that extends from the upper left to the bottom right of the photo is nearly 200 miles long.

Mercury is, on average, almost 36 million miles (58 million kilometers) from the Sun. At times, however, the planet can be much closer—as close as 29 million miles (47 million kilometers). At other times, Mercury can be as far away as 44 million miles (70 million kilometers). Why does the planet get so up close and personal to the Sun, then so aloof? The path it follows around the Sun is not circular; rather, it is stretched out into an elongated ellipse, with the Sun far from the center. In fact, Mercury's orbit is much more "eccentric" than Earth's or that of any other planet except Pluto.

In keeping with its namesake, the "winged messenger," Mercury is the fastest-moving of all the planets. It whips around the Sun in only 88 days, while the Earth takes 365 days, and lazy Pluto about 248 *years*! The speed at which Mercury moves in its orbital path varies according to its distance from the Sun. When Mercury is farthest from the Sun, it travels at 23 miles (37 kilometers) per second. When it is nearest to the Sun, however, it speeds up to a velocity of 35 miles (56 kilometers) per second—a speed that enables Mercury to escape from being sucked into the fiery solar furnace.

Slow-Moving Rotation

While swift in its race around the Sun, Mercury rotates slowly on its axis. The planet takes almost 59 Earth days for one complete rotation.

Mercury's sluggish twirl means that while one hemisphere of the planet faces the scorching Sun for long periods of time, the other side is exposed to the chill of interplanetary space. Surface temperatures on the sunlit side can reach 950° F (510° C); on the dark side, temperatures plummet to -346° F (-210° C).

One would imagine that Mercury's daytime side might be blindingly bright. Actually, the surface remains fairly dim, more like twilight than noon on Earth. This is because Mercury's surface reflects only about 6 percent of the sunlight that hits it. Because Mercury is so near the Sun, however, this has been enough to keep the planet veiled in mystery—until we went for a close look.

MARINER 10 FLYBY

In 1973 the United States launched the space probe Mariner 10 to Mercury. In March 1974, it flew to within 435 miles (700 kilometers) of the planet. The probe then circled the Sun and returned to within about 29,827 miles (48,000 kilometers) in September of that same year. On March 16, 1975, Mariner 10 made its last useful flyby, coming within 125 miles (200 kilometers) of the planet's surface. The wealth of photographs and other data collected by special instruments aboard the space probe have revealed much about Mercury.

In the photographs, Mercury can be seen to be covered with hundreds of thousands of impact craters—holes made when space rocks smash into the planet's surface. Some of the craters appear to have been smoothed over by ancient lava flows. This suggests that the planet was volcanically active at one time in its early history. Temperature measurements indicate that Mercury is covered with a fine-grained, porous material. This fine dust, together with the many craters, gives Mercury an appearance much like Earth's Moon. As expected, Mariner 10 revealed not a single Mercurial moon.

Mercury is crisscrossed with long cliffs, or scarps, up to 1.2 miles (1.9 kilometers) high and 930 miles (1,500 kilome-

Mercury's closeness to the Sun has made it difficult for astronomers to study the planet from Earth. Scientists soon hope to land a spacecraft on the surface of Mercury to study the planet more extensively.

© Pamela Lee

ters) long. Scientists think the scarps formed after Mercury's hot early history. As the planet cooled, the surface contracted and formed the scarps.

Scientists were surprised to discover that Mercury has a very thin helium atmosphere. Further research revealed the presence of sodium and oxygen atoms and traces of potassium and hydrogen. The word "atmosphere" may actually be an exaggeration, since Mercury's gases are so sparse—a million billion times less dense than Earth's atmosphere.

Another surprise discovery was a weak magnetic field around Mercury. This suggests that the planet has a hot molten core, perhaps mostly iron and some nickel. (The magnetic field would be generated by a core of molten metal mixed by streams of electrically charged materials.) New evidence, however, shows that Mercury's core may not be molten after all, but cool instead. Scientists used the Very Large Array radio telescope in New Mexico to measure the heat coming off the surface of Mercury. Analysis of the heat waves indicates that the heat is coming, not from the core, but only from sunlight that has reflected off Mercury's surface.

So what is responsible for Mercury's magnetic field? It may be the solar wind—the stream of charged particles flowing from the Sun. Scientists are still debating this issue.

Mercury has been the center of another, perhaps more passionate debate. Naming its major craters, some of which are 50 miles (80 kilometers) in diameter, caused an uproar in the scientific community. The International Astronomical Union (IAU), which decides the names of various geographical features on planets, had planned on naming the craters after birds. But some scientists were outraged at the thought of using bird names on an airless planet.

Astronomer Carl Sagan of Cornell University wrote to the IAU: "If the present inclination is followed, I suppose we will find other solar system objects sporting the names of fish, minerals, butterflies, spiders, and salamanders. I have nothing whatever against salamanders; but from the perspective of a millennium hence, it will be thought interesting that no features larger than a kilometer or so across have been named after Shakespeare, Dostoevsky, Mozart, Dante, Bach. . . ."

After bitter debate, the IAU settled on names of famous artists, writers, musicians, and other great contributors to the humanities. Major craters on Mercury now hold such distinguished names as Tolstoy, Mark Twain, Bach, Michelangelo, Shakespeare, and Renoir.

MERCURY IN TRANSIT

Since the planet Mercury lies inside Earth's orbit, it can pass directly between our planet and the Sun. Such a passage is known as a *transit*. If Mercury's orbit were in the same plane as that of the Earth, there would be three transits of Mercury each year. It would revolve about four times around the Sun in our year, and the Earth would revolve only once during the same period of time. However, the orbit of Mercury is inclined by about 7 degrees to the ecliptic—the plane of the Earth's orbit around the Sun. Mercury therefore crosses the ecliptic twice every 88 days, but generally not in a line with the Earth and Sun.

A transit of Mercury can take place only when the planet is between the Earth and the Sun, and is at the same time crossing the ecliptic. These conditions are satisfied from time to time, but not at any definite intervals. All of Mercury's transits occur in May and November. Dates of upcoming Mercury transits include November 5, 1993; November 15, 1999; and May 6, 2003.

As the planet makes its transit, the astronomer peering through a telescope can see the tiny black disk creeping across the dazzling solar background. This cannot be called an eclipse, because only an insignificant amount of the Sun's surface is obscured. Carefully observing the transit, astronomers can determine the exact position of Mercury in the heavens, and can also obtain added information about the planet's orbit.

Venus, the planet second from the Sun, is often called Earth's sister for its similar size and mass. The clouds that shroud the planet reflect enough light to make Venus the second-brightest object (after the Moon) in the night sky.

VENUS

by Jeffrey Brune

The elegant white planet orbiting the Sun between Mercury and Earth is called Venus after the Roman goddess of beauty. Although similar to the Earth in size (with a diameter of about 7,500 miles—12,100 kilometers—compared with Earth's diameter of 7,900 miles—12,750 kilometers) and mass (Venus is a bit more than four-fifths the mass of Earth), the Venusian surface and atmosphere are far different from ours.

Like Mercury, Venus is at times an evening star and at other times a morning star, depending on whether it is to the east or west of the Sun as viewed from the Earth. The planet may rise as much as four hours before the Sun, and may set as much as four hours after it.

Venus revolves around the Sun once every 225 days in an orbit that is very nearly circular. And as it revolves, it slowly rotates about its axis—about once every 243 Earth days, in a direction opposite that of Earth and most other planets.

When Venus is on the far side of the Sun from the Earth, it is quite far indeed. But at its closest, when it is between the Sun and the Earth, it is only some 26 million miles (41,840,000 kilometers) away.

Venus, like Mercury, shows a complete cycle of "phases," or shapes, to an observer on Earth armed with a small telescope or good binoculars. When the planet is at the farthest part of its orbit from the Earth, it appears as a disk. When Venus is between the Sun and the Earth, it is seldom visible. About 35 days before and after this time, it appears as a crescent and is at its brightest—two and one-half times brighter than when it is seen as a disk.

BENEATH THE GODDESS' VEIL

The surface of Venus is obscured by the planet's thick clouds, and so is invisible to optical instruments. For centuries, astronomers could only guess what lay beneath this veil. Some conjured tales of swamps, forests, and strange creatures.

Starting in 1962 the United States and the Soviet Union sent more than 20 probes to the planet for a true view. The U.S. Pioneer-Venus 1 craft and the Soviet Venera 15 and 16 orbiters used radar to pierce the thick clouds and make low-resolution maps of the planet's surface. The mapping revealed mainly rolling upland plains, some lowland plains, and two highland areas. One, called Aphrodite Terra, is about half the size of Africa. The other highland, Ishtar Terra, is about the size of the United States, and contains a 7-mile (11-kilometer)-tall mountain named Maxwell Montes.

Other Venera probes actually traveled through the clouds, landed softly on the surface, and transmitted the first color pictures of the planet's surface. Chemical analyses of the Venusian crust, conducted by the probes, showed that it contained basaltic rock similar to that associated on Earth with recent volcanic activity.

The U.S. spacecraft Magellan, launched in May 1989, has radar-mapped most of the Venusian surface with far better resolution than previous craft. Magellan sent sharp images of a world strewn with cracks and fissures, rugged mountains, and bizarre "pancake domes" formed by hot lava welling up from beneath the surface.

Magellan also photographed giant craters formed when large hunks of rock from space crashed into the planet. Interestingly, even the smallest craters found on Venus are quite huge—over 2.5 miles (4 kilometers) in diameter. Only very large hunks of space rock can survive the fiery passage through Venus' thick atmosphere and reach the surface.

Perhaps the biggest surprise from Magellan's pictures is the apparent lack of erosion on Venus. The mountains, craters, and other surface features appear rough and unweathered, almost as if they were newly formed. A major reason for this is that Venus is bone-dry. Water on Earth smooths down surface features. Rivers, for example, gradually change mountains into valleys. But with surface temperatures of 896° F (480° C), Venus is too hot for water. If Venus did have water in its past, it must have quickly evaporated.

Though there appear to be no signs that water ever flowed across Venus' sur-

NASA

Much like the Moon and the planet Mercury, Venus exhibits an entire cycle of phases. Venus appears brightest when it is at the crescent phase.

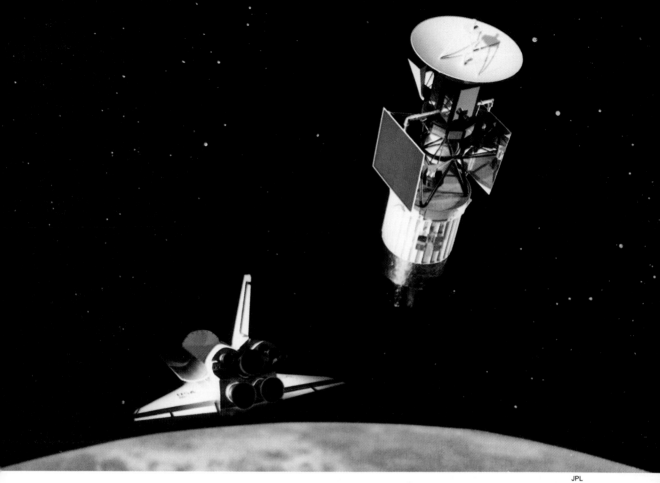

The Magellan space probe, pictured above after its 1989 deployment from the space shuttle Atlantis, *used radar to map the cloud-shrouded surface of Venus.*

face, Magellan's pictures show long channels cut by rivers of hot lava. Volcanoes must have resurfaced Venus in its recent past, because the planet has far fewer craters than do Mars or the Moon. Scientists suspect that Venus' volcanic face-lift came within the past few hundred million years.

Scientists are also trying to understand the geologic forces responsible for producing the volcanoes, mountain belts, and other surface features of Venus. On Earth the crust is divided into separate moving plates. Their movement, a process known as plate tectonics, is responsible for mountain building, seafloor spreading, and earthquakes. The crust of Venus does not appear to work in this way. Instead, upwellings of hot, light material and downwellings of cooler, denser material may pull and shove the surface crust.

Hot, Dense Atmosphere

About 95 percent of Venus' atmosphere consists of carbon dioxide, with some clouds of sulfuric acid. Traces of oxygen, hydrogen, nitrogen, neon, and ammonia have also been found.

The thick cloak of carbon dioxide is responsible for surface temperatures high enough to melt lead. In a process called the "greenhouse effect," sunlight passes through the thick atmosphere and heats up the planet's surface, which in turn gives off mainly infrared heat rays. But unlike sunlight, the infrared rays cannot penetrate the carbon dioxide of the atmosphere. The rays are thus trapped, and the planet heats up like a giant greenhouse.

Air pressure at the base of Venus' cloud cover is up to 100 times as high as it is at the Earth's surface. This massive pressure is a great obstacle to human exploration. At ground level the atmosphere of Venus is so dense that it is probably very slow-moving. To walk on the surface of the planet would be somewhat like walking through a furnace full of boiling oil—not a very pleasant thought!

At the equatorial region, bands of air as hot as 396° F (202° C) spiral upward to a height of 43 miles (70 kilometers). The ascent cools the air to about 55° F (13° C), causing clouds to form. These clouds whip around the planet at 224 miles (360 kilometers) per hour, veering constantly poleward. At the polar regions, the air descends and moves toward the equator, where it heats up again.

In 1986 two Soviet Vega spacecraft dropped weather balloons into the Venusian atmosphere. Floating about 34 miles (54 kilometers) above the planet's surface, the balloons revealed that the planet is buffeted by powerful east-west winds of up to 155 miles (250 kilometers) per hour.

Severe winds and storms probably occur as the air circulates around the planet. One enormous, long-lasting storm, somewhat akin to Jupiter's Great Red Spot, has been observed. This storm, called the Venusian "Eye," is the size of the United States.

The multiprobe Pioneer-Venus 2 found several hundred times more argon and neon in the Venusian atmosphere than is found in Earth's. This has caused astronomers to rethink one theory about the formation of the solar system. According to that theory, Venus would have less of these gases than does the Earth.

All the unmanned probes seem to indicate that Venus, unlike the Earth, has no appreciable magnetic field. As a result, there is no zone of trapped radiation in a magnetosphere, such as surrounds our planet. However, Venus does have an ionized atmospheric layer, or ionosphere, above its surface, due to the reaction between its atmosphere and the stream of particles and radiation coming from the Sun. This ionosphere is much thinner and lower than Earth's. But it does protect Venus' surface to some extent from fierce solar radiation.

Images of Venus transmitted from Magellan show a planet of towering mountains and volcanoes. The black lines in the image below are areas that lack radar data.

JPL

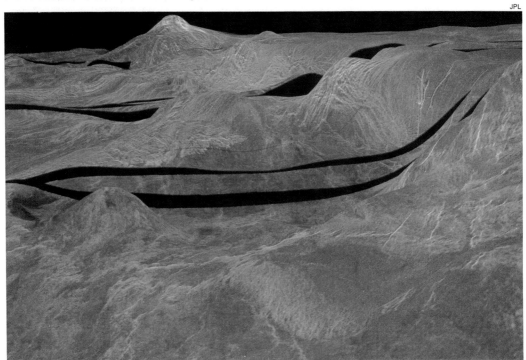

From space, Earth might be called the blue planet, since three-quarters of its surface is covered with water. The white swirls in the photo above are clouds—water vapor that has condensed in the atmosphere.

THE EARTH

by Franklyn M. Branley

Four planets in the solar system are smaller than the planet Earth. Four are considerably larger. The Earth in terms of mass, therefore, is not an outstanding member of the vast solar family—except, of course, as our home, and the vantage point from which we view the universe.

Astronomers have proposed many explanations of how the Earth and the other planets in our solar system originated. They are little more than ingenious conjectures, however, since they are based on insufficient data. Though we know little about the beginnings of the Earth, we know a great deal about our planet's shape, structure, properties, and motions.

THE SHAPE OF THE EARTH

Photographs of the Earth taken from rockets, satellites, and other spacecraft far above the surface show its distinct curvature, indicating it is nearly round. Before this evidence became available, the roundness of the Earth could be inferred from certain facts. It was known, for example, that as a lunar eclipse advances, the Earth casts a curved shadow on the Moon.

The Earth is not a perfect sphere, but slightly flattened. This shape is probably caused by our planet's force of rotation, which deforms the somewhat plastic Earth into a form that is in balance with the forces of rotation and gravity. The diameter of the Earth is 7,900 miles (12,700 kilometers) from pole to pole, and 7,920 miles (12,750 kilometers) around the equator. Recent measurements also indicate that the Earth is slightly more flattened on the South Pole than on the North Pole, which makes it slightly pear-shaped.

THE EARTH'S MASS AND DENSITY

The mass of an object represents the concentration of matter in it. It is a constant value, as opposed to weight, which is actually a measure of gravity that changes from place to place. Various methods were used to determine Earth's mass. In 1735 mathematician Pierre Bouguer, while in Ecuador, measured the extent to which a plumb line was deflected by the gravitational pull of a mountain, the peak called Chimborazo. Since he could estimate the mass of the mountain, he was able to estimate the mass of the Earth after the deflection was measured.

Today a sensitive instrument called a torsion balance is generally used to determine the Earth's mass. The attraction of a large ball of known mass to a small ball is compared with the attraction of the Earth to the small ball. According to a recent estimate, the mass of the Earth is 6.59×10^{21} tons (5.98×10^{21} metric tons).

To determine the Earth's density, we divide the mass in grams by the volume in cubic centimeters. The volume of the Earth is 1.083×10^{27} cubic centimeters. If we divide 5.98×10^{27} by 1.083×10^{27}, we get approximately 5.5 grams per cubic centimeter as the figure for the density of the Earth. That is, if we mixed together the air, water, and rock of our planet, the mixture would weigh about 5.5 times the same quantity of water. The Earth is the densest of all the planets.

GRAVITY AND MAGNETISM ON EARTH

Gravity and magnetism are still mysterious forces in many respects, and yet we have gathered considerable information about them. In the 17th century, Sir Isaac Newton clarified our understanding of gravity when he formulated his famous law of gravitation. It states that every particle in the universe attracts every other particle with a force that varies directly as the product of their masses, and inversely as the square of the distance between them. This is the statement of universal gravitation.

The term gravity (or, more accurately, terrestrial gravity) is applied to the gravitational force exerted by the Earth. Gravity is the force that pulls all materials toward the center of the Earth. This force becomes smaller as we move away from the center. You are really measuring the force of gravity every time you weigh yourself. If your weight is 143 pounds (65 kilograms), you are pulled toward the center of the Earth with a force of 143 pounds. Since weight decreases farther from the center, you will weigh slightly less on the top of a mountain than in a deep valley.

In 1600 Sir William Gilbert, an English physician, advanced the idea that the Earth behaves like a huge magnet, with north and south poles. This idea is now universally

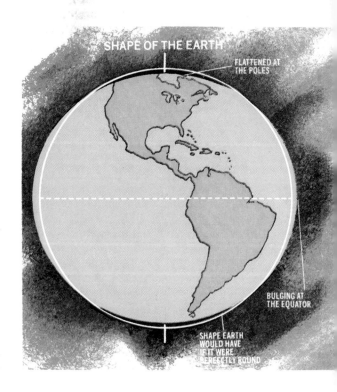

SHAPE OF THE EARTH
FLATTENED AT THE POLES
BULGING AT THE EQUATOR
SHAPE EARTH WOULD HAVE IF IT WERE PERFECTLY ROUND

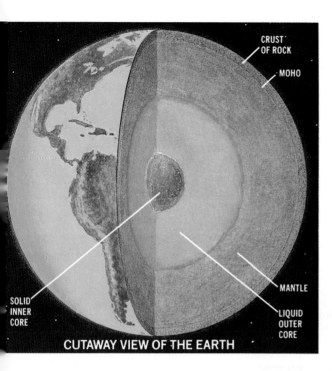
CUTAWAY VIEW OF THE EARTH

accepted. When you use a compass, the needle falls along the lines of force that run from one magnetic pole to the other. The magnetic poles do not correspond exactly to the geographic poles.

THREE PARTS OF THE EARTH

The Earth is made up of air, water, and solid ground—or, as a scientist would say, the atmosphere, hydrosphere, and lithosphere. The scientific terms are derived from Greek roots: *atmos* means vapor; *hydro*, water; and *lithos*, stone.

The atmosphere. The air surrounding the Earth is composed of about 78 percent nitrogen, 21 percent oxygen, and 1 percent other gases, including water vapor and carbon dioxide. Dust is also present.

The lower layer of the air envelope is the *troposphere*. *Tropos* means "change" in Greek, and the troposphere is the region where great changes take place in the temperature, pressure, and water-vapor content of the air. It is the part of the atmosphere where our weather occurs. Although most atmospheric changes take place relatively close to the Earth, the troposphere extends to an altitude of about 6 miles (10 kilometers). At the outer limit of the troposphere, there is a zone of division between the troposphere and the next sphere, called the *tropopause*.

The next atmospheric layer, extending from 6 to 25 miles (10 to 40 kilometers) above the Earth's surface, is the *stratosphere*. It is the zone of the strange winds known as jet streams—fast-moving currents of air that may reach velocities of 250 miles (400 kilometers) per hour. Temperature in the stratosphere rises from a low of −76° F (−60° C) at an altitude of 6 miles (10 kilometers), to a high of 32° F (0° C) at about 25 miles (40 kilometers).

At this point the stratosphere gives way to the *mesosphere*, which reaches from 25 to 44 miles (40 to about 70 kilometers) above our planet's surface. Temperature in the mesosphere ranges from a high of 32° F (0° C) at 25 miles (40 kilometers) elevation, to a low of −130° F (−90° C) at about 50 miles (80 kilometers) up. The air of the mesosphere is much thinner than that of the stratosphere.

The region of the atmosphere extending from 44 to 250 miles (70 to 400 kilometers) above the Earth is the *thermosphere*, where the air is extremely thin. Because of exposure to radiation from space and the Sun, many of the molecules and atoms are electrically charged, or ionized. Scientists often call these layers the *ionosphere*.

From 250 miles (400 kilometers) altitude and higher is the *exosphere*, the outermost fringe of the atmosphere. The extremely thin gas there consists chiefly of hydrogen. The exosphere eventually merges with the Sun's atmosphere.

In the late 1950s, it was discovered that the Earth is encircled by a large belt of radiation. This belt is now known as the Van Allen radiation belt, after its discoverer, American scientist James A. Van Allen. The belt begins about 400 miles (650 kilometers) above the Earth, and extends out some 25,000 miles (40,000 kilometers) into space. It is made up of charged particles radiated by the Sun and trapped by the Earth's magnetic field.

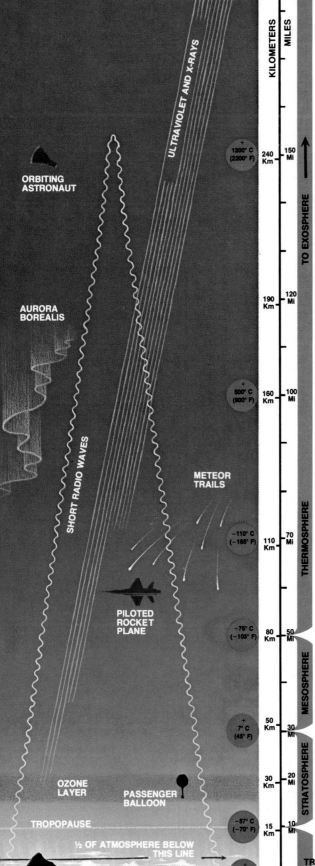

Layers of the atmosphere differ in the way they absorb radiation from the Sun, and thus differ in temperature. The bottom 10 miles of atmosphere—the troposphere—is where all weather takes place. Pollution has led to the formation of holes in the ozone layer.

The hydrosphere. The Earth appears to be the only planet that contains large amounts of liquid water. About three-fourths of the surface is covered by the oceans. These bodies of water, together with large inland lakes, contribute great amounts of water vapor to the air. They play a large part in the atmospheric changes that we call weather.

Of the Earth's surface water, almost 96 percent by weight is made up of hydrogen and oxygen. Sodium, chlorine, and many other elements are also found in oceanic waters. Traces of all the elements would probably be revealed if instruments with enough sensitivity were used.

The plants and animals found in the sea are an immensely valuable resource. They provide us with food, fertilizers, and industrial materials. The ocean is also a vast storehouse of minerals, such as common salt (sodium chloride), magnesium, manganese, gold, iron, copper, uranium, and silver. Some of these, such as salt and magnesium, are obtained from the sea in quantity. Others will no doubt be made available to us as more-efficient methods for extracting them are developed.

The lithosphere. The solid part of the Earth is made up of three types of rock—*igneous*, *sedimentary*, and *metamorphic*—and soil. Soil consists of rock debris combined with organic materials. Igneous rock is derived from the molten, rock-producing matter called *magma*. Sedimentary rock consists principally of rock fragments that have accumulated through untold millennia and have been pressed together. When igneous rock or sedimentary rock becomes altered through changes in temperature and pressure and other forces within the Earth, it gives rise to metamorphic rock.

We do not know for certain just what we would find if we were to dig to the center of our planet. However, Earth scien-

"The Living Earth" (above) is a composite of satellite photos taken over three years. The colors indicate biological activity both on land and in the sea, and thus help to convey the interactive nature of Earth's environments.

tists, gathering evidence from various indirect sources, have developed a more-or-less clear picture of the Earth's interior.

One of these scientists, Keith E. Bullen of the University of Sydney in Australia, has presented the following view: the Earth has a solid outer *mantle* about 1,740 miles (2,800 kilometers) thick. The Earth's crust makes up only a small part of this mantle. It extends only 25 miles (40 kilometers) or so below the Earth's surface. Beneath the mantle is the *core,* with a radius of some 2,200 miles (3,500 kilometers). It is divided into a solid inner core and a molten outer core.

AGE OF THE EARTH

The most-exact method known to science for determining the age of the Earth is based on the study of the radioactivity of certain minerals. In these minerals, one or more chemical elements decay radioactively, that is, their atoms give off very small particles and other radiation. During this process the radioactive elements are changed into other elements. A given element may also have different forms, or *isotopes*, which have different atomic weights. Some of these isotopes may be radioactive and undergo change.

Each radioactive-decay process takes a fixed length of time, regardless of external circumstances, depending on the isotope and its atomic weight. As the element decays, its quantity in the rock or mineral becomes smaller, while the amount of the element it is changing into becomes greater. Knowing the decay times of elements and the proportions of these elements and of their end products, scientists can calculate the age of the rock or mineral.

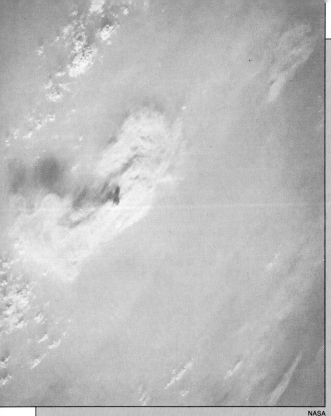
NASA

OUR FRAGILE WORLD: THE MISSION TO PLANET EARTH

When Apollo astronauts traveled to the Moon, they looked back on Earth and saw a tiny blue-and-white ball floating in space. That view, captured in pictures, conveyed as never before just how fragile our planet really is. We learned that we must take care of it.

But before we can intelligently care for the Earth, we must understand its complexities. How do the oceans affect weather? How do our life-styles—driving cars, using electricity, and disposing of garbage—affect the land, sea, and air around us?

In its home-observatory program called the "Mission to Planet Earth," the National Aeronautics and Space Administration (NASA) plans to monitor the overall health of the planet. Throughout the 1990s the agency will launch a series of observatory satellites to monitor such things as the Earth's ozone levels, changing temperatures, rain-forest destruction, volcanic eruptions, ocean currents, ice cover, sea levels, the extent of pollution, ocean life, and the increase in gases that may contribute to the "greenhouse effect." With such information, scientists will be able to make more-accurate conclusions about the Earth's health. These, in turn, can help all of us make the appropriate changes in our life-styles to save our fragile blue world before it is too late. The following problems, among the most important threats to our environment, are now under intense study:

• **The ozone hole.** Ozone is a three-atom molecule of oxygen. Formed high in the atmosphere, it shields life on Earth from the Sun's deadly ultraviolet rays. These rays can cause skin cancer, eye damage, immune deficiencies, and other health problems. The rays can also harm crops and aquatic life.

Ozone is very scarce in the upper atmosphere, making up less than one part per million of the gases there. In 1985 scientists announced that the springtime amounts of ozone over Antarctica had decreased by more than 40 percent between 1977 and 1984. With less ozone in the atmosphere, more harmful ultraviolet radiation reaches the Earth's surface. Alarmed, the scientists rushed to find out the causes of the

There are several radioactive elements, or isotopes, that are commonly used to date ancient objects, including rocks and minerals. These elements are carbon 14 (carbon isotope with atomic weight 14), rubidium 87, potassium 40, strontium 90, and uranium 235 and 238. The quantities of isotopes in a sample are measured by radiation detectors and other methods. The age of a rock may be based on one or more isotopes. However, these methods are not exact—and there may be an uncertainty of as much as several hundred million years, or there may be disagreement in ages measured with different isotopes. The oldest known rock on Earth was formed nearly 4 billion years ago. The Earth itself is thought to be about 4.5 billion years old.

"ozone hole." One contributing factor is chlorofluorocarbons (CFCs), a class of chemicals used in refrigerators, air conditioners, and other industrial devices. It has been estimated that one CFC molecule can destroy thousands of ozone molecules.

The ozone layer over the United States and other parts of the world is also thinning. Steps are being taken by industrialized countries to reduce the use of CFCs. But even if CFCs were no longer used, full recovery of the ozone layer could take decades or even centuries. In the meantime, scientists continue to monitor ozone levels around the world, using spacecraft, aircraft, and balloons.

• **Global warming.** Without carbon dioxide gas in our atmosphere, our planet would be covered with ice. Like the glass panes of a greenhouse, carbon dioxide allows sunlight to penetrate the atmosphere, and then traps some of the heat rays produced when sunlight reflects off the Earth's surface. The Earth is kept comfortably warm. But the presence of too much carbon dioxide, such as in the atmosphere of the planet Venus, causes surface temperatures to become hellish.

Human activities, such as the burning of fossil fuels, have caused a steep increase in carbon dioxide levels and other "greenhouse" gases in the past century. As a result the Earth may be getting warmer. According to NASA's Goddard Institute for Space Studies, at least seven of the warmest years in the past century have occurred since 1980.

No one really knows what climatic changes, if any, may occur in the future. But if the climate does change, the effects could be devastating. In a worst-case scenario, ice caps could melt, inundating coastal cities. Rainfall patterns may change, causing droughts in some areas and floods in others.

Scientists are using the vantage point of space to study the influence that oceans, ocean life, and land vegetation have on carbon dioxide levels. In the meantime, some nations are already taking steps to reduce consumption of fossil fuels.

• **Rain-forest destruction.** The fact that life exists on Earth makes our planet special and different from all others in the solar system. There's no better example of the rich diversity of life on Earth than the countless birds, flowers, monkeys, and insects of the rain forests. Tragically, humans are destroying the world's rain forests and other woodlands at an astonishing rate.

Trees naturally absorb carbon dioxide, a greenhouse gas. Cutting down trees destroys a natural defense against the threat of global warming. Burning trees adds carbon dioxide into the air. Worst of all, by destroying the forests, we destroy the homes of many plants and animals. In Brazil, species have been lost even before they have been identified.

In 1988 astronauts aboard the space shuttle *Discovery* photographed a massive smoke cloud hovering over South America's Amazon River basin (see photo, facing page). The cloud, measuring over three times the size of Texas, was literally and figuratively part of a lush rain forest gone up in smoke.

Jeffrey Brune

MOTION OF THE EARTH

In ancient times and throughout the Middle Ages, many people believed that the Earth was motionless. They explained the succession of day and night and the changing position of the stars by saying that the Earth stood still, and that the sky moved around the Earth. We now know that the apparent daily movement of the stars in the heavens is due to the rotation of the Earth about its axis. It makes a complete rotation in 23 hours, 56 minutes, and 4.09 seconds.

One of the best proofs we have of the rotation of our planet is a pendulum experiment first performed in 1851 by a French physicist, Jean-Bernard-Léon Foucault.

He suspended a heavy weight at the end of a steel wire, which was suspended from the dome of the Pantheon, a public building in Paris. A pin attached to the end of the weight rested on a circular ridge of sand. Foucault set the pendulum swinging. The pendulum moved to and fro, in the same plane, and the pin at the end of the weight began to trace lines in the sand. As the pendulum continued to swing, the lines followed different directions. There could be only one explanation. The pendulum did not change direction. Therefore, it must be the ridge of sand that was turning. Since the sand rested on the floor of the Pantheon, and since the Pantheon itself rested on the Earth, Foucault concluded that the Earth itself must be rotating.

The device with which Foucault proved the rotation of the Earth is called the Foucault pendulum. One of these pendulums has been erected in the General Assembly Building of the United Nations in New York City.

The alternation of day and night is due to the rotation of the Earth about its axis. As our planet turns, a given place on its surface will be in sunlight or in darkness, depending upon whether it is facing the Sun or facing the part of the sky on the other side of the Earth from the Sun. The Earth's rotation also causes air currents to be turned toward the right in the Northern Hemisphere, and to the left in the Southern —a phenomenon called the *Coriolis force* (see Volume 2, page 114).

An interesting effect, due to the rotation of the Earth, can be produced by focusing a camera on the North Star and leaving the shutter open for several hours. The stars will appear, not as points, but as curved lines. This is because the Earth, on which the camera rests, has been rotating on its axis.

Yearly Motion

At the same time that it rotates, the Earth revolves about the Sun. It completes a revolution in 365 days, 6 hours, 9 minutes, and 10 seconds, when reckoned relative to the position of the stars in space. This is called a *sidereal* (star) *year*.

The orbit of the Earth around the Sun is an ellipse, with the center of the Sun at one of the two foci. The definition of an ellipse is that it is the path of a point, the sum of whose distances from two fixed points—the foci—is constant. An ellipse can be drawn by the method described below.

Place two pins on a piece of paper resting on a flat surface. Prepare a piece of string more than twice as long as the distance between the two pins, and splice the ends of the string. Place the string around the two pins, and set a pencil in position. Stretching the string to the fullest extent, pass the pencil point over the paper, going around the pins. The figure drawn by the pencil is an ellipse. The pins will represent the two foci, F^1 or F^2. If this ellipse represented the orbit of the Earth around the Sun, the Sun would be at F^1 or F^2. Actually, the ellipse described by the Earth in its movement around the Sun is very nearly a circle.

The distance of the Earth from the Sun will vary according to its position in its elliptical orbit. At *perihelion*—its nearest approach to the Sun—the Earth is some 3 million miles (4.8 million kilometers) closer to the Sun than at *aphelion*, when it is farthest away. According to a commonly accepted figure, the mean distance is 93 million miles (149.6 million kilometers). This is often used as a unit of length—the astronomical unit (A.U.)—by astronomers for measuring large distances. For example, instead of giving the distance of the planets from the Sun in miles or kilometers, an astronomer could give it in A.U. To Mars, for instance, it would be 1.5 A.U.; to Jupiter, 5.2.

As the Earth revolves about the Sun, its axis is tilted at an angle of 23.5 degrees from the perpendicular to the plane of its orbit. As a result the Northern Hemisphere will be tilted toward the Sun in one part of the orbit and away from it in another part, as will the Southern Hemisphere. This accounts for the fact that the days are longer in summer than in winter in the Northern Hemisphere. In the summer months, this hemisphere is tilted toward the Sun. Hence, a given spot on the hemisphere—

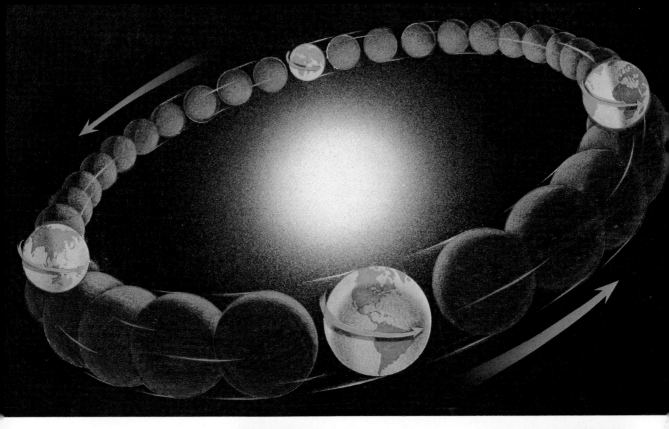

The Earth's motions (above) give our planet its days and its years. Every 24 hours, the Earth completes one rotation on its axis, a fact proven by the Foucault pendulum (right) in 1851. Every 365¼ days, the Earth completes one orbit around the Sun.

Buhl Science Center

say, Chicago—will be in sunlight more than it will be in darkness in the course of a single rotation of the Earth. In the winter the Northern Hemisphere is tilted away from the Sun. Hence, Chicago will be in the shadow longer than it will be in sunlight. The situation will be reversed in the Southern Hemisphere.

The motion of the Earth about the Sun and the tilting of its axis are the principal causes of seasonal changes. As we saw, the Northern Hemisphere is tilted toward the

EARTH 111

Sun during its summer months and away from it in the winter months. When it is summer in the Northern Hemisphere, the Sun's rays fall more directly upon the hemisphere, heating it more effectively, and areas in the Northern Hemisphere remain longer in the sunlight than they do in the shadow. For a time, more heat is therefore received from the Sun than can be radiated away. During our winter months, the Sun's rays fall more obliquely upon the surface of the Northern Hemisphere. The more the rays slant, the less effectively they heat the surface of the Earth. Likewise, because the days are shorter than the nights, the heat received from the Sun has more time to radiate away. That is why temperatures are lower during the winter.

The Southern Hemisphere is tilted away from the Sun during the time that the Northern Hemisphere is slanted toward it. Hence, the winter months south of the equator correspond to the summer months north of it. It is winter in January in Chicago, while it is summer in the same month in Buenos Aires.

Our planet moves at an average speed of about 19 miles (30 kilometers) per second along its path around the Sun. Sometimes the Earth moves slower, sometimes faster, in its orbit. It is also traveling through space at a much faster speed as it follows the Sun in its wanderings through the heavens. The Sun revolves around the center of the galaxy called the Milky Way, completing a turn in about 2 million years. As a satellite of the Sun, the Earth takes part in this journey, maintaining a speed estimated at 118 to 168 miles (190 to 270 kilometers) per second.

Wobbly Motion

In addition to the motions that we have just described, the Earth also wobbles, or *precesses*, as an astronomer would say. Precession is due to the combined effects of gravitational attraction and the Earth's rotation. The Moon (and, to a lesser extent, the Sun) is constantly pulling upon the Earth. This effect, combined with the Earth's rotation, causes the axis of our planet to wobble about its center. As it does so, it traces out two cones in space. These cones have their vertexes, or tips, at the Earth's center, and their bases in space above the geographic poles.

We might compare the effect with the spinning of a top at a slant. If we extended the axis of the top, say, to the ceiling of the room in which it is rotating, the axis would describe a cone with its vertex at the point of the top and its base at the ceiling. It would take the top a fraction of a second to complete a single spin. It takes the axis of the Earth about 25,800 years.

Because of precession, different stars become our North Star—the one most directly above the north geographic pole—in the course of the years. Right now, Polaris is the North Star. Alpha Cephei will be nearest the pole in A.D. 7500; Vega, in A.D. 14,000. Eventually the full cycle will be completed, and Polaris will be the North Star again.

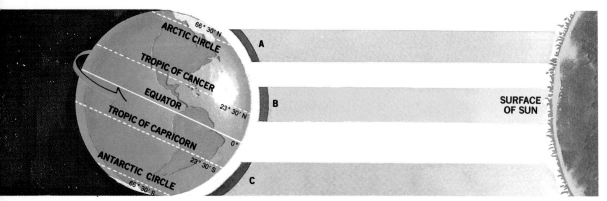

Areas A, B, and C receive equal amounts of solar energy on June 21. A's and C's energy is spread over larger areas than is B's energy, which is concentrated over the Tropic of Cancer. B's area is thus the warmest on June 21.

The Moon is the only natural satellite of the Earth. When astronauts landed on the Moon's surface in 1969, it was the realization of a centuries-old dream.

THE MOON

by Cecelia Payne-Gaposchkin
and Katherine Haramundanis

The Moon, circling the Earth under the pull of our planet's gravity, passes across our sky once every 24 hours, exerting its own considerable gravitational pull on all of Earth's oceans and seas. Since the dawn of history, the peoples of the world have used the regularly changing shape of the Moon's face as a calendar. But the Moon remained more an object of legend than of science until 1609, when Galileo first focused his telescope on the details of its surface. He recognized mountains and large, flat, dark areas, which he called *maria*, the Latin word for "seas." Astronomers now know that there is no water on the Moon, although the term is still used.

The light and dark areas we see from Earth are, respectively, the Moon's uplands and its low-lying flat regions. Shadows thrown by some of the Moon's features can be used to estimate their heights. Among the most striking features is the 146-mile (235-kilometer)-wide crater Clavius, its rim surrounded by mountains 3.2 miles (5.2 kilometers) high. Telescopes also reveal bright streaks radiating from some craters. Because these rays do not throw shadows, we know they are not elevated. Other dark, riverlike features, called *rills*, are probably dry cracks cutting through the Moon's surface.

As astronomers built more-powerful telescopes, more lunar details have been noted and mapped. Beginning in the 1960s, artificial satellites and manned spacecraft were launched to pass near the Moon and eventually land on it. Pictures from these missions revealed the surface in extraordinary detail, even making visible small boulders about a foot across. Moon probes have provided us with detailed close-up photographs of the far side of the Moon, which is always hidden from an Earthbound observer's view.

ORIGIN AND HISTORY

The history of the Moon is bound up with the history of the whole solar system, which is believed to be more than 4.5 billion years old. One theory holds that the Moon formed out of the Earth, perhaps tugged away by the pull of a passing star. But the most accepted theory seems to be that the Earth and the Moon were formed almost simultaneously from the agglomeration of cold material then circulating around the Sun. Similar processes are suggested for the origins of the other planets and their satellites. A third theory states that the Moon is a former planet that was captured by the Earth's gravity.

Close observation of the Moon has revealed very few signs of activity in recent times. However, signs of volcanic activity and possibly radioactive glows seem to indicate that the Moon is not really such a dead world after all. The present topography, plus the damage that has apparently been inflicted on its features, indicate that the Moon has had a tumultuous past.

If the Moon once did have an atmosphere and surface water, as some experts believe, then a history somewhat like the Earth's must have taken place until air and water vanished. But even with air and water gone, erosion is still possible. Extreme heating and chilling of rock will crack

The Moon's phases depend on the part of its sunlit side facing Earth. Its apparent shape changes nightly, depending on the positions of the Sun, Earth, and Moon.

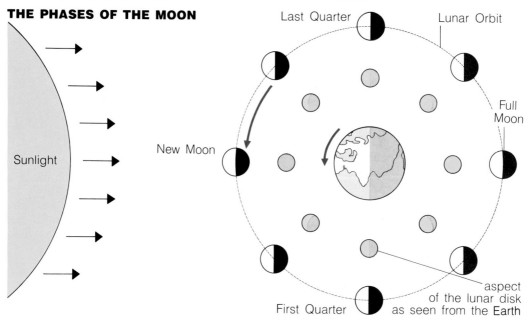

THE PHASES OF THE MOON

and flake it. Collisions with asteroids and meteorites pulverize the surface and may release volcanic forces. Movements and sliding of rock masses down slopes cause damage. Even the strong radiation from space may disrupt the lunar crust.

LUNAR STATISTICS

By measuring the time it takes for a laser beam to reflect off a set of mirrors left on the lunar surface by U.S. astronauts, astronomers can gauge the exact distance between the Earth and the Moon to within less than an inch. This distance is not constant, however, as the orbits of the Earth and Moon are elliptical. The yearly average distance is around 240,000 miles (386,000 kilometers)—at least for now. The Moon is slowly but surely pulling away at a rate of about 1.5 inches (4 centimeters) a year.

The Moon's diameter is 2,163 miles (3,480 kilometers), about a quarter of the Earth's diameter. Some planets in our solar system have larger moons, but the moons of Earth and Pluto are the largest with respect to the size of their parent planets. In fact, the Earth-Moon system may actually constitute a double-planet system. The mass of our Moon, as measured by its gravitational effect on the Earth, is $1/81$ of the Earth's; its volume is a somewhat larger proportion—$1/50$—meaning that the Moon is less dense than our planet. Actually, the Moon is about as dense as the rocks on the Earth's surface, but, unlike our planet, it may lack a dense metallic core.

BRIGHTNESS OF THE MOON

The Moon produces no light of its own, but shines by reflected light. The percentage of light reflected by the Moon is known as its *albedo*. On average the Moon reflects only 7 percent of the sunlight that falls vertically upon it, with some areas reflecting more light than others. All the light by which we see the Moon comes from the Sun, either directly or after reflection from the Earth. As the second-brightest object in the sky, the Moon sends us only two-millionths as much light as the Sun.

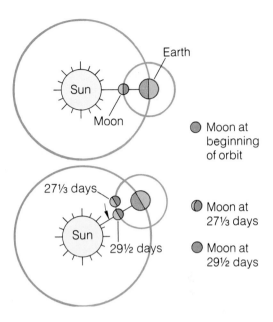

The period in which the Moon completes an orbit around the Earth and returns to the same position in the sky is 27⅓ days. But because the Earth is also moving in its orbit around the Sun in the same direction as the Moon, it takes two more days before the Moon is again between the Earth and the Sun. So the interval from one new moon to the next is 29½ days.

PHASES OF THE MOON

We all have noticed that the Moon's apparent shape changes from night to night, and it runs through a complete cycle in about a month. These shape changes, or phases, are caused by the relationship between the positions of the Moon, Sun, and Earth.

When the Moon is directly in line between the Sun and the Earth, the Sun is shining on the far side of the Moon; this phase is known as the *new moon*. As the Moon travels around the Earth and moves out of the Sun-Earth line, the illuminated part becomes visible to us as a thin crescent, which increases, or waxes, night after night. When the line from Earth to Moon makes an angle of 90 degrees with the line from Earth to Sun, we see half the Moon's face illuminated. This phase is called the *first quarter*. When the Earth, Moon, and Sun are again in line, with Sun and Moon

Lunar craters range in size from 684 miles across (Mare Imbrium) to only inches across.

on opposite sides of the Earth, we see the whole face of the Moon illuminated. This phase is the *full moon*. Thereafter the Moon wanes, and the illuminated surface grows smaller. When the direction from Earth to Moon again makes an angle of 90 degrees with the direction from Earth to Sun, we again see half the Moon's face illuminated in the *last quarter* phase. The crescent continues to wane until new moon is reached again. Between the quarters and new moon, the shape of the illuminated portion of the Moon is called a *crescent*. Between the quarters and full moon, the shape of the illuminated disk is described as *gibbous*—that is, not fully circular.

The line that separates the illuminated and dark portions of the Moon is known as the *terminator*. The crescent moon, whether waxing or waning, has horns, or cusps, which always point away from the Sun. Near the time of new moon, it is often possible to see the whole disk of the Moon faintly illuminated. The light by which we see this phenomenon is light that has been reflected from the bright surface of the Earth.

Although the illuminated area changes, we always see virtually the same side of the Moon. This is because the Moon is gravitationally locked to the Earth and makes one rotation on its own axis in nearly the same time it takes to make one revolution around the Earth. However, it should not be forgotten that the Sun shines on all its sides in turn.

The Months: Synodic and Sidereal

Astronomers recognize more than one kind of month. The simplest is the *synodic* month, or *lunation*, the interval from one new moon to the next new moon—$29\frac{1}{2}$ days. This, however, is not the time taken by the Moon to make one complete orbit around the Earth. The Moon falls behind because the Earth's motion around the Sun carries our planet about $\frac{1}{12}$ of the way around its orbit between lunations. The

A unique view of a portion of both the near and far sides of the Moon as captured by the Galileo probe.

true orbit of the Moon, known as the *sidereal* month, is 27⅓ days. Gravitationally locked to the Earth, the Moon rotates on its axis once in a sidereal month.

THE TIDES

Because of its close proximity to the Earth, the Moon's gravity pulls powerfully on our oceans, seas, and lakes, and even upon solid land. The Sun, too, tugs at the planet's waters, but, owing to its greater distance, the effect is much smaller. These forces, working together, are always creating two maximum tidal bulges at diametrically opposite sides of the Earth. These bulges, or high tides, draw water from all points between, creating two areas of low tides along meridians 90 degrees removed from them. As a result, every shoreline experiences two high tides and two low tides each day.

Twice a month the Sun, Moon, and Earth fall into perfect alignment, exerting their gravitational force in a mutual or additive way. At the full moon, the Earth moves between the Sun and the Moon; and at the new moon, the Sun and Moon lie on the same side of the Earth. These higher high tides and lower lows are called *spring tides*.

When the Sun and Moon make an angle of 90 degrees in relation to the Earth, they reduce each other's effect, producing semimonthly *neap tides*, when the difference between high and low is at a minimum. The exact times of high and low tides at a given location vary, however, depending on variations in coastlines, physical barriers such as reefs, strong winds, and even changes in barometric pressure.

ORBIT OF THE MOON

The motion of the Moon is far from simple. Its shape and position relative to the Sun and Earth are continually changing. For these reasons, the part of the

THE MOON 117

Moon seen from the Earth varies slightly, so that, over a period of time, we can view 59 percent of the Moon's surface from a place of observation on the Earth. Because the changes in the Moon's orbit run in cycles, its visible surface seems to undergo rocking motions, or librations, which bring small areas near the edges of the observable disk into view. Several decades elapse before all possible areas are visible to viewers on Earth.

Eclipses of the Sun and Moon

The relative sizes of the Moon and the Sun—as seen from the Earth—are almost exactly the same. This is an extraordinary coincidence, because the more-distant Sun is about 64 million times the volume of the Moon.

If the orbit of the Earth around the Sun and the orbit of the Moon around the Earth were in exactly the same plane, the Moon would pass directly across the face of the Sun at every new moon, producing a monthly solar eclipse. Similarly, at every full moon, the shadow of the Earth would fall on the Moon, producing a total eclipse of the Moon.

But the Moon's orbit is inclined by about 5 degrees to that of the Earth. This means that the Moon can come in front of the Sun only near the position where the two orbits intersect, points called the *nodes* of the Moon's orbit. When a new moon occurs very near the node, the Moon will pass exactly across the face of the Sun, and there will be a *total eclipse* of the Sun. Farther from the node, the Moon will cover only part of the Sun's face, resulting in a *partial eclipse*. Because neither the Moon's nor the Earth's orbit is circular, the distances from Earth to Moon and Sun are not constant. As these distances vary, so do the apparent sizes of the Sun and Moon. At times the Moon may not cover the entire solar disk, allowing a thin rim of sunlight to be visible around its edge—an *annular eclipse* of the Sun.

An eclipse of the Sun is visible only from the small portion of the Earth on

As the world watched in awe on July 20, 1969, Edwin E. Aldrin, Jr., the second man to step on the Moon, heralded this major accomplishment of space exploration by planting the flag of the United States firmly on the Moon's surface.

A lunar module stands on the cratered surface of the Moon, highlighted by the Sun. The lunar surface consists of a layer of fine particulate material and rock fragments extending down from the surface about one to three yards. The forbidding landscape is never marred by erosion, since the Moon lacks wind, rain, or other weather.

which the Moon's shadow is cast. The shadow moves rapidly across the surface of the Earth, and total eclipses last only a few minutes, as seen from one station.

As for lunar eclipses, they take place when the Moon passes through the Earth's shadow. Therefore, lunar eclipses always occur at the time of a full moon, with partial eclipses seen when the Moon passes partially within the Earth's shadow. Unlike solar eclipses, eclipses of the Moon are seen from every part of the Earth where the Moon would ordinarily be visible at the time (see also page 175).

LUNAR TEMPERATURE

When it is midday on the Moon, with the Sun directly overhead, the temperature reaches 212° F (100° C). At lunar midnight the temperature drops to about −177° F (−116° C). It should also be remembered that the Moon rotates only once in about 27.3 of our Earth days—giving it long periods of daylight and darkness during which it heats and cools.

But the Moon's astounding temperature extremes are due primarily to the lack of atmosphere, which would act as an insulating blanket. You can directly see that the Moon has no atmosphere by watching an *occultation*, that is, the passing of a star behind the Moon. If the Moon had an atmosphere, the star would twinkle and fade gradually, but it always vanishes abruptly, and abruptly reappears at the other edge.

It is not surprising that the Moon has no appreciable atmosphere, because gravity at its surface is only one-sixth that of the Earth's. This is not enough to retain most gases. Small amounts of gas have been seen to exude from certain points on the Moon's surface, but they seem to quickly dissipate into space.

LUNAR GEOGRAPHY

As mentioned earlier, the largest features on the Moon, readily seen with the naked eye, are the dark maria. These flat areas are strewn with small boulders and pocked with craters. We now know that, remarkably, all the maria but one are on the side of the Moon facing the Earth. The largest is Mare Imbrium, about 684 miles (1,100 kilometers) across. It is generally thought that the maria are hardened lavas that flowed in the distant past, perhaps due to the impact of huge extraterrestrial objects or due to some internal processes.

The Moon's surface is also heavily pocked with circular craters. They have clearly raised rims, and some have central peaks. Many craters overlap with each other, and may occur within the maria and on the mountain chains. Some craters appear to be filled with hardened lava, others to be partially buried in dust flows. In the photographs of the great crater known as Aristarchus, both types of flow may be seen. Aristarchus has a central peak, and its floor seems to be filled with lava. The crater nearby is very shallow and has been nearly obliterated by a dust flow. A number of much smaller, bowl-shaped craters dot the area.

Very large craters are often called "walled plains," because they enclose fairly level surfaces. These walled plains may be light in color, like lunar uplands, or dark, like maria. The crater wall, or rampart, is roughly circular and has often been designated as a circular mountain range. However, the wall is often low in comparison with the surrounding land, and even in comparison with the enclosed surface, or crater floor. The diameters of the largest craters reach more than 185 miles (300 ki-

lometers). Some craters, however, are so deep that their bottoms are always in shadow. Still other craters are mere pits in the lunar surface, with little or no raised rim surrounding the central depression.

On the Earth, there are two major types of craters: the volcano and the impact crater. Volcanic craters have steep sides and often a central rise or knob. Crater Lake in Oregon is a good example. Impact craters, such as the Meteor Crater in Arizona and the Chubb Crater in Canada, are shallower relative to their diameters, and do not usually possess central peaks. Impact craters can be very much larger than volcanic craters: the southeastern shore of Hudson Bay may be the remainder of a great impact crater.

Craters similar to Earth types can be recognized on the Moon. There is little doubt that the Moon's surface has been heavily bombarded by meteoric bodies over a long period of time. This bombardment has most likely produced not only a large percentage of the craters, but also the maria. The great dust splash extending halfway around the Moon from the huge crater Tycho is clear evidence that it is an impact crater. A large impact crater usually has many small impact craters near it, probably formed by the debris thrown out by the original impact. The boulders found near impact craters have a similar origin.

A possible example of a volcanic crater on the Moon is Aristarchus. Many small changes have been recorded near it, and gases may emerge from it on occasion. It has a pronounced central peak and what appears to be a lava-filled floor.

Lunar *rays* are Moon features that extend radially outward from certain craters

The Apollo 17 lunar landing (left) was the last of six manned Moon missions. The data collected from these missions have helped explain mysteries about the Moon that have haunted humanity for centuries. Scientists now conclude, for instance, that most of the Moon's craters (below) were created by meteorites bombarding the lunar surface, while a smaller portion were formed by Earth-style volcanoes.

Astronauts left a variety of instruments on the Moon's surface to collect data. The Laser Ranging Retro-Reflector (left of lunar module) is a set of mirrors that helps astronomers gauge the exact distance between the Earth and the Moon.

NASA/Johnson Space Center

such as Tycho and Copernicus. They appear lighter than the surrounding terrain principally because they reflect light better. This is probably because they are made of very finely divided particles. The reflectivity of particles depends largely on their average size, finer particles reflecting light more brilliantly under vertical illumination than do coarse ones.

Rills are narrow, riverlike valleys made visible by shadows cast into them. They may be cracks in the surface, or possibly canyons formed by ash flow from volcanoes. Some are twisted and tortuous, and some are associated with chains of craters. Some of the straight rills seem to be associated with natural settling around the maria. Others may have been furrowed by rolling boulders.

The mountains of the Moon form large, rugged chains principally concentrated around the maria. Their heights can be measured by the length of the shadows they cast. The highest, the Leibniz Mountains, attain a height of 5 miles (7.9 kilometers). Lunar mountains seem to have been thrust up as a result of impact rather than formed by folding of rocks, as has been the case with many mountains on the Earth. A mountainous feature known as the Straight Wall, in Mare Nubium, south of the lunar equator, appears to be a 68-mile (110-kilometer)-long line of cliffs produced by faulting.

Naming Lunar Features

The early astronomers examining the Moon by telescope gave the principal maria chains fanciful Latin names, and many mountain chains were named after well-known European mountains. Since that time, advances in observational techniques have revealed more and more detail, and thousands of features have been named. The International Astronomical Union (IAU) now has the responsibility for assigning names, a task that has been enormously increased by observation of the far side of the Moon. Typically, major craters have been named after astronomers or scientists. Other prominent individuals, such as Jules Verne, a 19th-century author, and Plato, a Greek philosopher, have also been commemorated.

LUNAR EXPLORATION

The first photograph of the Moon was made by American scientist John Draper in 1840. In 1897 the Paris Observatory completed a photo atlas of the Moon. By the 19th century, astronomy and geology had sufficiently advanced so that explanations of the Moon features could be seriously offered and the origins of the features discussed.

In the early 20th century, astronomers began to be able to view many never-before-seen details of the lunar surface. These discoveries inspired scientists to seriously consider means for traveling to the Moon. The vivid imaginations of fiction writers, in the meantime, were not idle. Authors such as Jules Verne and H. G. Wells kept popular interest in our solar system alive with their tales of interplanetary and lunar adventures.

Although rockets had been in use for fireworks and as weapons since the 13th century, few individuals ever considered them as a safe or reliable means of travel. But the 20th century saw the fruition of the rocket-ship dream. In the 1920s and the 1930s, the groundwork for rocket-ship technology was laid. During the later 1940s and the 1950s, the Soviet Union and the United States sent rockets aloft to explore the upper atmosphere and near space. They then decided to launch artificial satellites to circle the Earth and later to orbit or land on the Moon.

TO THE MOON AND BACK

In 1957 the Soviet Union launched Sputnik, the first artificial satellite to orbit the Earth. The United States followed suit with Explorer I in early 1958. The Soviet Union then sent unmanned probes to the Moon in its series of Luniks. In 1959 Lunik 2 crashed on the Moon. Later the same year, Lunik 3 circled the Moon and relayed the first pictures of the lunar far side. It was not until 1966 that Lunik 9 made a soft landing on the Moon and sent pictures of the lunar surface back to Earth.

In the meantime, the United States prepared to send unmanned probes to the Moon. Its Ranger series of mooncraft in 1964 and 1965 were designed to crash on the Moon, but before a Ranger would crash, it took thousands of photos of the approaching surface of the Moon. The photographs were transmitted to Earth electronically. Three Ranger craft, 7 through 9, showed details of the lunar surface about 1,000 times greater than could be seen with the best Earth telescopes. Even small craters only 1 or 2 yards across could be seen.

From 1966 to 1967, the United States launched unmanned artificial satellites to orbit the Moon. Known as Lunar Orbiters, five of these vehicles swept close to the Moon's surface and took thousands of photographs, which were then relayed to Earth. These vehicles also took photos of the Earth as seen from the Moon. Another series of unmanned craft were designed and launched with the purpose of soft-landing on the Moon. Known as Surveyors 1 through 7, five of them settled gently on various spots on the Moon to test its surface. They determined that the Moon was safe for landing. They also carried out various tests of the Moon's surface, and excavated some of it for study. Surveyor 5 made a chemical test of lunar rocks by irradiating them with atomic particles and recording how the rocks reacted. In this way, scientists on Earth deduced something of the chemical composition of lunar material. They concluded it was like Earthly basalt, a volcanic rock. The Surveyors showed that the Moon's surface was covered with pulverized rock and larger pieces of stone. The spacecraft also sent many photos of the Moon to Earth, as well as pictures of the Earth as seen from the Moon.

In the 1960s and 1970s, the Soviet Union continued its program of unmanned lunar exploration. Probes circled the Moon or landed on its surface. Luna probes took samples of lunar soil and sent them, by means of rockets, back to Earth for study. Two wheeled vehicles—Lunokhods 1 and 2—were landed on the Moon. They were driven across the lunar surface by remote control from Earth.

As early as 1961, the United States had decided to land astronauts on the Moon, and soon inaugurated Project Apollo. From late in 1968, a series of Apollo missions sent U.S. astronauts into orbit around the Moon, and, in July of 1969, the first humans landed on the lunar surface. The astronauts brought back many specimens of rock that have added immensely to our knowledge about the Moon. They also took thousands of photographs of the Moon and the Earth, while on the Moon itself and from the Apollo spacecraft. They set up scientific experiments on the lunar surface, the results of which were telemetered to Earth.

Travel on the Moon

The Moon presents many obstacles to travel on or above its surface. Conventional modes of Earth transportation—airplane, automobile, train, and ship—would be very difficult or impossible there. The Moon has no air, so combustion engines, such as gasoline, diesel, steam, and jet engines, cannot operate, unless a supply of air or oxygen as well as fuel is provided. The absence of an atmosphere on the Moon also rules out aircraft. Only rockets or nuclear power would make flight through the lunar skies possible. There are no water bodies for boats.

Locomotion on the lunar surface in a ground vehicle involves many problems. Motors need to be powered by electricity. The low lunar gravity—one-sixth the gravity at the Earth's surface—makes control of the vehicle difficult, so special suspension, steering, and axle systems needed to be designed for the Apollo astronauts.

Traction, or grip, on the soft and yielding lunar dust is difficult to achieve. Engineers knew that air-filled tires would not work—they would simply explode in the Moon's near vacuum of an atmosphere. Treads or tracks, as on a tank, were one solution. But drivers and engineers had learned that wheels with thick, flexible tires offer the best means of locomotion through soft dirt and sand. Airless tires of flexible, springlike metal or other material were the best answer.

Other requirements for a lunar vehicle are lightness and compactness, for easy transport to the Moon; a tough, flexible chassis and axles, for movement over rough lunar ground; an ability to withstand the airlessness and great temperature extremes of the Moon; and enough speed for traveling significant distances in a reasonable time. The vehicle had to have room for at least two astronauts with their equipment, experimental packages, and specimens of lunar rock. Navigational and communication systems were also vital.

The United States incorporated all these features into its first lunar roving vehicle (LRV). On its first mission, the LRV carried two U.S. astronauts on several short journeys of exploration across the Moon's surface.

When the aluminum LRV was unfolded from the lunar module on the Moon's surface, it was 10 feet (3 meters) in length. Each of its four wheels was powered by a one-quarter-horsepower electric motor. Flexible wire-mesh tires provided excellent traction. Power was supplied by electric batteries. Top speed was about 10 miles (16 kilometers) per hour; maximum range, around 40 miles (65 kilometers). The LRV could cross cracks in the ground 24 inches (60 centimeters) wide and move up slopes of 25 degrees. The astronauts steered the LRV with a control stick much like those in airplanes. All four wheels could be turned, for sharp maneuvers.

The vehicle carried two astronauts and their equipment. Computerized navigation instruments told the astronauts where they were. A color-television camera on the LRV relayed to Earth, via a high-gain antenna, views of the Moon.

The LRV's Earth weight was 440 pounds (200 kilograms); on the Moon, somewhat over 75 pounds (34 kilograms). The vehicle was designed to operate even if some of its vital parts broke down, which in fact did happen at first.

EXPERIENCING THE MOON IN PERSON

The Apollo astronauts brought back vivid descriptions of the lunar world. Its forbidding landscape comes in various shades of gray. Even the sky is black. Without an atmosphere, there can be no

Scientists expect it will not be too long before an astronaut again views the Earth over the lunar horizon.

blue sky and no weather. There can be no sound and no life. In the sunshine the temperature is that of boiling water, but without an atmosphere, a step into the shadow brings one to a temperature far below freezing. There are no beautiful sky colors at sunset and sunrise, and no twilight. Like the blue sky, these Earthly beauties depend on our atmosphere. On the airless Moon, they are absent.

For these reasons a visitor to the Moon needs protection, not only from extremes of temperature and lack of air, but also from the incessant bombardment of cosmic rays and particles, and from the effects of ultraviolet radiation. On the Earth, we are shielded from the deleterious effects of strong solar and cosmic rays by our atmosphere and the Earth's magnetic field. But the lunar explorer has no such natural protection, since the Moon is without an atmosphere and has an extremely weak magnetic field.

Since gravity at the Moon's surface is one-sixth that on the Earth, all weights on the Moon are diminished by a factor of six. A man who weighs 175 pounds (80 kilograms) on Earth weighs about 29 pounds (13 kilograms) on the Moon, and, with the same muscular effort, would be able to jump six times as high or lift six times as great a weight as on Earth. This compensates to some extent for the weight of the equipment that lunar explorers have to carry to protect themselves from the severe conditions. Similarly, the launching of a rocket from the surface of the Moon would require only one-sixth of the thrust required to launch the same rocket from the Earth's surface. This makes the return journey of an astronaut from the Moon and its vicinity easier.

In some cases, this diminished weight works to the disadvantage of the lunar visitor—for instance, when he or she must push a lever or dig with a shovel.

Explorers Confirm Theories

The Moon probes and U.S. Apollo missions have confirmed the existence of at least some lunar features that strongly resemble volcanic features on Earth—domes and lava flows. The question remains whether volcanic activity was ever widespread and important on the Moon. Extensive rock melting was undoubtedly caused by impacts of large meteorites on the lunar surface. They probably also produced the large craters and the maria, or plains.

Lunar rocks brought back to Earth by the Apollo astronauts generally resemble the Earth rocks known as basalts, which are finely grained, and gabbros, which are coarser-grained. These are dark, dense, chemically related rocks of igneous origin.

Among the youngest of Moon rocks are the mare basalts. These rocks formed from dark lavas that poured out 3 billion to 3.8 billion years ago, to form the floors of many large craters and maria. The rocks also confirm that the Moon has had a long, complex history, beginning about 4.6 billion years ago or even earlier, when the original crust is believed to have solidified.

EARTH AS SEEN FROM THE MOON

A person standing on the Moon sees the Earth as a disk two and one-half times the size of the Moon seen from the Earth. Because of its high albedo, the surface of the Earth has five times as much reflecting power as the surface of the Moon. Also, because of its greater apparent size, the full Earth sends about 30 times as much light to the Moon as the full Moon sends to us. To the lunar explorer, our clouds look brilliantly white, the oceans dark blue, and the continents almost all purplish brown.

Because the Moon and Earth are gravitationally locked, lunar observers do not see the Earth rise or set. From the near side of the Moon, the Earth is always visible. If explorers venture to the far side of the Moon, they will not see the Earth at all.

When the observer on the Moon is at lunar midnight, he or she sees a *full earth*, completely illuminated, unless the Moon is directly in line between the Sun and Earth. In that case, he or she observes an eclipse of the Earth by the Moon at the same time as Earth observers see a solar eclipse. When the observer is at lunar midday, and the Sun is directly behind the Earth, he or she observes *new earth*. If the Sun is completely eclipsed by the Earth, the Earth would be seen surrounded by a bright halo caused by the sunlight scattered in the Earth's atmosphere. Conditions for such an eclipse are very similar to those for a solar eclipse seen from the Earth, but the track of totality on the Moon would be two and one-half times as wide.

Like the Moon seen from the Earth, the Earth seen from the Moon passes from "new earth" through crescent, gibbous, and full phases, and back again to new earth. These changes take the same time as the corresponding changes in the Moon seen from the Earth, and run through a full cycle in one lunation.

THE MOON AS AN OBSERVING STATION

From the point of view of the astronomer, the Moon would be an ideal site for a telescope. The useful size of telescopes on Earth is limited by our atmosphere, which blurs optical images and limits the kinds of light waves reaching the Earth. In the absence of an atmosphere—as on the Moon—the detail that a telescope can record is limited only by the optical properties of the instrument.

The smaller value of gravity on the Moon would also make it possible to operate larger instruments than are feasible on the Earth. A lunar observing period for optical instruments would be one-half month long, followed by a similar period in sunlight. Weather would never interfere with observing schedules, although instruments would have to be well protected to withstand the lunar temperature extremes.

The National Aeronautics and Space Administration (NASA) does indeed have plans in place for a manned base on the Moon—both to provide astronomers with the observatory for which they've longed, and to serve as a launching pad for flights to Mars and other planets.

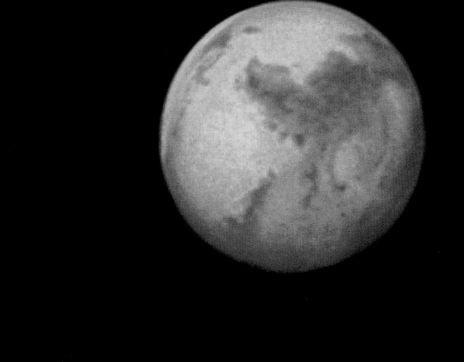

Of all the planets in our solar system, Mars most closely mimics the life-sustaining conditions on Earth. Scientists are trying to determine if the planet has ever supported life or could do so in the future.

MARS

by Jeffrey Brune

Mars, the fourth planet from the Sun, has long fascinated us. In many ways, Mars is much like Earth, and the apparent similarities between the two planets have raised the possibility that life might exist there. The planet assumes a bricklike color in the night sky, a phenomenon that has led to its nickname—the Red Planet. Mars is named for the Roman god of war.

Except for the Moon, and perhaps Mercury, Mars is the only solid body in the solar system whose surface we can view from Earth with ground telescopes. The planet has been under telescopic observation for over three centuries. Telescopes show Mars as a small disk with red, dark, and white markings. The red areas, which cover nearly three-fourths of the surface, are called *continentes* (Latin for "mainlands"). The dark regions are the *maria* ("seas" in Latin). White polar caps cover the planet's geographic pole.

For many years, astronomers thought Mars to be much like the Earth, with oxygen, water, and polar ice. Somewhat later they realized that Mars is dry, with no large or visible bodies of water at all.

During the late 1800s, some astronomers claimed they saw many fine lines crisscrossing much of the Martian surface. These lines were named *canali*—Italian for "channels" or "grooves." The Italian word can also be translated as "canals" or "waterways." This translation led to the widespread belief that there were intelligent beings on Mars who had built the canals. Perhaps coincidentally, this myth arose shortly after the building of the Suez Canal, when plans were being drawn for constructing the Panama Canal.

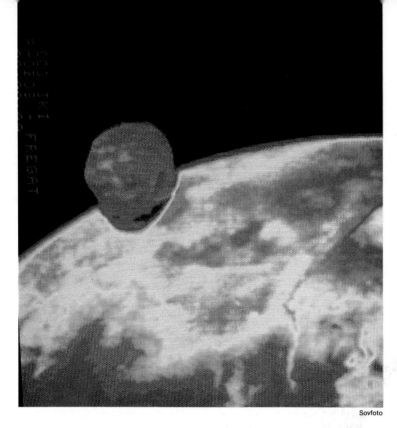

A Soviet satellite probe traveled within 200 miles of Mars to capture this infrared photo of the planet and one of its moons, Phobos.

Sovfoto

In the 20th century, improved telescopes and cameras revealed that these apparent Martian canals were mere optical illusions. However, the search for life on Mars continued undaunted, as did the quest to understand the planet better.

SOME BASIC STATISTICS

Mars lies about one and one-half times farther from the Sun than does the Earth. A good deal smaller than Earth, Mars' equatorial diameter—4,200 miles (6,787 kilometers)—is a little more than half that of our planet. The surface area of Mars is a little more than one-quarter of Earth's; its volume is about one-seventh of Earth's; its mass, 11 percent; and its density, 71 percent of Earth's.

Due to its greater distance from the Sun, the amount of light and heat that Mars receives is also diminished—less than one-half that received by Earth. As one would expect, Mars is therefore quite a bit colder, with temperatures ranging as low as $-305°$ F ($-187°$ C), enough to freeze carbon dioxide gas into "dry ice." Temperatures rarely rise high enough to melt ordinary ice into water.

The mean distance of Mars from the Sun is about 140 million miles (228 million kilometers), some 48 million miles (78 million kilometers) farther away than is the Earth. At the perihelion of its orbit—the nearest point to the Sun—Mars lies about 126 million miles (203 million kilometers) away. At aphelion—the farthest point from the Sun—the distance is about 155 million miles (250 million kilometers). This means that its elliptical orbit is quite eccentric. (The more an ellipse departs from the shape of a circle, the more eccentric it is.)

The distance between Mars and our own planet varies widely. When Mars is in conjunction—that is, on the other side of the Sun from Earth—its distance from us averages 234 million miles (377 million kilometers). When it is in opposition—that is, on the other side of the Earth from the Sun—its distance from the Earth ranges from about 35 million miles (56 million kilometers) to about 61 million miles (98 million kilometers), depending on the point in the Martian orbit where the opposition occurs.

Mars is in opposition about every 26 months. In opposition, its disk appears far larger to us than at any other time, and therefore provides astronomers wonderful opportunities for carefully examining the Martian surface. When nearest to the Earth, Mars has three times the brilliancy of Sirius, the brightest star in our galaxy (with the exception of the Sun). When farthest from the Earth, its brightness is reduced to that of a star of the second magnitude.

Mars completes its orbit around the Sun in 687 of our days, traveling along its path at an average rate of 15 miles (24 kilometers) per second. The orbit is inclined to the ecliptic—the plane of the Earth's orbit—by less than 2 degrees, and the planet rotates on its axis. The clear markings on its surface have made it possible to determine the speed of that rotation with great accuracy. The Martian day is 24 hours, 37 minutes, and 23 seconds long—almost the same as an Earth day.

Like the Earth, Mars is somewhat flattened at its poles. Its equator is inclined about 25 degrees to the plane of the planet's orbit. In comparison, the inclination of the Earth to its orbital plane is 23.5 degrees—only a little less. This tilt accounts for seasonal changes on the Earth. In the case of Mars, too, the corresponding tilt brings about changes of season. Since it takes Mars almost twice as long as the Earth to complete an orbit around the Sun, each of the Martian seasons is nearly six months in length—that is, almost twice as long as the corresponding season on the Earth.

Mars reflects 15 percent of the light received from the Sun. Area for area, the disk of Mars is a better reflector than that of our Moon and Mercury, but far inferior to that of Venus, owing primarily to thick clouds that shroud Venus' surface. Therefore, it does not appear as bright as Venus.

A CLEARER PICTURE

Detailed studies of the surface of Mars began with the images provided by Mariner 9 in 1969. An even clearer picture was revealed with the landing of the U.S. space probes Viking 1 and 2. Viking 1 landed on Mars on July 20, 1976. Viking 2 followed,

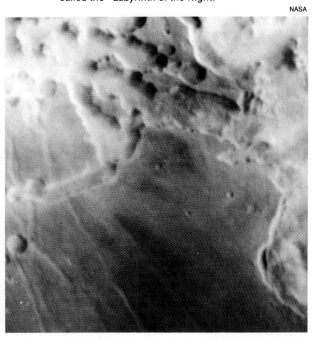

As the Sun rises over Mars, bright clouds of ice crystals rest above the canyons of this high plateau region called the "Labyrinth of the Night."

NASA

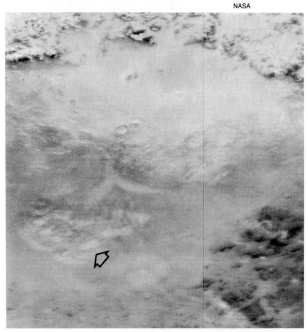

Winds of up to several hundred miles per hour trigger a Martian dust storm (arrow), strong enough to erode terrain structures with a sandblasting effect.

NASA

landing 4,600 miles (7,400 kilometers) from the Viking 1 site on September 3 of that year.

These Martian probes revealed that the surface of Mars is very similar but more rugged than the Earth's. There is hardly a land formation on Earth that does not have an equivalent on Mars. Mars has high mountains and plateaus, craters many miles across, broad plains, valleys, steep cliffs, jagged ridges, canyons deeper than the Grand Canyon, sand dunes, long scratches, and faults extending for great distances.

While Earth's crust is made of a number of plates that shift and move, the crust of Mars is believed to be made of a single, very thick plate that hardly budges at all. That may be why Mars has so many giant volcanoes. The largest, Olympus Mons, is taller than any surface feature on Earth. While volcanoes on Earth eventually shift away from their underground source of lava and peter out, Martian volcanoes seem to stay in the same place, continuing to grow ever larger. The absence of moving plates could also help explain why the seismometer on the Viking 2 lander didn't detect a single tremor in 1,200 hours of testing. On Earth, quakes are caused by plates bumping against each other.

If the Earth were drained of all its water, it would be at least as rugged as Mars. Empty ocean basins would form deep depressions, with long rifts and jagged ridges along their bottoms. Certain Martian cracks and ridges look like those found on the floors of Earth's oceans.

Some scientists think that about 3.5 billion years ago, parts of Mars were covered with an ocean of water, much of which has since evaporated. As evidence, they point to the scarcity of impact craters—large holes formed by falling meteorites—in some regions. While the southern hemisphere of Mars is pockmarked with impact craters, parts of the northern hemisphere have many fewer craters. A deep blanket of water may have covered these northern parts, shielding them from meteorites.

There is other evidence that large amounts of water once existed on Mars.

Lowell Observatory Photo; NASA

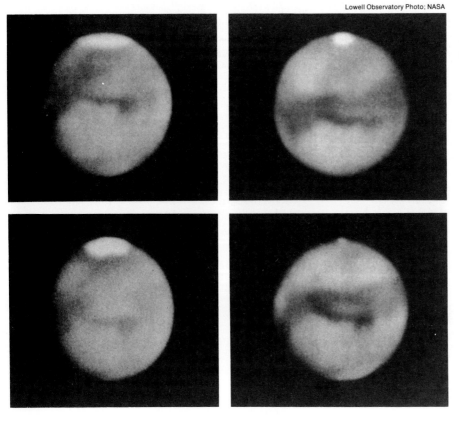

The surface features of Mars change with the seasons. The polar cap expands in winter, sometimes reaching 45 degrees latitude, and then melts away during the Martian spring and summer.

Certain Martian valleys, gullies, and deposits seem as if they could have been caused only by flowing water. Some plateaus are also marked by long grooves and scratches resembling those made by glaciers on Earth.

There may still be lots of water on Mars. Some scientists think the entire planet may have a thick layer of permafrost (permanently frozen water) under its apparent surface of dust and rock. Soil samples taken by the Viking probes yielded a good deal of water upon heating. The presence of water is especially exciting to planetary biologists searching for life. Water would also be extremely useful for human outposts on Mars, serving as a source of drinking water, oxygen for breathing, and hydrogen for rocket fuel.

Changes in the shape and color of Martian surface markings have been observed through ground-based telescopes and in the Mariner and Viking pictures. Many are undoubtedly caused by the action of meteorites, wind, ice, or volcanoes. Other changes, however, cannot yet be explained in these ways, suggesting there may be forces at work on Mars that have no counterpart on Earth. The Hubble Space Telescope will likely solve some of these mysteries during its long-term program to monitor the changes in the Martian surface and atmosphere.

Past scientific studies show that Martian atmospheric dust and surface rocks contain a wide range of minerals. They resemble Earth rocks chemically, thus showing that Mars, like the Earth, has had a long, complex development. And like Earth, Mars probably has a dense core at its center, which produces a weak magnetic field.

The Viking landers revealed many interesting facts about the chemistry of Mars' soil. As suspected, the soil's rusty red color is due to a high concentration of iron. It also contains large amounts of oxygen. Some preliminary Viking tests showed that the soil contained organic compounds—the carbon-based building blocks of life—at a concentration of about one part per million. But later tests seemed to contradict this

In 1976, the Viking Lander 2 transmitted back to Earth photos of the Martian surface—rock-strewn fields covered with red iron-oxide dust.

finding. Further tests showed that oxidants —chemicals that break down organic matter —are present in Martian soil and air. So even if there were organic compounds on Mars, they would be quickly destroyed through the action of oxidants.

Despite the above evidence, some scientists still believe that life may exist on Mars—in habitats the Viking landers did not explore. After all, these scientists contend, it was never believed that life could exist in hostile places on Earth, such as Antarctica, until we found rugged microorganisms eking out a living there. It may be that only a human mission to Mars and a thorough search for life can settle the question. Such a mission may, at the very least, uncover fossils of past life. Scientists may find that life evolved on Mars during its wetter, warmer past, then died out when conditions changed for the worse. Perhaps scientists will never find life—past or present—on Mars.

THE MARTIAN ATMOSPHERE

That Mars is surrounded by air has been known for a long time. What could be explained only as atmospheric hazes, clouds, and dust storms have often been observed. Mars' blanket of air is too thin, however, to protect much of its surface from the radiation of space. The atmospheric pressure on Mars is only about 6.1 millibars, much less than the average 1,013 millibars found on Earth.

Whereas Earth's atmosphere is dominated by nitrogen and oxygen, the Martian atmosphere is 95 percent carbon dioxide, with only faint traces of oxygen, nitrogen, and water vapor. Thin clouds—composed perhaps of dust, dry-snow crystals, and frozen water—often form. Clouds rise daily over high elevations. They may come from volcanoes or may simply be condensed atmospheric vapors.

The most spectacular and puzzling features of the Martian atmosphere are the gigantic dust storms that periodically sweep the entire planet. The Mariner 9 probe arrived during the height of such a tempest. Winds averaging about 174 miles (280 kilometers) per hour whip up enormous clouds of dust from the desertlike surface. A storm of this kind may last for weeks or months, during which the planet is one vast dust bowl. By comparison the worst Sahara sandstorm on Earth would seem like a mere breeze.

MOONS OF MARS

Mars has two satellites, or moons. They were discovered in 1877 by the American astronomer Asaph Hall. He named them Deimos and Phobos ("Terror" and "Fear"), after the two mythical sons and attendants of the ancient god Mars.

Both moons are irregular in shape and very small. Deimos, the outer one, measures about 6 to 7 miles (9 to 11 kilometers) in diameter. Phobos, the inner one, measures about 10 to 14 miles (16 to 22 kilometers). Pictures show them to be dark, rocky bodies, pitted with craters. Results from the Soviet Phobos space-probe mission in 1989 support the widely held view that Phobos and Deimos are asteroids that strayed from the nearby asteroid belt, only to be captured by Mars' gravity.

Deimos and Phobos both revolve around the equatorial region of Mars in nearly circular orbits, in the same direction that Mars spins on its axis—west to east. Deimos orbits Mars at an average distance of about 12,000 miles (19,300 kilometers) above the planet's surface. It takes 30 hours and 18 minutes to go once around Mars completely. On Mars, Deimos would be seen to cross the sky from east to west once in about two and one-half Martian days.

Phobos is only some 3,700 miles (6,000 kilometers) above the Martian surface. It orbits the planet once every seven hours and 39 minutes. Because it circles Mars faster than the planet rotates, Phobos is seen there to rise in the west and set in the east. It makes two crossings of the sky in one Martian day.

NASA

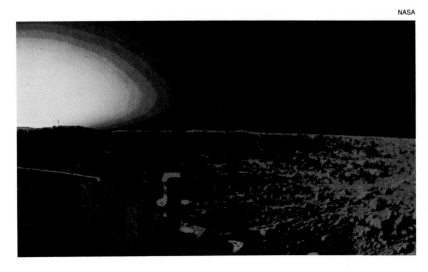

Martian atmospheric dust causes scattering and absorption of sunlight, producing the white-to-red-to-blue variation in sky color seen in this image of a sunset taken by the Viking Lander 1.

Jupiter, named for the Roman king of the gods, has a diameter 11 times that of Earth. The Great Red Spot—the vortex of a storm—is Jupiter's foremost feature.

JUPITER

by Jeffrey Brune

Jupiter is the largest planet in our solar system, bigger than many stars, though not as large as our Sun. Jupiter's volume is 1,300 times that of the Earth, with a diameter of 88,770 miles (142,860 kilometers). In comparison the Earth's diameter is less than 8,000 miles (13,000 kilometers)—scarcely $1/11$ of Jupiter's.

Despite its huge bulk, Jupiter spins on its axis much faster than our planet, making one complete turn in slightly under 10 hours. This fact explains why Jupiter is noticeably flattened at its north and south poles, and bulges around its equator. Jupiter's fast spin also draws out the planet's clouds into colorful horizontal bands.

For all its enormous size, however, Jupiter is only 318 times more massive than the Earth and has only a quarter of the density of the Earth. This suggests that Jupiter is made mostly of materials considerably lighter than the rock, soil, and iron that make up our planet. Jupiter is thought to consist mostly of gases—primarily hydrogen, some helium, and traces of methane, water, and ammonia.

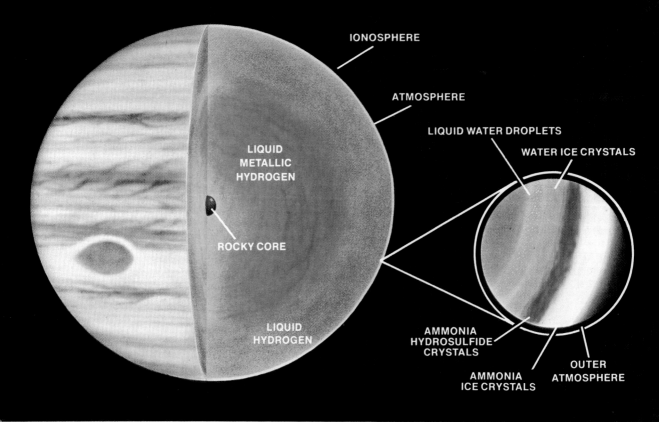

A model of Jupiter's interior reveals a small rocky core surrounded primarily by a ball of liquid hydrogen. Jupiter's atmosphere (close-up at right), about 620 miles deep, consists of layers of ammonia, hydrogen sulfide, hydrogen, helium, methane, and water. Despite its huge size, Jupiter is much less dense than Earth.

Jupiter lies between the orbits of Mars and Saturn, at an average distance from the Sun of nearly 484 million miles (779 million kilometers). Because of this vast distance, Jupiter takes nearly 12 years to go once completely around the Sun. At its apogee (closest approach to Earth), Jupiter is about 370 million miles (591 million kilometers) away. At its perigee (farthest point), it is 600 million miles (966 million kilometers) away.

Jupiter's magnetic field, which is the strongest of all the planets' in the solar system, traps charged particles that make up the "solar wind" that flows out continuously from the Sun. These particles produce extremely high radiation levels around the planet. The Pioneer and Voyager probes, which flew past Jupiter, revealed that its magnetic field extends out some 5 million miles (8 million kilometers).

Though Jupiter was previously thought to be ringless, Voyager 1 revealed a relatively thin, gossamer ring surrounding the planet. Some 29,800 miles (48,000 kilometers) above the Jovian cloud tops, the ring is only about 19 miles (30 kilometers) thick and 5,600 miles (9,000 kilometers) wide. It is too faint to be seen from the Earth.

JUPITER'S APPEARANCE

Despite its great distance from us, Jupiter is easily visible to the unaided eye. It is very large, and reflects more than 70 percent of the sunlight falling on it. As a result, among the planets of our solar system, only Venus and Mars appear brighter.

Just a pair of binoculars shows the planet as a disk, sometimes surrounded by four spots of light. These are the four "Galilean satellites," the largest of Jupiter's moons. They were discovered in 1610 by the Italian scientist Galileo Galilei, who was the first to observe Jupiter with a telescope.

Viewed through a large telescope and in the Pioneer photographs, Jupiter is a spectacular sight. Its wide disk is crossed by many bands of color—pastel shades of blue, brown, pink, red, orange, and yellow. These bands—or belts, as astronomers call them—run parallel to Jupiter's equator.

Jupiter's belts, despite changes in their colors and other features from time to time, are mostly permanent. Jupiter's face is also marked by shifting patches and spots, the most obvious of which is the famous Great Red Spot.

PHYSICAL CONDITIONS

Jupiter is thought to be composed of three basic layers: a gaseous outer layer, a liquid middle layer, and a solid core at the center of the planet.

The multicolored belts we see on Jupiter's "face" are actually part of the planet's gaseous outer layer, or atmosphere. These belts are layers of dense clouds, composed of liquid drops and frozen particles. Just how deep the atmosphere and its clouds extend is not certain, but the depth is probably about 620 miles (1,000 kilometers). Scientists will likely be more certain in 1995, when the Galileo spacecraft is expected to hurl a probe into Jupiter's atmosphere. In addition to gauging the atmosphere's depth, the probe will analyze its chemical makeup, clock wind speeds, and search for lightning.

As one approaches the interior of the planet, its clouds probably become denser and denser with increasing pressure and eventually assume a liquid form. In Jupiter's middle layer, there may be great oceans of liquefied gas such as hydrogen. This ocean may be some 12,400 miles (20,000 kilometers) deep, and can be thought of as Jupiter's "surface."

Beneath the great oceans of liquid gases, the pressure becomes so enormous that the liquid gases turn solid. In this third of Jupiter's layers, scientists suspect there may be a gigantic ball of metallic hydrogen. Within that ball, at the very center of the planet, may be an iron-and-silicate rock mass about the size of Earth.

Because of Jupiter's rapid spin, its great distance from the Sun, and its deep clouds, the planet's weather must be very strange indeed. Yet certain cloud formations in the Earth's atmosphere resemble the belts of Jupiter. And Jupiter, like Earth, has great cyclonic storms, or regions of low atmospheric pressure, as well as anticyclonic flows with high pressure.

Jupiter radiates two or three times as much energy as it receives from the Sun. This means that the planet has a great internal source of energy—probably left over from the time when it was first formed. Some astronomers think of Jupiter as a star that simply didn't "make it" because of its relatively small mass. That is, its mass was too small to produce the internal pressures and temperatures needed to set off the nuclear reactions that take place inside a star.

If Jupiter were 100 times bigger than it is now, astronomers have calculated, it may have become a star instead of a big gas ball. Imagine our solar system with two stars! There would be a sun on both sides of Earth, and we'd be bathed with nearly continuous sunlight. We'd rarely have nighttime, only when the two stars neared each other in our sky, and Earth blocked the double source of light.

Voyager 1 captured this image of a volcanic eruption faintly silhouetted on the horizon of Io, one of Jupiter's 16 known moons.

JPL

Jupiter also radiates energy in the form of radio waves, which scientists detect with radio telescopes. The radio waves come from solar-wind particles trapped in belts around Jupiter by the planet's large magnetic field. These radiation belts are thought to produce the strongest radio signals in the solar system.

CHEMICAL COMPOSITION

The substances in Jupiter's atmosphere would probably poison or suffocate most living things found on the Earth. These dangerous substances include the elements hydrogen and helium and the hydrogen-carrying compounds methane, ammonia, and possibly hydrogen sulfide. Jupiter lacks the chief ingredients of the Earth's atmosphere: free oxygen, nitrogen, and carbon dioxide. But the planet probably has water vapor, which constitutes about 1 percent of Earth's atmosphere.

If you somehow could survive breathing Jupiter's atmosphere, you would probably say it stinks—literally. Ammonia is a pungent compound of hydrogen and nitrogen. On Earth, it is produced by the decay of certain proteins. Hydrogen sulfide, another protein-decay product, contains sulfur and smells like rotten eggs.

The other chief constituents of Jupiter's atmosphere—hydrogen, helium, methane, and water—are odorless. Methane is a chemical compound of hydrogen and carbon. It is formed on Earth by the decay of organic matter in the absence of atmospheric oxygen. It is also called marsh gas, because it frequently arises in swamps and bogs.

A color-enhanced view of Jupiter's Great Red Spot, captured by Voyager 1. Believed to be a major storm greater than the size of Earth, the spot has raged for centuries with no signs of a letup.

NASA

THE GREAT RED SPOT

In the three centuries that astronomers have been peering at Jupiter through the telescope, they have been puzzled by one of the strangest features ever observed on any body of the solar system: Jupiter's Great Red Spot. We now know that it is a swirling storm in the upper atmosphere that is so large it could hold several planets the size of Earth.

Located at 20 degrees latitude in Jupiter's southern hemisphere, the Great Red Spot has an oval shape and measures about 31,000 miles (50,000 kilometers) at its widest. But its size varies. It also changes hue over periods of many years—from a bright, brick red to an almost invisible pink and then back.

Although the spot maintains more or less the same latitude, it may change its longitudinal position markedly. Also, the Great Red Spot has a speed of rotation around Jupiter's axis averaging a few miles per hour slower than those of the areas close to the spot. It may move slightly faster or slower from time to time.

Interestingly, the storm that is the Great Red Spot has been raging since before it was first observed 300 years ago. The reason the storm can continue unabated is that Jupiter has no true solid surface to slow it down. A hurricane on Earth dies down soon after it leaves the ocean for land.

The close-up photographs taken by Pioneer 10 and 11 show the "pinwheel" structure of the Great Red Spot. It is spinning so rapidly that it displaces the clouds of the south tropical zone. Temperature measurements indicate that the storm rises some 8 miles (13 kilometers) above the surrounding cloud deck.

The photographs also reveal the presence of other, smaller storm centers, such as the Little Red Spot in the northern hemisphere of Jupiter. In fact, the face of Jupiter as a whole is a scene of dramatic weather activity. The grayish white zones are warm, upward-rising weather "cells" that carry heat from the interior of the planet to the surface. The lower, reddish brown areas are downward-sinking cells.

MOONS OF JUPITER

Jupiter has 16 known moons—four of which are 1,860 miles (3,000 kilometers) or more in diameter. Some of the small, outer moons may be former asteroids captured

Both illustrations: JPL

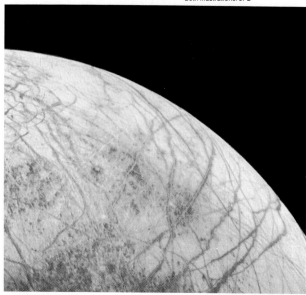

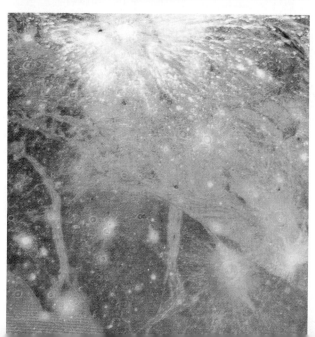

The Voyager probes have provided a wealth of information about Jupiter's many natural satellites. Europa (top), Jupiter's brightest moon, has an icy crust resembling a cracked eggshell. The mountains and valleys of Ganymede (right) show evidence of continental drift.

by the planet's vast gravitational field. The densities of the moons decrease outward as the spacing of their orbits increases. The inner moons are rocky, and the outer ones may be frozen liquid or gaseous.

Perhaps Jupiter's most interesting moon is Io, the most volcanically active body in the solar system. During Voyager 1's flyby, it photographed Io's gigantic volcanoes spewing sulfurous debris more than 90 miles (150 kilometers) above its surface. The spacecraft's infrared sensor detected enormous hot lava lakes, ranging in temperature from 172° to 199° F (78° to 93° C). "It's a pepperoni pizza," said one scientist, in describing Io's many volcanoes and lava-strewn surface.

The volcanoes of Io are responsible for producing the largest visible object in the solar system, a huge sodium cloud detected by telescopes from Earth. Besides spewing sulfur, the volcanoes also kick out sodium high above its surface. The sodium is then accelerated to over 155,340 miles (250,000 kilometers) per hour by Jupiter's magnetic field and slung into interplanetary space. The sodium forms a crescent-shaped cloud around Jupiter that is 50 times bigger than the Sun.

Europa is Jupiter's brightest moon, suggesting that its surface is highly reflective. Scientists say that this smooth moon is probably covered with ice. Its icy crust and the dark, linear features on its surface give Europa the appearance of a giant, cracked egg.

There are few impact craters on Europa, suggesting that its crust is slightly soft. Beneath the icy crust is thought to be a giant ocean of water, kept from freezing by the heat of radioactive elements in the moon's core. Scientists believe this watery world resembles the aquatic environment of Antarctica on Earth, and may have given rise to life.

Ganymede is Jupiter's largest moon. Indeed, it is the largest moon in the solar system. At 3,275 miles (5,270 kilometers) in diameter, Ganymede is even larger than the planets Mercury and Pluto. Voyager 2 came within 37,280 miles (60,000 kilometers) of Ganymede, close enough to photograph its Earth-like surface. The surface shows evidence of continental drift, especially in its numerous parallel faults (cracks), which appear similar to those on the Pacific and Atlantic Ocean basins. There are also cratered areas that have been pounded by meteorites.

The moon Callisto, almost as large as Ganymede, has an icy surface riddled with impact craters from heavy meteorite bombardment, probably dating to an early period of the solar system. Nothing appears to have happened to Callisto since its final formation. All its craters appear to be old, rounded, and smooth, rather than jagged and fresh. One of Callisto's most striking features is a bull's-eye formation called Valhalla. This probably formed when the moon was blasted with a large object, producing waves that froze in place.

Scientists view Jupiter's largest moons as worlds unto themselves, not mere satellites. The Galileo spacecraft promises to shed more light on these worlds, flying in 100 times closer than Voyager.

LIFE ON JUPITER?

For many years, astronomers were convinced that nothing could live on Jupiter. They pointed to its brutal climate, lack of oxygen, and the presence of poisonous or suffocating chemicals. But that attitude has changed. These very conditions are now thought to be a breeding ground for substances that could develop into living matter.

As we have said, methane, ammonia, and hydrogen sulfide, which are abundant on Jupiter, often arise on Earth from decay of organic matter. Not that this means that Jupiter is filled with decaying organic matter. But suppose we could reverse the process of decay. What if we could take ammonia, methane, hydrogen sulfide, water, and other chemicals, combine them in different ways, and get proteins and other biological compounds? Such processes likely took place on Earth 4 billion to 5 billion years ago. Many scientists believe the same process could in fact be happening on Jupiter today.

Saturn, the sixth planet from the Sun, is distinguished by its magnificent rings. The rings are made up of everything from dust particles to huge boulders.

SATURN

by Jeffrey Brune

For beauty and interest alike, there are few objects in the starry heavens to compare with Saturn. This magnificent planet, the sixth from the Sun, provided an unforgettable spectacle when viewed by the U.S. spacecraft Voyagers 1 and 2, in 1980 and 1981. The Saturnian system also includes numerous rings and at least 18 moons.

Saturn is so far from the center of the solar system that, viewed from its orbit, the Sun would appear as a brilliant pinpoint rather than as a disk. Saturn receives only one-ninetieth of the heat and light that we receive on Earth.

The mean distance of Saturn from the Sun is 887 million miles (1,428,000,000 kilometers), or about 9.5 times the distance of the Earth from the Sun. It completes its orbit once every 29.5 years. Our own planet, the Earth, overtakes it and comes in line between it and the Sun once every 378 days. Saturn is inclined to the ecliptic, the apparent path of the Sun among the stars, by 2.5 degrees. Its orbit, as it revolves around the Sun, is also much more eccentric than that of the Earth.

The distance of Saturn from the Earth varies according to the position of the two planets in their orbits: from 744 million miles (1,197,000,000 kilometers) to 1,027,000,000 miles (1,654,000,000 kilometers)—a variation not sufficient to cause any great difference in the planet's brightness.

SWIFTLY SPINNING EGG

The globe of Saturn is greatly flattened, so that when the planet is in such a position that the plane of its equator passes through the Earth, its profile appears distinctly egg-shaped, or elliptical. The planet's polar diameter is nearly nine-tenths that of its equatorial diameter—67,100 miles (108,000 kilometers) and 74,600 miles (120,000 kilometers), respectively. These dimensions show the vast size of the planet. Its volume is more than 750 times that of the Earth's and its superficial area is over 80 times that of our globe.

The density of Saturn is much lower than that of any other planet and, in fact, only three-quarters that of water. (If Saturn were the size of a tennis ball, it would float in a bucket of water.) The reason Saturn has such a low density is that it is composed largely of gas. There may be liquid beneath its thick atmosphere, and, deeper yet, a small, rocky core. Saturn's flattened shape is due to its mostly gas composition and to its rapid rotation.

Saturn's swift rotation on its axis was first observed in 1794 by English astronomer Sir William Herschel. By timing how long it took Saturn's cloudlike markings to make one complete rotation, Herschel calculated a rotation period of 10 hours and 16 minutes. In 1876 American astronomer Asaph Hall noticed a brilliant white spot on Saturn's equator. Using this spot as a point of departure for his calculations, he found that the equatorial period of rotation was 10 hours and 14 minutes, two minutes less than the period determined by Herschel.

In 1981 radio astronomers using data from Voyagers 1 and 2 were finally able to pin down Saturn's rotation rate to 10 hours, 39 minutes, 24 seconds. The inaccuracies of earlier measurements were due to their being based on the planet's shifting cloud cover.

STORMY WEATHER

Saturn's atmosphere is composed mostly of hydrogen with some helium and traces of methane and ammonia. And this atmosphere, like Jupiter's and Neptune's, is turbulent. Near Saturn's equator, there is a wide band of extremely high winds called a jet stream. Voyager probes clocked clouds traveling in this stream at 1,100 miles (1,770 kilometers) per hour—three times faster than the jet streams on Jupiter.

Scientists don't yet know why Saturn's equatorial winds are so strong or why they happen to blow eastward.

Over the past two centuries, astronomers have observed nearly two dozen "white spots," short-lived storms in the top layers of Saturn's atmosphere. In September 1990, amateur astronomers first detected what was to become one of the largest of these storms. This white spot was first seen near the equator as a relatively small, oval-shaped storm. About a month later, the National Aeronautics and Space Administration (NASA) pointed the Hubble Space Telescope toward Saturn and photographed a monster storm, one that had grown to the size of 10 Earths placed end to end. The huge system of swirling clouds and eddies was called the "Great White Spot," similar to, but much larger than, Jupiter's "Great Red Spot." This storm system probably developed when warm ammonia gas bubbled up from deep within the atmosphere. Once at the top of the cool cloud deck, the ammonia gas crystallized and was swept away by high winds.

The Pioneer 11 probe, which flew past Saturn in 1979, discovered that the ringed planet has a strong magnetic field. The field traps high-energy electrons and protons spewed from the Sun and keeps them in belts around the planet. Saturn's rings absorb and wipe out many of these electrons and protons, making the radiation that surrounds the planet much less intense.

SURROUNDED BY PARTICLES

Saturn is surrounded by a vast swarm of solid objects, ranging in size from dust to car-sized boulders. Each of the particles, composed mostly of ice and frosted rock, follows an individual orbit around the planet. Together, these particles form many rings, some of which are quite complex in structure.

Saturn's rings were long thought to be unique in the solar system. We now know that the other giant planets—Jupiter, Uranus, and Neptune—all have rings, though not nearly as extensive nor as beautiful as Saturn's. Study of Saturn's rings pro-

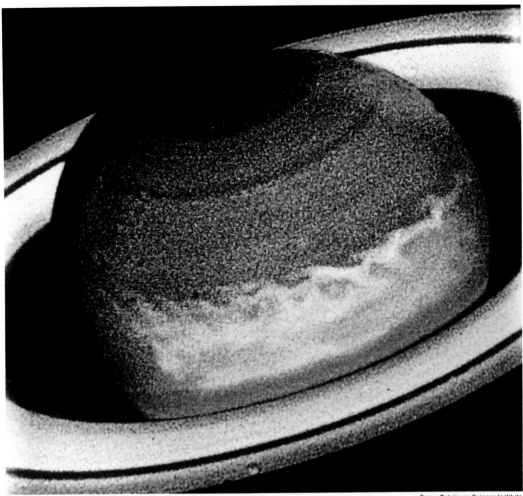

Space Telescope Science Institute

The whitish area above the rings is a great storm on Saturn detected by the Hubble Space Telescope. The Saturnian storm was large enough to envelop the entire Earth.

ceeded slowly over the centuries and then took a giant step forward with the surprise findings of the Voyager probes.

It was not until the telescope was discovered, in the first decade of the 17th century, that we even had an inkling that the rings existed. When Galileo examined Saturn in 1610 with a telescope that he had made with his own hands, he came to the conclusion that the planet had a triple form. "When I observe Saturn," he wrote to a friend, "the central star appears the largest; two others, one situated to the east, the other to the west, and on a line that does not coincide with the direction of the zodiac, seem to touch it. They are like two servants who help old Saturn on his way and always remain at his side. With a smaller telescope, the star appears lengthened, and of the shape of an olive."

Continuing to watch this strange performance month after month, Galileo was amazed to see Saturn's attendants becoming smaller and smaller, until they finally disappeared altogether. He doubted the evidence of his telescope. "What can I say," he wrote, "of so astonishing a metamorphosis? Are the two small stars consumed like sun spots? Have they vanished and flown away? Has Saturn devoured his own children? Or have the glasses cheated me, and many others to whom I have shown these appearances, with illusions?" Discouraged, he abandoned his quest.

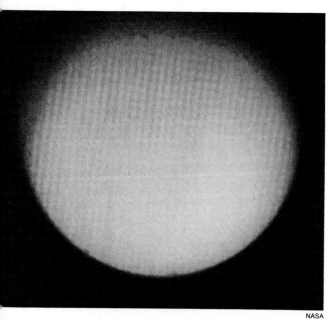
Titan, Saturn's largest moon, is larger than either Mercury or Pluto. Titan's substantial atmosphere contains molecules necessary for the evolution of life.

Saturn's moon Iapetus is composed almost entirely of water ice. The Voyager probes detected a large "stain of organic matter" on the moon's surface.

Others, however, watched the planet whenever it was in view, and they gradually established the fact that Saturn's unusual appendages underwent regular changes. They appeared first as bright, straight lines stretching outward on either side of the planet's elliptical disk. For the next seven years, these mysterious lines would expand into two luminous crescents attached to the planet like handles to a dish. For yet the next seven years, the crescents would flatten until they were again lines projecting from Saturn. Finally they would disappear altogether. For as the planet pursues its vast orbit, slanting always in the same direction, its rings, at opposite points in its orbit, appear in an edgewise position to observers on Earth. Between these two edgewise appearances, the north and south faces of the rings are visible in turn, always foreshortened. Each is seen for a period of about 15 years.

In 1655 Dutch mathematician, physicist, and astronomer Christiaan Huygens invented an improved method of grinding lenses. As a result, he was able to construct a powerful telescope that showed details more clearly than any earlier instrument. Using his new telescope, Huygens observed that the rings of Saturn cast shadows on the planet and were separated from it. From this observation, he deduced the true nature of Galileo's "appendages," which had long puzzled astronomers.

SATURN'S MANY RINGS

The idea that Saturn had rings was accepted in time. Huygens' fellow astronomers, including the renowned Giovanni Domenico Cassini, began to study the rings more carefully. In 1675 Cassini observed a dark band in what was then believed to be the single ring of Saturn. This band divided the ring into two separate rings. The dividing "band," which was really a gap, has since been labeled the Cassini division.

A third ring was observed in 1838 by German astronomer Johann Gottfried Galle. But Galle's report was ignored by his contemporaries. It was not until the ring was observed and reported simultaneously in 1850 by W. C. Bond at Harvard and W. R. Dawes in England that its existence became an accepted fact. In the third ring, there is apparently much less material that can reflect the Sun's light back to us. It has been likened to translucent crepe paper or gauze. As a matter of fact, it is often called the crepe, or gauze, ring.

In 1837 Johann Franz Encke, the director of the Berlin Observatory, saw what he believed to be another division in the outer ring. It was not complete and was not equally distinct at all times. This indistinct and transitory division has been called the Encke division.

In 1969 French astronomer Pierre Guerin discovered a fourth faint ring of Saturn, long undetected because it is so close to the bright globe of the planet. Scientists later found two more rings, for a total of six. Then came the Voyager flybys in 1980 and 1981 and a wealth of new discoveries.

As the Voyager probes approached Saturn, 100 or so ringlets were seen dividing up the major rings. When the rings came between Voyager 2 and a bright star —and when the rings passed in front of a star as seen from Earth in 1989—the probe was able to detect hundreds of thousands of rings. Voyager even found ringlets within the Cassini and Encke divisions, once thought to be empty space. Luckily, scientists didn't follow an early plan that would have sent Voyager through a division; the craft might have collided with one of the boulders that make up the rings.

Ring Dimensions

The major rings of Saturn, going from the outermost to the innermost, are designated by the letters E, F, A, B, C, and D. The E, or extended, ring lies outside the other major rings and may be as much as 62 miles (100 kilometers) thick.

The F ring has generated much interest —and confusion. Voyager 1 pictures showed a very narrow, wispy ring composed of three ringlets. The ringlets appeared to be braided, perhaps because they were orbiting in different planes. But when Voyager 2 took higher-resolution pictures of the same F ring, it found not three, but five ringlets; and the braids, for the most part, were gone. Perhaps the braiding— when it is seen—is caused by two tiny moons found straddling the F ring. These "shepherd" moons are thought to compress the ring into its narrow path by a kind of gravitational pinching action. Perturbations in the moons' gravity field might somehow cause the ringlets to twist.

Ring A is the second-brightest ring. It is 9,940 miles (16,000 kilometers) wide and has an outside diameter of 170,000 miles (273,000 kilometers). It may be no more than 328 feet (100 meters) thick. In this ring appears the narrow Encke division, containing multiple ringlets. Rings on either side of the division have wavy ripples like those produced in the wake of a speedboat, suggesting the presence of a moon that pushes material away from its orbit. In

NASA

The Voyager 2 space probe discovered that Saturn is actually circled by hundreds of thousands of rings. Saturn's many moons all lie outside the planet's elaborate ring system.

1990, after years of analyzing Voyager pictures, astronomers found this moon, the 18th confirmed Saturnian satellite, measuring a mere 12 miles (20 kilometers) in diameter.

The Cassini division, which lies between Rings A and B, is about 2,200 miles (3,500 kilometers) wide. As mentioned, astronomers now know that it contains a series of ringlets, some tightly packed together.

Ring B is the brightest ring, with a width of 16,156 miles (26,000 kilometers) and an outside diameter of 146,000 miles (235,000 kilometers). Voyager studies revealed that the B ring comprises some 300 ringlets, each of which may be made up of some 20 to 50 subringlets. The biggest surprise involving the B ring, however, was the appearance of "spokes" across it. The spokes move with the ring as spokes do on a bicycle wheel, and are thought to be clouds of dust-size particles raised above the plane of the ring. Meteor collisions might be responsible for producing the spokes, as might lightninglike electrical discharges observed in the B-ring plane.

Ring C is the crepe or gauze ring. It is separated from Ring B by only about 1,000 miles (1,600 kilometers). It is 11,500 miles (18,500 kilometers) wide. Its outside diameter is 121,800 miles (196,000 kilometers). Ring D is the faintest of all. It extends from the surface of Saturn to the inner edge of the C ring.

How Did the Rings Originate?

In 1850 Edouard Roche proved mathematically that the gravitational forces of a planet would tear apart any satellite within a certain distance of it. The gravitational force exerted by one body on another is inversely proportional to the distance between them. As they approach one another, the force becomes greater. Roche calculated that the exact distance at which the force of gravity of a planet would be great enough to tear apart its satellite is 2.44 times the radius of the planet. This distance is now known as Roche's limit.

A line drawn from the center of Saturn to the outside edge of the visible ring system would be approximately 2.35 times the radius of the planet. This places the visible ring system inside Roche's limit. All of the known moons of Saturn are outside Roche's limit. Even the moons first observed by Pioneer and Voyager probes and believed to be closest to the planet are at distances at least 2.55 times the radius of Saturn. Roche's calculations thus delineate a specific region within which satellites do not appear.

So here we have a clue to the origin of the rings. According to one theory, the rings are the remains of "wandering moons" torn apart by the planet's gravitational forces. The fragments of the former satellites now form the material of the rings. There is another, quite similar, theory of the origin of the rings. It holds that when the gases in the vicinity of the planet cooled and formed the various solid balls, or satellites, the gases inside Roche's limit were prevented from combining into satellites by the strong gravitational forces of the planet. Therefore, instead of merging into balls, they cooled as small fragments. These fragments revolve around Saturn separately in the form of the rings. The fact that the bona fide satellites of Saturn are all outside Roche's limit seems to support these interesting theories.

SATURN'S SATELLITES

Titan

Of the 18 known satellites that orbit Saturn, Titan is the largest. Larger in diameter than either Mercury or Pluto, Titan is a planet in its own right. It is the only moon in the solar system known to have a substantial atmosphere. The Voyager probes found that Titan's atmosphere is composed mostly of nitrogen, the main ingredient of Earth's atmosphere.

Titan's atmosphere also has a small percentage of methane, carbon monoxide, acetylene, ethylene, and hydrogen cyanide. The presence of hydrogen cyanide is of particular interest to scientists, for this molecule is believed to be a necessary precursor to the chemical evolution of life. It is doubtful, however, that life as we know it exists on Titan, since the moon's surface temperature is a frigid −292° F (−180° C). Some scientists, though, have suggested that Titan's interior may be hot enough to produce thermal vents at the surface, which would provide vital heat. Scientists hope that further study of Titan may unveil clues about how life began and evolved on Earth.

What lies beneath Titan's smoggy orange atmosphere has been the subject of much scientific debate. Some scientists think its surface is dry and rocky, while others think it is covered with oceans of liquid ethane mixed with a bit of liquid methane. In 1989 and 1990, American astronomers managed to penetrate Titan's murky veil with radar signals sent from an Earth-based antenna. The signals traveled through space, bounced off Titan's surface, and returned to Earth, where they were measured. The study revealed for the first time that Titan's surface does indeed have liquid on it, though not enough to cover it entirely. Most likely, Titan has scattered lakes of liquid ethane, not vast oceans.

A far better understanding of Titan's surface geology should result from the Cassini spacecraft, scheduled to arrive at Saturn in 2002. Radar aboard the craft will map Titan's surface, much as the Magellan probe has radar-mapped the cloud-covered surface of Venus. A small probe sent to Titan by the Cassini craft will study the moon's atmosphere, take pictures of the surface, and search for complex organic molecules.

Other Moons of Saturn

Mimas, Tethys, Enceladus, Rhea, and Dione are all nearly perfect spheres. All appear to be composed mostly of frozen water ice.

Mimas and Tethys are perhaps best known for their heavily cratered surfaces. Voyager pictures of Mimas showed an impact scar one-third the moon's diameter, with a central mountain almost as large as Mount Everest on Earth. Halfway around Mimas is a large canyon believed to have been formed by the same impact that

Saturn appears differently from Earth at different times of year. Sometimes (bottom right), the rings seem to disappear, an illusion that confused early astronomers.

Lowell Observatory

formed the crater. Pictures of Tethys revealed an even larger crater than the one on Mimas.

Enceladus, on the other hand, appears much smoother. Some scientists think that in the moon's recent past, slush may have gurgled up, filled in impact craters, and froze. The moon's icy surface reflects nearly 100 percent of the sunlight that strikes it, making it appear very bright.

Rhea is brownish in color and has a crater-saturated surface. The craters are of different sizes, suggesting to some astronomers that Rhea may have undergone two stages of impact: one series of impacts from the formation of the Saturnian system, and another from debris from the rest of the solar system. Rhea and Dione both have dark, wispy marks on their icy surfaces, suggesting some material leaked from their interiors along fractures in their crusts and then froze.

The Voyager probes also studied the moons Iapetus and Hyperion. The composition of Iapetus was found to be 80 percent ice, with at least half of its surface covered with "a stain of organic material." The moon appears to have a light and dark side, the dark side perhaps resulting from a coating of dust particles.

Hyperion held some surprises. Resembling a "battered hockey puck," this strange, misshapen moon, about 118 miles (190 kilometers) by 214 miles (345 kilometers), may have been smashed into its irregular shape by impact with another body. Hyperion tumbles erratically when it passes Titan and is tugged by the larger moon's gravity.

Phoebe, Saturn's outermost moon, orbits in a retrograde direction—opposite in direction to that of the other satellites. For this reason, scientists think Phoebe is a captured asteroid.

The beauty of Saturn has made it a natural subject for science-fiction artists. Below is a painting of how Saturn might look from the surface of the moon Titan.

Chesley Bonestell, Griffith Observatory

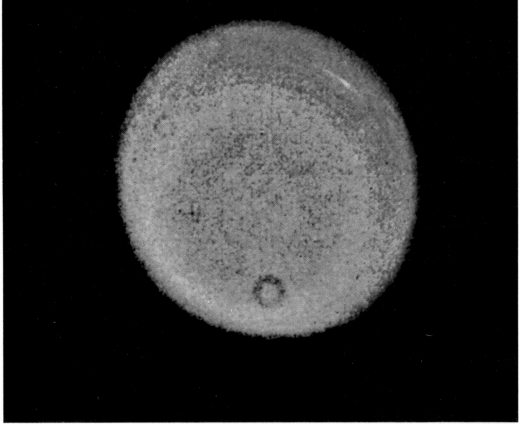

Voyager 2 unveiled much new information about Uranus, including immense clouds in its atmosphere (bright streak near the planet's upper edge above).

URANUS

by Jeffrey Brune

Of all the planets, Uranus may be the most bizarre. The seventh planet from the Sun, Uranus' orbit around the Sun is like a rolling ball instead of a spinning top like the other planets. Its poles, which take turns facing the Sun, receive a great deal more sunlight than its equator, and yet the whole planet has about the same temperature. Its magnetic field is off-center and tipped 60 degrees from its rotational axis.

Tipped on its side, Uranus looks like a bull's-eye, especially with its moons and rings orbiting roughly in the planet's equatorial plane. In 1977 scientists launched the space probe Voyager 2 toward this floating target in space. Most of what we know today about the planet was revealed by Voyager 2.

DISCOVERY

Uranus was not even known to exist until the late 18th century. It was first recognized by German-born English astronomer William Herschel. On the night of March 13, 1781, Herschel was observing the stars in the constellation Gemini with a 7-inch (18-centimeter) reflector telescope he had just made. Suddenly he observed what seemed to be a star shaped like a disk. Herschel was puzzled, since all true stars (with the exception of the Sun) appear as mere points of light, even when viewed with the most powerful telescope. Observing the unknown body night after night, he noted that it changed its position among the stars. He then came to the conclusion that his "moving star" was a comet, a celestial chunk of rock and ice that orbits the Sun.

The supposed comet was carefully followed by astronomers. They noted that it followed an almost circular orbit far beyond the orbit of Saturn, then thought to be the outermost planet. As time went on, they realized that the new body was a planet, and Herschel was hailed as its discoverer.

Herschel named the new planet Georgium Sidus (the Georgian Star), after the reigning monarch, George III. English astronomers called the planet the Georgian Star until about 1850; to others, it was known as Herschel. The name finally given to it—Uranus—was proposed by German astronomer Johann Elert Bode, who pointed out that all the other planets had been named after ancient gods. In Roman mythology, Uranus was the god of the heavens.

THE URANIAN ATMOSPHERE

Uranus can just barely be made out by the naked eye on a clear, moonless night. Through a telescope, it appears as a light blue-green disk. The planet has a diameter of about 31,750 miles (51,100 kilometers) at its equator, about four times that of the Earth. But despite its large size, Uranus is only one-fourth as dense as the Earth. That's because the planet is made mostly of the light element hydrogen.

The Voyager 2 probe found Uranus' atmosphere to contain 83 percent hydrogen, 15 percent helium, 2 percent methane, and scant amounts of ammonia and water. Beneath the atmosphere lie deep oceans of water, ammonia, and methane. Uranus' temperature rises with increasing depth. The planet probably has a rocky core of iron and silicon.

As Voyager 2 approached Uranus, it found little activity in the planet's atmosphere, certainly nothing like the storms found raging on Jupiter, Saturn, and Neptune. Scientists were rather bored with a featureless planet that stubbornly refused to change in appearance. They called the planet "the fuzzy blue tennis ball."

Just 10 days before Voyager 2's closest approach, however, it found a few small clouds in the upper atmosphere. These clouds were enough to help scientists determine wind speeds on the planet—from 130 to 535 feet (40 to 160 meters) per second. On Earth, jet streams race along at about 160 feet (50 meters) per second.

Despite external differences, Uranus and Neptune have similar internal structures. Both have a rocky core with a mantle of liquid water, methane, and ammonia. Both also have an atmosphere of hydrogen and helium gas.

© Paul DiMare

Umbriel is the darkest of the Uranian moons. Astronomers think that Umbriel may contain large amounts of the same carbonaceous material that makes up the rings and the smaller moons.

Titania, the largest moon of Uranus, is about half the size of Earth's moon. Although heavily cratered, Titania's surface has been much modified by geological processes that have occurred since its formation.

THE MOTION OF URANUS

Uranus revolves around the Sun once about every 84 Earth years. A day on Uranus lasts about 17 hours and 15 minutes—the time it takes the planet to make one complete rotation on its axis. As mentioned, the planet rotates on its side; its axis of rotation is tipped about 82 degrees from the vertical, so that it is nearly horizontal to the plane of the planet's orbit around the Sun. Scientists believe that the planet was knocked off-balance by a cosmic collision with a planet-sized space rock.

Because it rotates on its side, Uranus has the longest seasons of any planet—each of its seasons lasts about 21 Earth years as different hemispheres take turns facing the Sun. Any given point on the planet is lit up about half the Uranian year. For example, the planet's north pole points toward the Sun and receives almost constant sunlight for about 42 Earth years. For the next 42 years, the north pole is in darkness, while the south pole receives almost constant sunlight.

Because Uranus orbits at an average distance from the Sun of 1.78 billion miles (2.87 billion kilometers), it receives only 1/400th as much sunlight as reaches Earth. Still, with one pole receiving continuous sunlight for half a Uranian year, it should be much warmer than the other pole, which is receiving no sunlight. But Voyager found that Uranus has a uniform temperature of −353° F (−214° C). Some unknown mechanism is spreading out the heat in Uranus' atmosphere.

ENCIRCLED BY RINGS

When Uranus passed in front of a distant star in 1977, astronomers saw that the planet has at least nine rings. Just before Uranus blocked out the star, it winked on and off nine times. On the other side of the planet, at nearly the same distances, the star winked on and off again nine times. Each wink—before and after Uranus blocked the star—occurred when a ring around the planet blocked the starlight.

During its flyby, Voyager 2 detected two more rings, for a total of 11. The spacecraft also sent back measurements of the rings, which appear remarkably narrow and neat. While the rings are about 155,000 miles (250,000 kilometers) in circumference, most are less than 6 miles (10 kilometers) in width. The exception is the outermost ring, epsilon, which spans 60 miles (100 kilometers).

The particles that make up the rings may be remnants of a moon crumbled by Uranus' gravity or broken up by an impact with a high-speed asteroid. Instead of spreading out, the particles seem to be herded together into close-knit bands by "shepherding" moons, as is the case with Saturn's rings. Voyager 2 discovered two small moons, Cordelia and Ophelia, whose gravity seems to shepherd the particles of Uranus' epsilon ring.

Though dust and tiny particles are common in Jovian and Saturnian rings, they are uncommon in Uranian rings. Radio measurements of the epsilon ring, for example, show that it is composed mostly of ice boulders no smaller than beach balls. It is thought that Uranus' extended hydrogen atmosphere sweeps smaller particles and dust from the ring.

Scientists say that the rings are younger than the planet itself. The rings are thought to be short-lived as well. In time the boulders in the rings will collide with each other and grind down to dust. As the planet's atmosphere sweeps away the dust, the rings will vanish, appearing again only with the breakup of another moon.

URANIAN MOONS

The 15 Uranian moons are named after characters in the plays of Shakespeare and the poems of Alexander Pope. There are five major moons—Ariel, Umbriel, Titania, Oberon, and Miranda. Voyager 2 discovered at least 10 minor moons, the largest of which is named Puck, the mischievous sprite in Shakespeare's *A Midsummer Night's Dream*. Craters dot the dark surface of Puck, which is just over 100 miles (160 kilometers) in diameter.

Titania, Uranus' largest moon, is about 1,000 miles (1,600 kilometers) in diameter, roughly half the size of Earth's moon. Like Umbriel and Oberon, Titania is heavily cratered, suggesting that it has a long history. Ariel appears to have a geologically younger surface, with many fault valleys. Many of the craters formed in its early history appear to have been erased by extensive flows of icy material. All of the five major moons appear to be composed mostly of water ice and rock, with some ammonia and methane ices as well. The surfaces appear gray and dark, probably because they contain sooty, carbon-based materials.

With its bizarre collection of terrains, Miranda is probably the most fascinating Uranian moon. In some regions, there are cratered plains. In others, there are canyons some 13 miles (20 kilometers) deep. Miranda is also marked by glacial flows, huge fractures, broad terraces, high cliffs, and grooves that look like claw marks made by some mythical cosmic monster. All these strange features are mixed together in a seemingly haphazard way.

What could have formed such a jumbled mess of geology? Scientists think Miranda was blasted apart as many as five times by massive meteorites. Each collision cracked the moon into smaller pieces, which later fell back together by the force of gravity.

Thanks to Voyager 2, we were able to zoom in on Miranda's alien landscapes, measure the tidy Uranian rings, and survey the topsy-turvy world of Uranus itself. If one craft revealed so much on just a quick flyby, imagine what we would have learned had we sent a craft to orbit the planet for a long period of time. Imagine what we may someday learn by hitting the Uranian "bull's-eye," sending a probe plunging past the blue-green veil that covers this bizarre, mysterious planet.

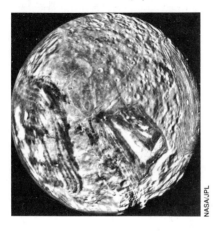

Miranda, the innermost of Uranus' large moons, was not discovered until 1948. Miranda has likely been subjected to meteoric bombardment in its past.

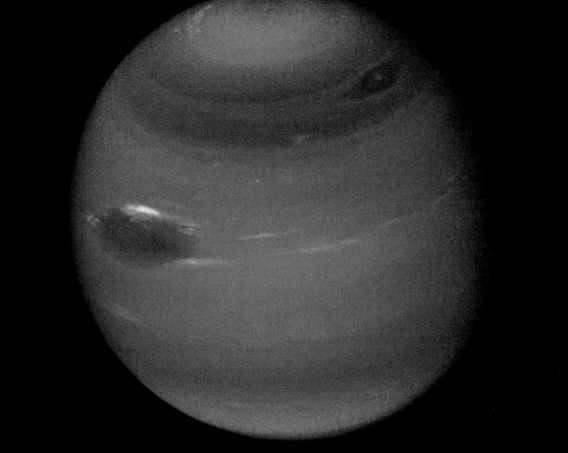

ASP Neptune Kit, William Kaufman & JPL

Neptune, the eighth planet from the Sun, is named for the Roman god of the sea. The planet's Great Dark Spot (center left, above), an Earth-sized storm system, was just one of the outward signs of Neptune's turbulent atmosphere discovered by Voyager 2 in 1989.

NEPTUNE

by Jeffrey Brune

Nearly 2.8 billion miles (4.5 billion kilometers) from the Sun lies a pale blue planet named Neptune. This, the outermost of the giant gas planets, was discovered in 1846. Yet Neptune remained largely a mystery until 1989, when the Voyager 2 probe flew by for a close look.

Voyager revealed that Neptune is alive with activity. In its turbulent atmosphere, a storm the size of Earth rages, with winds fierce enough to make our hurricanes seem like gentle breezes. On Triton, Neptune's largest moon, icy geysers spew forth freezing gas.

Voyager also discovered six new moons orbiting Neptune, giving the planet a known total of eight. Neptune and its moons orbit the Sun very slowly—once in about 165 years. That means the planet has not yet completed a single orbit around the Sun since it was first discovered in 1846! Neptune's elliptical orbit is only slightly eccentric; that is, the Sun lies nearly at its center. The plane of Neptune's equator is inclined by about 29 degrees to the plane of the planet's orbit. It rotates once in about 16 hours. Neptune's equatorial diameter is about 31,000 miles (49,500 kilometers); its volume is large enough to swallow nearly 60 Earths.

INVISIBLE TUG

Neptune might never have been discovered had neighboring Uranus followed its expected orbit. In 1790 French astronomer Jean Baptiste Delambre mapped the orbit of Uranus. For a few years, the ob-

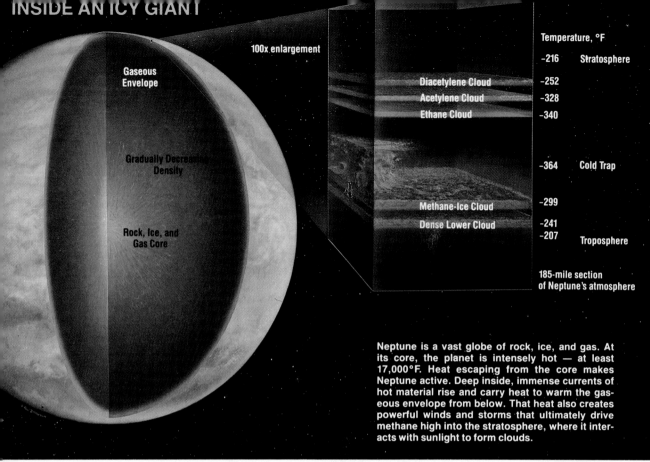

Neptune is a vast globe of rock, ice, and gas. At its core, the planet is intensely hot — at least 17,000°F. Heat escaping from the core makes Neptune active. Deep inside, immense currents of hot material rise and carry heat to warm the gaseous envelope from below. That heat also creates powerful winds and storms that ultimately drive methane high into the stratosphere, where it interacts with sunlight to form clouds.

© Paul DiMare

served positions of Uranus generally agreed with Delambre's astronomical tables. Over time, however, the discrepancy between the tables and the actual orbit of Uranus increased. Astronomers agreed on the need for revision.

French astronomer Alexis Bouvard undertook this task. His calculations of Uranus' orbit seemed accurate at first, but soon they, too, failed to agree with the planet's actual positions. Bouvard suspected that an undiscovered planet lying beyond the orbit of Uranus must be pulling it away from its calculated course.

On July 3, 1841, John Couch Adams, then an undergraduate at St. John's College, Cambridge, entered the following memorandum in his notebook: "Formed a design, in the beginning of this week, of investigating as soon as possible after taking my degree, the irregularities in the motion of Uranus, which are as yet unaccounted for; in order to find whether they may be attributed to the action of an undiscovered planet beyond it, and if possible thence to determine approximately the elements of its orbit, etc., which would probably lead to its discovery."

By 1845 Adams had completed some ingenious calculations that showed that a heavenly body—probably a planet—beyond the orbit of Uranus was indeed exerting a gravitational tug on the latter planet. The next step was to carefully search the region of the skies in which the undiscovered planet might be found. In September 1845, Adams shared his results with the eminent astronomer Sir George Biddell Airy, and asked him to undertake the search.

For a time, Airy did nothing, perhaps because he was not interested in the request of the unknown Adams. In July 1846, however, Airy asked astronomer James Challis to search for the planet. Challis observed Neptune on August 4, 1846, but he failed to recognize it at the time.

DISCOVERY OF NEPTUNE

In the meantime, French astronomer Urbain-Jean-Joseph Leverrier had calculated the unknown planet's orbit, quite independently of Adams. Leverrier sent a series of three memorandums to the French Academy on this subject—in November 1845 and in June and August 1846. Then he wrote to Johann G. Galle, chief assistant at the Berlin Observatory, enclosing his calculations and urging Galle to look for the planet. The letter reached Galle on September 23, 1846. That same night he turned his telescope toward the quarter of the heavens suggested by Leverrier, and found the planet less than one degree from the place where Leverrier said it would be. Galle's discovery of Neptune was made possible because the Berlin Observatory had just received a new and complete map of the stars in that region of the skies.

There was a good deal of rather unpleasant controversy for a time concerning who would receive credit for Neptune's discovery. It is now generally agreed that Adams and Leverrier deserve equal credit for the abstruse calculations that led to the discovery. To Galle goes the distinction of having been the first to identify the planet in the heavens.

The new planet was named Neptune, after the Roman god of the sea. Its discovery was a fine example of what has been called the "astronomy of the invisible," that is, the detecting of heavenly bodies before they are actually observed in the skies, through the attraction they exert on known bodies.

Since Neptune's discovery, little else was revealed about the mysterious planet until Voyager 2 passed to within 3,000 miles (5,000 kilometers) of it on August 25, 1989.

STORMY BLUE WORLD

Neptune's atmosphere is composed primarily of hydrogen and helium with a dash of methane. It is the methane that gives the planet its lovely blue color. Methane absorbs longer wavelengths of sunlight (near the red end of the visible-light spectrum), and reflects shorter wavelengths (near the blue end of the spectrum).

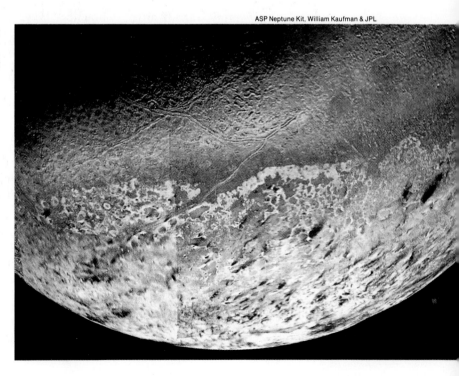

ASP Neptune Kit, William Kaufman & JPL

Triton, Neptune's largest moon, has the unusual distinction of following an orbit in a direction opposite to that of the planet's rotation.

Voyager 2 found Neptune ravaged by furious weather, the most violent area being a phenomenon dubbed the Great Dark Spot, an Earth-sized, pulsating swirl laced with white clouds. This hurricanelike storm lies in the planet's southern hemisphere, spins in a counterclockwise direction, and from time to time spawns smaller dark spots in a trail to the west. In many ways the Great Dark Spot is similar to Jupiter's Great Red Spot, though it changes in size and shape far more often. Near Neptune's south pole is another major spot, shaped like an almond, called Dark Spot 2. It has distinctive white clouds in its center that bubble up from below and spread out.

Though high winds streak the entire planet, the fastest occur near the Great Dark Spot. Here winds blow over 1,500 miles (2,400 kilometers) per hour—the strongest winds measured on any planet, including blustery Saturn. The winds of Neptune push wispy, cirruslike clouds of frozen methane gas around the planet. One particularly prominent cloud whips around the planet so fast that scientists call it "Scooter."

Scientists were surprised that Voyager found Neptune so stormy. On Earth, storms are driven by heat energy from the Sun. But Neptune receives 1,000 times less sunlight than the Earth receives. What's more, three years earlier Voyager found Neptune's planetary neighbor, Uranus, to have relatively little atmospheric activity; cold Uranus was thought to be too far from the warm Sun to have stormy weather. Surely Neptune would be equally as calm as Uranus, if not calmer, given that Neptune is a billion miles (1.6 billion kilometers) farther away from the Sun.

The existence of high winds and hurricanes on Neptune suggested that the planet was getting its heat from a source other than the Sun. But where? Scientists say that Neptune must have an internal heat engine, which probably fuels the planet's storms. As proof, Voyager showed that Neptune enjoys temperatures similar to those of Uranus, even though Uranus is much closer to the Sun. In fact, Neptune, unlike Uranus, emits more heat than it receives. The heat may come from the intensely high temperatures and pressures of the planet's innermost rocky core.

MAGNETIC FIELD

Voyager 2 confirmed that Neptune has a strong magnetic field. The probe also showed this field to be tilted 47 degrees from the planet's rotation axis. The magnetic field of Uranus has a similar tilt, suggesting that the two planets are magnetic "twins."

Because magnetic fields originate deep within a planet, they reveal much about a planet's interior. Neptune and Uranus are each believed to have a hot, rocky core, surrounded by a huge slushball of water ice, frozen gases, and liquid gases. Covering all that is the atmosphere.

Studies of Neptune's magnetic field allowed scientists to determine more precisely the length of a Neptunian day—how long it takes the planet to complete one rotation. Before Voyager, the best estimate was 18 hours. This was determined by using Earth-based telescopes to track cloud features in the Neptunian atmosphere. The practice, however, can be inaccurate because clouds speed up and slow down with the changing wind. How fast the part of the planet below the cloud layers is moving is best determined by studying the magnetic field generated by the planet's interior. An instrument aboard Voyager measured the pulse of radio waves created by Neptune's magnetic field. From these radio heartbeats, scientists determined that the planet spins once every 16 hours, seven minutes.

RINGS AND MOONS

As astronomers on Earth watched Neptune pass in front of distant stars, those stars sometimes winked out briefly, while other times they did not. On yet another occasion, one telescope captured an image of a star winking out behind Neptune, while, at the same time, a nearby telescope did not. It appeared from these observations that Neptune had partial rings, or "ring arcs," instead of complete rings.

Voyager 2 discovered a system of rings revolving around Neptune. The rings encircle the planet at distances between 17,000 and 23,000 miles.

NASA

Voyager 2, however, found at least five rings that go all the way around the planet, though they are thin in some areas and clumpy in others.

Astronomers have long recognized two of Neptune's moons: Triton, discovered in 1846; and Nereid, discovered in 1949. Voyager 2, however, discovered six more: Naiad, Galatea, Thalassa, Larissa, Proteus, and Despina.

Triton, Neptune's largest and most fascinating moon, orbits in a direction opposite to the planet's rotation. Triton's retrograde motion suggests that it did not originate out of Neptune. Some astronomers think that Triton was once a small planet until it came too close to Neptune. Then Neptune "harpooned" Triton with its strong gravity, keeping the moon close to its side ever since.

After Voyager's flyby of Neptune, the probe took a series of close-up photos of Triton. Ironically, this moon ended up capturing far more scientific attention than did Neptune itself. Scientists had expected Triton to be a cold, drab world. The moon did indeed prove cold at $-391°$ F ($-235°$ C). But it turned out to be far from drab. Voyager 2 revealed an unexpectedly varied and colorful terrain. The frosty white-, blue-, and pink-splotched moon was covered with cliffs and faults and sprinkled with craters. In some areas, dimples and dark spots ringed by whitish bands gave the moon the look of cantaloupe skin. Elsewhere, methane and nitrogen glaciers flowed slowly, icy slush oozed out of fractures, and ice lava formed frozen lakes.

Probably the most intriguing features on Triton are its geyserlike eruptions. It is thought that pressurized nitrogen gas spurts out from beneath Triton's frozen nitrogen surface. This gas spray, or plume, rises vertically about 5 miles (8 kilometers) into the moon's thin nitrogen atmosphere. Each plume creates a cloud of dark particles and possibly ice crystals that drifts some 93 miles (150 kilometers) westward. Energy to drive these ice geysers may come from solar heat building up in the ice. Another theory is that hot radioactive elements in the moon's core warm and expand the nitrogen gas. When enough pressure builds up, the geyser spouts off. What bizarre worlds Voyager 2 revealed at the edge of the solar system!

Pluto is the ninth and smallest planet of the solar system. Astronomers think that Pluto (foreground) and its moon Charon may constitute a binary planet system.

PLUTO

by Jeffrey Brune

Mysterious Pluto is the only planet that has not been studied closely by a space probe. Still, what scientists have been able to see using Earth-based telescopes is very odd indeed.

In its corner of the solar system, Pluto is a dwarf amid four giants. It has a solid surface, while those four giants—Jupiter, Saturn, Uranus, and Neptune—are made mostly of gas. Pluto's orbit is the most elongated and tilted of all the planets. And, probably oddest of all, it has a huge moon about half its size. Because Pluto and its moon, Charon, are so similar in size, astronomers consider the pair to be virtually a "double planet."

Pluto is only dimly visible because of its vast distance from the Sun—on average, about 3.7 billion miles (5.9 billion kilometers). Even through a powerful telescope, Pluto appears only as a very faint, yellowish point of light. Discovering this dimly lit planet in a background of countless bright stars was indeed a challenge.

DISCOVERY OF PLUTO AND CHARON

The road to Pluto's discovery began with another planet, Uranus, and followed the discovery of Neptune. The quirkiness of the orbit of Uranus led scientists to surmise that the gravity of some other planet was pulling Uranus out of its orbit. That planet turned out to be Neptune. But even taking Neptune's gravity into account, Uranus did not follow the course that astronomers had expected.

A number of astronomers, including Percival Lowell, director of the Lowell Observatory near Flagstaff, Arizona, decided to investigate the matter. In 1915 Lowell published *Memoir on a Trans-Neptunian Planet*, in which he presented his mathematical calculations of a ninth planet's probable position. He believed that it would be found in one of two areas in the sky. The search continued for years.

At last, in February 1930, Clyde W. Tombaugh, an assistant on the Lowell Observatory staff, found what seemed to be the long-sought planet as he studied a series of photographs he had taken. In March the observatory announced the discovery. The planet was named Pluto, after the Greek god who rules the underworld, because of its position in the distant part of the solar system.

It turns out, however, that Pluto, with a mass only 1/450th that of Earth, is not big enough to distort the orbit of Uranus with the pull of its gravity. This suggests the possibility of yet another planet beyond Pluto!

The discovery of Pluto's moon came much later. In 1978 James W. Christy of the U.S. Naval Observatory was methodically examining photographic plates to measure Pluto's position throughout its orbit. On one plate, Christy noticed Pluto had a slight bump. Looking at other plates, he noticed the bump moved around the planet. Christy quickly realized he had discovered a satellite, or moon, orbiting Pluto. He named the new moon Charon, after both his wife Charlene and the boatman in Greek mythology who ferried souls across the River Styx to the underworld.

SMALLEST AND SOMETIMES FARTHEST

With an equatorial diameter of 1,430 miles (2,300 kilometers), Pluto is the smallest of the known planets. It is not, however, always the farthest from the Sun, as is commonly thought. Sometimes Pluto orbits closer to the Sun than does Neptune. This has been the case since 1979. In 1999 Pluto will regain its distinction as the most distant planet.

The reason for this planetary flip-flop is Pluto's highly elliptical orbit around the Sun. While Neptune follows a roughly circular path, with the Sun in the center of its orbit, Pluto follows a very stretched-out path, with the Sun far removed from the center. At its closest approach to the Sun (perihelion), Pluto is 2.74 billion miles (4.41 billion kilometers) away; at its farthest approach, it is 4.57 billion miles (7.36 billion kilometers) away. So when Pluto approaches perihelion, it comes closer to the Sun than Neptune does.

Pluto's orbit is also the most tilted of all the planets'. It is inclined over 17 degrees to the plane in which Earth revolves around the Sun.

STUDY FROM AFAR

Only twice during Pluto's 248-year trek around the Sun can the orbit of its moon be seen edge-on from Earth. Astronomers were thrilled when one of those rare chances for study occurred in 1987 through 1988. During that time, Charon appeared to pass completely in front of Pluto, then completely behind it, then in front again, and so on every 3.2 days.

When Charon fell behind Pluto, only the light from the planet reached Earth. Scientists studied the colors in the light and concluded that Pluto's surface is frozen methane. When Charon was in front of Pluto, scientists measured the light from both bodies. They then subtracted Pluto's light to get only the light from Charon. Analysis of this light showed that the moon is covered mostly with frozen water. Pictures taken by the Hubble Space Telescope support these findings. Hubble shows Pluto to be much brighter than its moon. We know that methane ice (Pluto's covering) reflects more sunlight than water ice (Charon's covering).

Twinkling starlight confirmed that Pluto has only a thin atmosphere. In 1988 Pluto passed in front of a star as seen from Earth. Instead of turning on and off like a light switch, the star gradually dimmed as its light passed through Pluto's hazy covering. Scientists think the heat from the Sun turns Pluto's methane ice into gas, which may fall as snow from time to time.

Though scientists managed to unveil some of Pluto's secrets, many still remain. Only a spacecraft sent to the planet will answer the endless questions: What is Pluto's surface like? Are there other moons, or even rings? What is Pluto's weather like? How did this planetary snowball ever come to be?

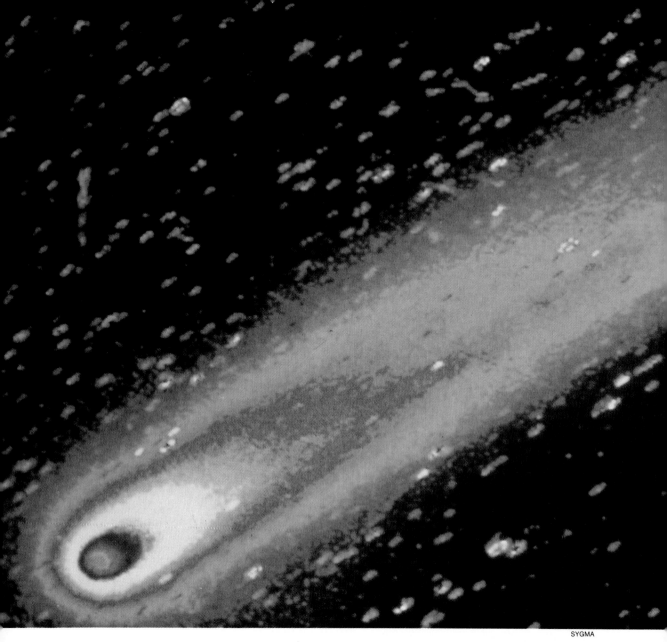

As a comet nears the Sun, its surface glows brightly and a tail forms. The best-known comet, Halley's (above, in a computer-enhanced photo), appears every 86 years.

COMETS

by Dennis L. Mammana

As the ancients gazed at the night skies, they were occasionally startled to see strange celestial objects intruding upon the familiar pattern of stars, Moon, and planets. These mysterious apparitions looked like fuzzy stars with long trains of light, moving from one constellation to another and cutting across the paths of the planets at every conceivable angle. The trains of light suggested a woman's tresses; hence, the celestial intruders came to be known as "long-haired stars," or *komētēs*, the Greek word for "long-haired."

A bright comet was a terrifying spectacle in antiquity. It was thought to foreshadow some dire catastrophe—plague, famine, war, or perhaps the death of a ruler. Today we realize that comets are simply another member of the solar system, and that their coming is no more portentous than the appearance of the first stars at twilight.

HOW COMETS APPEAR

When a particular comet is first discovered, it usually appears as a faint, diffuse body with a dense area near its center. This dense part, which sometimes looks like a tiny star, is known as the *nucleus*. The nebulous, or veil-like, region around it is the *coma*. Nucleus and coma together form the *head* of a comet.

In a certain number of cases, however, a spectacular transformation takes place as the comet approaches the Sun. The coma changes from a diffuse, round mass to sharply defined layers, called *envelopes*. Nebulous matter streams away from the comet's head in the direction opposite to the Sun and forms an immense tail. Most comets of this type have only one tail. A very few have two or more. The bright comet of 1744 had six in all. Some comets occasionally also have forward spikes. As a comet recedes from the Sun, the tail (or tails) can no longer be seen, the coma becomes diffuse again, and, in the great majority of cases, the comet itself disappears from view.

ORIGIN AND STRUCTURE

How do comets originate? According to one theory, they represent celestial building blocks left over after the formation of the planets. According to another, they are remnants of shattered worlds. All this is pure conjecture, as are the various theories that attempt to explain how comets are launched on their journey around the Sun. One theory, proposed by Dutch astronomer J. H. Oort in 1950, holds that there is a vast storehouse of comets—as many as 100 billion, perhaps—in the icy reaches beyond the farthermost planetary orbit. According to Oort's theory, a given comet would normally remain entirely inactive in the "deep freeze" of space unless the passage of a star disturbed it. The comet then would swing into the sphere of gravitational attraction of one of the major planets, such as Jupiter or Saturn, and would revolve around the Sun a few hundred or a few thousand times until it disintegrated.

We are on more solid ground when we try to analyze the structure and composition of comets. The general belief is that the nucleus consists of a vast number of small, solid bodies held together by mutual attraction. The nuclei of certain comets that have ventured close to Earth have been measured with considerable precision. The tail of the great comet of 1861 stretched across two-thirds of the sky and was bright enough to create shadows on the ground. Yet it had a nucleus less than 100 miles (160 kilometers) in diameter. When Pons-Winnecke's comet came within 4 million miles (6.5 million kilometers) of the Earth in 1927, its nucleus was found to have a diameter no greater than 1 or 2 miles.

As the nucleus of a comet approaches the Sun, the solar heat vaporizes the material on the outer surface of the nucleus. Escaping gases, carrying fine dust with them, diffuse into the coma. They are then swept away by the force of the Sun's radiation to form the tail. The gases and the dust they transport are illuminated partly by reflected sunlight and partly because they absorb ultraviolet light and re-emit it in the form of visible light.

The tail, which always flows away from the Sun, increases in breadth as the distance from the head increases. The tail does not form an exact line between the Sun and the comet's head. The greater the distance from the comet's head, the more the gases and dust that make up the tail lag behind. Hence, the tail often has the shape of a curved horn, with its tip at the comet's head.

When the comet turns away from the Sun, the material that formed the tail is swept off into space. In time, comets gradually lose all their substance, unless it can be replenished by dust and by gas molecules swept up in the course of their journeys through space.

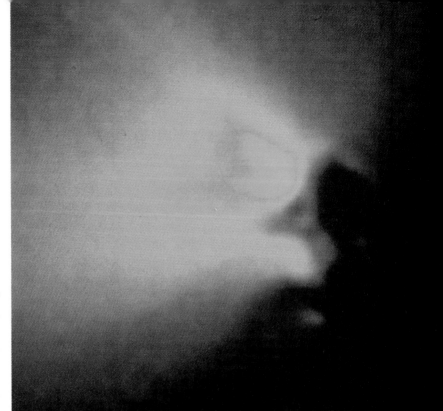

Astronomers think that the green jets in the nucleus of Halley's comet (right) represent the emission of gas and dust from the comet. Until relatively recently, comets were thought to hold supernatural significance. Below, the Bayeux tapestry records the passage of Halley's comet just prior to the Norman conquest of England in 1066.

Max Planck Institute for Astronomy, Lindau Harz, FRG

Musée de Bayeux

COMPOSITION

Analysis has revealed that comets contain various gases, including cyanogen (CN), carbon (C_2), carbon monoxide (CO), nitrogen (N_2), hydroxyl (OH), and nitrogen hydride (NH). The presence of such highly poisonous gases as cyanogen and carbon monoxide gave rise to a certain uneasiness in 1910, when Halley's comet passed between the Earth and the Sun. Astronomers had announced that our planet was certain to pass through at least a part of the comet's tail, triggering fears that the Earth's inhabitants would be subjected to gas poisoning. After the comet had passed, however, it was realized that such fears were groundless. The gases in the tail did not produce the slightest effect as they swept over the surface of the Earth.

Undoubtedly the Earth has passed through the tails of comets many times, and, as far as we know, without causing any harm to living organisms. The reason is that there are too few molecules in the tails to contaminate the Earth's atmosphere. It has been calculated that the best vacuum obtainable in the laboratory contains millions of times more matter per unit of volume than does the tail of a comet.

Small Amount of Matter

The total amount of matter in a comet, nucleus and all, is so small that it does not exert any significant gravitational pull on planets or moons. Through indirect methods, astronomers have surmised that the entire mass of Halley's comet cannot be more than one one-billionth that of the Earth.

Despite its small mass, however, a comet may be very large. The head of the great comet of 1811 was considerably larger than the Sun! As a rule the head of even a very small comet has a diameter larger than the Earth's. The tail may stretch over several hundred million miles.

Every time a comet approaches the Sun, the strong gravitational pull exerted by that huge body subjects the comet to a tremendous strain. As a result the comet may be broken up into two or more smaller bodies. That is what happened to Biela's comet, which split into two comets in the winter of 1845–46. These two comets were next observed in 1852, traveling not far apart from each other in what had been Biela's original orbit. They were no longer visible at the predicted returns of 1859 and 1866. On what would have been the next approach of these comets to the Sun—in 1872—a dazzling meteor shower was observed, indicating that the comets had disintegrated. We know that the breathtaking spectacle produced by other meteor showers, including the Perseids each August and the Leonids each November, have been due to the disintegration of comets.

The Dirty-Snowball Theory

In the early 1950s, U.S. astronomer Fred L. Whipple of the Harvard Observatory proposed a novel theory of the nature of comets—the so-called icy-conglomerate, or "dirty-snowball," theory. Whipple maintained that 70 to 80 percent of the mass of a comet is made up of icy particles, consisting of such hydrogen-containing compounds as methane (CH_4), ammonia (NH_3), and water (H_2O). As a comet approaches the Sun, according to Whipple, the ice particles on its outer surface sublimate, or pass directly from the solid to the gaseous state. The resulting gases, together with fine dust particles, form the tail. The remaining 20 to 30 percent of the comet's mass—consisting of compounds of the heavier elements—do not vaporize appreciably. These comparatively heavy compounds are the particles that produce spectacular meteoritic showers when a comet finally disintegrates.

Whipple's dirty-snowball theory was strengthened in 1986 when Soviet spacecraft Vega 1 and 2 flew within 6,000 miles (9,650 kilometers) of Comet Halley's nucleus. Astronomers found the nucleus to be a very black, potato-shaped object about 10 miles (16 kilometers) long and 5 miles (8 kilometers) wide. According to astronomer Horst Uwe Keller of the Max Planck Institute for Astronomy, the comet's nucleus is "like velvet. As dark as the darkest objects in the solar system." Data from space probes also indicate that the gas and dust from the comet's core are emitted in focused streams.

Even five years after its closest approach to the Sun, astronomers were still watching Halley's from more than 2 billion miles away. The comet is still full of surprises. For instance, though it has reached the coldest, darkest regions of the solar system, Halley's continues to erupt, occasionally shooting off jets of gas. The energy source behind the explosive display remains a mystery.

Coupled with the remarkable Hubble Space Telescope high in Earth orbit, large, ground-based telescopes continue to help astronomers keep an eye on Halley's as it completes its 76-year orbit around the Sun. Such continuous observation of the long-familiar comet is a first for science.

But scientists want an even closer look at comets. Now in planning stages are several international spacecraft that will journey out toward these mysterious, primordial bodies and send back a wealth of information. One of the most ambitious probes—the Comet Rendezvous Asteroid Flyby (CRAF)—will pay a close visit to Comet Kopff sometime in the mid-1990s.

MOTION OF COMETS

The motion of the comets baffled astronomers for a long time. In contrast to the sometimes complex but always reliable movement of the planets, comets emerge into view with a flourish, only to disappear for years on end. They traverse all regions of the sky at various angles to the plane of the solar system. No wonder that, until the late 16th century, comets were thought to be phenomena of the Earth's upper atmosphere, like the aurora borealis.

It was not until 1705 that the true nature of cometary motion was established.

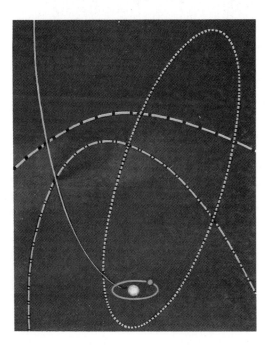

Periodic comets like Halley's follow an elliptical orbit (dotted line) through the heavens. Comets that appear only once follow open curves, either parabolic (dot-and-dash line) or hyperbolic (dashed line). Still others (solid line) disintegrate in the solar system.

In that year, Edmund Halley, a friend of Sir Isaac Newton, applied the law of gravitation to the observations of a number of comets. He found that they traveled in space in accordance with that law. He noted, too, that the comets of 1682, 1607, 1531, and 1456 had moved in much the same way. He came to the conclusion, therefore, that these supposedly different comets were really one and the same body, which reappeared every 75 or 76 years. Halley predicted that the comet would return in 1758, and his prophecy was fulfilled. Halley's comet appeared also in 1835 and in 1910. When it became visible again, early in 1986, spacecraft launched the previous year by the European Space Agency (ESA), the Soviet Union, and Japan rendezvoused with the comet. These spacecraft transmitted close-up pictures of the comet as well as data on its composition and temperature.

Historical investigation revealed that the comet had been observed at every appearance as far back as the year 240 B.C. It should have been visible in 315 B.C. and 391 B.C., but there are no actual records of its appearance in those years. However, the comet of 467 B.C., the first ever recorded, was undoubtedly Halley's comet.

When Halley demonstrated that comets move around the Sun according to the law of gravity, he dispelled forever the idea that they are signs of divine wrath. Since Halley's time the orbits of many comets have been traced. To understand the nature of these orbits, it is necessary to point out that, according to the law of gravitation, comets must move around the Sun in conic sections. A conic section is a curve obtained when a plane cuts through a cone. There are only two such curves that are closed—the circle and the ellipse. A body moving in either one of these curves ultimately returns to the place from which it started. As astronomers would say, the body has a definite period of revolution. The point in a comet's orbit when it's nearest to the Sun is called perihelion. The opposite point in the orbit, farthest from the Sun, is called aphelion.

A comet cannot move in a circle. If it started by tracing a circular path, the gravitational attraction of planets in the solar system would soon distort the comet's path into an ellipse. All periodic, or regularly returning, comets move in elliptical orbits around the Sun.

Other comets move in conic sections that are open curves—parabolas and hyperbolas. A comet moving in one of these curves would travel around the Sun and off into space, never to be seen again.

NAMING COMETS

Astronomers must be able to identify comets in order to calculate their orbits. Each comet, therefore, is distinguished by a special name. This contains the year when a particular appearance was first discovered, and a letter indicating the order of discovery among the comets of that year. But this is only a temporary name. Once the orbit has been computed and the time of perihelion worked out, the same comet is designated by the year of perihelion passage, and a Roman numeral, indicating its order among the comets that have reached perihelion that year.

Halley's comet will next appear in the year 2062. Astronomers have fully mapped its orbit around the Sun.

For example, Halley's comet in its last apparition was first called 1909c and then 1910-II. This means that it was the third comet discovered in 1909, and the second comet to pass perihelion in 1910. In the case of more recently sighted comets, the name of the discoverer is added in parentheses. Thus, the bright comet temporarily designated as 1936a is now referred to as 1936-II (Peltier). Thus, it was the first comet discovered in 1936, it was the second comet to pass perihelion in that year, and the discoverer was Peltier.

News of a comet's discovery in the Eastern Hemisphere is immediately sent to the observatory in Copenhagen, Denmark, the international clearinghouse for information of this kind. When the Harvard Observatory receives word of the discovery from Copenhagen, it telegraphs the information to all the observatories in the Western Hemisphere. News of a discovery in the Western Hemisphere is sent to the Harvard Observatory, which then forwards the news to other observatories.

DETERMINING THE COMET'S ORBIT

After a comet is discovered, astronomers begin to measure the apparent position of the comet in reference to known stars. Its position, measured at intervals of a few days, gives the astronomer enough information to calculate a preliminary orbit and to predict the comet's motion in the future. Some of these measurements may be in error, or the comet may be too diffuse to be measured precisely. As a result the first calculation of the orbit is usually somewhat inaccurate. It is amended as further observations become available. When the comet has disappeared from view, all observations are collected, and the definitive or final orbit is computed.

The main difficulty in determining an orbit is that at first we have no idea how distant the comet is. It is presumably "near" the Earth—(that is, within 100 million miles, or 160 million kilometers)—but it appears to be as remote as the stars. It is only when it moves against the background of the constellations that we can begin to calculate its actual distance from the Earth. There is an added complication. The apparent motion of the comet among the constellations is due principally to its real motion around the Sun, but it is also due in part to the motion of the Earth during the interval between observations. The astronomer must disassociate, or separate, the motion of the Earth from that of the comet.

At best, even with modern calculating machines, the computation of the orbit is very tedious, and even the best computers make mistakes. Therefore two computers generally work independently and check their results from time to time. After the orbit has been computed, the *ephemeris* is

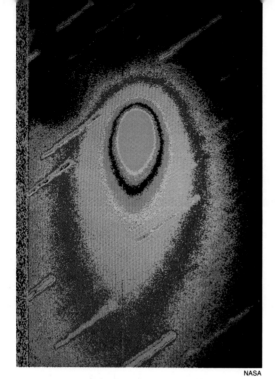

The colors in the computer-enhanced image of Comet Giacobini-Zinner above indicate levels of brightness. Comets are initially named for the year of their discovery; later, they take the names of their discoverers.

set up. An ephemeris is a statement, in the form of a table, of the assigned places of a celestial body for regular intervals.

It may take a long time to work out the definitive orbit. Each observation must be checked, the best-known position for each comparison star must be obtained, and the attraction, or perturbation, by other members of the solar system must be calculated. From all these computations, various quantities are derived, which are called elements of the orbit. One of the elements is the time of perihelion. Another is the distance of perihelion from the Sun, measured in astronomical units. An astronomical unit (abbreviated A.U.), used by astronomers for great distances, equals the distance from the Earth to the Sun, or approximately 93 million miles (149.6 million kilometers).

Short and Long Periods

The time that elapses between two returns of a comet to perihelion is called its period. Some comets move in comparatively small ellipses, so that they come back to perihelion every few years; that is, they are short-period comets. Encke's comet has the shortest known period—about 3.3 years. It was first discovered in 1786, and has since been seen returning to perihelion more than 40 times. None of the short-period comets are bright. Under the best conditions, Encke's comet is barely visible to the naked eye.

If the orbits of a certain group of about 30 short-period comets are plotted, it is seen that all of their aphelia, or positions farthest from the Sun, fall near the orbit of Jupiter. Obviously this giant among the planets exerts an important influence upon these comets. It can be shown mathematically that they were deflected by Jupiter from their original orbits and made to move around the Sun in small ellipses. Therefore, these short-period bodies are called Jupiter's family of comets.

Other planets have also deflected comets in much the same way, and now have their own comet families. Saturn's family is small. It includes the remarkable Comet 1925-II (Schwassmann-Wachmann), which has a period of 16 years and an almost-circular orbit. It was the first comet to be observed in every part of its orbit, from perihelion to aphelion. This strange body is subject to remarkable changes in brightness, the cause of which is unknown. Sometimes it flares up and becomes more than 500 times as bright as it had been a few days previously. Unfortunately, this comet is always so far from both the Earth and Sun that, even at its brightest, a good telescope is necessary to see it at all.

Halley's comet belongs to Neptune's family, which, like Saturn's family, has only a few members. Some of these, including Halley's comet and Comet 1884-I (Pons-Brooks), are quite bright.

For some comets, periods of 100, 200, or 500 years have been calculated. The longer the period, the more uncertain it is. The trouble is that we can observe such comets only in the small part of their orbit near perihelion. In the case of comets with periods of more than a few thousand years, the orbit that can be observed shows exceedingly little curvature. This often makes it impossible to figure out the shape of the orbit or the length of the period. According to Estonian astronomer E. Opik, some of the longer orbits may extend out to the nearest stars, more than 4 light-years away. A light-year is approximately 6 trillion miles (9.6 trillion kilometers).

Most astronomers believe that all comets are periodic, although the periods may be many thousands or even millions of years. It is thought, for example, that Comet 1914-V (Delavan) has a period of 24 million years. It is possible, however, that the orbits of some comets may form an open curve, whose ends can never meet. In such cases the comet may have been traveling in an elliptical orbit when it entered the planetary region of the solar system, and may have been thrown into an orbit forming an open curve because of the gravitational attraction of the major planets.

Collision with a Planet

Since comets move in all directions in the solar system, it is conceivable that one could collide with a planet like the Earth. We have already seen that no damage is done when the Earth passes through a comet's tail, but what would happen if the Earth were struck by a comet's nucleus?

The chances of this occurring are very small. Yet in the Earth's long history, there must have been a few collisions with comets. In such an event, there would certainly be a brilliant display of meteors as the comet's nucleus disintegrated upon coming in contact with the Earth's atmosphere. The larger pieces of the comet's nucleus might cause considerable destruction if they were to fall into a populated region.

APPARENT BRIGHTNESS

As we have observed, the illumination of comets is due, directly or indirectly, to the Sun. The nearer to the Sun a comet is, therefore, the brighter it is. The distance of the comet from the Earth will also naturally affect its apparent brightness. In 1910 Halley's comet appeared brightest, not at perihelion, but a month later, about the middle of May, when it came closest to our planet. The brightness of comets is measured by the same scale of apparent magnitudes used to measure the apparent brightness of the stars.

One of the brightest comets ever seen in the skies was that of 1577. It was carefully observed by Tycho Brahe, who then proved that it was not traveling in our atmosphere. This demonstration disproved the old theory that comets are vapors from the surface of the Earth, ignited in the upper atmosphere of our planet.

Only a few truly bright comets have appeared in the 20th century, though there have been many easily visible to the naked eye. We have already mentioned Halley's comet, which appeared in 1910, and again, in less spectacular fashion, in 1986. Comet Skjellerup in 1927 was very bright, but it was almost entirely ignored by the general public because it appeared to be so close to the Sun. Another extremely bright comet was Ikeya-Seki. Discovered by two ama-

Halley's 1986–87 passage disappointed many comet watchers in the Northern Hemisphere. Still, many vivid displays were captured on film, including the photo below of Halley crossing the Arkansas night sky.
© Greg Polus

teur Japanese astronomers, K. Ikeya and T. Seki, in September 1965, it created quite a sensation. After passing within 310,700 miles (500,000 kilometers) of the Sun, the nucleus of the comet broke into three pieces, which then sped back into space.

The brightness of a comet is difficult to predict. In the winter of 1973–74, Comet Kohoutek failed to appear as brightly as had been predicted, while in early 1976 Comet West surprised many, becoming one of the brightest comets observed since the beginning of the 20th century.

AMATEURS DISCOVER MANY

Many amateur astronomers actively search for new comets. As a matter of fact, most new comets are discovered by amateurs. A professional astronomer using telescopes follows a rigid program of observation, and, for the most part, observes individual stars or small parts of the sky. Once in a while, a hitherto-unknown comet will come within this field of view, but not often. The amateur who makes a deliberate search for a comet by scanning every quarter of the sky is more likely to make a discovery.

The amateur comet hunter should use a small telescope that has a wide field of view—four or five times the diameter of the Moon—and should work out a definite system of observation. One excellent method is to sweep the sky from east to west and to change the position of the telescope with every sweep so that the new field partially overlaps the old one. The most promising areas for search are the western sky after nightfall, and eastern sky before dawn.

The comet-hunting amateur will find many celestial objects that look like comets. Many of these, however, will prove to be nebulas and globular clusters. The only way to distinguish between such permanent celestial bodies and comets is to memorize the positions of the nebulas and globular clusters or to consult a sky atlas, such as *Norton's*. The real comet will betray its nature by changing its position among the stars within a few hours. Only occasionally will an observer come upon a comet with an unmistakable tail. In most cases the comet will be faint, diffuse, and tailless.

Only a fairly skillful amateur astronomer can search for new comets with any hope of success. The comet hunter must be able to operate a telescope efficiently, must know how to distinguish a comet from a nebula, and must be able to determine with what speed and in what direction the new comet is moving. Besides a certain amount of proficiency in astronomy, an amateur comet hunter must also have a great deal of persistence, and must realize, too, that the best of efforts may be fruitless. Luck plays an important part in every discovery. Even so, there are plenty of encouraging stories.

A comet, reportedly as bright as Halley's, was discovered in the summer of 1986 by a 24-year-old student, Christine Wilson. Wilson, working at a summer job at Mount Palomar Observatory, noticed the comet on a photographic plate. Comet Wilson takes more than 10,000 years to complete an orbit around the Sun.

The champion discoverer of comets was Jean-Louis Pons, who had been a janitor at the Marseilles Observatory. Between the years 1802 and 1827, he discovered 28 comets—more than one a year—with his homemade telescope. American amateur comet hunters W. R. Brooks and E. E. Barnard attained such fame for their discoveries that they became professional astronomers. Brooks discovered 25 comets between 1883 and 1911, and Barnard found 22 in about the same period.

STILL PROBLEMS TO SOLVE

Much remains to be done in observing the spectra of comets and in determining their brightness, shape, and motions. The observation of comets from space—away from interference of the Earth's atmosphere—provides a new approach to the study of comets. In 1986 the study of Comet Halley by Soviet, Japanese, and European probes added much new information. Future observations from space will no doubt further advance our understanding of these unusual members of our solar system.

ASTEROIDS

by Dennis L. Mammana

In the region between the orbits of Mars and Jupiter, a vast number of small celestial bodies known as asteroids, or minor planets, make their home. About 2,000 of these bodies have already been cataloged, and there may be over 100,000. Some of them have orbits that swing beyond Jupiter, and some come inside the orbit of Mars.

The asteroids were discovered as a result of an apparent flaw in the law astronomers used for estimating the relative distances of the planets from the Sun. This law, formulated by German astronomer Johann Elert Bode in 1772 and known as Bode's law, was based on the fact that as the distance from the Sun increases, the planetary paths are more and more widely separated. Bode claimed that the increasing distances of the planets from the Sun followed a more or less regular ratio. Today we realize that the law is based upon coincidence and, in fact, does not apply at all to the two outermost planets, Neptune and Pluto, which were not known in Bode's day. However, for a number of years after it was proposed, the law was accepted by astronomers.

At first the only flaw in Bode's law was an apparent gap between Mars and Jupiter, one too great to be explained by the law. Astronomers came to the conclusion that another planet, hitherto undiscovered, must lie within this belt. Toward the end of the 18th century, an association of astronomers was formed for the express purpose of hunting for the missing planet. They realized that the object of their search must be very small, or else it would not have been able to escape observation for such a long time.

DISCOVERY OF A "MISSING PLANET"

When Giuseppe Piazzi, an Italian astronomer, announced on January 1, 1801, that he had discovered a new heavenly

© William K. Hartmann

Astronomers think that asteroids are chunks of cosmic matter that somehow never united to form a planet. A collision of two asteroids creates vast amounts of debris, some of which may fall to Earth as meteorites.

body in the zone between the orbits of Mars and Jupiter, astronomers assumed that this was the planet for which they had been searching. Curiously enough, Piazzi came upon the supposed planet more or less by accident. He was engaged in making a catalog of the fixed stars. He had developed a very exact method for mapping the sky by determining the relative positions of the stars within a given area on a number of successive occasions. If any "star" moved in relation to its neighbors, it was obviously not a star at all, but some other sort of heavenly body, such as a planet or a comet. The moving body would not be included, therefore, in Piazzi's catalog.

Piazzi had already mapped out more than 150 areas of the sky without incident.

However, when he compared four successive observations of the constellation Taurus, he discovered that a certain small star within the constellation had changed its position from one observation to another. The Italian astronomer suspected that the "star" was really a comet. But when Bode heard of its movements, he decided that it was the planet for which so many astronomers had been searching.

Three More "Planets"

Piazzi fell ill before he had time to make many observations of the new heavenly body, and it was lost to view for a time. But word of the discovery had come to Karl Friedrich Gauss, a young mathematician from Göttingen, Germany. Using a new method for determining planetary orbits, Gauss was able to calculate the path of the supposed planet from the few observations made by Piazzi, and, as a result, the new heavenly body was rediscovered in December 1801. It received the name of Ceres, after the Roman goddess of agriculture. Not long after Ceres had been rediscovered in the heavens, astronomers discovered three other "planets"—Pallas (1802), Juno (1804), and Vesta (1807). It was realized by this time that these newly discovered heavenly bodies were too small to rank as full-fledged planets. Rather, they were minor planets, also dubbed asteroids or planetoids.

Ceres, Pallas, Juno, and Vesta are considered the Big Four among the asteroids, the only ones with definite diameters and which appear as disks when viewed through a telescope. Ceres is about 480 miles (770 kilometers) in diameter; Pallas, 304 miles (490 kilometers); Vesta, 240 miles (385 kilometers); and Juno, 118 miles (190 kilometers). Though Vesta is the third in size, it is the brightest, perhaps because its surface reflects sunlight the most effectively. Vesta, alone among the asteroids, can be seen with the naked eye.

Many More on Film

It was not until 1845 that the fifth asteroid, Astraea, was discovered. Astronomers were finding the search for the comparatively tiny asteroids painstakingly slow and very laborious. But in 1891 a new asteroid-hunting technique was developed by Max Wolf, a German astronomer. He attached a camera to a telescope that was moved by clockwork in such a way that it continually pointed to the same fixed stars. A photographic plate was set in the camera and exposed for a certain period of time. When the plate was developed, the stars in the photograph appeared as white points. If there were any asteroids within the field of the telescope, they would appear as short white lines, because the asteroids would have moved in their orbits during the exposure of the plate. After adopting Wolf's technique, astronomers found so many new asteroids that it became difficult to keep track of them.

The larger asteroids have measurable disks, but in the great majority of cases, the diameter can be determined only by indirect means. As mentioned earlier, the largest asteroid, Ceres, is only 480 miles (770 kilometers) across. The smallest ones may be only 1 or 2 miles (1.6 or 3.2 kilometers) in their longest dimensions. Probably they are not spheres, but irregularly shaped solids.

VARIED ORBITS

The asteroids cover a very wide belt in space. Hidalgo, at its farthest distance from the Sun, approaches the orbit of the planet Saturn. Icarus, discovered in 1948, moves in an orbit that passes beyond Mars and then closer to the Sun than the planet Mercury itself—a distance of only 19 million miles (30 million kilometers), which is about half that of Mercury from the Sun. The asteroid Hermes, only about a mile (1.6 kilometers) across, sometimes comes as close as 198,840 miles (320,000 kilometers) to the Earth's orbit—closer than the orbit of the Moon. In 1937 Hermes moved to within 500,000 miles (800,000 kilometers) of the Earth itself. Evidently the orbits of the asteroids are highly varied and distorted compared to those of the planets. They may intersect each other, and often lie at high angles to the planes of the Earth's path and the paths of the other planets.

Space probes venturing beyond Mars encounter the asteroid belt, where thousands of the tiny bodies orbit the Sun. So far, no probe has collided with an asteroid.

CLOSE APPROACHES AND IMPACTS

The asteroids have never been seen to collide with each other or with any of the planets, but undoubtedly they have in the past. At times, some asteroids actually cross Earth's orbit. Perhaps as many as 1,000 asteroids larger than a half mile wide have crossed our planet's path. In 1990 we came as close to an asteroid as we have in half a century. The chunk of rock cataloged as 1989 FC streaked by only 500,000 miles (800,000 kilometers) from Earth, about twice the distance to the Moon.

In the past, many asteroids have actually collided with our world and the Moon. A quick look at the lunar surface with a small telescope shows thousands of impact craters. While lunar craters are preserved from time immemorial by the lunar vacuum, the evidence of impacts on Earth is gradually but continually erased by weather and erosion. Some of the meteors that create such impact craters are fragments of asteroids.

One of the most famous of all such collisions may have caused the extinction of the dinosaurs some 65 million years ago. The subsequent explosion may have blanketed the Earth with dust, cutting off sunlight for long periods of time and wiping out a significant number of species—including the dinosaurs.

ORIGIN OF THE ASTEROIDS

Astronomers have long speculated on the origin of the asteroids. A once-popular theory held that they are the remains of an exploded planet that had previously circled the Sun between the orbits of Mars and Jupiter. The planet approached Jupiter too closely and was broken up by the gravitational pull of that giant planet. The fragments then collided, accounting for their varied orbits.

Astronomers now favor another explanation—that asteroids are chunks of cosmic matter that somehow never finally united to form a planet during the time that the solar system was coming into being. Here, too, the gravitational attraction of Jupiter would be the decisive factor. It would prevent the chunks from drawing together and ultimately forming a single body.

A meteor is matter from outer space that enters the Earth's atmosphere. As gravity pulls it toward Earth, the meteor vaporizes, leaving behind a glowing trail (above). A meteor that survives its plunge to Earth is called a meteorite.

METEORS AND METEORITES

by Ray Villard

From time to time, you might glance into the sky to see a point of light trailed by a fleeting, luminous train as it races against the background of stars. It may be barely visible or as bright as the full Moon. This object is often called a "shooting star."

Actually, what you're watching is not a far-off star in motion, but something quite close to home. A fast-moving body from outer space has penetrated Earth's atmosphere and has become so heated from air resistance that it begins to glow.

The term *meteor* refers to solid bodies that burn up into vapor when they scorch through our atmosphere. If the celestial visitor survives its fiery plunge and reaches the ground, it is called a *meteorite*.

Meteorites provide us with an unusual opportunity to study specimens from outer space. From these specimens we can determine the composition of matter from beyond the Earth and Moon. We can also draw conclusions about the conditions under which this matter originated and evolved, as well as estimate its age.

Scientists believe most meteorites coalesced from the same cloud of dust and minerals that condensed to form all the planets and moons in the solar system. Others are no doubt fragments of shattered asteroids or debris thrown off from comets, moons, or even planets. Radioactive-dating techniques show that most meteorites date back to what is thought to be the beginning of the solar system, 4.6 billion years ago. So meteorites also provide scientists with a cosmic clock by which to estimate the age of our solar system.

METEOR SHOWERS

On a typical clear night, you can see the occasional meteor blazing across the heavens. But meteors also appear in groups at more or less regular intervals. These showers may consist of no more than five meteors in an hour, but may reach as many as 75. On rare, fantastic occasions, the sky may be lit by thousands of meteors per hour. For example, some 35,000 or so meteors fell in one hour during the Leonid meteor shower of November 1833. This spectacular display terrified most everyone who witnessed it.

It seems more than coincidence that the timing of meteor showers corresponds to the known orbits of certain comets. Astronomers conclude that the showers come from a stream of debris through which Earth passes as a comet shoots by.

Spectacular Impacts

While meteor showers dazzle, meteorite falls can pack a wallop. In the late 1940s, there were two spectacular meteorite falls. On the morning of February 12, 1947, a solid metallic meteorite fell and shattered as a "shower of iron" in the Ussuri taiga in the Soviet Union. The pieces created more than 120 craters, some roomy enough to hold a two-story house.

About a year later, on February 18, 1948, tens of thousands of people in the Midwest watched a stony meteorite fall. Its main mass, now known as the Furnas County stone, weighs more than a ton. Before it struck the atmosphere, it probably carried a mass of more than 11 tons (10 metric tons).

A few other meteorites come close to the size of the meteorites in the Ussuri and Furnas County incidents. They include a huge, iron-rich meteorite of perhaps 77 tons (70 metric tons) near Hoba, Namibia, and a 33-ton (30-metric-ton) iron meteorite brought back from Greenland by the American explorer Robert Peary.

Shortly after midnight on February 8, 1969, a 30-ton meteor broke up over the town of Pueblito de Allende in northern Mexico. Thousands of fragments rained out

A huge meteor striking the ground can leave an enormous crater. One such crater, the Canyon Diablo crater in Arizona, also called the Barringer Crater, resulted from a meteor impact more than 50,000 years ago.

Doddy & Zeller/*Discover* magazine, Time, Inc.

Workers needed a crane to move the 33-ton Anaghitto meteorite into New York's American Museum of Natural History. The meteorite fell in Greenland; had it fallen in a populated area, the meteorite would have caused great damage.

of the sky, the largest weighing nearly 5 tons (4.5 metric tons).

On August 10, 1972, a 54-yard (50-meter)-wide meteor (weighing 1 million tons) skipped across the Earth's atmosphere as a fireball and sped back into space. Had it struck Earth, it would have exploded with the force of an atomic bomb!

Scientists today can spot meteors with radar and estimate how many bombard our planet. Some figure that 110 tons (100 metric tons) of meteorites reach the Earth every day. Most of these bodies are too small to see—so tiny that they can pass through the atmosphere without vaporizing. Eventually these particles drift down to the surface, virtually unaltered.

Meteors that actually flash across the night sky range from the size of a sand grain to that of a peanut shell. Visiting less frequently are the larger meteorites, weighing 10 pounds (4.5 kilograms) or more. These don't completely burn up during their atmospheric travel, so that small but recognizable portions reach Earth's surface. As it traverses the atmosphere, the meteorite puts on a light-and-sound show that can be startling or terrifying. And at the moment of impact, high-speed "meteoric shrapnel" has cut off tree branches as smoothly as would a sharp ax and shot through roof slate without cracking it.

Statistically speaking, a meteor will probably hit someone in North America once every 180 years. The only such documented case occurred on November 30, 1954, in Sylacauga, Alabama. A 13-pound (6-kilogram) meteorite crashed through the roof of a house and struck Mrs. Hewlett Hodges in her bedroom. She was bruised, but not seriously injured.

WHAT'S IN A METEORITE?

Meteorite composition generally falls into one of the three following groups: (1) irons, or *siderites*; (2) stones, or *aerolites*; and (3) stony irons, or *siderolites*.

Meteorites are composed of the same elements contained in the solid portion of the Earth, but in different relative abundance. Whereas oxygen (49.4 percent), silicon (25.8 percent), aluminum (7.5 percent), and iron (4.7 percent) make up almost 90 percent (by weight) of the rocks of the Earth's crust, these elements show up in differing amounts in the three classes of meteorites.

Irons, or siderites, are about 91 percent iron and 8.5 percent nickel. They also contain cobalt, phosphorus, and minute quantities of other elements. Stones, or aerolites, contain, on the average, 41 percent oxygen, 21 percent silicon, 15.5 percent iron, 14.3 percent magnesium, and smaller percentages of other elements. The aerolites are much less dense than the siderites and break more easily. Consequently, they shatter on their way through the atmosphere and rain down in widely scattered showers of small masses, only a few of which are recovered. People see aerolite falls about 10 times more frequently than siderite meteorites.

Stony irons, or siderolites, show characteristics of both the iron and the stony meteorites. Much less common in collections than either siderites or aerolites, stony irons remain objects of mystery.

What Prizes Inside?

It was once thought that meteorites carry living spores and thus propagate life in different parts of the universe. Today's experts no longer hold this theory. But scientists have found the chemical building blocks of life—amino acids and other organic compounds—in meteorite samples. Over a long period of time, organic compounds could have formed in space, where the elements carbon, hydrogen, nitrogen, and oxygen are known to exist.

Within some meteorites are microscopic "pieces" of other stars, which mixed with meteors when the solar system formed. Meteor researchers can recognize these interstellar grains because they contain carbon, nitrogen, and other elements in odd combinations of atomic forms, or isotopes. These combinations are different from those found in the solar system. Supernova explosions could have blasted some of these grains through space. This suggests that the radioactive debris from a nearby exploding star may have "seeded" our very early solar system. Researchers have also found microscopic diamonds within meteorite dust.

PIECES OF PLANETS?

Antarctica is meteorite country. There, scientists have found nearly 10,000 meteorites preserved under the ice. A few of these samples appear to be pieces of the planet Mars. Unlike most meteorites, these pieces appear to have once been molten rock that solidified 1.3 billion years ago. The Mars meteorites also have different isotopes than do those found in most other meteorites. The crystalline structure of the Mars meteorites indicates that they underwent a severe compression and shock, as if they'd been struck with some incredibly powerful sledgehammer.

With all this evidence, scientists think a 7.5-mile (12-kilometer)-wide asteroid crashed into Mars about 200 million years ago. Some of the debris was blasted into space, and, after no more than about 10 million years, a small fraction of that shrapnel fell to Earth. The Martian meteorites contain organic compounds, meaning that life might exist on Mars.

Scientists have also found golf-ball-size meteorites that evidently came from the Moon, no doubt through similar cosmic collisions. These meteorites match exactly the chemical signature of Moon rock samples brought back to Earth by the Apollo astronauts.

© Roger Ressmeyer

The presence of the element iridium in certain dinosaur deposits has led Walter Alvarez (at right in photo) and his son Luis to speculate that a giant, iridium-rich meteorite collided with the Earth, causing the extinction of dinosaurs and other species.

Craters as Calling Cards

A truly gigantic meteorite would probably be virtually destroyed upon striking the Earth's surface. That's because it would be too heavy to slow down as it tore through our atmosphere. Experiments have shown that when a projectile hits a target at high speeds, the projectile and a considerable part of the target are simply blown up, and an explosion crater is formed. Although such a crater would be relatively shallow, its volume would be surprisingly large compared to the size of the projectile that created it.

If a large enough meteorite were to hit Earth, both the meteorite and the chunk of Earth it struck would vaporize almost immediately. The resulting explosion would carve out an enormous crater.

Earth bears scars from at least 130 such impacts. More and more ancient, buried craters are turning up as scientists learn how to recognize them. The best-preserved evidence of such an explosive impact is the Barringer Crater, near Winslow, Arizona. The crater has a circumference of nearly 3 miles (5 kilometers) and is nearly 328 yards (300 meters) deep. The meteorite that blew out this vast hole had a mass estimated at between 2 million and 8 million tons (1.8 million and 7.2 million metric tons). Evidence indicates that this huge meteorite collided with Earth 50,000 years ago.

In 1908 a mysterious, atom-bomb-type explosion devastated 193 square miles (311 square kilometers) of forest in Siberia. Because there is no large impact crater from the event, the destruction may not have been due to a monster meteor, but perhaps a small comet or asteroid that exploded in the atmosphere.

Death of the Dinosaurs

The dinosaurs dominated Earth for 250 million years. Yet they mysteriously vanished 65 million years ago. A number of competing theories try to explain what happened. One possibility is that a 1-mile (1.6-kilometer)-wide meteorite hit Earth. Such an impact would have destroyed everything within hundreds of miles. More important, the dust raised by the explosion could have severely altered Earth's climate, devastating plant life. Such a disruption in the food chain would have been particularly lethal to larger animals, like dinosaurs.

There is evidence for this catastrophe in the geologic record. About 65 million years ago, a layer of iridium, an element rare on Earth but common in meteors and asteroids, was deposited over the entire world. At the same time, a layer of soot fell, possibly from forest fires. Grains of shocked quartz, the best evidence for a tremendous crash, have also been found at this same geologic layer.

Such an impact should have left behind a huge crater at least 93 miles (150 kilometers) across. But where is it? Experts believe they have found at least three craters dating back 65 million years: a 111-mile (180-kilometer)-diameter ring buried under Mexico's Yucatán Peninsula, a 62-mile (100-kilometer)-diameter multiring impact basin in Siberia, and a 22-mile (35-kilometer) crater near Mason, Iowa. Perhaps all three craters were made simultaneously by pieces of an asteroid or comet that broke apart before hitting Earth.

If a giant meteorite killed the dinosaurs, their extinction likely resulted from the debris hurled into the atmosphere by the force of the impact, debris that ultimately shielded Earth from sunlight.

© Don Davis

A solar eclipse occurs when the Moon, passing between the Earth and the Sun, briefly blocks the Sun's rays. Scientists can then study the solar atmosphere, the only part of the Sun visible during totality.

ECLIPSES

by Dennis L. Mammana

An eclipse of the Sun or Moon is truly an awe-inspiring spectacle. The word "eclipse" comes from the Greek word *ekleipsis*, meaning "forsaking" or "abandonment," indicating how the ancients dreaded this celestial drama. As the Sun or Moon disappeared from view, it seemed indeed to be deserting mankind. Eclipses, like comets, were held to be portents of war, pestilence, the death of princes, or even the end of the world. To this day, certain primitive peoples come to the aid of the Sun or Moon, as it is being eclipsed, with solemn rites and loud entreaties.

We know now that there is a perfectly logical explanation for eclipses—they are caused by the enormous shadows of the Earth and of the Moon. Both of these bodies are opaque. Hence, when they are lit by the Sun, each has a shadow extending out into space, away from the Sun.

The shadow cast by the Earth or Moon has several parts (see *Diagram 1*). There is a region of complete shadow, which is known as the *umbra* (the Latin word for "shadow"). Since both the Earth and the Moon are smaller than the Sun, the umbra of each is conical in shape. It diminishes in diameter as it extends farther out in space until, finally, it comes to a point. No light comes directly from the Sun to any object within the umbra. Surrounding the cone of complete shadow, there is a region of partial shadow, called the *penumbra* (Latin for

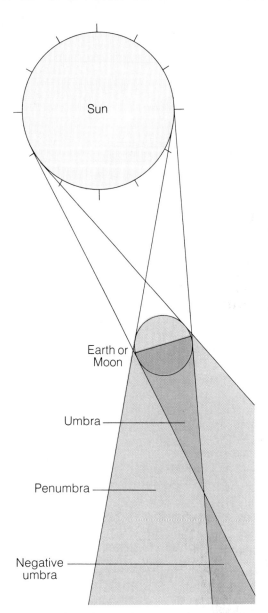

Diagram 1. *Scientists use the terms* umbra, penumbra, *and* negative umbra *to describe the shadows cast by the Earth or by the Moon during an eclipse. The* umbra *is the area of complete shadow, which appears on the Earth as a solar eclipse. The area of partial shadow is called the* penumbra. *The* negative umbra *represents the continuation of the lines that bound the complete shadow.*

"almost a shadow"). Any object within this area receives light from a portion of the Sun. If the lines bounding the conical region of complete shadow are extended outward, as shown in *Diagram 1*, an inverted cone is formed. It is called the *negative umbra* and, as we shall see, it is an important factor in certain eclipses of the Sun.

LENGTH OF THE SHADOW

It is evident from the diagram that the length of the cone of complete shadow depends on three factors: the diameter of the source of light—the Sun; the diameter of the Earth or Moon; and the distance between the Sun and the Earth or Moon.

It is important to bear in mind that while the diameters of the Sun, Earth, and Moon are constant factors, the distances between the Earth and the Sun and between the Moon and the Sun are variable. For this reason, the umbra of the Earth or of the Moon varies in length. The average length of the Earth's umbra is about 870,000 miles (1,400,000 kilometers); the average length of the Moon's umbra, about 233,000 miles (375,000 kilometers).

When the Earth enters the Moon's shadow, an eclipse of the Sun takes place. A solar eclipse can take place only at the time of new moon, when the Moon is between the Sun and the Earth. A lunar eclipse always takes place at full moon, when the Earth is between the Sun and the Moon (see *Diagram 2*).

There would be an eclipse of the Sun at every new moon, and an eclipse of the Moon at every full moon, if the Moon's orbit were in exactly the same plane as the Earth's orbit around the Sun. However, this is not the case. The Moon's orbit is slightly inclined (about 5 degrees) to that of the Earth (see *Diagram 3*).

The Moon passes through the plane of the Earth's orbit around the Sun twice every month, at points called the *nodes* of the Moon's orbit. Generally, the Moon is at one side or the other of the Earth's orbital plane at new moon or at full moon. If it is not in the plane at new moon, its shadow does not fall upon the Earth, and the Sun is not eclipsed. If the Moon is not in the plane of the Earth's orbit at full moon, it remains outside of the Earth's shadow, and the Moon is not eclipsed. From time to time, the Moon is at full or new moon at about the time when it crosses the plane of the Earth's orbit. When that happens, there is a lunar eclipse at full moon and a solar eclipse at new moon.

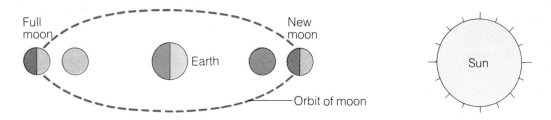

Diagram 2. *Solar eclipses only occur during a new moon, when the Moon is between the Earth and Sun. Lunar eclipses only occur during the full moon, when Earth is between the Sun and Moon.*

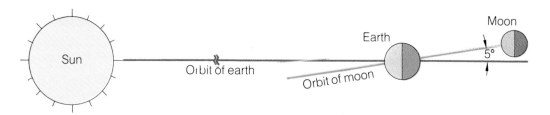

Diagram 3. *The plane of the Moon's orbit around the Earth is slightly inclined to the Earth's orbit around the Sun. If the planes were the same, there would be a solar and a lunar eclipse each month.*

ECLIPSES OF THE MOON

For an eclipse of the Moon to take place, then, it must be full and it must be near an orbital node. When the Moon plunges into the cone-shaped area of complete shadow, it is much nearer to the base of the cone than to its tip. The diameter of the cone, where the Moon passes through it, is about two and one-half times the diameter of the Moon.

If the path of the Moon happens to pass through the center of the shadow, the Moon may remain totally eclipsed for about an hour. The shadow may cover part of it for about two hours. A lunar eclipse begins when the Moon enters the penumbra, and ends when it leaves the penumbra. There is little significant darkening, however, until the Moon enters the umbra.

If the path of the Moon takes it near the edge of the umbra shadow, the total phase of its eclipse may last only a few minutes. If the Moon's path is such that only a portion of its disk, and not the whole of it, enters into the conical shadow of the umbra, the eclipse is partial and not total. Sometimes the Moon in its path passes, not through the cone of complete shadow, but through only the penumbra. If this happens, so much light is still received from a portion of the Sun's disk that there will be no marked obscuring of the Moon unless it passes very close to the true shadow. The Moon is usually not altogether lost to view even in the midst of a total eclipse. It shines with a strange, copper-colored glow.

Eclipses of the Moon occur less frequently than solar eclipses. Each year, there are at least two and as many as five solar eclipses. But there are years when there are no lunar eclipses, and only rarely is there more than one. This may seem to run counter to our experience. After all, it seems that more people have seen lunar than solar eclipses. This is because, during a solar eclipse, the Moon's shadow covers only a small part of the Earth's surface. As a result, solar-eclipse enthusiasts may have to travel thousands of miles to stand in the narrow track where the full eclipse is visible. During a lunar eclipse, however, the Earth's shadow covers the entire face of the Moon—so that the eclipse is seen over half the Earth, the half where it is night.

ECLIPSES OF THE SUN

There are three kinds of solar eclipses: *total*, *annular*, and *partial*. We have seen that the average length of the Moon's shadow is about 233,000 miles (375,000 kilometers), and it never extends beyond about 236,000 miles (380,000 kilometers). The average distance of the Earth from the Moon, however, is about 240,000 miles

The Moon gradually moves through the Earth's shadow and reappears (above, left to right), during a total lunar eclipse. Because of light refracted by the Earth's atmosphere, the Moon is never wholly dark, even during totality.

(385,000 kilometers), so that in general the Moon's umbra—its true shadow—is not long enough to reach the Earth. At times, however, the Moon is only about 221,500 miles (356,500 kilometers) from the Earth's surface. The true shadow then falls upon a small part of the Earth's surface, causing a total eclipse of the Sun over an area that never exceeds 168 miles (270 kilometers) in diameter. At other times the Moon may be more than 252,277 miles (406,000 kilometers) from the Earth. If it is then interposed between the Sun and the Earth, its negative umbra will partially obscure a small area of the Earth's surface, causing an annular eclipse of the Sun at that point. In an annular eclipse, the Sun's rim appears as a "ring" of light around the dark Moon. (*Annulus* means "ring" in Latin.)

Around the area where there is a total eclipse or an annular eclipse of the Sun, there is always a much larger area where there is a partial eclipse. This area is in the Moon's penumbra. It generally extends for more than 1,865 miles (3,000 kilometers) of the Earth's surface on each side of the path where the total eclipse can be seen. Sometimes the area of partial eclipse extends nearly 3,100 miles (5,000 kilometers) on each side of the path of totality.

The Moon's shadow passes along this path at great speed—about 1,000 miles (1,700 kilometers) an hour. The longest period of total eclipse, under the most favorable conditions, is about 7.5 minutes.

Few spectacles in the heavens are as startling as a solar eclipse. The approach of a total solar eclipse is particularly impressive and, to some persons, alarming. The sky darkens ominously; birds fly to shelter. Finally the dark shadow of the Moon, like a vast thundercloud, advances with awe-inspiring rapidity from the western horizon and covers the land. Usually, just before the last rays of the Sun are obscured, one can see swiftly moving bands of light and shade, which are probably due to uneven refraction in the atmosphere. Then the day becomes like an eerie night.

As the eclipse approaches totality, the Sun is seen as a very narrow crescent of brilliant light. The crescent then becomes a curved line, and finally breaks off into irregular beads of light known as Baily's beads. The beads are caused by irregularities in the surface of the Moon.

But even at totality, the atmosphere that surrounds the Sun extends so far out in space that it is never completely covered by the Moon. We see the eclipsed Sun surrounded by a beautiful glow. While observing a solar eclipse, remember that the eyes must at all times be protected from the direct rays of the Sun!

SUMMER ECLIPSE OF 1991

One of the most spectacular solar eclipses on record occurred on July 11, 1991. On that day the Moon's 150-mile- (241-kilometer)-wide shadow drifted eastward from Hawaii, across the Pacific Ocean, across Baja California and mainland Mexico, and into Central and South America, before trailing back into space. Astronomers were able to make unprecedented observations of this eclipse from one of the most celebrated observatories—

that on Mauna Kea in Hawaii. An international crew of astronomers trained seven of Mauna Kea's 10 large telescopes on the Sun's corona.

During its trek across the Earth's surface, the Moon's shadow also passed over some of the most popular tourist spots in the world and five Latin American capitals. According to estimates, some 50 million "eclipto-philes" drove, sailed, flew, or hitchhiked to witness this cosmic event. In Mexico City alone (where the total eclipse lasted nearly seven minutes, longer than any until the year 2132), some 17 million people were swept into the eerie darkness of the Moon's shadow—more than had ever before witnessed a single, total solar eclipse in recorded history.

PREDICTING ECLIPSES

Inasmuch as eclipses of both the Sun and the Moon depend upon the regular movements of the Sun, Earth, and Moon, they can be calculated with great accuracy. In the 1880s, for example, Austrian astronomer Theodor Oppolzer published a book called the *Canon of Eclipses*, in which he gave a table of 8,000 solar eclipses and 5,200 lunar eclipses taking place between 1207 B.C. and A.D. 2162. His *Canon* has now been extended to A.D. 2510 and revised with the aid of computers. In the case of solar eclipses, he indicated the areas of the Earth from which the eclipses would be visible.

Our knowledge of when eclipses occurred in the past helps modern historians pinpoint the exact time of a surprising number of important events. Ancient writers who associated eclipses with battles, royal births or deaths, and other occurrences inadvertently provided us with a foolproof method of assigning precise dates (often down to the hour and minute) of many significant occasions.

SERIES OF ECLIPSES

The interval between an eclipse of the Sun or Moon and the next one in a given series is called a *saros*. Each saros is 18

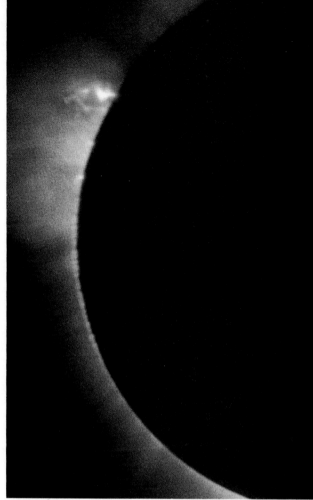

© Alberto Levy

During totality, solar prominences become visible, extending like tongues from the obscured Sun. Prominences dramatically appeared during the sensational total solar eclipse of July 1991 (above).

years and 11⅓ days long (or 18 years and 10⅓ days long if there are five leap years instead of four in a saros). While a given series of solar eclipses may run through 70 saroses and last for 1,250 years, there are only 48 or 49 saroses in the average series of lunar eclipses. A given series of lunar eclipses lasts somewhat under 900 years.

The first eclipse in a series of solar eclipses is partial, the Moon encroaching but slightly on the Sun's disk. At the next eclipse, the Moon obscures a somewhat larger area of the Sun. Each subsequent eclipse becomes more extensive, until there is an annular or total eclipse. Then there follows a succession of partial eclipses, each one obscuring a smaller area of the Sun's disk than the one before.

OPPORTUNITY FOR STUDY

Total solar eclipses offer astronomers exceptional opportunities to study the atmosphere of the Sun; the distribution of its material; the chromosphere, or Sun's inner layer; and the splendor of the corona, or outer layer. Furthermore, only at the time of total eclipse can astronomers photograph celestial objects close to the Sun.

Unfortunately, few total eclipses of the Sun take place in areas where well-equipped astronomical observatories are located—the eclipse of July 1991 being a notable exception. Astronomers must generally undertake expensive expeditions and erect temporary research camps to install costly scientific instruments within the narrow band of shadow where the eclipse can be viewed—usually for a matter of just a few minutes.

The simplest observations made during a total eclipse of the Sun are the precise moments of each "contact." The first contact takes place when the Moon first encroaches on the Sun's disk. In the second contact, the Sun disappears behind the Moon, and the solar corona becomes visible. In the third the Sun's rim is seen again. In the fourth the Moon passes completely off the Sun's disk. If the times of contact are observed at widely separated stations along the path of the eclipse, scientists get detailed information about the Moon's position and motion.

In the period of total eclipse, observers search for possible undiscovered planets within the orbit of Mercury, the planet closest to the Sun. Thus far, no new planet within the orbit of Mercury has been discovered. The search for hitherto-unknown comets during solar eclipses has been a more rewarding exercise. A number of new comets have been tracked at perihelion—the part of their orbit where they pass closest to the Sun.

On his fourth voyage to the New World, Columbus used his foreknowledge of a lunar eclipse to avoid a conflict with Indians. Modern astronomers have pinpointed the eclipse to February 19, 1504, shortly after 6 P.M.

The New York Public Library Picture Collection

The Milky Way is a giant spiral of several billion stars that includes our Sun. Its true shape, size, and nature were not discovered by astronomers until the 20th century.

THE MILKY WAY

by Dennis L. Mammana

The Milky Way, the part of our galaxy visible to the naked eye, is one of the most striking sights in the night skies. It is too faint to be seen in bright moonlight or amid the myriad lights of our large cities, but on moonless nights in the country, the outlines of its cloudy track of light across the heavens are easily discernible. If viewed through a powerful telescope, the Milky Way is unveiled as the combined light of vast numbers of stars, none of which can be made out individually without a telescope.

We know today that when we look at the Milky Way, we are looking into the heart of a vast system of stars that includes our Sun—lying as it does in the system's outer suburbs. In the past, however, the Milky Way was a celestial puzzle. Many explanations were provided in Greek and Roman mythology. Some writers called it the highway of the gods, leading to their abode on Mount Olympus. Others held that it sprang from the ears of corn dropped by the goddess Isis as she fled from a pursuer. Still others believed that the Milky Way marked the original course of the sun god as he sped across the skies in his chariot.

In medieval times, pilgrims associated the Milky Way with their journeys to various sanctuaries. In Germany, for example, it became known as Jakobsstrasse, or James' Road, leading to the shrine of St. James at what is now Santiago de Compos-

In the southern hemisphere of the Milky Way lies a strikingly large, dark area called the Coalsack. This inky black void, really a mass of obscuring matter, is surrounded by bright stars.

tela, in Spain. In England, it was called the Walsingham Way. It was associated with the pilgrimages to the famous shrine of Walsingham Abbey. The pilgrims of those days did not seriously believe that the Milky Way had anything to do with their travels. Rather, they saw it lying overhead, a misty path in the heavens. Their belief in the universal kinship of all things caused them to find comfort in its presence.

The best time to view the Milky Way is on an autumn or winter evening. It is then highest in the heavens, and therefore its light is least affected by our atmosphere. It is seen to stretch like a vast, ragged semicircle over the skies of the Northern or the Southern Hemisphere. Actually, it traces a rough circle that continues through both hemispheres.

IRREGULAR SHAPE

The path traced by the Milky Way is full of irregularities. It is by no means a simple stream of stars. Its average width is about 20 degrees, but it varies considerably, both in width and in brightness. Even with the naked eye, one can make out something of its irregular detail when the atmosphere is unusually clear and the Moon is new. When viewed under such conditions through a good telescope, the Milky Way is a truly exciting spectacle.

Its general effect has been likened to that of an old, gnarled tree trunk, marked here and there with prominent knots. As details become clearer in a telescopic view, we see that at one point the Milky Way may consist of separate stars scattered irregularly upon a dark background. Elsewhere, there are numerous gorgeous star clusters. In many places the track is engulfed in a nebulous blur in which a great many stars are embedded.

A powerful telescope reveals a great many dark bands in the Milky Way. Sometimes these dark bands are parallel; sometimes they radiate like spokes from a common hub; sometimes they are lined with bright stars. In certain places, they are quite black, as if utterly void of content. In others, they are slightly luminous, as if powdered with small stars.

THE COALSACK AND OTHER APPARENT VOIDS

Large, dark areas occur here and there. The most famous of these is the so-called Coalsack, which is near the constellation called the Southern Cross, visible in the Southern Hemisphere. Just before the Milky Way divides into two branches in the southern constellation Centaurus, it broadens. It becomes studded with a collection of brilliant stars, making it one of the most resplendent areas in its whole course. Right in the center of this host of bright stars, near the four stars that form the Southern Cross, is the inky black cavern known as the Coalsack.

The Coalsack is by no means unique. There are many similar black areas in the Milky Way, though they are generally less clearly defined and less striking in appearance. American astronomer Edward E. Barnard described one of these, in the constellation Sagittarius, as "a most remarkable, small, inky-black hole in a crowded part of the Milky Way, about two minutes in diameter, slightly triangular, with a bright orange star on its north northwesterly border, and a beautiful little star cluster following."

STUDYING THE SYSTEM

The starry band seen extending across the heavens is really part of our own galaxy, or system of stars. Every star that can be seen with the naked eye belongs to this vast system, including our Sun and the other members of our solar system.

An infrared image of the Milky Way produced by NASA's Cosmic Background Explorer (COBE), which collects data about the origin of the universe.

NASA

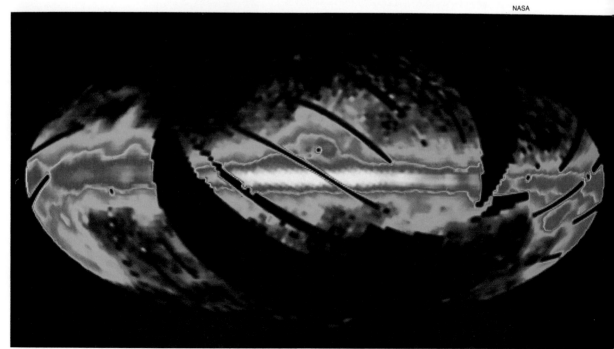

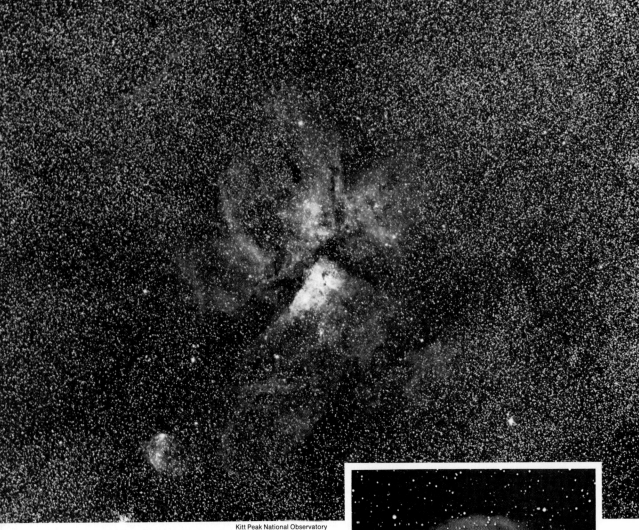

The apparent luminosity of many nebulas, such as the Eta Carinae nebula (above), is actually the reflection of light from neighboring stars. Planetary nebulas, like the Dumbbell nebula in Vulpecula (right), are gaseous envelopes surrounding hot stars.

Various methods have been used to solve the mystery of its structure. The eighteenth-century German-English astronomer William Herschel decided to attack the problem by making a survey of the stars. He called his method "star gauging." It consisted of counting all the stars visible in a reflecting telescope that had an 18-inch (45-centimeter) mirror and a field 15 inches (38 centimeters) in diameter.

Dutch astronomer J. C. Kapteyn developed an even more elaborate method for determining the distribution of stars in the heavens. He selected 206 areas distributed uniformly over the whole sky, and he urged astronomers to determine the apparent magnitudes and other data for all the stars in these regions. A number of the great observatories of the world took part in the cooperative venture suggested by Kapteyn. Two outstanding contributions to the program are the Mount Wilson Catalog and the Bergedorfer Spectral-Durchmusterung

(Bergedorfer Spectral Catalog). The latter gives not only the apparent magnitudes of the stars, but also the spectral classes of the stars.

Another way of determining the dimensions and structure of our galaxy is to observe the positions and distances of some of its members, such as the Cepheid variables. We can find the approximate distances of these star groups from the Earth if we know their apparent magnitudes.

The chief difficulty in determining the extent and structure of our galactic system is that the Earth is so deeply embedded in it that a comprehensive idea of the system is impossible to obtain through mere observation. It is almost as if we were required to make a map of New York City, for example, from a vantage point somewhere in a crowded section of the Bronx. However, we do have a clear overall view of a number of other galaxies, such as the magnificent spiral galaxy in Andromeda (a galaxy called Messier 31) and the Whirlpool in Canes Venatici (Messier 51). Our knowledge of these other galaxies, and radio telescope and infrared astronomy, enable us to draw a number of plausible conclusions about our own galaxy.

A Spiral Galaxy

An analysis of our galactic system by the methods described above indicates that most of the stars it contains are crowded into a sort of wheel with a pronounced hub. When viewed edge on, the wheel and hub look something like the illustration on page 189. Of course, since the Sun, our own star, is located within the wheel, we cannot see the wheel edge on. We infer its shape from viewing other galaxies whose form we can make out and think are similar. Our Sun does not lie anywhere near the center, or hub, of the wheel. It is at a distance of about two-thirds of the way from the center to the outer rim. The wheel as a whole is inclined at an angle of 62 degrees to the plane of the celestial equator.

The Rosette nebula. The gas surrounding the central stars is thought to be contracting to form new stars.
© California Institute of Technology and Carnegie Institution of Washington

The gas and dust near the Pleiades, or "Seven Sisters," star cluster scatter starlight to form diffuse nebulas around each of the stars.

The stars that make up the wheel show a distinct spiral pattern. Astronomers therefore classify our galaxy as a spiral galaxy. The spiral arms spring from a nucleus or core at the center of the galactic system. In the spirals are found individual bright stars, star clusters, bright nebulas, and a great deal of obscuring matter. This obscuring matter is made up of dust particles and various gases. The general haze it causes has been detected through the dimming and reddening effects that it produces.

Dark Nebulas

Astronomers have given the name "dark nebulas" to the masses of obscuring matter that have no nearby stars to illuminate them. Some of these dark masses can be seen quite clearly with the naked eye, and have been known to astronomers for a long time. The earlier astronomers thought of them as gaps in the starry firmament. The dark nebulas, or masses of obscuring matter, cut off our view of vast numbers of stars that lie beyond them. To obtain an idea of the form of the galactic system, certain observers have counted the stars in various parts of the sky, and have then made allowances for the obscuring matter and estimated the form of the galaxy.

Obscuring matter is responsible for the large, dark areas such as the Coalsack. It was once thought to be an optical illusion, but this hypothesis never seemed too convincing, since its huge size, its utter darkness, and the similar brightness of the starry edge surrounding it could never be explained. The existence of obscuring matter, now proved, offers a simple explanation of what was once a mystery.

STAR CLUSTERS

The clouds and wisps of bright nebulosity mentioned earlier are the bright nebulas, whose atoms have either been excited by the hot stars in the vicinity or reflect the light of nearby stars. It is interesting to note that the most luminous stars are all to be found in the wheel of our galaxy. The so-called galactic clusters are also confined to this area. They are sometimes called "open clusters." They consist of many groups of hot stars, each group consisting of several hundred stars. These stars excite the atoms of dust or gases in their vicinity; hence, they are embedded in bright nebulas. Among the best known of the galactic clusters are the Pleiades, the Hyades, and Coma Berenices.

The globular clusters are distributed through a roughly spherical region bisected by the plane of the galaxy. American astronomer Harlow Shapley held that the main aggregation of stars in the galactic system is arranged in the form of a wheel and that this is enclosed in a roughly globular haze of stars—globular clusters and others. Globular clusters differ in many respects from the galactic clusters. For one thing, each one contains hundreds of thousands of stars, or perhaps even millions, compactly and symmetrically grouped. They are comparatively free of gas or dust.

GALACTIC CORE

The core or nucleus of the galactic system is so heavily obscured by clouds of dust that it was not until 1983, when the InfraRed Astronomical Satellite (IRAS) was launched, that astronomers were able to obtain a comprehensive view of its structure. They have since begun to map the heavy concentration of stars in this area. It has been estimated that these stars account for something approaching one-half of the total mass of our galactic system.

Measurements of infrared radiation, combined with information on high-energy radiation such as X rays and gamma rays, suggest that the very core of our galaxy may contain a black hole. It may be as small as a star, but thousands or millions of times more massive. The object that astronomers suspect to be a black hole—named Sagittarius A—may be gobbling up nearby stars and gas clouds at an incredible rate. Simultaneously it would be releasing tremendous amounts of energy back into space.

THE ROTATION OF OUR GALAXY

The entire galactic system is rotating around an axis, which is at right angles to the wheel of stars. Outside the dense nucleus of the galaxy, the speed of rotation decreases, and the period increases with greater distance from the center. Corresponding differences in period and speed can also be noted in the planets that revolve

Our Milky Way galaxy is a member of what astronomers call the Local Group. This group may contain as many as 100 galaxies arranged in an oval shape at least 6,000,000 light-years across.

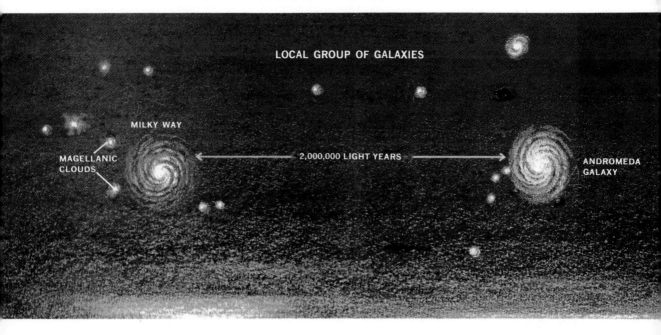

Our galaxy, when viewed on edge, appears like a wheel and hub, with the Sun lying about 28,000 light-years from the hub (below). NASA's COBE satellite collected data from both the far-infrared wavelengths of our galaxy, capturing radiation from cold dust (top), and from the near-infrared wavelengths (above), showing radiation from the stars. Scientists believe that a black hole may exist at the center of our galaxy.

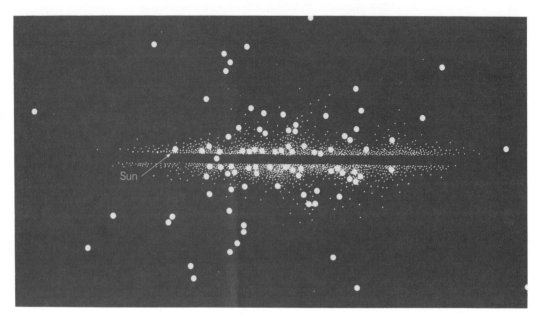

about the center of our solar system—that is, the Sun. The nearer a planet is to the Sun, the faster it revolves around it, and the shorter the period of revolution. For example, in the case of Mercury, the planet nearest the Sun, the period is only 88 days, compared with the period of 248.43 years of the planet Pluto, whose orbit is farthest from the Sun.

Our solar system orbits the Milky Way's galactic center every 225 million years, which astronomers call one "cosmic year." Obviously, the stars closer to the galactic core travel around its center more rapidly than the Sun does, and eventually overtake and pass it. On the other hand, the stars between the Sun and the outer rim of the wheel revolve around the center of the galactic system more slowly than the Sun does. As a result, they lag farther and farther behind our star in their voyaging.

THE DIMENSIONS OF OUR GALAXY

In the foregoing pages, we have given you some idea of the form of our galaxy. Astronomers have also sought to discover its dimensions in space. Recent research based on the study of special types of variable stars (the Cepheid variables) and on the visual, photographic, and spectrographic analysis of globular clusters has revealed how extraordinarily great these dimensions are. This is natural enough when we consider that our own solar system, vast and complex though it is, is an infinitesimally small part of the galactic system. The estimates of the dimensions of our galaxy vary considerably, but even the smallest are almost overpowering when we seek to grasp their significance.

According to one of the more conservative estimates, made by American astronomer Heber D. Curtis, a pulse of light, starting from one edge of the galactic system and traveling at the speed of over 186,400 miles (300,000 kilometers) a second, would take from 20,000 to 30,000 years to reach the other edge. According to Harlow Shapley, Curtis' estimate, staggering though it may seem to us, is far too modest. To accept Shapley's figures, we must tax our imagination still further. He maintains that it would take about 100,000 years for a pulse of light to travel from one confine of the galaxy to the other. In other words, the diameter of the galactic disk is approximately 100,000 light-years.

In the mid-1970s, astronomers studying the rotational speed of the Milky Way and its speed toward the Great Andromeda Galaxy, some 2 million light-years away, came to a startling conclusion. To account for the motions they observed, the Milky Way must be significantly larger and more massive than they had believed—perhaps as much as five to 10 times greater. But where was all this material? Scientists have speculated that it may consist of "dark matter"—clouds of dark particles, neutrinos, brown dwarfs, unseen planets, even black holes. Astronomers have yet to account for all the suspected missing mass, much less hazard a good guess as to where to look for it.

As remarkably huge as our galaxy is, it is only one of billions of other isolated galactic systems. Each of these "island universes" contains hundreds of billions of stars. The nearest—two irregular galaxies called the Magellanic Clouds—lie between 160,000 and 185,000 light-years away and orbit our galaxy.

THE "EDGE" OF THE UNIVERSE

Whether we shall ever see the "limit" of the universe is extremely doubtful. For if the universe is 10 billion years old, as some scientists believe, our horizon will be limited to a distance of 1 billion light-years from the Earth, even if the universe actually extends far beyond that distance. The reason is that the light from galaxies beyond that mark will not have been able to reach us since the beginning of time. The same is true of the radio waves emanating from such galaxies, since light waves and radio waves travel at the same speed—roughly 186,400 miles (300,000 kilometers) per second. Such galaxies, therefore, will remain a deep mystery to astronomers, unless some hitherto-unsuspected method of detecting them is discovered.

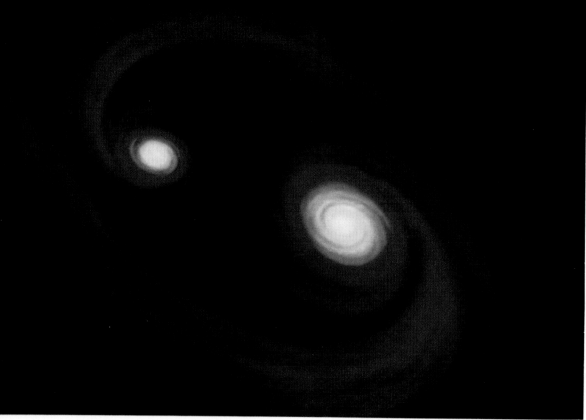

A star begins life as a dense cloud of gas that condenses and contracts to produce a protostar.

© Dana Berry/STSCI

THE STARS

by Charles Beichman

The thousands of millions of stars in the heavens are so far away that, with the exception of our Sun, they are visible only as twinkling points of light. How far away are the stars? What are they made of? What powers them? How old are the stars, and how are they formed? These questions are among the most basic in astronomy. To develop a fundamental understanding of stars, let's first examine a typical star, and the closest one to Earth: our own Sun.

THE SUN AS A STAR

The Sun is a globe of glowing gas. Despite its sharply defined disk, it has no solid surface. Its diameter is about 864,000 miles (1,390,000 kilometers), or 109 times that of the Earth. However, the gaseous Sun is less dense than the rocky Earth, and its mass is only 330,000 times as large. The temperature at its surface is about 10,800° F (6,000° C). A new technique called *helioseismology* has enhanced our understanding of the interior of the Sun. Astronomers study the vibrations of the Sun much like geologists, who learn about the Earth's depths by measuring the sound waves carried by its rocky interior after an earthquake. A global network of telescopes monitors the Sun around the clock, measuring physical waves roiling across its surface. The intervals between these waves and sets of waves may range from a few minutes to several weeks. The amplitude, or fullness, of these waves gives astronomers much information about the physical conditions deep within the Sun. From such measurements and sophisticated computer models, we now know that the temperature and pressure rise swiftly toward the center of the Sun, reaching values of 18,000,000° F (10,000,000° C) and 100 billion atmospheres, respectively.

Even though the Sun is millions of miles away, we have also been able to determine its composition using a technique called *spectroscopy*. Every element emits or absorbs specific colors of light; these patterns can be used as fingerprints to identify an element, to determine its abundance, and to determine the temperature and pressure in its surroundings. By measuring the Sun's spectrum, we have learned that the lightest element, hydrogen, is by far the most common, accounting for more than 92 percent of its atoms. The second most abundant element in the Sun, and in the universe as a whole, is the lighter-than-air gas helium. It makes up about 8 percent of the Sun. Roughly speaking, the heavier the element, the less of it exists in the Sun. Certain elements particularly common on Earth—such as carbon, nitrogen, and oxygen—total less than 0.1 percent of the Sun's material.

The basic properties of the Sun are common among all stars. Although all other stars are so distant as to appear only as points of light, astronomers have gained a surprising amount of information about these heavenly bodies by careful observation and the application of physical laws.

DISTANCE TO THE STARS

The apparent brightness of stars belies their true nature. A star may appear bright because it is nearby, or because it is intrinsically luminous. It is impossible to determine the intrinsic brightness of a star unless we also know how far away it is.

Astronomers measure the distances to the closest stars using the same principles of triangulation that a surveyor might use. Suppose we observe a star on a given date and again six months later, when the Earth is on the opposite side of its orbit. The position of the star will be displaced slightly against the background of more-distant stars. The shift in the star's position, called its *parallax*, can be combined with the size of the Earth's orbit to yield the distance to the star.

Unfortunately, even the closest stars are very far away, so that the angles used to calculate their distances are small, a second of arc (1/3600 of a degree) or less. Only the distances to the closest few hundred stars have been determined using the parallax method. The distances to stars farther away must be inferred indirectly.

Once the distance of a star is known, we can calculate its absolute brightness. With all stars compared on a common scale, we see that our Sun is quite ordinary. Some stars are intrinsically a million times brighter; others are a thousand times fainter. The brighter stars are quite rare, while fainter stars are very common and outnumber those like our Sun.

TEMPERATURE, SIZE, AND MASS

The laws of radiation, familiar to blacksmiths and physicists alike, permit us to calculate the temperature and size of a star. Stars, like horseshoes being forged, first glow red-hot, then yellow-hot, then blue-hot, and finally white-hot, as their temperatures increase. Thus, a star's color reveals its temperature. Measurement of the star's temperature and absolute bright-

The shift in a star's position in relation to Earth, known as its parallax, is used to calculate the star's distance.

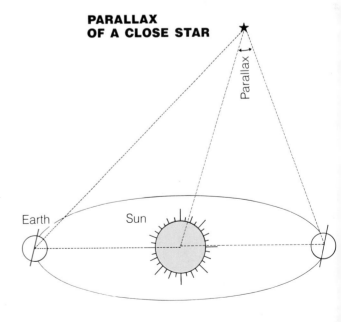

PARALLAX OF A CLOSE STAR

THE STARS 191

TEMPERATURES AND DIMENSIONS OF TYPICAL STARS

	STAR	SURFACE TEMPERATURE IN DEGREES C	RADIUS IN SUNS	MASS IN SUNS	DENSITY IN SUNS	ABSOLUTE MAGNITUDE
SUPER-GIANTS	Beta Lyrae	12,000	19.2	9.7	.0014	0.1?
	Rigel	11,300	78	20?	.00004	− 7.0
	Deneb	10,200	96	20?	.00002	− 7.0
	Gamma Cygni	4,100	67	20?	.00007	− 5.0
	Betelgeuse	2,700	1000	10?	.0000005	− 5.8
	Antares	2,900	776	20?	.00000004	− 4.0
GIANTS	Capella	4,800	13	2.1	.00096	0.3
	Arcturus	3,800	35	8	.00018	− 0.1?
	Aldebaran	2,700	87	4	.000006	− 0.3
	Beta Pegasi	2,000	40	9	.00014	− 1.4
MAIN SEQUENCE STARS	Hadar	21,000	22	25	.0023	− 5.1
	MU-1 Scorpii	20,000	5.2	14.0	.1000	− 5.1
	Sirius A	10,200	1.9	2.3	.335	1.2
	Altair	7,300	1.6	1.7	.415	2.4
	Procyon A	6,800	2.6	1.8	.102	2.1
	Sun	5,900	1.0	1.0	1.0	4.86
	61 Cygni A	2,600	0.7	0.58	1.69	7.65
	Krueger 60	2,800	0.35	0.27	6.30	11.9
	Barnard's Star	2,700	0.15	0.18?	53.3	13.2
WHITE DWARFS	Sirius B	5,000?	0.022	0.99	90,000	11.4
	40 Eridani B	5,000?	0.018	0.41	71,000	11.2
	Van Maanen's Star	5,000?	0.007	0.14?	47,000	14.2

ness establishes its surface area and its diameter. The table above compares the size and other properties of the Sun relative to other stars.

The mass of a star can be determined only by the gravitational effects of one star on another. Since about half of all stars are found in systems where two or more stars orbit one another, we have been able to estimate the masses of many hundreds of stars. The most luminous stars are typically the most massive.

TYPES OF STARS

The properties of stars fall into four distinct categories, or sequences. Most of the stars in the neighborhood of our Sun belong to the *main sequence*, which is a regular series that runs from hot, bright stars, down through cool, faint ones. There are many more stars at the faint end of this sequence than at the bright end. For every luminous, massive star, there are millions of faint ones.

Red giants form another distinct group of stars that are cooler, but much larger in radius, than the main-sequence stars. The *supergiants* are even larger and more luminous than giant stars, but are very rare. In contrast, stars called *white dwarfs* are so faint that they are extremely difficult to find, even though they may be more common than stars like our Sun.

SOURCES OF STELLAR ENERGY

The Sun radiates energy at the rate of 3.8×10^{23} kilowatts. Ordinary processes of combustion or gravitational contraction cannot account for the continuous release of so much energy over the 4.5 billion years that the Sun has been shining. Scientists have learned, however, that the energy that streams out of the Sun is due to the fusion of hydrogen atoms into helium atoms.

A variety of similar nuclear reactions play a role in the energy production in stars. While some stars have nuclear reactions involving heavier elements such as

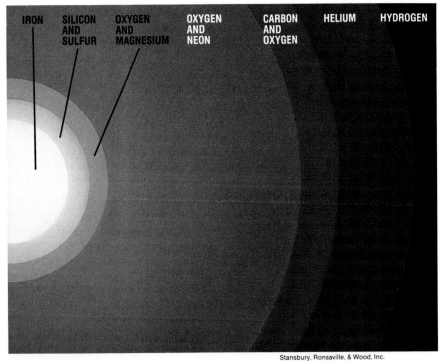

IRON SILICON AND SULFUR OXYGEN AND MAGNESIUM OXYGEN AND NEON CARBON AND OXYGEN HELIUM HYDROGEN

A giant star begins as a sphere of mostly hydrogen atoms. For most of the star's 10-million-year life, the hydrogen atoms fuse to form helium, releasing energy that makes the star shine. In its dying stages, the star builds up layers of other elements. Helium in the core is converted to carbon and oxygen; further conversions produce still heavier elements until at the very end, the core is pure iron. Since iron cannot produce energy, the stellar fires in the core eventually die out.

Stansbury, Ronsaville, & Wood, Inc.

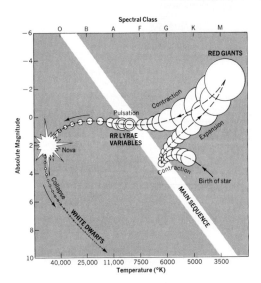

This Hertzsprung-Russell diagram illustrates the changing spectra and magnitude of the various stages of a star's life. When cosmic material condenses to form a star, it falls in the main sequence range. Near the end of its life, a star expands to a red giant, then contracts again, sometimes erupting as a nova. Eventually it becomes a white dwarf and dies.

carbon and nitrogen, the principle is the same: light elements—such as hydrogen, with an atomic mass of 1—combine into heavier ones—such as helium, with an atomic mass of 4—liberating energy.

THE EVOLUTION OF STARS

A star's life is a constant fight between the power of gravity—trying to pull all the star's material toward its center—and various outward forces, which resist that pull. Stars begin their existence as condensations of cool gas within dense clouds of gas. Gravitational forces overpower the weak gas pressure, and the clumps contract into protostars. When the interior of a contracting star reaches the temperature at which the hydrogen-to-helium fusion begins, its contraction is stopped by the resulting internal pressure of the heated gas or the outpouring of radiation.

The star then takes its place on the main sequence. If it is of high mass, it will have already become very hot at the surface, perhaps 90,000° F (50,000° C). A star like the Sun has an initial main-sequence temperature around 10,000° F (6,000° C). The important fact is that the basic characteristics of a star—its temperature, absolute brightness, and lifetime—depend primarily on its original mass.

The lifetimes of stars vary widely. The mass-luminosity law states that a star's luminosity is proportional to nearly the fourth power of its mass. Thus, a main-

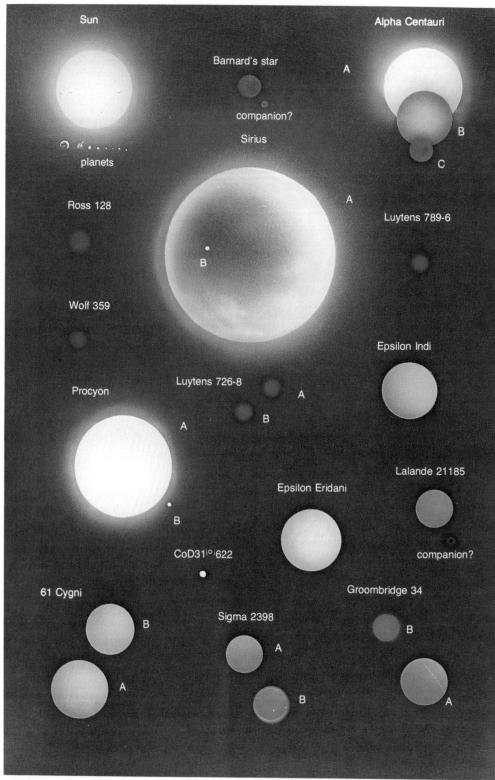

Very few stars are completely isolated. Scientists estimate that half of all stars have planetary or stellar companions. The farther away a star is, the harder it is to identify its companions. Pictured above are the relative sizes and spectral differences of our Sun and its closest neighbors.

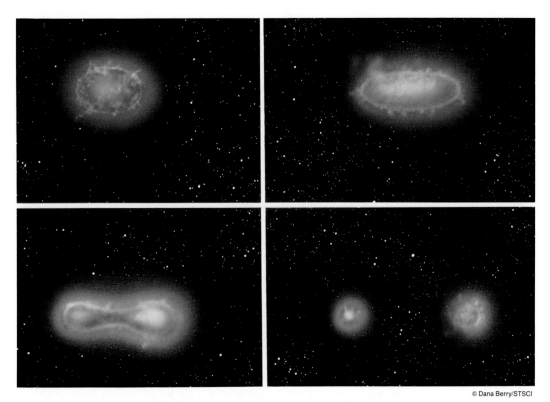

One theory of binary-star origin suggests that gravity causes a star to shrink and increase its rotation speed. Eventually the star, its equator drawn out into an ellipse, assumes a dumbbell shape that finally divides.

sequence star with 10 times the Sun's mass uses up its initial supply of hydrogen about 10,000 times more rapidly than does the Sun. At the Sun's rate of consumption, calculations suggest that its store of hydrogen will last about 10 billion years; about half this interval is already past. The most massive stars known have about 50 times as much mass as the Sun, are about 1 million times more luminous, and will live only 100,000 years.

When a star has used up its available supply of hydrogen, it can no longer generate energy by hydrogen-to-helium reactions. However, the star can release gravitational energy by contracting. As it does so, its internal temperature rises again. And at this higher temperature, the star can use helium as fuel, fusing it into still heavier molecules, such as carbon. It has a new lease on life. A new equilibrium is established, with the star's surface being cooler, but much larger in size. No longer on the main sequence, the star has become a red giant. After a few million years as a giant star, the nuclear-energy resources of the star are almost entirely depleted. What happens next depends upon the star's mass.

Relatively small stars whose mass is no more than 1.4 times the mass of the Sun will gradually contract and grow fainter and denser. In these stars the pull of gravity is balanced by the pressure of electrons resisting being squeezed too closely together. These stars wind up as *white dwarfs*, which will continue to glow until they can contract no longer. Then they will cease to give off light. More-massive stars suffer a more violent fate, ending their lives in massive explosions called *supernovas*.

IS THE NUMBER OF STARS INFINITE?

How many stars are there? The answer is a complex one, for space is not filled uniformly with stars. The stars that we see with the unaided eye, and most of those that we are able to photograph, are parts of a system of stars known as the Milky Way system—our own galaxy. The Milky Way is so large that light, speeding at 186,000 miles

(300,000 kilometers) per second, may take as long as 100,000 years to cross from one end of the galaxy to the other. Yet the Milky Way is finite. It is a flattened, disk-shaped system, whirling like a pinwheel. It consists of a densely populated center wreathed with coiled spiral arms, rich in stars, gas, and dust. It contains about 100 billion stars.

However, there are billions of other galaxies outside our own—each one containing tens of billions of stars. The collective light of the stars in distant galaxies is observable by telescopes only as a faint blur. The total number of stars is indeed enormous, numbering perhaps more than a trillion million. Yet, as far as we know, the number is finite.

THE EVOLUTION OF THE MILKY WAY

From careful observations, we know that some stars are very young, only a few million years old, while others are much older, perhaps 10 billion to 15 billion years old. Since spectroscopic data show that the oldest stars contain only small quantities of the heavy elements, we believe that the first stars in the Milky Way formed out of the simplest materials in the universe, hydrogen and helium. The carbon, oxygen, silicon, iron, and other elements created within these stars were strewn about the galaxy by stellar winds and by supernova explosions, enriching the interstellar gas out of which subsequent generations of stars have formed. Later stars added still more heavy elements. As this process continued over billions of years, enough heavy elements were eventually manufactured so that stars like our Sun, along with its surrounding planets, and the living creatures found here on Earth, could come into being.

Most modern scientists dismiss the ancient notion that the positions of the stars and planets influence the course of human events. However, scientists have discovered a deeper bond between us and the stars. The carbon, nitrogen, oxygen and all the other elements necessary for life as we know it were processed through the nuclear furnaces within distant stars. Thought of in these terms, we are really only stardust.

A star is born when the density of interstellar material is significantly increased and is pulled inward toward the center by gravity. It typically takes several hundred thousand years to produce a star of one solar mass.

© Rob Wood/Stansbury, Ronsaville, & Wood, Inc.

Astronomers believe that stars continuously form, evolve, expand, and, eventually, contract and die out.

COLLAPSED AND FAILED STARS

by Charles Beichman

When a star is born, it is essentially a huge ball of hydrogen atoms mixed with a smattering of other elements. The strong pull of gravity in the star's core creates levels of heat and pressure high enough to cause the hydrogen atoms to fuse, forming helium. These nuclear fusion reactions generate the energy that balances the inward pull of gravity. Eventually, however, the supply of hydrogen atoms runs out. When this happens, the star, depending upon its size, either contracts and dies or violently explodes.

SUPERNOVAS

The explosion of a star, called a supernova, is among the most violent events imaginable. What causes a single star, erupting unexpectedly, to outshine 100 billion ordinary stars, and then disappear?

Novas

The most common—and least violent—stellar eruption is called a *nova*, from the Latin word for "new." Novas are thought to occur in binary systems in which a white-dwarf star orbits a red-giant star. Material from the red giant is pulled by gravity onto the white dwarf. The material slowly accumulates on the white dwarf's surface for years or thousands of years, until the pressure and temperature surge high enough to trigger a sudden outburst of nuclear reactions. The star throws off a shell of incandescent material and, in doing so, increases its brightness a thousandfold. After a few months, the star fades back to its former anonymity. Since many novas recur again and again, astronomers believe that the process does not destroy either star. Nevertheless, one wouldn't want to live too close to such a celestial volcano.

Supernovas

A star comes to the end of its life when the energy thrown off by its nuclear furnace can no longer resist the inward pull of gravity. At this point, stars up to 1.4 times the mass of the Sun contract into white-dwarf stars that cool and slowly fade from view. More-massive stars, however, feel a greater gravitational pull than can be resisted; instead of fading out, they erupt in explosions of unimaginable violence.

Numerous supernovas have been seen in historical times. The "guest star" observed by Chinese astronomers in 1054 created the Crab nebula now visible in the constellation Taurus. European astronomers Tycho Brahe and Johannes Kepler observed supernovas in 1572 and 1604. Rock paintings suggest that American Indians also marveled at one or more of these events. Judging by the frequency of supernovas in our own and other galaxies, we infer that a typical galaxy experiences a supernova explosion about every 100 years. The most recently observed was a bright supernova sighted in a neighboring galaxy in 1987.

When the Center Will Not Hold

With the use of supercomputers, astronomers have developed models that detail the last minutes of a dying star. As nuclear fuel is depleted, the star burns its own nuclear ashes, breeding and then con-

A supernova occurs when a massive star can no longer resist the inward pull of gravity and explodes, increasing the star's brightness by a factor of many billions. Below, photos of the same portion of the sky, taken 13 years apart. The photo on the right shows the emergence of a supernova just below the galaxy at center.

Hale Observatories

Supernovas are relatively rare phenomena in terms of human life spans. So astronomers rejoiced when a supernova blazed into view in February 1987 (above).

suming progressively heavier elements such as oxygen, magnesium, sulfur, and silicon. Each element requires higher temperatures and higher pressures to fuse into still heavier elements. But the star is doomed—these nuclear fires can stave off collapse for only a short time. The nuclear furnace eventually converts the star's core into iron, which can neither fuse nor fission to produce additional energy.

Suddenly the final and devastating collapse begins. In an instant the temperature at the core rises billions of degrees, and the iron core breaks up into elementary particles like neutrons, protons, and neutrinos. The collapse ends when the inner core reaches densities characteristic of an atomic nucleus. The sudden rebound, or bounce, of the infalling core sends some of the inner core outward, to crash into the still collapsing outer core. The collision makes the explosion even more violent. Heavy elements in the outer skin of the star and those created in the blast wave are strewn throughout space at speeds of nearly 1,000 miles (1,600 kilometers) per second. These heavy elements are the final legacy of the dying star, an inheritance for subsequent generations of stars that will ultimately shine in the heavens.

Type I and II Supernovas

Astronomers have observed two types of supernovas. Type I explosions may simply be violent versions of the nova explosions described earlier, in which a companion star dumps material onto a white-dwarf star. In some cases the amount of material falling onto the compact white dwarf may be so great, and the rate of infall so rapid, that the star suddenly exceeds the 1.4 solar masses permissible for white

Although not in the same league as a supernova, the light output of a nova explosion can still exceed 100,000 suns for several days. In the photo above, the arrow points to the star, now known as Nova Aquilae 1918, as it appeared before the outburst that made it a nova. The facing page photo shows how much brighter the star became after the explosion. Experts estimate that about 40 novas occur a year.

Some astronomers speculate that the fabled star of Bethlehem, which heralded the birth of Jesus Christ (left), may have actually been the magnificent display of a supernova.

dwarfs. The star collapses and explodes. Type I supernovas are mostly found in parts of galaxies where older stars are located.

Type II explosions come from the collapse of a massive star at the end of its short life. Such supernovas are found in the disks and spiral arms of galaxies containing massive, young stars. These stars consume their resources so rapidly that they live for only a few million years before meeting violent deaths.

Supernova Remnants

The blast from a supernova rushes outward into space, sweeping up surrounding material, heating and compressing it. These telltale signs of an explosion, recognized by their peculiar optical, radio, or X-ray patterns, are known as supernova remnants. Astronomers have identified more than 100 remnants of supernovas in the Milky Way alone. Careful examination of the light from supernova remnants shows they are abundant in hydrogen, helium, and heavy ele-

Yerkes Observatory

ments. The abundance of these elements in our own solar system has led some astronomers to suggest that the blast wave from a nearby supernova explosion triggered the collapse of the cloud of gas that eventually became our Sun.

Our ideas about supernovas were dramatically and unexpectedly confirmed in 1987, when a star exploded in a nearby galaxy, the Large Magellanic Cloud. Telescopes sensitive to infrared radiation and energetic gamma rays confirmed the presence of elements such as iron and cobalt in roughly the amounts expected from a supernova. Scientists also detected a wave of elusive subatomic particles, which were picked up by giant underground instruments. The number of neutrinos detected was consistent with theories about the physical conditions within the core of a supernova at the moment of its collapse.

PULSARS

Surprisingly, a star is not completely destroyed in a supernova. Some of the core survives, held together by the nuclear forces that bind atoms. These supernova cores are known as *neutron stars*. A neutron star is born when a supernova explosion compresses the subatomic particles in a star so tightly that a cube the size of a sugar lump would weigh 100 million tons! Such stars can be as small as 20 miles (32 kilometers) in diameter.

Rapidly spinning neutron stars are often wreathed in strong magnetic fields that give rise to intense, narrow beams of radiation that sweep across the sky like lighthouse beacons. These periodic bursts of radiation were first detected in 1967, when a British graduate student, Jocelyn Bell, and her colleagues detected regular pulses of radio waves sweeping toward Earth every second or so from specific directions in space. They announced their discovery only after ruling out interference from radio stations and the possibility of messages from extraterrestrial life.

Other astronomers soon explained the signals in terms of a swiftly rotating neutron star, and coined the name *pulsars*. Since that time, some 500 pulsars have been found. Their rotation periods range from a few seconds to a few thousandths of a second. Some pulsars are highly accurate clocks, their cadence hardly varying.

BLACK HOLES

The most massive stars may collapse so completely that not even a neutron star can survive. These dying stars may become black holes—stars so dense that even light cannot escape from their surface. The surface of a black hole containing the mass of

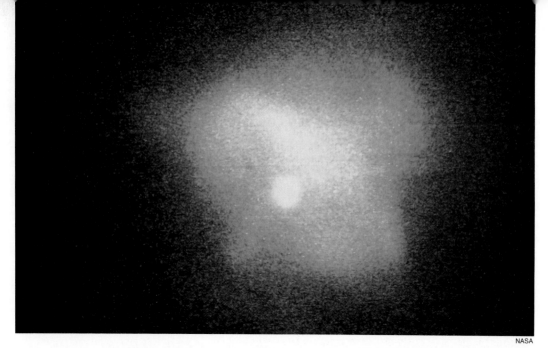

The Crab nebula supernova that exploded in A.D. 1054 contains a stellar remnant known as a pulsar (large white spot, above). Spinning 30 times a second, this pulsar emits regular signals once per revolution.

the Sun would be only 2 miles (3 kilometers) across. The physics of such objects is poorly understood, but it is thought that the X-ray emissions detected from some celestial objects come from material streaming into black holes. As the black hole forcefully sucks material away from nearby stars, this material is heated to millions of degrees, emitting more- and more-energetic radiation as the gas spirals into oblivion (see also page 204).

BROWN DWARFS

Astronomers reserve the name "star" for an object that is massive enough to convert hydrogen into helium as its primary source of energy. The planet Jupiter, which contains only one-thousandth of the mass of the Sun, may have almost become a star. Even now the planet gives off more energy than it receives from the Sun, as it slowly shrinks and cools, converting gravitational energy into heat radiation.

But Jupiter is not a star, since its central core will never reach the extremes of temperature and pressure necessary for stellar birth. Computer calculations show that a star must be at least 0.08 the mass of our Sun before it will convert hydrogen into helium, and thus begin nuclear fusion. Objects smaller than this can never become true stars.

These "failed" stars are known as brown dwarfs. Astronomers have yet to actually see a brown dwarf, with the possible exception of Jupiter. But they have theorized about their appearance and number.

Brown dwarfs may start out being just as hot as the smallest stars, with temperatures about 3,600° F (2,000° C). But lacking an inner fire, brown dwarfs cool. Over tens of millions of years, they may drop to around room temperature! At this level of coolness, they would emit only one-millionth the power of the Sun.

In contrast to any other type of star, brown dwarfs don't even glow red-hot. So it's not surprising that they have remained invisible to telescopes operating in the visible-light spectrum. Astronomers expect that brown dwarfs, if and when they are sighted, will likely be faint sources of infrared radiation. They may be found orbiting larger stars, just as Jupiter orbits our Sun, or floating freely in interstellar space. Many astronomers continue to search for brown dwarfs, but nobody has yet found a convincing candidate. The coolest, faintest objects seen so far have proven to be hydrogen-burning stars with masses just around the 0.08-solar-mass cutoff.

Jupiter (right) may be a "failed" star known as a brown dwarf. It has a starlike composition and does radiate some energy, but its central core is not large enough for the nuclear fusion common to true stars.

Brown dwarfs may be an important but so-far-undocumented population within our own galaxy. Imagine how distorted a census of a city would be if the census takers ignored all people shorter than 6 feet tall! Some astronomers suspect that the number of known stars in the galaxy will remain seriously underestimated until we can detect brown dwarfs.

The greatest evidence of brown dwarfs is the estimated mass of our own galaxy. Astronomers estimate this mass much like accountants trying to balance a checking account. On one side of the ledger, astronomers estimate the gravitational pull in a region of space by measuring the motions of nearby stars. On the other side of the ledger, astronomers total up all the mass they can see: stars of all types, plus interstellar gas. But somehow the columns in this ledger don't balance, leaving scientists to speculate about "missing mass."

Judging from the movements of stars, which reflect the gravitational forces acting on them, the galaxy may contain twice as much mass as we can see! Such a conclusion remains controversial, however.

The missing-mass problem also exists on larger scales as well. The motions within many other galaxies demand the existence of 10 times more material than can be bound up by the stars and gases that astronomers have observed.

Brown dwarfs are not the only possible explanation, however. Large numbers of subatomic particles might provide the gravitational tug necessary to keep galaxies spinning properly. Plans for detailed infrared surveys of the sky, scheduled for the 1990s, may reveal many brown dwarfs and at last settle the argument over whether they are indeed the most numerous objects in the universe.

The supernova in the constellation Vela contains a pulsar that emits optical- and gamma-ray frequencies.

Top: JPL; right: NASA

BLACK HOLES

by Mark Strauss

"Abandon all hope, ye who enter here," would be an appropriate warning for any space traveler foolish enough to approach a black hole. Black holes are proposed by astrophysicists as regions of space where gravity is so strong that the black holes act like stellar vacuum cleaners, sucking in matter and energy from space and allowing nothing, not even light, to escape.

The American physicist John Wheeler coined the term "black hole" in 1969, but, in fact, the theory has been around for much longer. As far back as 1783, English astronomer John Michell suggested that if a star were massive enough, it would have such a strong gravitational field that any light leaving the star would immediately be dragged back to the star's surface.

Michell's theories were largely ignored until 1939, when physicists Robert Oppenheimer and Hartland S. Snyder demonstrated that, based upon Albert Einstein's general theory of relativity, it would be possible for a star to collapse to the point where it would become a black hole.

HOW A STAR AGES

In order to understand how a star could collapse into a black hole, it is first important to understand the life cycle of a star. A star is, essentially, a giant fusion reactor. At the central core of the star, swirling atoms of hydrogen gas collide with one another and merge to form helium. In the process of fusing together, these hydro-

Astronomers propose that when a massive star dies, it collapses to form a black hole. This tiny, dense area has a surface gravity strong enough to suck any nearby light and matter into the hole, never to escape.

gen atoms release a tremendous amount of energy in the form of heat.

At the same time, the star as a whole is continuously struggling against the inward pull of gravity. The inward gravity is from the central core of the star, which is surrounded by a massive envelope of gas. This inward pull is so immense that the star is always on the verge of collapsing under its own weight.

What prevents the star from collapsing? Tremendous internal pressure that is generated by the extreme heat at the star's core, which pushes outward, counterbalancing the inward pull of gravity. In our own Sun, for example, the temperature at the core is about 27,000,000° F (15,000,000° C), generating pressure 100 billion times the air pressure at sea level on Earth.

After thousands of millions of years, however, a star comes to the end of its hydrogen fuel supply. It starts to cool and contract. What happens next will depend entirely on the mass of the star.

Small stars, such as our Sun, will collapse to form objects called *white dwarfs*. About the size of the planet Earth, white dwarfs resist further collapse with internal pressure caused by electrons spinning at near the velocity of light. White dwarfs are very dense objects: 1 cubic inch of white dwarf weighs several tons. But they are considered lightweights when compared to *neutron stars*.

Neutron stars are the evolutionary end products of larger stars—those 1.4 to 2 times as large as the Sun. Electrons cannot resist the greater gravitational collapse of such stars, and are pushed into atomic nuclei, where they combine with protons to form uncharged, tightly packed neutrons. Neutron stars are only a few miles in diameter. They weigh about 1 million tons per cubic centimeter. They can resist further collapse only by invoking the strongest force in nature—appropriately called the "strong force"—the force that binds together an atomic nucleus.

X-ray binary stars are composed of a small, normal star (in orange, below) that orbits around an X-ray-emitting invisible companion—possibly a black hole.
Painting by Steven Simpson. Courtesy Sky & Telescope Magazine

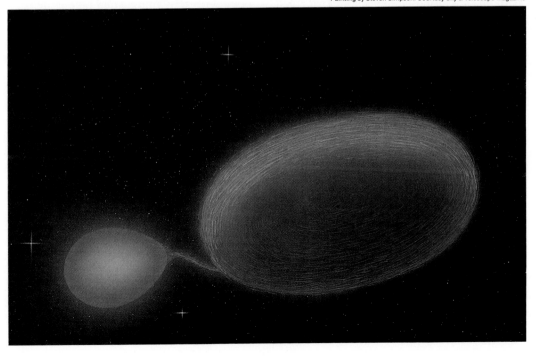

The strong force halts the imploding matter so abruptly—in a tenth of a second—that the collapsed stellar cores act as explosive charges. The resulting explosion in the star's outer regions is called a *supernova*. Such celestial fireworks, observed by Chinese astronomers in July 1054, produced the Crab nebula, a cloud of gas that still writhes and glows today, 6,000 light-years from Earth.

What happens to a dying star that is more than twice as large as the Sun? Even the strong force cannot halt its in-falling momentum. It collapses completely, beyond the neutron-star stage, to an even smaller, denser object. Back in 1939 Oppenheimer and Snyder calculated that the gravitational field at the surface of such an object would become so strong that even light (traveling at a speed of 186,400 miles —300,000 kilometers—per second) would be unable to escape. According to Einstein's theory of relativity, nothing in the universe can travel faster than light. Therefore, if light cannot escape, neither can anything else. The collapsed star becomes what we call a black hole.

The best way to visualize a black hole is to imagine, for a moment, that space is a flat rubber sheet. If you were to drop a steel ball on the sheet, the rubber would curve downward, forming a shallow hole. This, in a nutshell, is how Einstein interpreted gravity. According to Einstein, gravity exists because massive objects bend the fabric of space around them. If, for example, we rolled a small marble across our rubber sheet, it would roll around the top of the hole formed by the steel ball, much in the way that the Earth orbits around the Sun.

Now imagine that we could increase the weight of the steel ball that we dropped on our rubber sheet. As the weight increased, the ball would sag farther and farther downward, creating a deep "gravity" hole. Eventually the rubber would be stretched so tight that the top of the hole would pinch together, closing off from the outside world the region containing the steel ball. Similarly, a collapsed star could eventually become so dense that it would curve space completely around on itself, isolating it from the rest of the universe.

How far would a star have to collapse before it "disappeared" from the visible universe? Astronomers refer to that critical size as the "event horizon" (otherwise known as the "Schwarzschild radius," named after German physicist Karl Schwarzschild). The event horizon is the outer boundary of a black hole, the exact point at which light rays fail to escape. The horizon acts as a one-way membrane—light and matter can cross the horizon into a black hole, but once inside, the horizon can never be recrossed.

The size of the event horizon is proportional to the mass of the collapsing star. Typically, the event horizon of a star would be on the order of miles. (For example, a star 10 times as massive as our Sun would have a Schwarzschild radius of 18.6 miles— 30 kilometers.)

Yet, according to our current knowledge of theoretical physics, once a star starts collapsing, no known force can stop it. It will continue to shrink past its event horizon, smaller and smaller, until it becomes a "singularity"—a mathematical point with zero volume and infinite density. This singularity lies at the very center of a black hole.

EXPLORING A BLACK HOLE

If an astronaut were to attempt to visit a black hole, it would be a one-way trip. Long before the astronaut even arrived at the event horizon, he would encounter tremendous tidal forces exerted by the black hole. Imagine, for example, that the astronaut is falling feetfirst toward the hole. The gravitational force pulling on his legs would be considerably stronger than the gravitational force pulling on his head. The difference between those two forces would stretch the astronaut like a piece of taffy.

As if that weren't bad enough, every single atom in the astronaut's body would be pulled toward the singularity at the black hole's center. For the astronaut the sensation might be similar to being squeezed by a giant fist.

The Uhuru satellite (right) was the first satellite designed for X-ray astronomy. It found the star Cygnus X-1 (above), a binary system located about 6,000 light-years from Earth. Astronomers believe that Cygnus X-1 may contain a black hole.

After being stretched and squeezed by the black hole's gravitational forces, our intrepid space traveler would resemble a strand of spaghetti, and would likely not be in the mood for any further exploration.

Let's imagine for a moment that we could instead send a robot probe to investigate the black hole, one that could somehow stay intact despite the tremendous tidal forces. For the sake of discussion, we will mount a clock and a light source on the outside of our probe.

If we were watching the robot back on Earth, we would notice a curious phenomenon. The light source mounted on the side of the probe would start to change color. If the light, for instance, started out green, it would turn yellow, and then red as it got closer and closer to the event horizon of the black hole.

This is because light is composed of particles known as *photons*. As the photons move away from the black hole, they expend some of their energy as they try to escape from the hole's tremendous gravitational pull. The closer they are to the event horizon, the more energy they need to pull away.

BLACK HOLES

The energy of a photon is proportional to the frequency of its radiation. As a result, light that loses energy will have a reduced frequency, and therefore a longer wavelength. This effect is known as "gravitational redshift." When light has a long wavelength, it is red in color.

Eventually, as the robot probe moves closer and closer to the event horizon, the light source will seem to disappear from view. The wavelength of the light will have become so long that it can only be detected with infrared and radio telescopes.

Just above the event horizon of the black hole, the wavelength of the light will approach infinity. Theoretically, radiation from the light source would still reach us back on Earth, but by then the wavelengths would be so long that no known scientific instruments would be able to detect them.

Meanwhile, the clock mounted on the side of our robot probe would also be behaving rather oddly. According to Einstein's theory of relativity, time slows down in the presence of a strong gravitational field—at least as viewed by an outside observer. As the probe got nearer and nearer to the black hole, astronomers back on Earth would notice that the clock was ticking more and more slowly.

The clock would continue to slow down, until the probe arrived at the event horizon, at which point the clock would stop altogether. The probe would appear frozen in time, hovering at the brink of the black hole for the rest of eternity.

Relativity predicts, however, that from the perspective of the robot, time would not seem to be affected in any way. The probe would arrive at the event horizon and enter the black hole without the clock slowing down for even an instant. Yet our dutiful robot explorer would have only a fraction of a second to contemplate this peculiar law of nature, at which point it would be pulled toward the center of the black hole, where it would encounter the singularity and be crushed to infinite density.

PROVING THAT BLACK HOLES EXIST

All of this might sound very strange, and, in fact, for many years the majority of astronomers and physicists were reluctant to believe it. (The prominent English astronomer Sir Arthur Eddington even declared

English physicist Stephen Hawking theorizes that black holes can emit radiation in the form of subatomic particles that do not obey the traditional laws of physics. These "virtual" particles are created in pairs in outer space, where they collide and annihilate each other. But if a pair of virtual particles is created in the vicinity of a black hole, one particle is sucked into the black hole, while the other particle escapes into space as "Hawking radiation." Over time, as the black hole loses energy, it may shrink, perhaps eventually evaporating with an enormous burst of energy.

The physical processes that occur when a large star collapses to form a black hole (above) may be the same processes that would be involved in the collapse of the universe.

that there must be "a law of Nature to prevent a star from behaving in this absurd way!") If astronomers were to believe in black holes, they wanted more than just mathematical equations on a blackboard; they wanted hard, physical evidence.

Such evidence became available in 1967, when two British astronomers, Jocelyn Bell and Antony Hewish, discovered objects in space that were emitting regular pulses of radio waves. At first the astronomers thought that they had made contact with an alien civilization in a distant galaxy. They even named the objects "LGMs," for Little Green Men. Eventually, however, astronomers came to the conclusion that the objects were rotating neutron stars, emitting radiation in the form of narrow beams. Like a celestial lighthouse, each time the neutron star spun toward Earth, astronomers could detect a pulse. Hence, these objects were named *pulsars*.

This was the first hard evidence that neutron stars actually exist. If a star

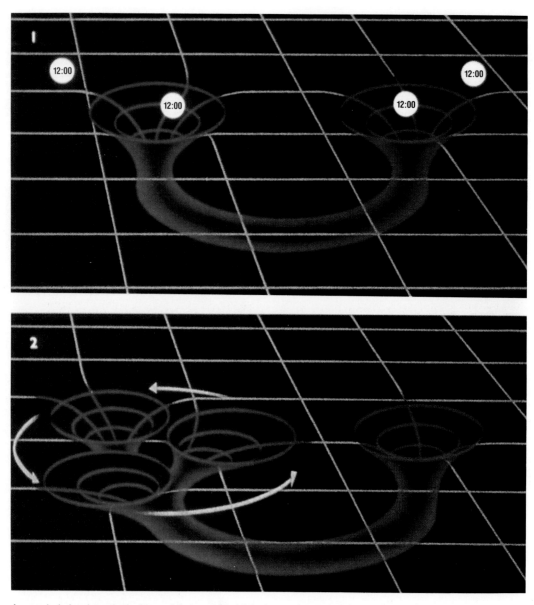

A wormhole is a hypothetical tunnel that could hold the key to time travel. It is envisioned as a black hole that has two "mouths" connected by a "throat." Picture a wormhole where the two mouths lie only one hour apart by spaceship (top). Before time travel can occur, one of the wormhole's "mouths" would have to be accelerated (above) to near the speed of light for one hour (starting at noon, for argument's sake).

could collapse into an object as small as a neutron star, it then seemed reasonable to assume that it could collapse to an even smaller size and become a black hole.

One problem remains. How do you find a black hole? They aren't as accommodating as neutron stars, in that they don't emit easily detectable beams of radiation. In fact, according to conventional theory, black holes don't emit anything at all.

Astronomers saw a way out of this dilemma. Black holes exert an enormous gravitational force on nearby objects. So although scientists can't see a black hole

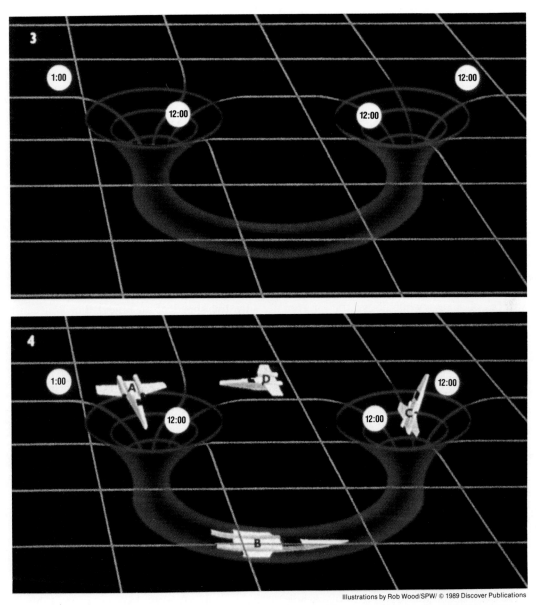

A clock outside the wormhole now reads 1:00 P.M., while clocks inside the wormhole and outside the stationary mouth are still at noon (top). The wormhole is now ripe for time travel. If a spaceship (above) plunges into the accelerated mouth (A), it can speed through the wormhole in an instant (B), emerge from the stationary mouth just seconds after noon (C), and return to the other mouth the long way around (D), in time to see itself going in.

"in the flesh," so to speak, they can observe how it would affect its surrounding environment.

To date, binary-star systems offer the best hope for locating a black hole. Astronomers have detected many such systems, where two stars orbit around one another. In some cases the astronomers have observed only one visible star, which seemed to be in orbit around an unseen companion. It is possible that the companion might be a star too faint to be seen from Earth. It is also possible, however, that the second object could be a black hole.

If a black hole were part of a binary-star system, its enormous tidal forces would pull gaseous material off the surface of the neighboring star. Like water draining out of a bathtub, the gaseous material would slowly spiral into the black hole, forming a swirling disk of gas around the event horizon, a phenomenon that astronomers refer to as an *accretion disk*.

Within the accretion disk, compression and internal friction would heat the gas to temperatures as high as 1,800,000° F (1,000,000° C). When gas gets this hot, it radiates tremendous energy in the form of X rays detectable by astronomers.

In 1970 a United States artificial satellite, the Uhuru, was launched off the coast of East Africa. (*Uhuru* is the Swahili word for "freedom.") Its purpose was to detect sources of X rays while above the interference of the Earth's atmosphere. Uhuru has found over 100 stars emanating X-ray pulses. One of the most powerful X-ray sources was Cygnus X-1, located about 6,000 light-years from Earth.

Closer examination of Cygnus X-1 revealed it to be a binary-star system, with a supergiant star orbiting around an unseen companion. By measuring the velocity and the orbital period of the supergiant star, astronomers were able to roughly calculate the mass of the unseen object. The object was estimated to be at least six solar masses (six times the mass of the Sun), far too massive to be either a white dwarf or a neutron star. By 1974 astronomers concluded that Cygnus X-1 must contain a black hole.

In recent years, other X-ray sources have been added to the roster of black-hole candidates. In 1983 astronomers discovered LMC X-3, a binary-star system in the Large Magellanic Cloud, one of our own galaxy's companion dwarf galaxies.

Another suspected black hole was found by the Astro Observatory, carried into orbit by the space shuttle *Columbia* in December 1990. The observatory's X-ray telescope detected high levels of X-ray radiation emanating from galaxy Markarian 335. Astronomers believe that the X rays might be coming from an accretion disk of superhot gases spiraling into a massive black hole at the center of the galaxy.

Perhaps the best black-hole candidate yet discovered is a binary-star system that goes by the uninspiring name A0620-00. Like Cygnus X-1, A0620-00 emits intense levels of X-ray radiation. The binary system has a visible orange dwarf star, which orbits around a dark, unseen mass. In the late 1980s, astronomers studied the motions of the orange dwarf and estimated that the star's dark companion was 3.2 times the mass of our Sun.

With a mass that large, the dark object was placed high on the black-hole suspect list. But in order to get a more accurate estimate, astronomers would have to measure the velocity of the dark object. At first that seemed impossible: how can you measure the speed of an object that you can't even see?

Then, in 1990, astronomers Carole Haswell and Allen Shafter found an accretion disk in A0620-00, surrounding the suspected black hole. By measuring the motion of the accretion disk—assumed to move every time the black hole moved—the astronomers concluded that the dark object was nearly four times the mass of the Sun, perhaps even greater. So far, no other suspected black hole has been measured with such accuracy.

HAWKING RADIATION

English physicist Stephen Hawking has suggested that radiation might not only exist in the vicinity of a black hole, but that it actually might be leaking from the black hole itself. Energy leaking from a black hole? It sounds impossible. But Hawking says that black holes emit radiation in the form of subatomic particles that do not obey the traditional laws of physics. Such "virtual" particles, as Hawking calls them, can be created in pairs in empty space, only to instantly collide and annihilate each other. But if such a pair were to come into being in the vicinity of a black hole, one particle would be sucked into the black hole, while the other would escape into space as "Hawking radiation."

As a black hole loses energy, it would also lose a proportionate amount of mass. Hawking's theory suggests that there might come a time when a black hole will lose so much mass that it will no longer be able to curve space around itself. The black hole would cease to be a black hole, and the remaining mass would likely explode outward, with a force equivalent to millions of hydrogen bombs.

But don't look up in the sky expecting to see a fireworks display of exploding black holes. A large black hole lives a very long time. More specifically, it would take trillions upon trillions of years for it to lose enough energy to explode outward. The universe itself has been around for only 20 billion years.

Yet it may be possible that very small black holes, formed in the early days of the universe, might be exploding just about now, releasing energy in the form of gamma rays, equivalent to about 100 million volts of electricity.

Astronomers are now searching the skies for just such bursts of gamma radiation. If found, then astronomers could verify what Stephen Hawking has been saying for the last 20 years: "Black holes ain't so black."

TIME TRAVEL

In 1895 H. G. Wells wrote a book about a device that could carry a man back and forth through time. The book was called *The Time Machine*, and for a century after it was published, the concept of time travel remained a favorite topic among writers of science fiction.

In 1988, however, science fiction moved closer to becoming science fact when American physicist Kip Thorne and his colleagues at the California Institute of Technology (Caltech) published a paper in the prestigious journal *Physical Review Letters*, titled "Wormholes, Time Machines, and the Weak Energy Condition."

Thorne didn't actually publish a blueprint for a do-it-yourself time machine. He speculated that an "arbitrarily advanced civilization" might be able to find a loophole in the laws of physics that would allow individuals to travel through time.

The loophole that Thorne had in mind is what physicists call a "wormhole." A wormhole is similar to a black hole, but with one noteworthy difference. At the bottom of a black hole, there is a singularity, a mathematical point of infinite mass through which nothing can pass.

A wormhole, by contrast, has no bottom. It has two "mouths" connected by a "throat." It is, essentially, a tunnel through space. A space traveler entering one mouth of a wormhole might emerge from the second mouth only a few seconds later, but halfway across the galaxy.

Time travel could leave the realm of science fiction if wormholes are found to really exist. Physicists are worried, however, about the possible ramifications if time travelers are able to tamper with events in either the past or the future.

Einstein's equations predict that wormholes exist, although nobody has ever found one. American physicist John Wheeler has suggested that a good place to look for one would be at a submicroscopic level, where random fluctuations occur in the fabric of space-time. In such an environment, wormholes would spontaneously appear and collapse, giving space a frothy, foam-like appearance.

Kip Thorne suggests that an advanced civilization could pull a wormhole out of this foam, enlarge it, and then move its openings around the universe until the wormhole assumed a desired size, shape, and location. Unfortunately, once such a wormhole was created, it would be highly unstable. If a space traveler entered the wormhole, the throat might instantly pinch shut. Even moving at the speed of light, the space traveler might be unable to reach the other side of the wormhole before it collapsed around him or her.

In order to avoid such a catastrophe, the Caltech physicists recommend that our hypothetical advanced civilization thread the throat of the wormhole with what they call "exotic material." In order to prop open a wormhole a half a mile or so across, the material would have to possess a radial (outward) tension comparable to the pressure at the center of a neutron star. Kip Thorne believes that there is a 50–50 chance that the laws of physics permit such a substance to exist.

Once our "arbitrarily advanced scientists" finished building a safe, traversable wormhole, they would be ready to convert it into a time machine. At this point they would rely upon Albert Einstein's general theory of relativity. According to Einstein, time slows down for a moving object when it is measured by a stationary observer.

This is often illustrated with what is known as the "twin paradox." Imagine that you have twin brothers, named Bill and Ted, each 20 years old. Bill takes off in a spaceship, while Ted stays back on Earth. Bill's destination is a star 25 light-years away. (A light-year is the distance that a beam of light can travel in one year.) His spaceship can attain a speed of 99.9 percent of the speed of light. From Ted's point of view, Bill will be gone for 50 years (25 years to reach the star, plus 25 years to return). However, from Bill's point of view on board the spaceship, the entire trip will last only one year. This effect is known as "time dilation." When Bill returns to Earth, he will be only 21 years old, but his brother Ted will be a 70-year-old man.

Now, instead of two brothers, imagine that we are dealing with two mouths of a wormhole. Our advanced civilization could move one end of the wormhole, perhaps by using a heavy asteroid or a neutron star as a kind of gravitational tugboat. If the mouth of the wormhole were accelerated to a high enough speed and then returned to its original position, it would behave just like our space-traveling twin brother. A clock fixed to the moving mouth would tick more slowly than one at the stationary mouth.

For instance, the clock outside the accelerated mouth might read 12:00 noon, but the clock outside the stationary mouth would read 1:00 P.M. By passing from one mouth to the other, a space traveler could move back and forth through time.

How far could our traveler move through time? That would depend upon how long and how fast the wormhole mouth is accelerated. If the mouth were moved at 99.9 percent of the speed of light for 10 years, the time difference between the two mouths would be nine years and 10 months. Theoretically, if you accelerated a wormhole mouth fast enough and long enough, the time difference between the two mouths could be stretched across several centuries.

There is, however, a limitation to Kip Thorne's time machine. Common sense tells us you cannot travel back to a time before you created the wormhole and accelerated one of the mouths through space. After all, what we're doing is exploiting the relative rate at which time passes under the effects of speed. So, unfortunately, you could not pop back through time to visit the dinosaurs. Unless, of course, you were lucky enough to find a time hole that had already been constructed by an advanced civilization several million years ago.

Quasars are very luminous objects; some shine 100 times brighter than the brightest known galaxy.

QUASARS AND ENERGETIC GALAXIES

by Charles Beichman

Normal galaxies consist of stars and various amounts of interstellar gas and dust. Some galaxies are large, containing 100 times the number of stars in our own galaxy, the Milky Way, while some are smaller, containing 100 times fewer stars than our galaxy.

For 100 million light-years in any direction, we see only normal galaxies, tens of thousands of spirals and ellipticals—some large, some small, some isolated, some in clusters—but all consisting of various populations of stars. At still greater distances, however, a few rare galaxies stand out as particularly bright. The luminosity, or total energy output, of such galaxies is 10 to 10,000 times greater than that of a typical galaxy such as our own. These distant objects, at least 100 times rarer than galaxies like the Milky Way, have some sort of additional source of energy that astronomers are still trying to understand. Intriguing possibilities include massive black holes or enormous amounts of star births.

So-called active, or energetic, galaxies are characterized by a compact nucleus that seems to be responsible for most of their tremendous energy output. Astronomers often detect strong radio, infrared, or X-ray emissions emanating from such galaxies. Such a phenomenon is rare around more-typical luminous objects.

Jets of energetic particles traveling at close to the speed of light are often seen

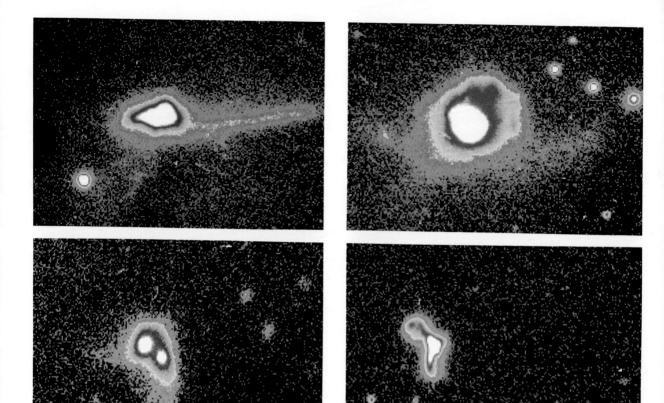

Ultraluminous objects are only found in merging galaxies. The collision and merger processes may be necessary to trigger the huge luminosities. The photos above show galaxies in various stages of merger.

D. Sanders and B.T. Soifer, Caltech

escaping from the nucleus of these galaxies. These particles are often highly ionized atoms being ejected from the galaxy at velocities of thousands of miles per second.

One way to explain many of these characteristics is to suppose that a black hole lies at the center of such an active galaxy. Black holes are thought to be the mysterious end result of stars and collections of stars that have become so massive and compact that no known force can resist the crushing force of their gravity.

A black hole only 4 miles (6 kilometers) across would still have more mass than our Sun. A black hole containing the mass of a billion suns would be no larger than our own solar system! Nothing, not even light, can escape from a black hole. But as gaseous material spirals into the mouth of a black hole, it is compressed and heated to millions of degrees, emitting copious amounts of energy, from X rays to radio waves.

Astronomers have seen the brightness of some active galaxies double in just a few days. Perhaps they were seeing a sudden burst of energy produced when a star falling into a black hole was shredded by the intense gravitational field. Some astronomers even speculate that all galaxies harbor black holes, some of which may occasionally burst into prominence to create an "active" galaxy.

STARBURSTS

Another type of luminous galaxy is the starburst galaxy, so called because most of its excess energy appears to be due to a large number of hot, young stars in the violent process of birth. Such a starburst galaxy may give rise to as many as 100 stars per year. In comparison, a galaxy like our own sees about one star formed per year.

Starburst galaxies are prominent at infrared wavelengths because star formation often occurs within dense clouds of gas and dust, which absorb visible starlight and reradiate the energy as heat. The InfraRed Astronomical Satellite (IRAS) identified more than 50,000 such galaxies. Many starburst galaxies appear to be in the process of colliding or merging with other galaxies. Such collisions between galaxies may actually trigger starbursts.

QUASARS

A "quasi-stellar object," or quasar, represents an extreme type of active galaxy. Quasars, even brighter than active galaxies, are up to 10,000 times more luminous and 1 million times rarer in a given volume of space than galaxies like the Milky Way. Their cores are in constant motion, shooting out plumes of gas at a velocity of several thousand miles per second.

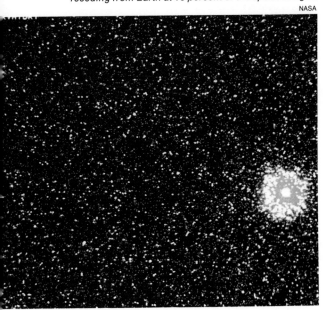

Comparison of the emission lines of the quasar 3C 273 (below) with those of a standard laboratory spectrum reveals that the quasar's lines are shifted to the red end of the spectrum. This indicates that the quasar is receding from Earth at 15 percent of the speed of light.
NASA

Quasars are so named because, when they were discovered in the 1960s, these apparently compact galaxies were hard to distinguish from individual stars. Now, however, we know that what astronomers are looking at are faint galaxies surrounding bright quasar cores. The same mysterious process that powers active galaxies may supply the even greater amounts of energy needed to produce quasars.

Since quasars are the brightest beacons in the night sky, they can be seen at enormous distances, billions of light-years away. But because light travels at a finite speed, telescopes act like time machines, permitting astronomers to view distant quasars as they once appeared billions of years ago. Thus, we see quasars as they appeared when the universe was new.

As we look deeper into the universe, and so, farther back in time, the number of quasars increases. Their population seems to have reached a peak around the time when the universe was only 2 billion or 3 billion years old, just a fraction of its current age. What happened in the universe's youth to produce such high-energy objects? Understanding this concept remains one of the primary challenges in astronomy.

PROTOGALAXIES

As we look backward in time through our most powerful telescopes, to a period when the universe was only a few billion years old, we should begin to see galaxies themselves forming out of intergalactic gas. Although such protogalaxies should be intensely powerful, they will probably appear as faint smudges—due to their tremendous distance from Earth.

Astronomers have yet to identify a protogalaxy with certainty. One of the infrared galaxies discovered with the IRAS satellite appears to be the most luminous object ever found in the universe. It is emitting more than 10,000 times the energy of the Milky Way. This object may be the first galaxy caught in the very process of formation. Astronomers may at last be viewing an entire generation of stars being born before their eyes.

Interstellar space is filled with tiny frozen crystals that condense to form a protostar (above), the first stage in the birth of a new star.

INTERSTELLAR SPACE

by Dennis L. Mammana

Modern astronomers are concerned not only with heavenly bodies of considerable size, such as the stars and the planets; they are also vitally interested in the space between the stars, areas filled with gases, solid particles of matter, and scattered atoms. In the early 20th century, this material was considered merely an obstacle to the study of the distant stars. Astronomers have come to realize, however, that interstellar matter is interesting in its own right. Today it is one of the greatest aids to our understanding of the structure and development of the stars and of the galaxies.

Perhaps as many as 200 billion stars make up the gigantic, flattened pinwheel that is our own Milky Way galaxy, in which the Sun and the planets of the solar system are located. It is estimated that light, which travels some 186,400 miles (300,000 kilometers) a second, may take as long as 100,000 years to cross the Milky Way. This gives some idea of its immensity. Vast as it is, our galaxy is only one of many. There are others, probably hundreds of millions, in the far reaches of space. Some of them are systems like our own. Others are quite different in shape and size. Interstellar materials are found in almost all of these systems of stars. We are beginning to suspect, too, that such materials also exist in the vast spaces between the different galaxies.

Our first knowledge of interstellar matter came from bright, diffuse nebulas—luminous clouds in the heavens. Perhaps the best known of these is the Great Nebula in

Orion, faintly visible to the naked eye as a greenish blur. Hundreds of other bright nebulas are known; most of them lie within or near the Milky Way. Other galaxies, especially those that are like our own, also contain bright nebulas.

THE NATURE OF NEBULAS

The early observers thought that nebulas like the one in Orion might really consist of faint stars. As more-powerful telescopes were built, astronomers tried to resolve the nebulas—that is, to illuminate individual stars. In certain instances, they were successful. Some of the nebulas proved to be galaxies or smaller star groups. However, a great number of them could not be resolved in this way. When astronomers viewed them even through the most powerful telescopes at their disposal, they could see only finely shredded, cloudlike structures. Late in the 19th century, it was proved through the use of the spectroscope, which measures electromagnetic radiation, that bright nebulas, such as the one in Orion, are not made up of stars, but of enormous balls of gas.

The instrument that enabled astronomers to make this observation—the spectroscope—gives us invaluable information about heavenly bodies by decoding the messages that their celestial light brings to us. White light is made up of all the colors of the rainbow. When the light of a star or of a distant galaxy passes through the spectroscope, the light is broken up into a band of colors—a spectrum. By the position and pattern of the lines of color in this band, we can tell what elements, in the form of incandescent gas, are present in the heavenly body being observed. Each element has its own distinguishing pattern of lines, with its own place in the long band of the electromagnetic spectrum.

A spectrum that shines with definitely separate, or discrete, colors is the earmark of a glowing, tenuous gas. The Great Nebula in Orion has just such a spectrum. Many other bright nebulas have spectra very much like this, and clearly also consist of glowing gas.

Solid Material

Not all the bright nebulas have gaseous spectra, however. The spectrum of the nebulous material that surrounds the Pleiades, a cluster of rather hot stars, resembles those of the bright stars within it. But we would be wrong in concluding that this nebula itself consists of stars. Actually, it is simply reflecting the light of the bright stars near it. We assume that it must contain solid particles. Otherwise, it would not be able to reflect starlight as well as it does. It is probable that the Pleiades nebula contains not only solid particles, but also gases.

We have another way of knowing that solid matter is present in interstellar space. Clouds of particles screen off, or obscure, the light of more-distant stars. Some clouds, like the Coalsack in the southern sky, are opaque and have well-defined edges. Others can be located by the general dimming of the stars behind them and, more important, by the reddening of the starlight that passes through them. If the observed color of a star is redder than astronomers expect it to be, they know that it is screened by a cloud that is absorbing some of the starlight. We can also tell how much of this obscuring material there is. Most of this matter lies in the plane of the Milky Way.

Solid particles are not the only obscuring bodies in the space between the stars. Individual atoms, too, can obscure starlight. Each atom absorbs radiation of the same wavelength as the atom itself emits. The radio telescope is a spectacular tool for discovering the existence of atoms in interstellar space. It detects electromagnetic radiation in the radio frequencies. The wavelength of red light (also a form of electromagnetic radiation), to which the eye is sensitive, is about 0.0000025 inch (0.000065 millimeter). The radio telescope can detect radiation with a wavelength of about 8 inches (200 millimeters). It happens that the atom of hydrogen, the most common substance, emits radiation of about this wavelength. After American physicists built a radio receiver to search for the predicted existence of radiation from hydrogen in the

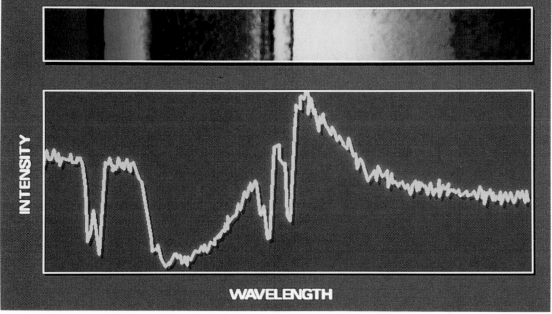

All elements produce a characteristic spectrum on a spectroscope. Experts analyze the spectra of celestial elements to identify the composition of interstellar space.

sky, they soon succeeded in picking up the signals sent out by interstellar hydrogen. The detection of radiation emitted by hydrogen marked a new era in the history of astronomy.

GLOWING GASES

The bright, gaseous nebulas are found only near stars burning at a temperature greater than about 25,000° F (13,900° C) at the surface. The light of these nebulas shows the colors of the spectrum characteristic of glowing hydrogen, and also helium, carbon, nitrogen, oxygen, sulfur, and other atoms. The Orion nebula contains over 300 atoms per cubic centimeter. Here, again, most of the material is hydrogen.

These atoms glow because they are close to a hot star. The hydrogen and helium atoms in the nebulas absorb extreme ultraviolet light from the star and re-emit it in the form of their particular radiations, or wavelengths. Oxygen, nitrogen, and sulfur atoms give out light after they collide with electrons that have been detached from other atoms—mainly hydrogen atoms—by the high-temperature radiation of the star. If the material had densities comparable to those in stellar atmospheres, collisions with other atoms would, in a short time, rob these atoms of the energy conferred by the electrons. But the density of the Orion nebula is so low that collisions may not take place for months. Under these conditions the atoms of oxygen, nitrogen, and sulfur will radiate the lines seen in the nebular spectrum, which are called the "forbidden lines." They are forbidden only in the sense that they are suppressed under stellar or laboratory conditions by constant collisions with neighboring atoms.

A fine gaseous haze, much fainter than these conspicuous knots of gas, pervades the Milky Way. Powerful instruments are required to photograph its spectrum, for it is 6,000 times fainter than the Great Nebula in Orion. On the average, this haze contains about one atom per cubic centimeter. The enormous majority are atoms of hydrogen. They shine because of the radiation that falls on them—the very faint light of many distant stars. There is evidence that atoms of oxygen, nitrogen, and sulfur are also present.

In addition to the glowing gases within the Milky Way, astronomers have also observed extremely hot, gaseous clouds that appear to be floating throughout the galaxy. According to Ralph Fiedler of the Naval Research Laboratory in Washington, D.C., these clouds may be responsible for temporary dips detected in the intensity of radio waves emitted by quasars. The Milky

Way may contain as many as 1,000 times more of these clouds than stars do. According to Fiedler, who based his findings on daily radio-wave observations from 1979 to 1985, other celestial clouds are at least 20,000 times wider than the newly discovered clouds. The observed dips are apparently the result of the deflection of radio waves as they pass between Earth and distant celestial objects.

OUTLINING GALAXIES

Interstellar hydrogen and solid particles help us to make out the details of some of the galaxies. These are divided into several types. In the spiral galaxies, huge curving arms are set around a nucleus, which may be spherical or in the form of a straight bar. Our Milky Way is an example of a spiral galaxy. Other galaxies, called irregular, have no marked symmetrical form. Still others, the elliptical galaxies, are distinguished only by their elliptical shape.

When astronomers used the radio telescope to observe the interstellar hydrogen in the faint nebulosity of the Milky Way, they found that the interstellar hydrogen gas outlines the spiral structure of our stellar system. The gas lies most densely in the spiral arms, and thins out between them. The hydrogen is accompanied by absorbing particles, which follow the course of the spiral arms. In the galaxy called the Great Spiral in Andromeda, the spiral arms are found to be studded with bright, gaseous nebulosities and associated with well-marked lanes of obscuring materials.

Yerkes Observatory

The Great Nebula, M42, located 1,500 light-years away in the Orion constellation, appears as a faint blur to the naked eye. Powerful telescopes, however, reveal a churning mass of glowing atoms, illuminated by the light of a nearby group of stars.

INTERSTELLAR SPACE

The chaotic, irregular galaxies are even richer in bright and dark nebulous material than are the spiral galaxies. Even many of the featureless elliptical galaxy systems, which contain few if any solid particles, are pervaded by a very diffuse haze of glowing atoms.

Astronomers have found that several distinct clouds, moving at different speeds, lie between us and many distant stars. Evidence has shown that these separate clouds are associated with different spiral arms of our galaxy. Each is moving at its own speed, and is silhouetted against the light of very distant stars.

The interstellar gases furnish us with enlightening evidence about conditions in space between the stars. For example, in the region of faint, hazy nebulosity, the temperature of a solid particle would be a few degrees above absolute zero, which is $-460°$ F ($-273°$ C).

SMALL CRYSTALS

Interstellar solids must be very finely divided to absorb as much starlight as they do. From studies of the motions of stars, scientists have determined that these solids cannot have a density greater than 0.00000000000000000000008 grams per cubic centimeter. At a density as low as this, the particles may have diameters of about 0.0000059 inch (0.000015 centimeter). But the size required to account for the observed absorption depends on whether the particles are metallic or not. If they are metallic, they need not be so small. Metallic atoms do not seem to predominate in the universe, however. In the Sun, which is a fairly typical example of cosmic material, there is less than one metallic atom per 6,000 atoms.

The extremely high reflecting power of nebulas—greater than that of snow—sug-

Views of the Milky Way using optical telescopes were long obscured by interstellar dust (below). Scientists, finding that radio waves pass through this dust, used radio telescopes to reveal the Milky Way's spiral structure.

Yerkes Observatory

The Milky Way is a slightly warped, scalloped disk. The elements it contains may be the building blocks from which life develops elsewhere in the universe.

gests that most of the solids in space are tiny frozen crystals of compounds of the most common chemical elements. Perhaps they are substances such as ammonia, methane, and even water, built of the cosmically common atoms of hydrogen, carbon, nitrogen, and oxygen.

Other evidence indicates that the interstellar particles must be elongated crystals, for the light of the distant stars is polarized. This means that the light vibrations tend to be lined up perpendicularly to the light ray. The polarization is associated with interstellar reddening. Spherical particles would not have this effect. Moreover, the particles seem to be lined up more or less parallel in any one region.

There is much controversy as to how these observations should be interpreted. The consensus is that the elongated particles are lined up by magnetic fields in interstellar space. One theory suggests that small, highly magnetic particles are embedded in the icy grains.

STELLAR NURSERIES

Interstellar matter does not serve merely as a space filler, but rather is important in the formation of stars and galaxies. It is believed that, under the right conditions, an interstellar cloud of gas and dust will condense and assume a spherical shape. This sphere is a protostar, or early stage, which will eventually lead to the birth of a new star.

While astronomers have long believed that such interstellar clouds were the birthplaces of stars, no one had ever seen the process take place—until just a few years ago. In 1986 radio astronomers found that a huge, cold cloud of gas named IRAS 162932422 in the constellation Ophiuchus was collapsing upon itself. At the rate they calculated for its collapse, astronomers figured that a sunlike star would be born within the next 100,000 years.

Radio telescopes have detected a variety of chemical elements and compounds in space. These substances exist as individual atoms, as molecules, or as parts of molecules. Many of these molecules are biological in nature. That is, they are chemically identical to substances found in or produced by living things on Earth. Among them are hydrogen, water, ammonia, and carbon-containing compounds such as formaldehyde and cyanogen. A number of scientists believe that these interstellar materials may be the building blocks from which life as we know it could develop elsewhere in the universe. Thus, the study of interstellar space, particularly by means of the radio telescope, is extending the science of astronomy into the realm of biology and the origin of life.

Cosmic rays are charged atomic nuclei that bombard the Earth constantly from every direction. Only about 1 percent of the cosmic rays that strike the atmosphere and interact with air molecules make it to the Earth's surface.

COSMIC RAYS

by Volney C. Wilson

Whizzing through our galaxy at all times, in every direction, and at fantastic speeds, are the charged particles that we call cosmic rays. They are really atomic nuclei—that is, atoms stripped of their electrons. Traveling at nearly the velocity of light, some cosmic rays have energies far greater than those produced with powerful atom smashers.

The cosmic rays that fly through outer space are called primary cosmic rays. A thin rain of these particles constantly strikes the upper atmosphere of the Earth. As they enter the upper part of the atmosphere, they collide with atoms of air. The fragments known as secondary cosmic rays result from these collisions.

About 1 percent of these secondary rays penetrate the remaining atmosphere and reach sea level. Roughly 600 rays pass through the human body every minute, day and night. This should not be alarming, though—a wristwatch with a radium-painted face exposes us to the same dose of radiation. Doctors estimate that the dose must be 30 to 300 times greater to produce the increasing rate of mutations that lead to harmful genetic effects on human beings.

If we could harness the energy of all cosmic particles striking sea level over the entire world, it would yield little more power than a modern automobile engine. Cosmic rays will never power our machines, but they do provide nuclear physicists with a natural laboratory for the study of super-high-energy phenomena. They may reveal the secrets of the forces that bind together the particles of matter in atomic nuclei. And they may tell us about the universe beyond our solar system.

DETECTION OF COSMIC RAYS

Cosmic rays made their presence felt long before they were identified. About the beginning of the 20th century, physicists discovered that electroscopes, which determine the electric charge on a body, and ionization chambers, which are used to measure the intensity of radiation, were affected by certain mysterious rays existing in the atmosphere. These rays were far more penetrating than X rays or any other form of radiation known at the time. X rays could be stopped by a lead plate only 0.06 inch (1.6 millimeters) thick; but 3.94 inches (10 centimeters) of lead would absorb only 80 percent of the "new" rays.

The first and most natural assumption was that these penetrating rays issued from some special radioactive material in the Earth's crust and atmosphere. Beginning in 1909, however, enterprising experimenters carried ionization chambers aloft in balloons and found that the rays grew more intense with increasing altitude. Between 1910 and 1914, Victor F. Hess of Austria and Werner Kolhörster of Germany made the first precise measurements, at altitudes up to 29,528 feet (9,000 meters). At this height the rays were 10 times more intense than at sea level. Clearly, the penetrating rays traveled downward through the atmosphere. Hess came to the conclusion that the mysterious rays must have their origin in outer space.

In 1925 a series of experiments by a team of physicists headed by Robert A. Millikan of the California Institute of Technology (Caltech) confirmed Hess' hypothesis. Millikan and his colleagues lowered ionization chambers into two snow-fed lakes at high altitudes in California in order to determine the absorption of the penetrating rays in water, as compared with their absorption in air. Measuring the intensity of the rays under different depths of water and at different altitudes, they came to the conclusion that the rays must originate outside the atmosphere. In the report of these experiments, the name "cosmic rays" was used for the first time.

In a few years, two newly invented instruments revealed important new facts about cosmic rays, and later became the cosmic-ray physicist's chief tools. In 1927 Russian physicist D. V. Skobeltsyn first adapted the Wilson cloud chamber to the study of the rays. With this instrument the path of an ionizing particle, or ray, is made visible as a trail of water droplets.

In a series of balloon-borne high-altitude experiments, Victor Hess (below) helped prove the existence of cosmic rays, an achievement for which he shared the 1936 Nobel Prize for physics.

Michael W. Friedlander, Washington University, St. Louis

To study cosmic rays, Carl D. Anderson (left) used a cloud chamber in which a series of lead plates (below) causes the rays to slow down and split. In the course of his cosmic-ray investigations, Anderson, almost by chance, confirmed the existence of positrons—positively charged electrons. He shared the 1936 Nobel Prize with Hess for his findings.

Collection of California Institute of Technology

In 1928 and 1929, Werner Kolhörster and another German, Walther Bothe, devised a research technique using sets of Geiger-Müller counters, devices that detect particles by electrical discharges. By arranging the counters in a straight line and providing them with the proper electrical circuits, one could trace the path of a single cosmic ray.

The bubble chamber, developed in 1952 by Donald Glaser, also proved helpful in the analysis of the rays. In the bubble chamber, a liquid is kept just below the boiling point. The pressure is suddenly lowered, and the liquid becomes superheated. When high-speed charged particles now pass through, their passage will be indicated by a series of bubbles.

These devices helped establish that most of the secondary cosmic rays are high-energy, electrically charged particles.

COSMIC RAYS AND EARTH'S MAGNETIC FIELD

In 1927 Dutch scientist Jacob Clay discovered that primary cosmic rays are affected by the Earth's magnetic field. During a voyage between Amsterdam (in the Netherlands) and Indonesia, he observed that the intensity of cosmic rays drops as one approaches the magnetic equator from higher latitudes. The existence of this "latitude effect" anticipated the 1952 discovery that primary cosmic rays are electrically charged particles.

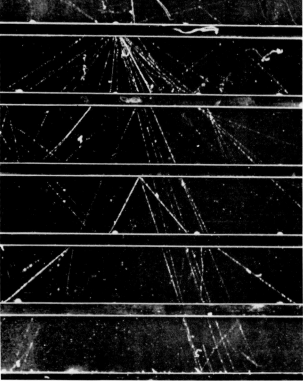

University of California Radiation Laboratory

A neutral (uncharged) particle moving in a straight line through a magnetic field is not influenced by the field. However, a charged particle in the same situation will have its path bent into a curve if it moves approximately at right angles to the magnetic lines of force.

The Earth's magnetic field is very weak, but it extends for thousands of miles into space. Any charged particle approaching the Earth must travel immense distances through the field, and will be appreciably deflected by the curving force. This force is greatest on particles that approach exactly at right angles to the lines of force, and weakest on those that travel parallel with the lines. The bending is more pronounced for slow-moving, or low-momentum, particles than for faster, or high-momentum, ones.

This is illustrated in the diagram on page 228. The dotted lines represent the Earth's magnetic lines of force. A, B, C, D, and E are all charged particles approaching the Earth with the same momentum or energy. A is only slightly deflected by the field, because it approaches near the pole and thus moves almost parallel with the lines of force. On the other hand, E approaches near the equator, moving exactly at right angles to the lines. The strong curving force on this particle eventually turns it back into the direction from which it came. The paths taken by B, C, and D will depend upon the angle formed by their direction as they approach the Earth and the magnetic lines of force. For a particle to penetrate the magnetic field and the atmosphere at the equator (particle F), it would have to start out with many times more momentum than the other particles.

Cosmic-Ray Intensity

If the particles are indeed electrically charged, we should expect many fewer rays to reach the Earth's surface at the equator than at the higher latitudes. This was conclusively shown in 1930. In that year a worldwide survey was begun, under the direction of Arthur H. Compton of the University of Chicago, to determine cosmic-ray intensities at different latitudes and altitudes all over the globe. The report of the survey in 1933 established that, from the geomagnetic latitudes of 50 degrees north (or south) to the equator, cosmic-ray intensity at sea level drops about 10 percent. It showed that primary cosmic particles are electrically charged.

Is the charge positive or negative? Manuel S. Vallarta, a Mexican mathematician, calculated in 1933 that a positive particle with low momentum could reach the Earth more easily when approaching from the west than from the east. The directions would be reversed for a negative particle. By 1938 experiments proved that the rays falling from the west were definitely more intense than those from the east. This led to the conclusion that the primary cosmic rays are positively charged.

American physicist Scott E. Forbush discovered that cosmic-ray intensities decrease during periods of high sunspot activity. The Sun is continuously throwing out tremendous quantities of protons moving at a speed of about 994 miles (1,600 kilometers) per second—a phenomenon referred to as the *solar wind*. During periods of sunspot activity, when a solar flare happens to be at a particular place on the Sun's surface, the solar wind in the direction of the Earth is greatly increased. This in turn increases the strength of the magnetic field in the vicinity of the Earth to such an extent as to shield out some of the low-energy cosmic rays. This is called the Forbush effect.

Rockets and artificial satellites have been used increasingly to measure cosmic-ray intensity above the Earth. They have also served to analyze the effect of the solar wind upon the Earth's magnetic field.

Cosmic-ray Showers

Italian-born physicist Bruno Rossi showed in 1932 that if three or more Geiger-Müller counters were spread out horizontally in an irregular pattern, they would sometimes be tripped simultaneously, indicating that showers of particles were traveling together. These showers could be produced as cosmic-ray particles passed through lead or other substances containing heavy atoms. Rossi concluded that each shower was produced from a single cosmic-ray particle as it passed close to the nucleus of a nearby atom. Showers could also be observed in a Wilson cloud chamber that had a lead plate across it. The track of a single cosmic-ray particle is seen entering

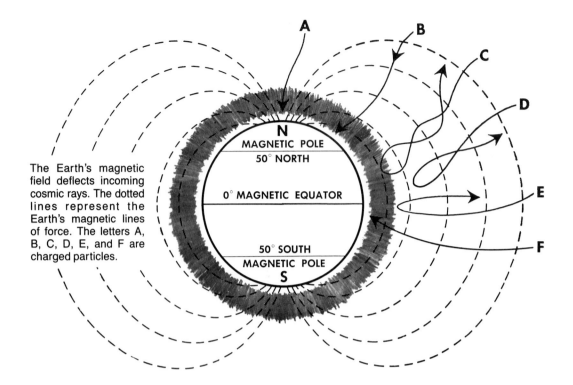

The Earth's magnetic field deflects incoming cosmic rays. The dotted lines represent the Earth's magnetic lines of force. The letters A, B, C, D, E, and F are charged particles.

the chamber. As it passes through the lead plate, the track splits in two. These tracks split again and again, producing a shower of particles.

In the same year in which Rossi revealed the existence of cosmic-ray showers, American Carl D. Anderson discovered a new fundamental particle in cosmic radiation—the positive electron, or *positron*. It appeared in a photograph of a cloud chamber containing a lead plate in a strong magnetic field. Anderson's historic achievement won him a share of the 1936 Nobel Prize in physics.

Cloud-chamber photographs soon showed that each time a shower track split, the two branches were oppositely curved, and an electron and a positron were created. A pair-production theory of cosmic-ray showers was then formulated. It was maintained that the process of shower formation represented the conversion of energy into charged matter and vice versa.

According to the pair-production theory, this is what takes place. As a gamma-ray photon—a fragment of light energy—penetrates the nuclear field of an atom, a portion of its energy is suddenly transformed into a pair of electrons, positive and negative. The remaining energy provides the velocities of the electron and the positron. Each of the daughter electrons may disappear in its turn, producing another gamma ray, if it is suddenly slowed down in the field of another nucleus. Two gamma rays may produce four electrons; four gamma rays may then arise, then eight electrons, and so on.

In 1938 such cosmic-ray showers were discovered in the atmosphere, up to .21 mile (0.33 kilometer) in diameter and with millions of particles each. The energy of one cosmic-ray particle causing a shower may reach 1 billion GeV (1 GeV equals 1 gigaelectron volt, or 1 billion electron volts). An electron volt is the energy gained by an electron when accelerated by 1 volt of electricity. In contrast, fission of a uranium atom releases only 200 MeV (1 MeV equals 1 million electron volts).

THE PRIMARY COSMIC RAYS

Primary cosmic rays contain 91 percent positive particles, or protons (the nuclei of hydrogen atoms); 8 percent alpha particles, or helium nuclei, with two protons apiece; and 1 percent nuclei of heavier elements, such as lithium (three protons per nucleus), beryllium (four protons), boron (five protons), oxygen (eight protons), and iron (26 protons). These particles are detected by photographic emulsions in high-altitude balloons.

A variety of heavier cosmic-ray nuclei are being discovered. Scientists have learned that moving heavy particles damage mica and plastics, leaving tracks, which, when etched, give an idea of the particles' sizes and charges. One nucleus has been found to contain 109 protons, representing an element more massively charged than uranium, which has 92 protons per nucleus.

Cosmic-ray composition is similar to the known distribution of chemical elements in stars, nebulas, and interstellar dust. Cosmic-ray composition and the distribution of the quantities of elements in the universe show sharp differences in the numbers of atoms having odd and even numbers of protons per nucleus, with a peak in regard to iron. "Even-numbered" nuclei are more common. These facts give a clue to the origin of primary cosmic rays.

TWO COMPONENTS OF SECONDARY COSMIC RAYS

Secondary cosmic rays are composed of two very different classes of particles, which can be separated by a piece of lead about 5 inches (13 centimeters) thick. One component is completely absorbed in this thickness. This soft component consists mainly of electrons, positrons, and gamma-ray photons—the typical particles of a cosmic-ray shower. The intensity of this soft component increases from the top of the atmosphere down to an altitude of about 11 miles (17 kilometers), where it makes up roughly four-fifths of the total radiation. From this altitude downward, the intensity decreases until it makes up only one-quarter of the total at sea level.

The other component passes through 5 inches (13 centimeters) of lead almost unobstructed. It is called the hard, or penetrating, component. It decreases in intensity continuously from the top of the atmosphere down to sea level. Approximately one-half of these rays at sea level can still penetrate 15 inches (38 centimeters) of lead. The difference in the absorp-

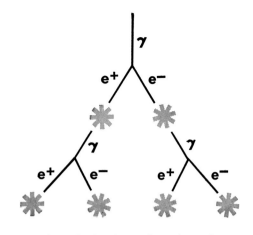

In the pair-production theory of cosmic-ray showers, a gamma ray (indicated by γ) yields a pair of electrons, positive and negative, indicated by e^+ and e^-, respectively. Each of these electrons may disappear, producing another gamma ray as it does. Each of the gamma rays in turn may yield two electrons. As the process is continued, a cosmic-ray shower is produced.

tion pattern and penetrating power of the two components represented a puzzling problem to physicists for many years.

In 1936 the riddle was finally solved by the teams of C. D. Anderson and S. H. Neddermeyer at the California Institute of Technology (Caltech), and J. C. Street and E. C. Stevenson at Harvard University. Working independently, these teams showed that all the evidence pointed to a new type of fundamental particle, intermediate in mass between the proton and the electron. This was the mesotron, now called the *meson*.

In the meantime, the author of this article (Volney Wilson of the General Electric Company) had been measuring the penetration of cosmic rays below the ground in a copper mine in northern Michigan. He reported that some rays were able to penetrate as much as 1,640 feet (500 meters) of rock. Further investigation showed that these rays were corpuscular, or made up of tiny particles, and that they ionized the rock and lost energy all the way down. Calculations showed that they must be the newly discovered mesons. It is now agreed that mesons make up the bulk of the hard component of secondary cosmic rays. The high energy of mesons accounts for the penetrating power of this component.

The existence of the meson had been predicted by Japanese nuclear physicist Hideki Yukawa in 1935. He also predicted

Solar flares (left) and other dramatic outbursts from the Sun's surface may serve as one nearby source of cosmic rays.

Big Bear Solar Observatory, photo courtesy of R. L. Moore

that the meson would be unstable outside the nucleus, decaying with the emission of an electron. In 1940 this decay was actually photographed for the first time in a Wilson cloud chamber.

In 1948 researchers at Berkeley, California, succeeded in producing mesons artificially by bombarding carbon atoms with alpha particles accelerated to 380 MeV. Both positive and negative mesons were produced. Some, known as *pi-mesons*, decay in about 200-millionths of a second into a lighter variety called *mu-mesons*. Mu-mesons live for roughly 2-millionths of a second before they disintegrate into electrons or positrons. In secondary cosmic-rays the heavier pi-mesons are created by tiny "nuclear explosions" in the upper atmosphere, which occur when primary cosmic rays collide with the nuclei of the atoms of the gases that make up the atmosphere. The pi-mesons usually decay immediately into mu-mesons, which live long enough to travel great distances. It is the mu-mesons that form the major component of secondary cosmic rays at sea level.

ORIGIN OF COSMIC RAYS

What is the source of cosmic rays, and how do they acquire their fantastic energies? Some astronomers have long held that the universe was created in a single primordial explosion, and that cosmic rays are tiny remnants of this colossal event. But the presence of heavy nuclei in the primary rays shows that their energies must be acquired gradually. In an explosive process, the larger nuclei would be completely shattered.

Present evidence suggests that most cosmic rays originate and remain largely within our own galaxy. Let us suppose that the distribution of rays observed in our galaxy extends far beyond its borders throughout the universe. We know the average energy of the rays in our region. From this, we can calculate the total energy carried by cosmic rays everywhere. The figure reached would equal the entire mass of the universe if this were converted into energy. Scientists consider such an idea impossible. Moreover, the solar system is constantly rotating along with the rest of our galaxy. If rays are entering the galaxy from the outside, certain complex effects due to the rotation of the galaxy should be visible. None have ever been observed.

At least 90 percent of the cosmic rays originate inside our own galaxy. The maximum lifetime of a cosmic ray is about 2 million years. The known characteristics of cosmic rays suggest they originated in thermonuclear (atomic) explosions in certain kinds of stars.

Old stars rich in iron atoms reach a stage when they just blow up. They then release vast amounts of particles and electromagnetic radiation, including light. These explosions occur by a process called neutron capture. Neutrons—heavy nuclear particles with no charge—can move freely outside the nuclei of atoms. They may, however, be absorbed, or captured, by the nuclei of atoms; this event makes the atoms unstable. The atomic nuclei therefore release energy in the form of alpha particles, protons, neutrons, and radiation of very short wavelengths, especially gamma rays, in a gigantic blast.

An exploding star is called a *supernova*. It is much brighter than an ordinary nova, which does not really explode, but shoots out matter instead. The best-known supernova was the event of A.D. 1054, recorded by Chinese astronomers. The remnant of this nova is the Crab nebula.

At present the Crab nebula is observed to emit radio waves and light that is markedly polarized. That is, the radiation waves vibrate along a single plane or along a few planes, instead of at many angles, as ordinary (unpolarized) radiation does. The polarization means that the Crab nebula's radiation originates from electrons moving at great speeds in strong magnetic fields.

This is called *synchrotron radiation*. A synchrotron is an atom smasher, or accelerator, that speeds up electrons and other particles by means of intense, changing magnetic fields. Polarized electromagnetic radiation is observed to arise from electrons accelerated in a synchrotron.

The Crab nebula, then, may act as a giant cosmic accelerator of particles—or cyclotron—with energies that dwarf those of man-made accelerators. The most powerful man-made accelerator to date reaches only 500 GeV. Some cosmic-ray energies reach the staggering figure of 1 billion GeV!

Within the Crab nebula, astronomers have identified a *pulsar*—a very small and very dense star composed mostly of neutrons, which emits radiation in very short bursts, or pulses. The pulsar itself is part of the exploding star that was seen by Chinese astronomers over 900 years ago. This pulsar may be the chief energy source of the Crab nebula. Not only does it radiate energy, but it ejects large numbers of particles from its surface. These particles are being accelerated in the very strong rotating magnetic field of the pulsar.

Scientists estimate that one to two supernovas occur in our galaxy per century. If they contain pulsars, the energy balances are such that these supernovas could just about produce the observed cosmic rays. Rays with energies of 1 billion GeV or less are trapped in the magnetic field of our Milky Way galaxy. They last an average of about 1 million years before striking an atom and being annihilated in the process.

The extremely few cosmic rays with energies greater than 1 billion GeV may have arrived at the Earth before escaping the galaxy, or may even have come from outside the Milky Way. Observations so far suggest that no more than 10 percent, and perhaps far less, of cosmic rays come from beyond our galaxy.

Today cosmic-ray telescopes aboard the space shuttle and some Earth-orbiting satellites permit scientists to probe the elemental makeup of cosmic-ray nuclei. The researchers are still searching for clues to the origin of the rays and to how they move through interstellar space.

THE MANNED SPACE PROGRAM

by Mark Strauss

"Tranquility Base here. The *Eagle* has landed."

And so, on July 20, 1969, at 4:17:40 P.M. eastern daylight time, two men landed on the Moon. More than 239,000 miles (384,000 kilometers) across space, the whole world was watching spellbound. In London, crowds of people clamored around a big-screen television in Trafalgar Square. In Prague the streets were deserted, as thousands of Czechoslovaks rushed to watch televisions in beer gardens and hotel lobbies. Several thousand people burst into applause at the Science Museum in Paris.

In a scant 30 years, space technology has exploded, advancing from the relatively primitive Sputnik to the sophisticated space shuttle. Scientists are confident that the shuttle's as yet untapped potential will ultimately extend our knowledge to the stars and beyond.

In Zambia, rural tribesmen sat around bonfires, listening to the news on transistor radios that had been distributed by the government.

"For one priceless moment in the whole history of man, all the people on this Earth are truly one," President Richard Nixon told the astronauts in a phone call from the White House.

The Moon landing was the culmination of a dream thousands of years old, made into reality by a space program that had existed for barely a decade. A new era of exploration had begun, and humanity seemed on the verge of moving out into the solar system, and beyond.

Since that historic day, manned spaceflight has become almost commonplace. Other crews followed the *Apollo 11* astronauts to the Moon, and both American astronauts and Soviet cosmonauts have lived and worked in space for extended missions on orbiting space laboratories. The space-shuttle program, employing the world's first fleet of reusable spacecraft, has sent astronauts into orbit more than 40 times.

But the space program has not been without tragedy. In 1967 three Apollo astronauts died when a fire swept through their command module during a ground test at Kennedy Space Center in Cape Canaveral, Florida. On January 28, 1986, the space shuttle *Challenger* exploded shortly after lift-off, killing its crew of seven.

Yet those setbacks, which at the time seemed insurmountable, proved temporary. Despite the Apollo tragedy, America continued on its path toward the Moon. Two years after the *Challenger* explosion, the shuttle program resumed with the successful launching of the *Discovery* shuttle. America's commitment to the space shuttle renewed itself in 1991, when the National Aeronautics and Space Administration (NASA) unveiled the latest addition to the shuttle fleet, *Endeavour*.

THE EARLY DAYS

The shuttle is the most complex machine ever created and piloted by man, a marvel of modern technology. But only with a long heritage of research into manned spaceflight could such a craft have come into existence.

It all started right after World War I. In 1919 an American scientist named Robert H. Goddard published a paper titled "A Method of Attaining Extreme Altitude," which contained the basic mathematics of rocketry. In that article he outlined a scheme by which a solid-propellant rocket could be launched with enough velocity to reach the Moon, and then signal its arrival by setting off a charge of flash powder.

These days the thought of astronauts sending up flash-powder flares in order to communicate with Earth seems hopelessly old-fashioned. But at the time there was little alternative. Radio transmitters and receivers were large and bulky, with limited range and power. Television was still only experimental.

Within the next 40 years, however, the field of electronics advanced in spectacular fashion. Radio and television had reached the stage where it was possible not only to use them to guide unmanned spacecraft, but also to transmit all sorts of scientific data, including photographs, to ground stations back on Earth.

It was that sort of technology that made it possible for the Soviet Union to launch Sputnik, the world's first satellite, into orbit in 1957. The news hit America like a slap in the face. The United States, which had always imagined itself the world's technological leader, was suddenly seen in the eyes of the world to be lagging very far behind.

In order to restore America's prestige, and to close the widening technological gap between the United States and the Soviet Union, President John F. Kennedy, in his State of the Union message to Congress on May 25, 1961, declared that, within the decade, the nation should commit itself to the goal "of landing a man on the Moon and returning him safely to Earth."

The Titan III hybrid rocket uses both liquid and solid fuel to create a lift-off thrust powerful enough to launch heavy payloads, such as the Viking and Voyager space probes.

ANATOMY OF A ROCKET

Sir Isaac Newton would have loved rockets. Rockets are working models of his famous third law of motion, which states that for every action, there must be an equal and opposite reaction. When a rocket fires, the fuel burns, and combustion gases quickly expand and push out the rear. As the gases thrust downward, the rocket is pushed upward. Although a rocket's motion is virtually independent of the surrounding air, it flies considerably faster in a vacuum where there is no air to resist the rocket's upward motion, or to impede the downward exhaust.

The actual thrust of the rocket depends on only two factors. The first is the quantity of combustion gases produced (which depends on the amount of fuel being burned). The exhaust blast resulting from the combustion leaves the exhaust nozzle, which is shaped in such a way that the gases push against the nozzle as they expand. The second factor is the ejection speed of these gases.

Currently all rockets fall into one of two categories: liquid fuel or solid fuel. A liquid-fuel rocket has two tanks, one containing fuel and the other containing an "oxidizer," which causes combustion when it reacts chemically with the fuel. Typically the fuel is either alcohol or refined kerosene, and the oxidizer is liquid oxygen. Pumps force the two liquids into a combustion chamber, where they are then ignited and burned.

In a solid-fuel rocket, the "fuel tank" and the "combustion chamber" are one and the same. The fuel tank contains hard, rubbery explosives into which an oxygen-rich chemical has been kneaded during the manufacturing process. The fuel feels to the touch like an automobile tire. It is shaped like a tube with very thick walls and a small central hole. The fuel burns outward from the central hole, which constitutes the combustion chamber. To bring about combustion, a pyrotechnic igniter is triggered. The igniter consists of an electric wire surrounded by a powder charge; when the wire is heated, it sets off the charge.

Liquid-fuel rockets, although more complicated, have one big advantage over solid-fuel rockets: pumps can control how quickly the fuel and oxidizer mix. That's why Saturn V, the rocket that launched men toward the Moon, was liquid-fueled. NASA's space-shuttle system relies on solid-fuel boosters simply to add extra thrust to get off Earth's surface. The liquid-fueled main rocket engines of the shuttle do the rest of the work once the shuttle is in space.

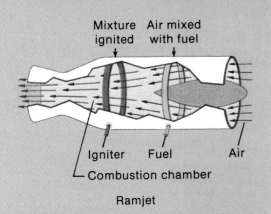

Ramjet

Jet-propelled planes and rockets both operate according to Newton's third law of motion, described in the text. However, in the jet plane, such as the ramjet shown above, air (which contains oxygen) is drawn in, mixed with injected fuel, and ignited. A rocket, by contrast, carries its own oxygen supply in the form of an oxidizer. In the liquid-fuel rocket (below left), the fuel is separate from the oxidizer. Fuel is mixed with the oxidizer and ignited in the combustion chamber. In the solid-fuel rocket (below right), the oxidizer is contained in the fuel itself. After the fuel is ignited, it burns outward from a central core. The fuel tank thus becomes the combustion chamber.

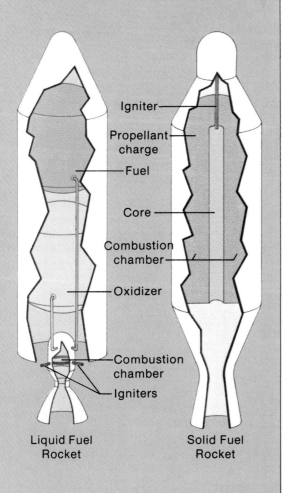

Liquid Fuel Rocket

Solid Fuel Rocket

Our First Manned Rockets

The early rocket researchers figured out one thing pretty quickly: any spacecraft would have to be launched into space by multistage rockets. A single-stage rocket could not achieve the necessary velocity. The U.S. rocket missions of the 1960s, which led to the Moon landing, all featured launchers with several stages.

At the top of these stages rode a manned capsule. As a precautionary measure, each manned capsule had, attached to its nose, an escape tower provided with its own rocket engines. If there appeared to be danger of fire or explosion at the end of a launch, the tower's rocket engines could fire—triggered either from the ground or by an astronaut in the capsule. This would cause the capsule to pull away from the booster-rocket stages. After reaching the proper distance from the landing site, the capsule could unfurl a parachute for landing. And if the launching went well, this escape tower would simply fall out of the way at the appropriate time.

The early space capsule was shaped like a cone with its point lopped off. Inside the capsule the atmosphere consisted of oxygen pumped from tanks. Extended flights relied on a three-gas system—oxygen plus nitrogen and helium. The Soviet Union has also used this mixture in its Soyuz spacecraft.

Each astronaut sat in a contour couch, specially constructed according to his measurements. Since the nose of the capsule pointed straight upward at the moment of launch, the astronaut had to lie on his back. Pipes connected his space suit to a network of tanks and air pumps. Once in orbit the astronaut could remove his helmet and breathe the oxygen in the capsule. If anything were to go wrong with the capsule's ventilating and filtering system, the astronaut would replace his helmet, and his space-suit system would take over.

The capsule had an elaborate control system connected to small rocket engines set around the craft. Controls were automatic during the launch. Once in orbit the pilot had the option of either taking control or leaving the craft on automatic. The pilot

THE MANNED SPACE PROGRAM

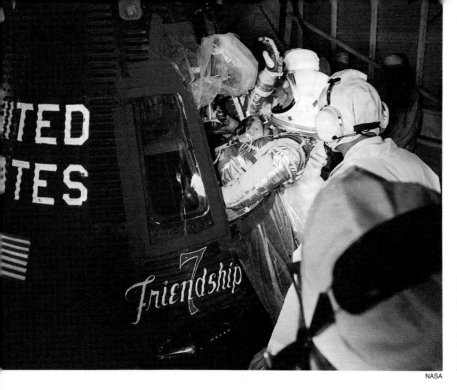

On February 20, 1962, John H. Glenn, Jr., became the first American to orbit the Earth. The 4-hour, 56-minute flight aboard Friendship 7 took him around the globe three times before he safely returned to Earth, splashing down in the Atlantic Ocean.

would also take control during crucial maneuvers such as space docking or nosing the craft out of orbit. A similar control system is used to position the space shuttle for the deployment of satellites and other work in space.

In the early spacecraft, no maneuvers were more vital than those required to bring about safe reentry through the atmosphere. When the time for reentry was at hand, the command pilot would have to turn his craft around so that it was moving with the blunt end—the base of the cone— foremost. Retro-rockets mounted in the blunt end then fired, slowing the craft. The heat shield, located on the craft's blunt end, absorbed much of the heat of friction as it collided with more and more molecules of air in the atmosphere.

Until the development of the space shuttle, all U.S. spaceflight landings took place at sea. In preparation for water landings, the capsule's speed first slowed to a few hundred miles per hour. When the craft was reasonably near the target area, small parachutes, called drogues, opened for further braking. Finally, when the capsule was about 9,900 feet (3,000 meters) above the target area, larger parachutes opened. The velocity dropped to about 22 miles (35 kilometers) per hour, and the craft "splashed down" into the sea.

Mercury and Gemini

Before the Apollo project, which led to a landing on the Moon, the U.S. space program went through two preliminary stages: Projects Mercury and Gemini.

In Project Mercury, conducted from 1959 through 1963, spacecraft carried a single astronaut aloft. The program was aimed at testing the physical reactions of the pilot and his ability to perform various tasks while in a state of weightlessness. It was also designed to solve basic problems of spaceflight, including reentry into the Earth's atmosphere.

Project Gemini ran from 1961 until late 1966. Named for the twins in the zodiac, a Gemini spacecraft housed two astronauts, who rehearsed the tasks needed to land an astronaut on the Moon. They also executed some vital maneuvers, such as taking the craft off-course and guiding it back again. They practiced rendezvous in space, in which two capsules might meet at a designated point in orbit. Likewise, they carried out space docking: connecting two craft in space.

Among the most spectacular feats accomplished in the course of Project Gemini were the "space walks." In a space walk, one of the astronauts floated free of the capsule, tethered to it by a lifeline known as an umbilical. The line furnished oxygen

SPACE FLIGHT: FROM VOSTOK TO THE SPACE SHUTTLE

Date of flight	Name of spacecraft	Name(s) of astronaut(s) or cosmonaut(s)	No. of orbits	Duration of flight	Remarks
Apr. 12, 1961	Vostok 1 (U.S.S.R.)	Yuri A. Gagarin	1	1.8 hrs.	First manned spaceflight
May 5, 1961	Freedom 7 (U.S.)	Alan B. Shepard, Jr.	–	15 min.	First American in space
July 21, 1961	Liberty Bell 7 (U.S.)	Virgil I. Grissom	–	16 min.	
Aug. 6–7, 1961	Vostok 2 (U.S.S.R)	Gherman S. Titov	17	25.3 hrs.	
Feb. 20, 1962	Friendship 7 (U.S.)	John H. Glenn, Jr.	3	4.9 hrs.	First U.S. manned orbital mission
May 24, 1962	Aurora 7 (U.S.)	M. Scott Carpenter	3	4.9 hrs.	
Aug. 11–15, 1962	Vostok 3 (U.S.S.R.)	Andrian G. Nikolayev	64	94.3 hrs.	Part of first Soviet "group flight"
Aug. 12–15, 1962	Vostok 4 (U.S.S.R.)	Pavel R. Popovich	48	71.0 hrs.	Came within 3 miles of Vostok 3 on first orbit
Oct. 3, 1962	Sigma 7 (U.S.)	Walter M. Schirra, Jr.	6	9.2 hrs.	
May 15–16, 1963	Faith 7 (U.S.)	L. Gordon Cooper, Jr.	22	34.3 hrs.	Last flight of Mercury program
June 14–19, 1963	Vostok 5 (U.S.S.R.)	Valery F. Bykovsky	81	119.1 hrs.	Part of second Soviet "group flight"
June 16–19, 1963	Vostok 6 (U.S.S.R.)	Valentina Tereshkova	48	70.8 hrs.	Came within 3 miles of Vostok 5. First woman in space
Oct. 12–13, 1964	Voskhod 1 (U.S.S.R.)	Vladimir M. Komarov, Konstantin P. Feoktistov, Boris B. Yegorov	16	24.3 hrs.	First 3-man crew in space
Mar. 18–19, 1965	Voskhod 2 (U.S.S.R.)	Pavel I. Belyayev, Aleksei A. Leonov	17	26.0 hrs.	Leonov spent 10 min. outside spacecraft; first "walk in space"
Mar. 23, 1965	Gemini 3 (U.S.)	Virgil I. Grissom, John W. Young	3	4.9 hrs.	First U.S. 2-man crew in space
July 3–7, 1965	Gemini 4 (U.S.)	James A. McDivitt, Edward H. White	62	97.9 hrs.	White spent 21 min. outside spacecraft
Aug. 21–29, 1965	Gemini 5 (U.S.)	L. Gordon Cooper, Jr., Charles Conrad, Jr.	120	190.9 hrs.	First extended U.S. manned flight
Dec. 4–18, 1965	Gemini 7 (U.S.)	Frank Borman, James A. Lovell	206	330.6 hrs.	Served as rendezvous target for Gemini 6
Dec. 15–16, 1965	Gemini 6 (U.S.)	Walter M. Schirra, Jr., Thomas P. Stafford	16	25.9 hrs.	Rendezvoused within 1 foot of Gemini 7
Mar. 16, 1966	Gemini 8 (U.S.)	David R. Scott, Neil A. Armstrong	7	10.7 hrs.	First "docking" in space. Docked with unmanned Agena target vehicle. Gemini 8 was forced down because of failure of maneuvering rocket
June 3–6, 1966	Gemini 9 (U.S.)	Thomas P. Stafford, Eugene A. Cernan	44	72.3 hrs.	Made triple rendezvous with orbiting target vehicle; Cernan spent 2 hrs. 5 min. outside spacecraft
July 18–21, 1966	Gemini 10 (U.S.)	Michael Collins, John W. Young	43	70.8 hrs.	Linked in space with orbiting Agena target; when latter's rockets fired, both vehicles reached record height of 475 miles. Gemini also rendezvoused with other Agena craft
Sept. 12–15, 1966	Gemini 11 (U.S.)	Charles Conrad, Jr., Richard F. Gordon, Jr.	44	71.3 hrs.	Docked with Agena target; reached record height of 851 miles; Gordon made brief space walk

SPACE FLIGHT: FROM VOSTOK TO THE SPACE SHUTTLE

Date of flight	Name of spacecraft	Name(s) of astronaut(s) or cosmonaut(s)	No. of orbits	Duration of flight	Remarks
Nov. 11–15, 1966	Gemini 12 (U.S.)	James A. Lovell, Jr. Edwin E. Aldrin, Jr.	59	94.6 hrs.	Docked with Agena target vehicle; Aldrin engaged in extravehicular activity for more than 5½ hours
Apr. 23–24, 1967	Soyuz 1 (U.S.S.R.)	Vladimir M. Komarov	17	26.7 hrs.	Komarov killed when vehicle crashed; first actual space fatality
Oct. 11–22, 1968	Apollo 7 (U.S.)	Walter M. Schirra, Jr. Donn F. Eisele R. Walter Cunningham	163	260.1 hrs.	Successful flight test of three-man Apollo command module
Oct. 26–30, 1968	Soyuz 3 (U.S.S.R.)	Georgi T. Beragovoi	60	94.8 hrs.	Rendezvous with unmanned Soyuz 2
Dec. 21–27, 1968	Apollo 8 (U.S.)	Frank Borman James A. Lovell, Jr. William A. Anders	10 (around Moon)	147 hrs.	First manned flight around the Moon
Jan. 14–17, 1969	Soyuz 4 (U.S.S.R.)	Vladimir A. Shatalov	45	71.2 hrs.	Rendezvous with Soyuz 5
Jan. 15–18, 1969	Soyuz 5 (U.S.S.R.)	Boris V. Volynov Aleksei S. Yeliseyev Yevgeni V. Khrunov	46	72.7 hrs.	Two men transferred to Soyuz 4 and landed with it
Mar. 3–13, 1969	Apollo 9 (U.S.)	James A. McDivitt David R. Scott Russell L. Schweikart	151	241 hrs.	First docking with lunar module
May 18–26, 1969	Apollo 10 (U.S.)	Thomas P. Stafford Eugene A. Cernan John W. Young	31 (around Moon)	192 hrs.	Descent in lunar module to within 9 miles of Moon
July 16–24, 1969	Apollo 11 (U.S.)	Neil A. Armstrong Edwin E. Aldrin, Jr. Michael Collins		195 hrs.	First manned landing on the Moon; Mare Tranquillitatis; lunar EVA time totalled 2 hrs. 13 min.
Oct. 11–16, 1969	Soyuz 6 (U.S.S.R.)	Georgi S. Shonin Valery N. Kubasov	80	118.7 hrs.	Together with Soyuz 7 and Soyuz 8, world's first triple launch of manned vehicles
Oct. 12–17, 1969	Soyuz 7 (U.S.S.R.)	Anatoly V. Filipchenko Viktor V. Gorbatko Vladislay N. Volkov	80	118.7 hrs.	Rendezvous maneuvers with Soyuz 8
Oct. 13–18, 1969	Soyuz 8 (U.S.S.R.)	Vladimir A. Shatalov Aleksei S. Yeliseyev	80	118.7 hrs.	
Nov. 14–24, 1969	Apollo 12 (U.S.)	Charles Conrad, Jr. Richard F. Gordon, Jr. Alan L. Bean		244.6 hrs.	Second manned lunar landing; Oceanus Procellarum; two lunar EVA's totalling 7 hrs. 39 min.
Apr. 11–17, 1970	Apollo 13 (U.S.)	James A. Lovell, Jr. Fred W. Haise, Jr. John L. Swigert, Jr.	—	142.9 hrs.	Planned lunar landing aborted after rupture of oxygen tank in service module
Jan. 31–Feb. 9, 1971	Apollo 14 (U.S.)	Alan B. Shepard, Jr. Stuart A. Roosa Edgar D. Mitchell		216.7 hrs.	Third manned lunar landing; Fra Mauro highlands; two lunar EVA's totalling 9 hrs. 19 min.
Apr. 23–25, 1971	Soyuz 10 (U.S.S.R.)	Vladimir A. Shatalov Nikolai N. Rukavishnikov Aleksei S. Yeliseyev	32	59.8 hrs.	Docking with Salyut orbital space station
June 6–30, 1971	Soyuz 11 (U.S.S.R.)	Georgi T. Dobrovolsky Vladislav N. Volkov Viktor I. Patsayev	359 (including Salyut)	569.7 hrs.	Longest stay in space; rendezvous with Salyut space station, which cosmonauts occupy; all three die of accidental depressurization as Soyuz returns to Earth
July 26–Aug. 7, 1971	Apollo 15 (U.S.)	David R. Scott James B. Irwin Alfred M. Worden		295.2 hrs.	Fourth manned lunar landing; three lunar EVA's for 18 hrs. 37 min.; use of lunar roving vehicle
Apr. 16–27, 1972	Apollo 16 (U.S.)	John W. Young Charles M. Duke, Jr. Thomas K. Mattingly		319.8 hrs.	Fifth manned lunar landing; three lunar EVA's for 20 hrs. 14 min.; lunar orbiter launched
Dec. 7–19, 1972	Apollo 17 (U.S.)	Eugene Cernan Harrison Schmitt Ronald E. Evans		301.8 hrs.	Sixth and last U.S. manned lunar landing; Taurus-Littrow Valley
May 25–June 22, 1973	Skylab I (U.S.)	Charles Conrad, Jr. Joseph P. Kerwin Paul J. Weitz	395	672.7 hrs.	Manned Earth-orbiting space station; test of human space endurance; studies of space, Earth

SPACE FLIGHT: FROM VOSTOK TO THE SPACE SHUTTLE

Date of flight	Name of spacecraft	Name(s) of astronaut(s) or cosmonaut(s)	No. of orbits	Duration of flight	Remarks
Nov. 16, 1973– Feb. 8, 1974	*Skylab III* (U.S.)	Gerald P. Carr Edward G. Gibson William R. Pogue		84 days, 1 hr., 16 min.	Manned Earth-orbiting space station; record test of human space endurance; studies of Sun, space, Earth resources
Dec. 18– 26, 1973	*Soyuz 13* (U.S.S.R.)	Pyotr Klimuk Valentin Lebedev		9 days	Test of spacecraft that serves *Salyut* space station
July 3– 19, 1974	*Soyuz 14* (U.S.S.R.)	Pavel Popovich Yuri Artyukhin		17 days	Docking with *Salyut 3* and crew transfer to it
May 24– July 26, 1975	*Soyuz 18* (U.S.S.R.)	Pyotr Klimuk Vitaly Sevastyanov		63 days	Docks with *Salyut 4* space station; experiments involving Sun, outer space
July 15– 24, 1975 July 15– 21, 1975	*Apollo* (U.S.) *Soyuz 19* (U.S.S.R.)	Thomas P. Stafford Donald K. Slayton Vance D. Brand Aleksei A. Leonov Valery N. Kubasov	136	9 days 6 days	*Apollo-Soyuz Test Project*, the first cooperative international space flight. *Apollo* successfully docks with *Soyuz*
April 9– Oct. 11, 1980	*Soyuz 35* (U.S.S.R.)	Leonid Popov Valery Ryumin		185 days	Docked with *Salyut 6* for scientific research and repairs to space station
Mar. 13– May 26, 1981	*Soyuz 38* (U.S.S.R.)	Vladimir Kovalyonok Viktor Savinykh		75 days	Docked with *Salyut 6*; made repairs; had visitors via *Soyuz 39* & *40* before returning to Earth
April 12– 14, 1981	*Columbia 1* (U.S.)	John W. Young Robert J. Crippen		54.5 hrs.	First flight of a reusable space vehicle (the shuttle)
For a listing of subsequent space shuttle missions, see the chart on pages 246–247.					
May 13– Dec. 10, 1982	*Soyuz 41* (U.S.S.R.)	Anatoly Berezovoy Valentin Lebedev		211 days	First to dock with new *Salyut 7* space station
Feb. 8– April 11, 1984	*Soyuz T-10* (U.S.S.R.)	Leonid Kizim, Vladimir Solovyev, Oleg Atkov		63 days	Ferried crew to *Salyut 7* space station
Apr. 3– Oct. 2, 1984	*Soyuz T-11* (U.S.S.R.)	Yuri Malyshev, Gennedy Strekalov, Rakesh Sharma		182 days	Ferried crew to *Salyut 7* space station
July 17– July 29, 1984	*Soyuz T-12* (U.S.S.R.)	Vladimir Dzhanibekov Igor Volk, Svetlana Savitskaya		12 days	Ferried crew to *Salyut 7* space station; first woman to walk in space
Aug. 29– Sept. 7, 1988	*Soyuz TM-5* (U.S.S.R.)	Abdullah Ahad Mohmand Vladimir Lyakhov		10 days	Ferried crew to Mir space station; malfunction strands crew in space for 24 hours
Nov. 26– Dec. 21, 1988	*Soyuz TM-3* (U.S.S.R.)	Jean-Loup Chrétien Alexander Volkov Sergei Krikalev		28 days	Ferried crew to Mir space station; Chrétien becomes first Frenchman to walk in space
Sept. 5, 1989– Feb. 19, 1990	*Soyuz TM-8* (U.S.S.R.)	Alexander S. Viktorenko Alexander A. Serebrov		168 days	Ferried crew to Mir space station; automatic docking system failed, requiring manually controlled linkup
Feb. 11– Aug. 9, 1990	*Soyuz TM-9* (U.S.S.R.)	Anatoly Solovyov Aleksand Balandin		180 days	Ferried crew to Mir space station; emergency space walk required to repair thermal blankets on *Soyuz* spacecraft
Dec. 2 1990– May 26, 1991	*Soyuz TM-11* (U.S.S.R.)	Viktor Afansev Musa Maranov		176 days	Ferried crew to Mir space station; journalist Toyohiro Akiyama is the first Japanese citizen in space

supplied by the capsule's pumping system and had an electrical system so the spacewalker could talk to the astronaut in the capsule.

MEN WALK ON THE MOON

All these adventures were merely dress rehearsals for the big show, Project Apollo, in which astronauts would land on the Moon. The early Apollo spacecraft were launched by the Saturn IB, a two-stage rocket with an initial thrust of 1,653,467 pounds (750,000 kilograms). Beginning with *Apollo 8*, the launch vehicle was the classic Saturn V rocket. This powerful, three-stage vehicle was 282 feet (86 meters) high and had a takeoff thrust of about 7,716,180 pounds (3,500,000 kilograms).

The Apollo spacecraft consisted of three basic parts: the command module, the service module, and the lunar module. In addition, a spacecraft-lunar module adapter attached the Apollo craft to the Saturn V rocket. Each Apollo flight also featured a launch-escape function, which could whisk the astronauts to safety in case of a malfunction during the initial launching stages.

The command module was the control center of the spacecraft. There three astronauts worked and lived during the mission (except when two of the men were in the lunar module and on the Moon). The service module contained electrical and propulsion systems as well as most of the spacecraft's oxygen supply. The lunar module served during the actual exploration of the Moon's surface. It consisted of an ascent stage and a descent stage.

The initial manned flights of the Apollo spacecraft were planned for 1967. But in January 1967, while testing their Apollo vehicle at Cape Kennedy, three astronauts were killed. Their capsule caught fire when uncovered wiring caused a spark in the pure-oxygen atmosphere. The ensuing investigations and redesign of the vehicle delayed the program for more than a year.

Between 1968 and 1969, several practice Apollo flights took place. Not only did they test the flight performance of the craft, but they also scouted ahead for potential lunar landing sites.

Apollo 11 launched for the Moon on July 16, 1969, and touched down four days later. Neil Armstrong and Edwin "Buzz" Aldrin bounced around on the lunar surface, while Michael Collins remained with the orbiting command ship. For more than two hours, the two Moon walkers collected rocks, photographed the area, and set up scientific experiments. The men moved easily in the low gravitational field, hampered only slightly by their bulky space suits. Their oxygen supplies, communications gear, and other equipment were contained in portable life-support systems strapped to their backs.

The astronauts placed several seismometers on the lunar surface. These have enabled scientists to monitor seismic events: both moonquakes and meteorite impacts. Meanwhile, a laser reflector that the astronauts left on the Moon has allowed scientists to measure the distance between the Earth and the Moon to an accuracy of 6 inches (15 centimeters). It has also provided invaluable information on the pos-

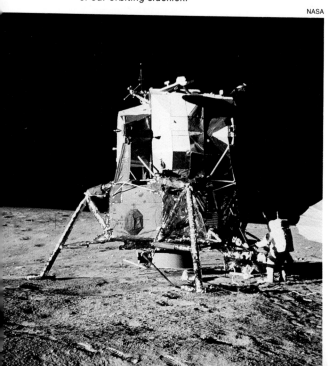

The world watched in amazement as Apollo 11 *made the first manned lunar landing on July 20, 1969. In the five lunar landings that followed, astronauts collected scientific data that added immensely to our knowledge of our orbiting sidekick.*

NASA

Prior to the space-shuttle era, all manned spaceflights ended with a splashdown in the ocean. This eliminated the need for a braking and landing system, but it also exposed the space capsule and its crew to certain environmental hazards. Here a flotation collar supports both the capsule and the astronauts until pickup by helicopter.

sible weakening of gravitational forces between the two bodies.

The astronauts also gathered velocity particles blown out from the Sun. By exposing a sheet of aluminum foil, they collected a small number of these atomic particles, which make up the solar wind.

Of great interest to both scientists and the general public was the precious cargo of lunar rocks brought back by *Apollo 11*. This material weighed about 44 pounds (20 kilograms) and included core samples as well as surface rocks and "soil." Some were igneous rocks, which suggested that the Moon might feature volcanic activity. The lunar soil itself was riddled with solar-wind particles.

In the years that followed, six other Apollo missions flew toward the Moon. *Apollo 13*, launched in 1970, had to return to Earth when an oxygen-tank explosion forced the mission to end prematurely. The final Moon flight, *Apollo 17*, returned to Earth on December 19, 1972.

Apollo's Legacy

Recently, the Soviet Union revealed that it had been planning to send a manned spacecraft to the Moon in 1968, but the program was canceled due to repeated failures of the booster rocket. Although the Soviet Moon program failed, the country continued its commitment to a manned space program. In 1971 cosmonauts began docking with their Salyut space station. In 1986 the

Skylab, the first U.S. space station, was launched in May 1973. For nine months, three successive crews used it to break human space-endurance records.

Buran, *the somewhat smaller Soviet equivalent of the U.S. space shuttle, is designed to carry crew and cargo back and forth from the space station. It is lifted into space by the Energiya rocket.*

SOVIET SPACE VEHICLES

Just as Apollo, Gemini, and the space shuttles *Challenger*, *Columbia*, *Discovery*, and *Atlantis* have become household words in this country, newspapers in the Soviet Union follow the adventures of cosmonauts working with these space vehicles:

- *Buran*—Its name means "blizzard," for its white color and the strong role the Soviets hope it will play in space exploration. This vehicle is almost identical to the American shuttle. It has been flown empty on tests, and should begin carrying crews late in the 1990s.
- *Energiya*—"Energy" is the rocket that boosts Soviet vehicles into space.
- *Icarus*—Named for the Greek mythological character who dared to fly on man-made wings, this is the Soviet version of NASA's manned maneuvering unit, used for space walks.
- *Mir*—"Earth" or "Peace" are both translations for this space station's name. Launched in 1986 and still in orbit, it is the world's first modular space station. The Soviets have added on a 20-ton materials processing lab and two similarly sized Kvant, or "quantum," modules, which hold telescopes and other scientific equipment.
- *Salyut*—This "salute" to Soviet astronaut Yuri Gagarin was a series of seven small space stations that remained in orbit between 1971 and 1991.
- *Soyuz*—Its name means "union," and it was first launched in 1967. It is still the workhorse spacecraft of the Soviet space fleet, transporting astronauts to and from the Soviet space station. It is complemented by a cargo ship called Progress.
- *Voskhod*—Its name means "ascent," and it put the second six cosmonauts into orbit in 1964.
- *Vostok*—The name means "East." This series of spacecraft launched the first six cosmonauts into space between 1961 and 1963.

Maura Mackowski

Soviets launched the Mir ("Peace") space station. Since then they have maintained an almost continuous presence in space. On December 21, 1988, a two-man cosmonaut crew returned to Earth after a 366-day stay aboard Mir.

The United States also put up a space station—Skylab—in 1973. Equipped to house a crew of three astronauts, Skylab carried an array of instruments for studying space, the stars, the Earth, and the biological effects of space on humans. Three separate crews of astronauts occupied Skylab. The space station was supposed to stay in orbit until 1983, but it entered into the atmosphere in 1979 and burned to a crisp.

Despite the tremendous success of the Moon-exploration program, space planners

In a space-shuttle lift-off, two external solid-rocket boosters blast the shuttle above the Earth's atmosphere. The boosters are then jettisoned and the shuttle's own engines take over, fueled by an external fuel tank (left). The boosters parachute into the ocean for recovery and reuse; the external fuel tank (right) disintegrates upon reentry into the atmosphere.

soon realized that the Apollo-Saturn system, based upon expendable rocket launchers, was simply too expensive to continue. The Apollo program lasted only four years, but cost over $20 billion.

MEET THE SPACE SHUTTLE

In 1969 President Nixon appointed a NASA task group to find a low-cost alternative to the Saturn rocket, and in 1972 NASA approved a design for the space shuttle. Nine years passed before the shuttle first roared into orbit, but today it is the workhorse of the space program. Comparing it with the Apollo spacecraft is like contrasting a Porsche to a horse-drawn carriage. It can launch satellites, carry a large science laboratory, and repair satellites in orbit or bring them back to Earth. At the completion of a mission, it reenters Earth's atmosphere and lands like a plane on a runway. After a few months, it's ready to be launched again.

Originally, NASA engineers envisioned the space shuttle as a spacecraft with two manned stages. The booster craft, about the size of a Boeing 747 jumbo jet, would carry the orbiter piggyback until it reached a high enough altitude, at which point the two vehicles would separate. The booster crew would turn around and fly back to base, while the orbiter would kick in its rocket engines and continue to fly into orbit.

But budget restrictions forced NASA to consider a simpler design. The space shuttle, as we know it today, consists of three separate elements—the orbiter itself, twin solid-rocket boosters, and an external fuel tank.

The solid-rocket boosters (SRBs) are the largest solid-fuel rockets ever developed for spaceflight. They give the shuttle two minutes of sturdy thrust to get it above the thickest layers of the Earth's atmosphere, to where the three main engines built into the orbiter can work most efficiently. After the SRBs burn out, they parachute into the ocean for recovery and are later reused. Each booster stands 141 feet (43 meters) high and weighs almost 1,322,774 pounds (600,000 kilograms) when fueled. A booster provides lift-off thrust of 3,306,935 pounds (1,500,000 kilograms) at launch (see also p. 234).

The external fuel tank carries liquid hydrogen and oxygen in separate compartments to feed the orbiter's three main rocket engines. The tank's diameter is 28 feet (8.5 meters), its height is 154 feet (47 meters), and its empty weight is 74,957 pounds (34,000 kilograms). At the base of the shuttle's launch platform are boxlike fueling stations for liquid hydrogen and oxygen. Machinery pumps these cold liquids into the external tank in the final hours before lift-off.

The only part of the shuttle system that goes into orbit is the orbiter itself. The vehicle is 121 feet (37 meters) long, spans 79 feet (24 meters) across the wingtips, and stands 56 feet (17 meters) tall on its landing gear. The crew cabin, located in the front part of the craft, carries an average of seven people, or 10 in case of an emergency. The mission commander and the pilot sit at the forward flight deck, surrounded by far more control lights and panels than you'd find in any Earthbound vehicle. The other crew members, technical and scientific personnel, work at the rear flight deck, called the "aft," or inside the payload bay. They are referred to as mission specialists and payload specialists, respectively.

Each orbiter is designed to last for 100 flights. But to accomplish this, NASA's engineers had to overcome a major obstacle: the amazingly high temperatures created by friction when the spacecraft reenters Earth's atmosphere.

The solution was to cover the undersurface of the orbiter with tiles. The current shuttle design uses 20,548 individually shaped silica-fiber tiles, varying in thickness from 1 inch (25.4 millimeters) to 5 inches (127 millimeters). Each tile is specially shaped to hug the contours of the orbiter, and can withstand temperatures up to 2,300° F (1,260° C). The nose cap and the edges of the wings, which are subjected to the most heat during reentry, are covered with Reinforced Carbon-Carbon (RCC), which can withstand temperatures up to 3,000° F (1,649° C), hot enough to melt steel. Engineers use lightweight thermal-insulation blankets woven from silica fibers on parts of the orbiter's upper surface, such as the shuttle's cargo-bay doors, where the temperature will not exceed 1,200° F (649° C).

At lift-off the space shuttle stands 185 feet (56 meters) tall and weighs more than 4,409,246 pounds (2,000,000 kilograms). All three engines and both solid-rocket boosters are burning at the moment of launch, producing well over 7 million pounds (3,175,144 kilograms) of thrust.

After two minutes of powered flight, at an altitude of about 28 miles (45 kilometers), the solid boosters burn out, separate from the orbiter, and parachute into the ocean for recovery. The orbiter and fuel tank hurtle into space for another 6.5 minutes. About 8.5 minutes after launch, at an altitude of about 68 miles (109 kilometers), the main engines cut off, and the external tank tumbles back into the atmosphere and burns up.

While the orbiter circles the planet, it travels upside down, so the payload-bay

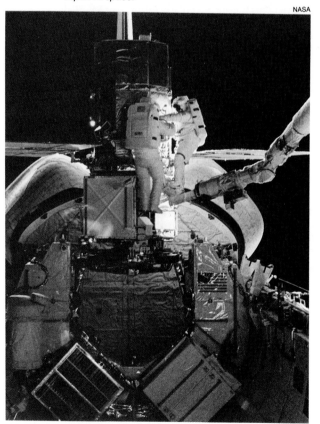

All shuttles are equipped with a robotic arm, an invaluable tool for a variety of repair purposes. Below, astronauts stand on the arm while performing satellite repair in space.

NASA

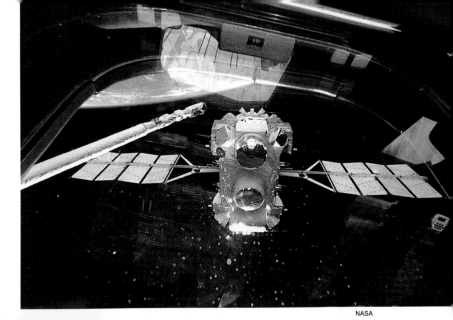

Space-shuttle cargo bays can hold up to 25 tons. Below, the shuttle Discovery releases the 11.5-ton Hubble Space Telescope into orbit in April 1990. A year later, the Atlantis deployed the 17-ton Gamma Ray Observatory (right), a mission that required an unexpected space walk to manually extend the satellite's antenna.

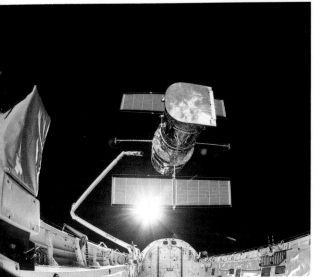

becomes a glider and makes a sweeping turn to line up with its landing runway. The pilot has just one shot at landing, before touching down on the shuttle's set of tricycle wheels.

The Shuttle at Work

Perhaps the most important feature of the orbiter is its payload bay. Stretching 59 feet (18 meters) long and 15 feet (4.5 meters) in diameter, the bay can tote up to 50,000 pounds (22,680 kilograms) into orbit. Unmanned space probes such as Galileo and Magellan were launched from a shuttle's payload bay.

At the edge of the bay stands a robotic arm, called the remote manipulator system (RMS). Astronauts control this device from the shuttle's aft flight deck. Not only does the arm seize and release satellites, it also provides an astronaut with a place to stand in space.

But if the payload doors do not close or some other problem arises, astronauts have to leave the shuttle and perform repairs while floating in space. This is known as extravehicular activity (EVA). On some flights, EVAs are scheduled for satellite-repair tasks. Other EVAs have involved astronauts piecing together large structures in space—practice construction projects for the day when the United States will begin building a permanently manned space station. Occasionally astronauts must perform unscheduled EVAs, and they must be prepared for such emergencies on a moment's notice.

doors face out to space. These doors generally remain open, so that the pipes lining their inner surface can radiate heat away from the spacecraft.

Two maneuvering engines give extra thrust to push the shuttle into orbit and help steer the orbiter in space. They also help it return to Earth. At the end of the mission, the orbiter fires its maneuvering engines, slowing the spacecraft so much that it dives into the atmosphere. Air in the atmosphere then acts like a brake, bringing the spacecraft from Mach 25 (25 times the speed of sound) down to 221 miles per hour (356 kilometers per hour) in only 30 minutes. Flying without engine power, the shuttle

SIGNIFICANT SPACE SHUTTLE FLIGHTS

Flight designation	Date of flight	Remarks
STS-1	Apr. 12–14, 1981	54-hour flight designed to verify that the shuttle could be launched, operate in orbit, then return to Earth safely. Because of the complexity of the system and the confidence that had been developed since the early manned spaceflight days, the first flight was manned rather than automatic. Veteran astronaut John Young and rookie Robert Crippen were selected to fly the mission on *Columbia*. It was, in almost all respects, perfect. Some concern developed when TV cameras showed small chips from the heatshield tiles on the aft thruster pods. The only cargo carried was an array of development flight instrumentation that measured all aspects of how the shuttle responded in flight. This instrumentation stayed aboard through the STS-5 flight.
STS-2	Nov. 12–14, 1981	The planned five-day mission was cut to 54 hours by a fuel cell problem. *Columbia* carried its first science payload, an Earth mapping radar and other observation instruments. Also aboard was the robot arm on its first test flight and a special contamination monitor.
STS-3	Mar. 22–30, 1982	The science payload included instruments for measuring the space environment around the shuttle and a mini-factory for making latex microspheres. It was flown by Jack Lousma and Gordon Fullerton.
STS-4	June 27–July 4, 1982	The first Defense Department payload; the first Getaway Special, assembled by Utah students; and the first biomedical separation facility were carried on this flight. The main military component was an infrared telescope to study the upper atmosphere. Its lid refused to open, however. It was flown by T. K. Mattingly and Henry Hartsfield.
STS-5	Nov. 11–16, 1982	The development instrumentation was still aboard and the satellites of the first two commercial customers—Telesat E for Canada and SBS for Satellite Business Systems—were deployed. The first attempt at a shuttle space walk was scrubbed when both space suits developed minor problems. The crew was made up of Vance Brand, Robert Overmyer, Joseph Allen, and William Lenoir.
STS-6	Apr. 4–9, 1983	First flights of *Challenger*, Inertial Upper Stage, and Tracking and Data Relay Satellite System (TDRS-1). First shuttle space walk.
STS-7	June 18–24, 1983	First "proximity operations" with a retrievable satellite (SAPS-01), first five-person crew, first American woman in space (Sally K. Ride).
STS-8	Aug. 30–Sept. 5, 1983	First night launch and landing, first American black to fly (Guion S. Bluford, Jr.), heavy-lift testing of robot arm.
STS-9	Nov. 28–Dec. 8, 1983	First flight of full Spacelab carrying 71 experiments. Largest crew (six men) and first flight of payload specialists. First around-the-clock science operations in space. Mission extended a day.
41-B	Feb. 3–11, 1984	First test flights of Manned Maneuvering Unit (MMU), first time humans took space walks untethered to their spaceship (up to 100 meters away). Both satellites deployed on the mission were lost when the nozzles on their PAM-D stages broke off shortly after ignition.
41-C	Apr. 6–13, 1984	First repair of a satellite by another spacecraft. Using the MMU and the robot arm, the crew picked up the ailing Solar Maximum Mission satellite and replaced the attitude control module and the electronics for one of the science instruments.
41-G	Oct. 5–13, 1984	The mapping radar flown on STS-3 was upgraded and reflown along with a large format camera and other instruments. Two astronauts (including Kathryn Sullivan, the first U.S. woman to do so) held a space walk to operate a test model of a satellite refueling station. The shuttle also deployed the Earth Radiation Budget Satellite. The seven-person crew was the largest ever.
51-A	Nov. 8–16, 1984	After deploying two satellites, the crew retrieved the Palapa and Westar satellites lost on the 41-B mission and returned them to Earth for refurbishment.
51-C	Jan. 24–27, 1985	The first totally military mission; it reportedly carried an electronic "ferret" satellite for eavesdropping on Soviet radio traffic.
51-D	Apr. 12–19, 1985	Two satellites were deployed, but one, Syncom IV-3, failed to fire itself into a higher orbit. An unplanned space walk was held to attach a flyswatter-type device to the robot arm in an unsuccessful attempt to trip Syncom's firing lever. Sen. Jake Garn, flying as a payload specialist, became the first elected official in space.

Flight desig- nation	Date of flight	Remarks
51-B	Apr. 29–May 6, 1985	The second flight of full Spacelab. It carried crystal growth and fluid mechanics experiments and special cages for holding rats and monkeys on future life-science missions.
51-G	June 17–24, 1985	A retrievable science satellite (Spartan) was flown, and the shuttle served as the target for a laser tracking experiment.
51-F	July 29–Aug. 8, 1985	Spacelab 2 carried an advanced instrument pointing system with four solar telescopes, three astrophysics instruments (X-ray and infrared telescopes and a large cosmic-ray detector), and a reflight of plasma instruments from STS-3.
61-A	Oct. 30–Nov. 6, 1985	Spacelab D-1, with materials and life-science experiments, was the first Spacelab mission to be purchased by another nation (West Germany).
51-L	Jan. 28, 1986	In the greatest disaster of the U.S. space program, the main external tank of the shuttle *Challenger* exploded less than two minutes after lift-off. The seven-person crew perished, including Christa McAuliffe, a high-school teacher and the first ordinary U.S. citizen in space.
STS-26	Sept. 29–Oct. 3, 1988	First flight after 32-month hiatus, shuttle deployed tracked data relay satellite.
STS-27	Dec. 2–6, 1989	Classified military mission.
STS-29	Mar. 13–18, 1989	Deployed tracked data-relay satellite.
STS-30	May 4–8, 1989	Deployed Magellan radar-mapping satellite on flight to Venus.
STS-28	Aug. 8–13, 1989	Classified military mission deployed advanced photo-reconnaissance satellite.
STS-34	Oct. 18–23, 1989	Launched the Galileo space probe on a six-year flight to Jupiter.
STS-33	Nov. 22–27, 1989	Classified military mission launched at night.
STS-32	Jan. 9–20, 1990	*Columbia* spent more than 10 days in space, setting a new shuttle record. The crew retrieved the Long Duration Exposure Facility satellite, used to test how various materials survive in a space environment.
STS-31	Apr. 24–29, 1990	*Discovery* launched the Hubble Space Telescope, achieving the highest shuttle orbit ever—380 miles (613 kilometers).
STS-41	Oct. 6–10, 1990	*Discovery* was the first shuttle to fly after a series of mechanical failures grounded the fleet for nearly six months. The crew launched the Ulysses probe, which is scheduled to arrive at the Sun in 1994.
STS-38	Nov. 15–20, 1990	Classified mission for the Department of Defense. The shuttle landed at the Kennedy Space Center, the first shuttle to do so since the *Challenger* explosion. Landing at Edwards Air Force Base was canceled due to bad weather.
STS-35	Dec. 2–10, 1990	Nine-day astronomy assignment for the *Columbia*. The shuttle released into orbit the Astro astronomy laboratory, consisting of three ultraviolet telescopes and one X-ray telescope.
STS-37	Apr. 5–11, 1991	*Atlantis* deployed the Gamma Ray Observatory. First EVA activity since the *Challenger* explosion.
STS-39	Apr. 28–May 6, 1991	*Discovery* tested equipment for the Strategic Defense Initiative (SDI). The 40th shuttle mission.

HOW SAFE IS SPACE?

Launching a few pounds of electronic hardware into space was one thing. But sending a human was something different altogether. Many problems had to be solved before manned spaceflight could be achieved. Among these problems were the physical effects of excessive acceleration, weightlessness, radiation in space, and the reentry into the Earth's atmosphere.

Anyone venturing into orbit on a spaceship, scientists calculated, would be subjected to acceleration forces as high as 8 G's, or eight times the force of gravity. Experts already knew that the top limit that a pilot could withstand without passing out was 4 G's. How, then, could future astronauts withstand even greater acceleration?

To test humans' ability to tolerate high acceleration, engineers spun them as fast as possible. A gondola was suspended from the end of a large arm, which in turn was attached to a rotary motor. A volunteer sat in the gondola. As the motor was set going, the gondola sped round and round with increasing speed. Volunteers who remained in a sitting position underwent alarming effects. Blood circulation slackened. When blood failed to reach certain vital areas of the brain, the volunteer would lose consciousness. But if he lay on his back with his head slightly raised and the knees slightly bent, the physical effects were not nearly so severe. It became evident that astronauts would have to lie almost supine, or on their backs, during moments of extreme acceleration.

Another concern was how astronauts would be affected by weightlessness. Here on Earth, investigators simulated a state of weightlessness by means of an airplane maneuver. The pilot would first make a shallow dive, then point the nose of the plane upward while throttling back, cutting power to the engine. The plane went through a curve, moving upward at first, then leveling out, and finally pointing downward. From the instant the pilot released the throttle until full-engine power was restored, the plane and everything it contained were weightless. The periods of weightlessness ranged from 25 seconds to about a minute.

During this brief period, people in an experimental chamber in the plane floated about freely. So did all loose objects. Weightlessness produced physiological side effects. It made humans disoriented, dizzy, nauseous, and unable to control their movements. The plane became known as the Vomit Comet, for obvious reasons. These effects varied widely in different individuals. They seemed to be slight in some, overpowering in others.

Clearly, astronauts would have to go through a period of physical conditioning to endure extreme acceleration and weightlessness. To test this idea, researchers put animals through training, and then sent them into orbit. The animals returned safely to Earth, apparently little worse for their experience. Scientists concluded that well-trained astronauts should be able to withstand intense acceleration and short-term weightlessness.

Investigators also knew that spacecraft outside Earth's atmosphere would be bombarded by radiation. Ultraviolet rays from the Sun would pummel the craft with undiminished force, because no air exists to absorb them. Solar-flare eruptions on the Sun's surface send out X rays and electrons into space. X rays are also produced when the free electrons hurling through space strike the metal of a spacecraft's walls. At the same time, spacecraft would encounter high-energy cosmic rays. The discovery of the Earth's radiation belt in 1958 laid bare another source of potential danger.

But studies carried out with unmanned satellites showed that radiation did not offer an insurmountable obstacle to spaceflight. Even very thin sheets of metal could stop most of the ultraviolet rays and the less powerful X rays that shoot from the Sun. Under normal conditions, the walls of the spacecraft would be thick enough to shield its occupants against radiation. It would not be practical to provide shielding against cosmic rays; but normally there would not be enough of this type of radiation to constitute a hazard. And the Earth's radiation

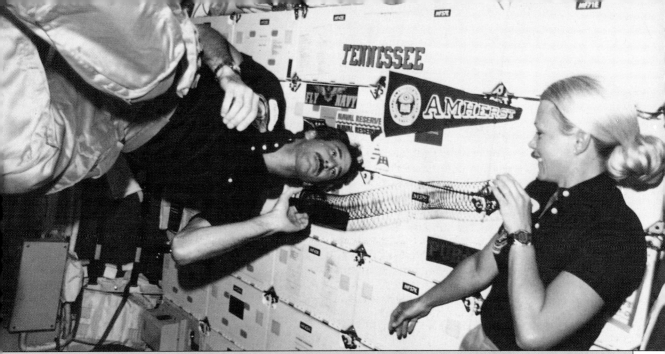

belt could be avoided with proper planning.

It was also necessary to protect manned spacecraft from the tremendous heat that would sear the craft as it plunged through the atmosphere on its return from space. The technique finally devised consisted of slowing down the craft with backward-facing braking rockets, called retro-rockets. Engineers also fitted the craft with a shield to absorb the heat produced during the period of reentry.

Astronaut safety is NASA's prime concern during spaceflight. Weightlessness (above) on long-term flights can cause a variety of ills, including cardiovascular problems and bone degeneration. Astronauts are closely monitored (below) for any changes in their medical conditions. Specially designed space suits (below left) protect astronauts on space walks.

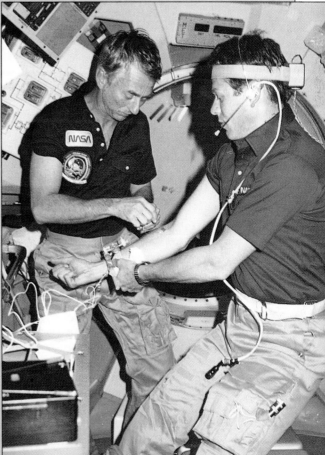

During an EVA the astronaut depends on his space suit, called the EVA mobility unit (EMU). The suit has a hard upper torso to which flexible pants are joined, plus gloves and a helmet. The life-support pack connects directly to the torso.

Before the shuttle first flew, astronauts who ventured into space were always tethered by a lifeline. But with the shuttle, the astronaut relies on the manned maneuvering unit (MMU). The unit snaps onto the back of the EMU and places two joysticks at the astronaut's fingertips for side-to-side motion (left hand) and rotation (right hand). Through a computer the joysticks command gas thrusters that push the astronaut, untethered, through space.

The Flying Scientists

As well as carrying unmanned probes and satellites for launch, the payload bay aboard the space shuttle is also equipped to carry Spacelab. Donated by the European Space Agency, a consortium of 14 countries, Spacelab is a pressurized laboratory module, featuring U-shaped pallets that expose experiments to the vacuum of space.

The pressurized module consists of two segments, each 9 feet (2.7 meters) long and 13 feet (4 meters) in diameter. The first segment, the core, contains life-support equipment, computers, and some working space. The other segment is a complete laboratory/workshop. The entire laboratory module connects to the crew compartment by means of a 15-foot (4.6-meter) flexible tunnel.

Packed with scientific instruments, Spacelab allows astronauts to perform a wide variety of experiments. Their science projects have ranged from physics and astronomy to botany and biology. For example, during one mission, investigators set up a fluid-physics module, in which they studied how fluids spread, change shape, and conduct heat without gravity. By studying fluids under these conditions, the researchers learned how liquid fuels behave in spacecraft tanks and how lubricants react to being in space.

Spacelab missions often involve crew members who are not professional astronauts, but scientists specializing in areas such as physics and chemistry. One advantage of the space shuttle over previous spacecraft is that during lift-off, acceleration is limited to about three times the force of gravity, or 3 G's. Reentry forces are typically less than 1.5 G's. As a result, scientists in good health can travel aboard the shuttle with only a minimum of training. Crews aboard American spacecraft are no longer made up exclusively of highly trained test pilots.

New and Improved

The space-shuttle program got off to a roaring start with the launching of *Columbia* on April 12, 1981. The program suffered a tragic setback, however, when, on January 28, 1986, the *Challenger* shuttle exploded two minutes after lift-off, killing its crew of seven. A commission appointed by President Reagan to investigate the accident concluded that the explosion had been caused by a faulty sealant ring, called an O-ring, in the right solid-rocket booster. Hot gases leaked out of the rocket booster through the faulty seal, burning a hole in the external liquid-fuel tank and setting off an explosion.

Following recommendations made by the commission, NASA began redesigning the booster joints and O-rings. The space agency also made a number of management changes and began developing a crew-escape system. The shuttle fleet remained grounded for 32 months until the program resumed with the successful launching of *Discovery* on September 29, 1988.

By 1991 it appeared that the shuttle program had completely regained its momentum. In April of that year, *Atlantis* sent the huge 35,000-pound (15,890-kilogram) Gamma Ray Observatory into low orbit. The mission marked the first time in over five years that astronauts had gone outside the spacecraft for an extravehicular activity. Only 17 days after *Atlantis* landed, *Discovery* streaked into orbit to conduct a series of military experiments for the weaponry program known as the Strategic Defense Initiative (SDI). The *Discovery* launching marked the 40th shuttle mission.

Despite a variety of setbacks early in its history, the shuttle has regained credibility as a reliable vehicle for space transportation and satellite deployment and repair. After the launch of the Endeavour in 1991, the latest addition to the shuttle fleet, NASA proposed plans to extend shuttle missions to as long as 28 days.

On April 25, 1991, NASA unveiled *Endeavour*, the newest addition to the shuttle fleet.

Another manned space shuttle might soon be orbiting the Earth, although it does not belong to the United States. In November 1988, the Soviet Union sent its first shuttle, *Buran*, into orbit for an unmanned test flight. The Soviet shuttle closely resembles the American craft in both size and appearance. One outstanding difference is that the Soviet orbiter has no large rocket engines of its own, but gets most of its propulsion from a massive disposable booster rocket similar to the Saturn V, the rocket used for the U.S. Apollo space program. The American space shuttle has three main engines, which are reusable.

Meanwhile, NASA is planning to upgrade the shuttle fleet, increasing the maximum mission time from 10 to 16 days, and ultimately perhaps to as long as 28 days. Some experts are concerned, however, that after several weeks in low gravity, pilots suffering from "space sickness" will be too disoriented to land the shuttle back on Earth. The prospect of longer missions has raised the possibility of starting automated, unmanned space-shuttle flights.

In theory the space shuttle could be equipped to fly into orbit and launch a satellite without the presence of even a single astronaut. When it comes time to return to Earth, radio beacons placed near the landing strip would tell the shuttle where it was in relation to that runway. Onboard computers could process the information and guide the spacecraft in for a landing.

For the time being, NASA is reluctant to start automating the shuttle fleet. The space agency argues that in the event of an emergency—if, for instance, the payload doors were to become stuck—only an astronaut would be able to save the mission. Also, NASA officials acknowledge that there is tremendous public appeal in manned spaceflight. More than 20 years after they walked on the Moon, astronauts are still considered heroes. NASA may need that popular support in the near future, if it hopes to gain sufficient funding for future projects. For instance, over the next decade, the United States hopes to launch a permanently manned orbiting space station, with the assistance of Canada, Japan, and the European Space Agency. The Soviet Union, once America's rival in space, has proposed a joint mission to Mars. On our own planet, bitter political conflicts continue, but in space, at least, it seems "the people on this Earth are truly one."

The extraordinary pace of innovative space technology makes it clear that the only limits to future space exploration will be the limits of our own imagination.

THE FUTURE OF THE SPACE PROGRAM

by Mark Strauss

How far should we venture into space? Should we build manned stations that orbit our own planet, send pioneers to live on the Moon, explore Mars, even push to the outer reaches of the solar system? The National Aeronautics and Space Administration (NASA), the U.S. space agency, has plans to do all these things, and they have people ready for the adventure. But the price will be steep: space exploration costs money and risks lives. Exactly how fast NASA can carry out its ambitious plans is hard to predict.

The first 25 years of the U.S. space program will probably be remembered as a golden age. From the launching of the first U.S. satellite, to the Mercury program that put our first man in space, to the unforgettable image of the American flag being planted on lunar soil, it seemed that the only limits to space exploration were the limits of our own imagination.

But the 1986 explosion of the space shuttle *Challenger* was a tragic reminder that the road to space will not always be smooth. In the aftermath of the accident, many people raised questions. Should we believe what NASA says it can do? And in what direction is our space program heading? After all, the heady days of the "space race" were apparently over. No longer was the space program driven by a single, all-encompassing goal: President Kennedy's vision to land a man on the Moon before the end of the 1960s. The space shuttle had promised to turn space travel into a routine event. But by the late 1980s, this seemed only a hollow promise.

Following the *Challenger* disaster, NASA's administrator, Dr. James Fletcher, sought to restore confidence in the space program and put America back on a path that would one day take us to the far edges of the solar system. Fletcher formed a task force to study potential U.S. space initiatives. After months of deliberation, the group outlined a list of proposals that, in the next century, would put America back in space for keeps.

MISSION TO PLANET EARTH

The first initiative calls for an in-depth study of the most important planet in our solar system: Earth. The proposed "Mission to Planet Earth" project will send a family of satellites into orbit to collect an unprecedented amount of data about our own world. Called the Earth Observing System (EOS), these satellites may start rocketing into space in 1998, finishing most of their work by 2015. They'll measure chemicals in the air and oceans, make maps of plant communities, take the Earth's temperature, map out the planet's magnetic field, and carry out many other tasks.

Why study our planet from space when we live on the ground? One pressing reason: scientists need to answer some troubling questions about Earth as a whole, and observation from space promises to answer many of them: How quickly is human pollution changing the makeup of the planet's atmosphere? What will that mean to the world climate? To what extent do the world's oceans, and the organisms that live in them, control climate? How fast are the rain forests really disappearing? These problems, and many others, can be answered by space observation.

NASA and other organizations already have orbited satellites to keep an eye on Earth. But the information picked up from the Earth Observing System will make such past observations seem insignificant. To process all the new facts, engineers will have to design computers bigger and faster than any currently available. Building such massive electronic brains is as important a part of the project as building the satellites.

PLANETARY EXPLORATION

NASA's second initiative is to expand the exploration of our solar system. Already, unmanned probes have voyaged through space, seeking out other planets. Among NASA's greatest successes were the Voyager spacecraft, which brought us our first close-up pictures of Jupiter, Saturn, Uranus, Neptune, and their mysterious moons. By today's technologically advanced standards, the Voyagers were simple, cheap, and remarkably reliable. Somewhat more complex probes are now on their way to their destinations. The Ga-

To traverse the terrain on other planets, NASA has specially designed a rover vehicle. Its navigation system allows a human operator to plan a general route, while built-in sensors steer it around local obstacles.

NASA

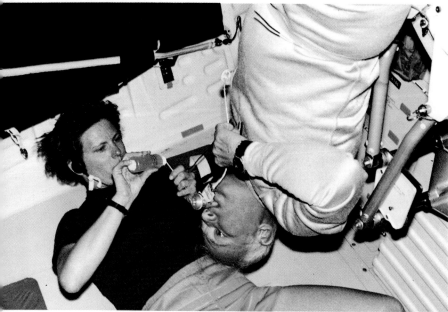

Medical tests conducted on space-shuttle missions measure human adaptability to space travel. The negative physiological effects that astronauts may encounter—and need to overcome—during a lengthy mission are an overriding concern of NASA's medical staff.

NASA

lileo probe will arrive at Jupiter in late 1995. The Ulysses probe will be peeking at the Sun in the summer of 1994.

Still another series of simple, hard-working probes is on NASA's drawing boards. Ideally, these little spacecraft will roll off assembly lines like cars, be fitted with the appropriate instruments, and head into space. The first of these may be the Comet Rendezvous/Asteroid Flyby (CRAF) probe. According to plan, CRAF will meet up with Comet Kopff sometime in the year 2000. CRAF will fire a dart loaded with instruments into the comet's very heart.

Next on the launching pad will be Cassini, designed to arrive at Saturn in 2002 and take a firsthand peek at the planet's rings. Other probes will target asteroids, the planet Uranus, and the big moons of Jupiter and Saturn.

MISSION TO THE MOON

But exploration of our solar system will not be limited to robot probes. Within 30 years, if all goes according to plan, human beings will stride across the surfaces of distant worlds.

For starters, NASA envisions a return trip to the Moon. This time, however, we're not going simply as tourists, just to take a few photographs and then go home. NASA intends to set up a permanently manned lunar base.

The first step to settling the Moon would be the launching of a Lunar Observer Probe. Despite the fact that the Moon is our closest neighbor, our knowledge of the lunar surface is sketchy. Previous Moon probes launched in the 1950s and 1960s had limited orbits, and the Apollo missions explored only a small portion of lunar territory. That's why a Lunar Observer Probe would have to scout ahead and locate the best spot suitable for a lunar base.

Once a base is established, astronauts must learn to live off the land. Moon pioneers could mine the lunar soil for helium-III, a form of helium that is very rare on Earth, but common on the Moon. Helium-III could be used to power nuclear-fusion reactors and might become a plentiful energy source in the future, perhaps even cheaper than oil.

It's even possible that lunar soil could be used to manufacture construction materials. Volcanic rock on the Moon contains a high concentration of silicon dioxide, which, when heated, melts into glass. By using giant mirrors to focus the Sun's rays, astronauts could melt down the soil, churning out tons of black lunar glass. Robotic machinery could stretch the molten-glass fibers, much like fiberglass. The material could be used for insulation, or it could be poured into molds and shaped into small, igloolike buildings.

An outpost on the Moon would also be a tremendous boon to astronomers. With almost no atmosphere to blur its vision, an optical telescope on the Moon could be 100,000 times more powerful than any telescopes here on Earth.

But perhaps most important, a Moon base would provide astronauts with invaluable experience on how to survive on the surface of a lifeless, alien world. It would be a stepping-stone to the next phase of NASA's solar-system exploration: a manned trip to Mars (see also page 264).

THE NATIONAL AERO-SPACE PLANE

In addition to the space-exploration initiatives outlined by Dr. Fletcher's task force, NASA officials have plans that are a little closer to Earth.

One of the most frustrating challenges NASA faces is finding an efficient way to get people and payloads into low-Earth orbit. When the space-shuttle program was introduced in the late 1970s, it promised routine access to space. From the very start, however, the space shuttle has suffered delays, mechanical failures, and constant budget problems. The *Challenger* explosion left the shuttle fleet grounded for nearly three years. And space officials grounded the fleet for five months in 1990, when engineers found a series of hydrogen leaks throughout the fleet's main engines.

The shuttle was intended to be a low-cost way to launch astronauts into orbit, but, in practice, it costs nearly $300 million per launch. What's more, NASA needs to employ a staff of nearly 12,000 people just to launch the spacecraft!

A future alternative to the space shuttle might lie in the National Aero-Space Plane, now being developed jointly by NASA and the Air Force. It's also known as the X-30, because it's No. 30 in a long

NASA envisions the X-30 aero-space plane (a proposed design illustrated above) to replace space shuttles as a more cost-efficient way to launch astronauts and payloads into orbit.

To collect astronomical data for future space exploration, the Kuiper Observatory, a modified jet, flies above the Earth's distorting atmosphere equipped with an infrared telescope.

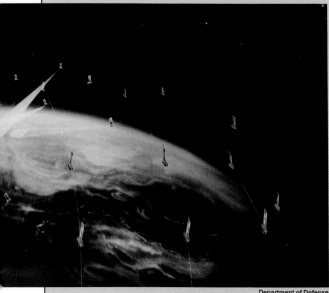
Department of Defense

THE STRATEGIC DEFENSE INITIATIVE

While NASA has its own plans for space, the nation's military has another plan. The popular press usually refers to it as the "Star Wars" program, but the government calls it the Strategic Defense Initiative, or SDI. On March 23, 1983, in a nationally televised address, President Ronald Reagan proposed an audacious new defensive system, which would be based primarily in space. The president envisioned an impenetrable shield that would render Soviet nuclear missiles "impotent and obsolete."

Back then, the proposed SDI arsenal consisted of a variety of high-tech gadgetry, most of which existed only on paper and in scientists' imaginations. One proposed orbiting device was a neutron-particle-beam weapon—a space-based accelerator that would aim atomic particles at a missile to destroy its electronics. Another potential weapon was a chemical laser that could aim intense beams of highly focused light at a missile's skin, causing its rocket fuel to explode.

In the years following Ronald Reagan's speech, the SDI program attracted many scientists, and much progress was made developing powerful laser beams and miniature rockets. But the technology needed to realize the program remained in the future. Moreover, the collapse of Communism in Eastern Europe and reforms within the Soviet Union greatly reduced tensions between the superpowers, making the frightening scenario of a full-scale Soviet nuclear attack suddenly seem very unlikely.

Still, some defense experts felt that the possibility of a small-scale missile at-

line of NASA/Air Force experimental planes. The X-30 will be the world's fastest plane, taking off from a runway like a normal jet, but building enough speed to escape gravity and shoot into orbit.

Unlike a normal jet, the X-30 will be equipped with a series of supersonic combustion engines, called "scramjets." These engines will do the difficult job of keeping a flame lit while air rushes through them at supersonic speeds. The entire front of the space plane will be shaped to scoop and compress air, channeling it into these scramjets. The X-30 will have to be able to take enormous gulps of air, because the plane will fly at altitudes where there's very little oxygen to feed its engines. Inside the combustion chambers of the scramjets, the air will mix with liquid or semifrozen hydrogen fuel, which must burn to power the plane.

The plane will race faster and higher, until there's not enough air left to keep its scramjets going. Then, at a speed of Mach 22 (which is 22 times the speed of sound, or 4 miles—6.4 kilometers—per second), the scramjets will shut down, and a rocket engine will kick in. The rocket will push the craft to Mach 25, the speed needed to propel the space plane into orbit.

Although the first two models will be strictly experimental, subsequent aerospace planes may become the taxicabs of space, making space travel easy and affordable. The space plane could cost just a little over $1 million per launch. Advocates claim that they could send the craft back into space a day and a half after it touches tack could not be entirely dismissed, especially in light of the rapid proliferation of long-range ballistic missiles in unfriendly countries.

In his 1991 State of the Union message, President George Bush announced a refocusing of SDI. The new system would be designated GPALS, for Global Protection Against Limited Strikes. Instead of an impenetrable shield designed to destroy nearly all incoming missiles, the proposed program has been scaled down to stop up to 200 warheads.

The centerpiece of the newly proposed SDI program is "Brilliant Pebbles"—an orbiting fleet of tiny satellites, each 2 feet (61 centimeters) long and weighing about 40 pounds (18 kilograms). Each satellite would fire nonnuclear rockets at an incoming missile, destroying it before impact. The GPALS plan would place between 750 and 1,000 of these weapons in orbit. The satellites would rely on infrared sensors to spot and track the hot exhaust plumes of flying missiles. Other parts of the GPALS picture include small, high-speed rockets that would leap from the ground. Research on neutron-particle-beam weapons and other machinery would continue more slowly.

down. Even a week would beat the shuttle, which can take several months between flights for refitting and repair. What's more, the aero-space plane will function more like a jet plane than it does a rocket. That means that if there are problems after take-off, it can turn around and land, an option that the shuttle doesn't have.

One drawback to the space plane is that it will likely carry a much smaller pay load than the space shuttle, which can lift as much as 50,000 pounds (22,680 kilograms). Proponents of the space plane argue, however, that it is very rare that the shuttle ever carries anything that heavy into orbit, and that most future payloads would not likely exceed 25,000 pounds (11,340 kilograms). Moreover, the space-transportation system of the future will most likely also include unmanned, heavy-weight rockets to lift the really big packages.

As an added bonus, since the space plane can be launched at a moment's notice, it can be used in cases of emergency, such as the sudden illness of a crew member aboard an orbiting space station. If America ever constructs a permanently manned space station for the 21st century, the space plane may serve as America's first ground-to-orbit ambulance.

NASA is developing technology for a small space-taxi system that will permit rapid human access to orbiting space shuttles or space stations. With its wings folded, the craft can fit within the shuttle's cargo bay.

Space station Freedom is an international effort to produce a permanently inhabited station by the year 2000. The station will serve as a scientific research base.

THE SPACE STATION

by Mark Strauss

In 1984 President Reagan, in a plea reminiscent of John F. Kennedy's pledge to put a man on the Moon, called on Congress to put a permanently manned space station in orbit by the early 1990s.

For space enthusiasts, President Reagan's request couldn't have come soon enough. Long before scientists seriously considered going to the Moon, many advocated building an orbiting space station. Early calculations suggested that chemically fueled rockets might lack sufficient thrust to hurl an astronaut all the way to the Moon. Instead, it was thought that a space station could act as a sort of stepping-stone, a refueling station from which a spacecraft could launch itself to the Moon and the other planets.

As far back as the late 19th century, Russian rocket scientist K. E. Tsiolkovsky suggested building a large, cylindrical space station, which would spin along its axis to simulate gravity. He imagined growing green plants on the station, to provide food and oxygen. The image of a rotating space station, spinning like a wheel high above the Earth, has endured for nearly a century. Rocket scientist Wernher von Braun proposed a similar space station, as did the British Planetary Society. Such a structure was stunningly depicted as "Space Station 5" in the film *2001: A Space Odyssey*.

In reality, America's first space station was rather ordinary-looking. Skylab, launched in 1973, was built from the third stage of a Saturn V Moon rocket. It was equipped to house a crew of three astronauts, and carried an array of instruments for studying space, the stars, the Earth, and the biological effects of living in space.

A proposed new space station, named Freedom, would go well beyond the simple design of Skylab. The longest amount of time that a crew of astronauts ever stayed on Skylab was 84 days. Freedom, by contrast, will be a permanently manned space station. Twenty-four hours a day, 365 days a year, a crew of astronauts will live, eat, sleep, exercise, and conduct experiments aboard the orbiting laboratory (see also page 260).

At first glance, the proposed space station looks like an enormous, floating Erector set. The scaffolding of the structure will be a truss of pipes molded out of lightweight, ultrastrong graphite-epoxy. Jutting from the truss are arrays of solar-powered cells. According to blueprints, each solar array will measure 118 feet by 39 feet (36

The proposed space station will likely be a series of docking platforms and laboratory modules constructed and repaired using robotic arms. Space shuttles will ferry materials and supplies to the station.

NASA

meters by 12 meters) and contain 32,800 solar cells. The station would orbit the Earth once every 90 minutes, exposing its panels to the Sun for 45 minutes of each orbit. Whatever generated power that isn't used right away will be stored in nickel-hydrogen batteries.

Modules to house the astronauts will cling to this scaffolding. One of the modules, more than 26 feet (8 meters) long, will be the crew-habitat area. This will provide the permanent crew members with gracious living accommodations (at least, by space standards). There will be a simple kitchen (complete with the world's first orbiting microwave oven); a small infirmary for minor injuries and routine medical checkups; and an exercise area that will include treadmills, stationary bicycles, and rowing machines to help crew members avoid muscle atrophy caused by long-term exposure to low gravity (see also page 262). There will be no designated sleeping area. The astronauts will simply use Velcro fasteners to attach their sleeping bags to the walls of the habitat area, where they will sleep like flies.

The habitat area will also be equipped with a docking port, allowing astronauts to move between the station and the space shuttle, or another visiting craft, without having to walk in deep space.

Also included will be laboratory modules, where the space-station crew will work and conduct their experiments. One of the labs will be constructed by the United States (it will have a shuttle docking port, just like the habitat module), and two other modules will be provided by the European Space Agency (ESA, a consortium of 14 countries) and Japan.

Inside the planned Japanese Experiment Module (JEM), experiment lockers will line the walls and ceiling. The module will have an open "back porch" to allow crew members to place experiments in the vacuum of space.

The European Space Agency plans to contribute both a manned laboratory module, which will be permanently attached to the space station, and a free-flying, unmanned laboratory. The manned laboratory will be carried into orbit aboard a space shuttle. The free-flying laboratory will likely be launched with a European Ariane booster rocket and placed in orbit 35 miles (56 kilometers) away from Freedom. The flying laboratory would then be used as

> **TO LIVE AND BREATHE IN SPACE**
>
> No matter what configuration the space station takes, it will show NASA just how challenging long-term space living can be. For example, on a permanently manned space station, a crew of four astronauts, in the course of one year, would consume 100,000 pounds (45,360 kilograms) of oxygen and water.
>
> Whenever astronauts visited Skylab, they brought with them all of the oxygen and water that they needed to sustain their lives during their temporary stay in space. But unless NASA plans to send continuous replenishing shipments of air and water into orbit, the space-station astronauts will quickly run out of oxygen and water.
>
> One possible alternative is to recycle some of the air and water consumed aboard the space station. Condensation loops, set in the ceilings of the work and habitation areas, could collect moisture that is released into the air as the astronauts sweat and breathe. The moisture would then drip into collection troughs to be purified and then stored in a tank for drinking.
>
> A separate system could recycle shower water, washing water, and the astronauts' urine. Since this water started off impure, once purified, it would probably be stored in a different tank and used again only for showers and washing.
>
> Recycling oxygen would be a much simpler process. The carbon dioxide (CO_2) exhaled by the astronauts could be collected from the air and sent into a combustion unit, where it would be burned with hydrogen (H_2) at 1,742° F (950° C). At this temperature the bonds binding carbon to oxygen and hydrogen

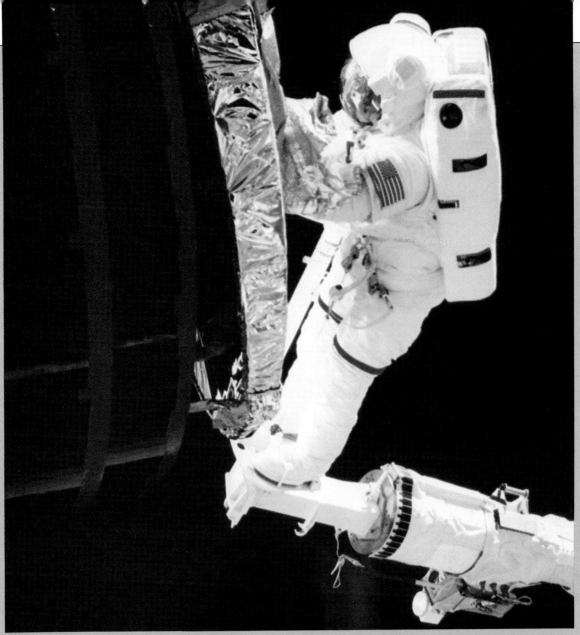

to hydrogen break and reform, creating a mixture of methane (CH_4) and water (H_2O). The methane would be vented from the space station, or possibly used as a rocket propellant. The water, however, would be sent on to an electrolysis unit. There an electric current would break down the water into hydrogen and oxygen. The hydrogen would be sent back to the combustion unit, but the oxygen would then be available for breathing.

a place to conduct delicate low-gravity experiments, which would otherwise be disrupted by the movements of the astronauts, or by the shaking caused by the occasional docking of the space shuttle. All the experiments will probably be automated. Once every few months, the space-station crew will bring the laboratory back to the station, remove its experiments, and load it up again for another flight.

The Canadians also plan to make a major contribution to the station: a highly

LOW-GRAVITY SCIENCE

Why build the space station at all? Just to hang around in microgravity is reason enough, say some space-station advocates. Scientists want to learn all they can about the strange sensation of weightlessness, the effect it has on everything from tiny crystals to the human body.

Once the crew arrives at the space station, they will also study ways in which a low-gravity environment can be used to manufacture goods. The resulting products may be far superior to any made on Earth, where the incessant pull of gravity causes materials of different densities and temperatures to break apart and deform under their own weight.

In space, astronauts could also create brand-new alloys, made from metals that resist mixing on Earth. Nearly flawless crystals could be grown, such as gallium arsenide crystals that could be a major component of ultrafast computer microchips of the future. Glass-fiber optics would have none of the internal-stress flaws that limit their use when they are made on Earth.

Space would also be an ideal location for medical research. Under low-gravity conditions, doctors would find it remarkably easy to separate and purify proteins, enzymes, cells, and cell components. Ultrapure drugs might be manufactured in space for only a fraction of their cost on Earth. What's more, since low gravity makes it possible to mix together substances that could never be mixed on Earth, researchers on board Freedom might develop a whole new generation of lifesaving drugs that have yet to be imagined.

Astronauts aboard the space station will also devote considerable time to research in the life sciences. In particular, NASA wants to learn whether human beings can endure living for long periods of time in low gravity. The research is especially important, in light of the fact that the space agency is hoping to send human beings to Mars by the year 2019 —a round-trip mission that may last two and a half years.

Some planners want to equip the space station with a centrifuge, a large chamber that would turn fast enough and create enough centrifugal force to simulate gravity. This centrifuge could be set to rotate at a velocity equivalent to the gravity on Mars, approximately one-third that of Earth's. Its chamber would

sophisticated robot manipulator arm, similar to the one used by the space shuttle to release payloads and repair satellites. The arm, nearly 56 feet (17 meters) long, will be able to lift objects so massive that on Earth they would weigh almost 100,000 pounds (45,360 kilograms). The robot arm will likely be one of the first segments sent into orbit, where it will help astronauts assemble the space station.

Once the space station is complete, the robot arm may rest on a sliding platform, which can travel along Freedom's truss. Astronauts will control the arm from a remote workstation. For delicate repair jobs, such as fixing a satellite, a Special Purpose Dexterous Manipulator (SPDM) would be slipped onto the end of the arm. According to its present design, the SPDM will have two smaller arms, about 6.6 feet (2 meters) long, to serve as a set of "fingers."

Crew members aboard Freedom will move from one module to the other by means of a system of connecting-tunnel adapters, also called nodes. Originally, designers imagined that the connecting tunnels would be little more than access crawlways, just large enough for a person to squeeze through. With the current design, the nodes will be roomy enough to store equipment or scientific instruments.

Encircling the outside of the nodes will be a series of docking ports. The ports will serve as attachment sites for future space-station modules. When NASA designed Freedom, it envisioned an evolving space station, one that could expand and adapt to the changing goals of the space agency. In-be large enough to house several lab animals.

By studying these animals, researchers might learn whether a weak gravitational force would be sufficient to prevent some of the dangerous side effects caused by prolonged exposure to weightlessness. We already know that human beings who have spent many months in space have suffered from nausea, calcium deficiency, muscle weakening, anemia, heart palpitations, and depression. If the centrifuge experiments prove successful, then engineers might even design a spinning Mars spacecraft that creates its own artificial gravity.

Experiments will involve not only animals, but plants as well. In the future, NASA scientists are hoping to put plants not only on spacecraft, but on Mars and lunar bases. The plants could provide both food and a good source of water and oxygen. Plants transpire water, releasing it through their leaves as vapor that can be condensed into drinking water. The oxygen that plants release could reduce the amount of oxygen brought from Earth.

The question still thwarting scientists is, will plants be able to grow without gravity? NASA researchers would like to set up a small greenhouse on the space station and test a variety of crops, including soybeans, wheat, and lettuce. If the plants grow too slowly in space, or don't yield as much food as they do on Earth, scientists aboard the space station might try different ways to improve the crop, such as introducing growth hormones, attempting genetic selection of plants, or even placing them in the centrifuge for a couple of hours every day.

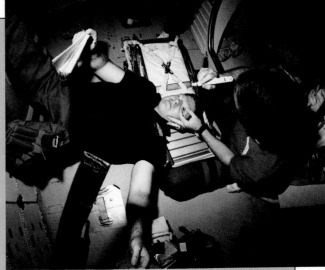

A shuttle crew member measures eye-pressure changes caused by exposure to microgravity. Such a test will be used to evaluate crew health on the long space missions of the future.

itially, Freedom will serve as an orbiting laboratory. Yet, 20 years from now, the station might become an orbiting shipyard, an assembly point for a manned spacecraft destined for Mars. Freedom might also one day serve as a sort of way station, a jumping-off point from which astronauts could begin to establish a permanent base on the Moon.

In the beginning the docking ports outside the connecting nodes may serve other purposes. The station escape vehicle (officially known as the Assured Crew Return Vehicle, or ACRV) will be permanently docked outside one of the nodes. The workstation that controls the Canadian mobile robot arm will stand in one of the nodes. An astronaut could monitor the movements of the robot arm with a series of TV screens mounted inside the station. Another possibility is that a cupola, a small dome, might be attached to one of the nodes. A workstation could be placed inside the cupola, where the astronaut would have a 360-degree view of his or her surroundings.

President Reagan had hoped that the space station would be completed by the early 1990s, but budget problems have pushed the program back indefinitely. Technicians, however, have already begun constructing prototype, or experimental, pieces of the space station here on Earth. In the years ahead, such pieces will be ferried into space by dozens of different space-shuttle missions, unmanned rockets, or perhaps a newly constructed aerospace plane.

THE MISSION TO MARS

by Mark Strauss

On the 20th anniversary of the first Moon landing, President George Bush announced that he was committed to "a journey into tomorrow, a journey to another planet, a manned mission to Mars." The president predicted that an American astronaut would set foot upon the surface of Mars in the year 2019, 50 years after Neil Armstrong walked on the lunar surface.

Mars has always tantalized the human imagination. At the turn of the century, American astronomer Percival Lowell turned his telescope on Mars and reported that he had discovered a network of canals crisscrossing the Red Planet. The canals, he suggested, had been dug by an advanced civilization to carry water from the planet's polar ice caps to parched cities sitting in the middle of the Martian desert.

When the National Aeronautics and Space Administration's (NASA's) Viking probes landed on the surface of Mars in 1976, the photographs and data that they sent back to Earth told a completely different story. Mars looked like a barren patch of the Arizona desert, strewn with rocks and craters, and, by all indications, completely lifeless. The temperature dropped as low as $-188°$ F ($-122°$ C). The air was almost entirely carbon dioxide. So light was the atmospheric pressure that it was the equivalent of Earth's atmosphere at an altitude of 100,000 feet (30,480 meters).

Yet Mars might not always have been a dry, lifeless planet. Photographs of the surface show deep, jagged canyons, perhaps etched into the rock by ancient rivers of liquid water. Moreover, there are very few impact craters in the northern hemisphere, suggesting that an ocean once covered part of Mars, shielding the surface from falling meteorites.

In fact, scientists believe that 3.5 billion years ago, Mars might have closely resembled ancient Earth, with oceans, rivers, and possibly a thick, dense atmosphere. Under those conditions, life might have thrived on the Red Planet. A future expedition to Mars could search for fossil evidence of those life-forms, perhaps proving to us for the first time that it is possible for life to exist on other planets.

An expedition to Mars would also answer many questions about the evolution of our own planet. Why did Mars, apparently once covered with water, suddenly dry up and die, while the Earth continued to thrive and grow?

THREE ROUTES TO MARS

But a trip to Mars is easier said than done. The Red Planet, at its closest approach to Earth, lies nearly 35 million miles (56 million kilometers) away, 150 times farther than our Moon. Although a round-trip to the Moon takes only a few days, a voyage to Mars might last as long as two and a half years. So a crew of eight astronauts aboard a Mars spacecraft would require some 200,000 pounds (90,718 kilograms) of water and oxygen a year. At least some of the ship's supplies, such as water and air, would have to be recycled, using a system similar to the one proposed for space station Freedom.

Then there is food. But carrying along enough supplies would be only half of the problem. The massive spacecraft would also have to bring fuel for a return trip back to Earth, perhaps several hundred thousand pounds' worth.

Obviously, the best way to get to Mars would be via the route that uses the least amount of rocket fuel. The most fuel-efficient route to Mars involves a direct trajectory when the planet is at its closest approach to Earth. Every 26 months, the orbits of Earth and Mars are aligned in such a way that it is possible to make such a direct journey in only seven months, using a minimum amount of fuel. One drawback looms, however. The astronauts would have to wait on the surface of Mars for a

The time and expense involved in a manned trip to Mars dictate that we establish a self-sustaining, permanently manned base there. Such a base would allow long-term, in-depth scientific studies of the Martian geology and meteorology.

Illustration by Pat Rawlings/Eagle Engineering, Inc./Courtesy of NASA

year and a half before the two planets were properly aligned again for an optimally short return trip.

A second possible scenario would be a "sprint mission," which would last about 15 months. In this case the astronauts wouldn't wait until the distance home was again at a minimum. But they would cut their mission to as short as 30 days in order to leave before the widening distance between the planets became too great. However, such a plan would require considerably more fuel, perhaps more fuel than a ship could carry. NASA might solve that dilemma by launching a craft to Mars, well in advance of the crew transport, with all of the fuel needed for a return trip.

A third plan would send a spacecraft to Mars by first launching it toward Venus. The astronauts would swing their ship by Venus, taking advantage of the planet's gravitational field to "slingshot" them toward Mars. The spacecraft would achieve the necessary velocity, yet it would conserve enough fuel for the return trip home. Such a mission would last about 22 months.

However, if lugging tons of fuel presents too much of a problem, why not carry along just enough fuel for a one-way trip? Once the astronauts arrive at Mars, it might be possible to manufacture fuel for the return trip, using the Martian atmosphere.

Almost 97 percent of Mars' atmosphere is carbon dioxide. The thin air could be drawn into a processor and heated to about 1,800° F (982° C). This would cause some of the oxygen atoms to break free from the carbon dioxide molecules, leaving behind a mixture of oxygen and carbon monoxide. The oxygen could then be separated, liquefied, and used as a propellant fuel. It might even be possible to send an automated fuel refinery to Mars, a year ahead of the manned expedition. When the crew finally arrived, a full tank of fuel would be waiting for them.

Another possibility is that NASA might avoid using chemical rockets altogether. Engineers at several universities and laboratories have been commissioned by the space agency to explore ways to produce nuclear-reactor-powered rockets. A nuclear-fission reactor, similar to the type used as a power source here on Earth, would be used to heat a fuel, such as hydrogen, to 4,940° F (2,727° C), before spewing it out the exhaust tubes. That would be considerably hotter than the liquid hydrogen in the space shuttle's main engine, and much more energetic. A nuclear rocket could one day allow astronauts to reach Mars in as little as two months.

Realistically, nuclear-propulsion technology is still many years away. For now, NASA will have to worry about keeping its astronauts alive and healthy on a trip that could last as long as two and a half years.

STAYING HEALTHY

Among NASA's biggest concerns will be the health problems that could result from prolonged exposure to weightlessness. Muscles, no longer straining against the tug of Earthly gravity, become weak and atrophied. In a weightless environment, the body begins excreting calcium, reducing the density of bones in the legs and feet. Loss of calcium makes bones brittle and prone to fractures. Meanwhile, the

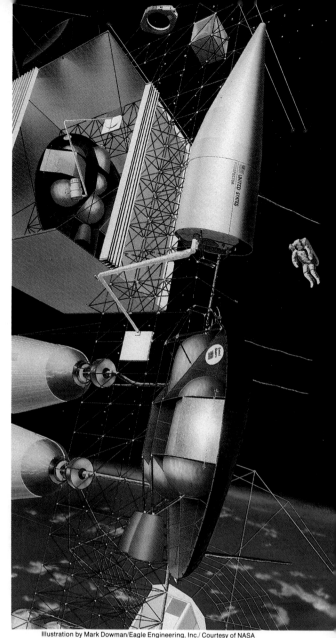

Illustration by Mark Dowman/Eagle Engineering, Inc./ Courtesy of NASA

An Earth-based launching of a ship large enough to travel to Mars would be impractical, if not impossible. More plausibly, a Mars ship would be assembled in space, with parts delivered via space shuttles.

heart, with no gravity to pump against, shrinks in size.

After arriving at Mars, the sudden transition from weightlessness to gravity one-third that of Earth would likely be a painful experience for the astronauts. Cosmonauts who returned to Earth after nearly a year aboard the Soviet space station Mir found it difficult to stand up, or perform simple tasks such as throwing a ball.

Exercise could thwart some of the side effects of weightlessness, notably muscular atrophy. But to combat other health problems, the only solution would be to set up an artificial-gravity environment aboard the Mars spacecraft. This might be accomplished by spinning the ship along its axis, creating enough centrifugal force to simulate the pull of gravity. Some engineers have proposed a spacecraft that would travel to Mars spinning at the end of a half-mile-long tether with an engine on one end and the habitat module on the other.

If the spacecraft itself doesn't spin, then engineers might consider placing a small centrifuge aboard the ship. Astronauts could spend a couple of hours a day in the centrifuge, as a sort of gravitational therapy.

Astronauts on their way to Mars also face the danger of prolonged exposure to radiation. Solar flares, which occasionally belch out from the Sun, produce radiation strong enough to kill an unshielded spacecraft crew in just a couple of days. Engineers would have to design a "storm cellar," a temporary shelter against solar flares, perhaps shielded by vast stores of oxygen, water, and hydrogen fuel—all of which are good energy absorbers.

Yet the health problems resulting from a two-and-a-half-year journey through space might pale next to the psychological problems. Soviet cosmonauts who spent several months at a time aboard the Mir space station eventually began to argue with one another, sometimes even refusing to live in the same area of the space station. After several months of isolation, work efficiency also suffered. Cosmonaut Yuri Romanenko, who spent nearly a year aboard the Mir space station, was able to work only four and a half hours a day by the end of his mission.

If matters got that bad aboard an orbiting space station, what would life be like aboard a Mars spacecraft? On board the space station, Earth was only a few hours away. On a trip to Mars, Earth would be only a dim point of light, millions of miles out of reach. The spacecraft crew would be confined together for more than two and a half years, unable to get away from one another for even the briefest vacation.

Behavioral psychologists at NASA have been studying how people in small groups interact with each other, and are trying to figure out ways to keep astronauts comfortable and happy on a long interplanetary trip. One unusual proposal came from former Apollo astronaut Michael Collins, who suggested sending only married couples to Mars, since they have already had several years' experience living together.

LANDING ON MARS

When the Mars spacecraft finally approaches its destination, it may be traveling as fast as 50,000 miles (80,450 kilometers) per hour. In order to slow down, the ship could fire massive retro-rockets, but that would be an enormous waste of fuel. A more sensible solution would be "aerobraking." The ship would rip through Mars' atmosphere, passing to within 25 miles (40 kilometers) of the planet's surface. A cone-shaped heat shield beneath the ship, called an "aeroshell," would prevent the astronauts from burning up, and the resulting friction would slow the ship down to the point where it would be possible to establish an orbit.

Once in orbit, some crew members would enter an excursion module and land on the planet's surface, using the module as a base for exploration. At the end of their stay, the landing party would rocket into orbit, hook up with the orbiter, and begin the return trip back to Earth.

THE ROAD TO MARS MISSION

In anticipation of a mission to Mars, NASA is planning to send a Mars Observer Probe to the Red Planet sometime in the 1990s. The probe will scout ahead for an ideal landing spot, sending back high-resolution photos of the entire planet's surface. A likely landing place would be an ancient riverbed that would have the highest probability of containing Martian fossils.

The next step will be to launch a Mars Sample Return Mission, a group of un-

After astronauts endure the two-and-a-half-year journey to Mars, they must then cope with the freezing temperatures and raging dust storms common on the Martian surface.

The Boeing Company

manned probes that will collect soil samples and return to Earth. Scientists will study the soil, looking for any toxic chemicals or viruses that could pose a threat to humans. Engineers will also want to understand how soft or strong the Martian soil is, so they can perfect the landing craft and rover vehicles. When the time comes to finally build the Mars spacecraft, the most difficult task might be getting it into orbit. Obviously, the craft, weighing more than 2 million pounds (1 million kilograms), would be too heavy to launch all at once. It would have to be launched in segments and then assembled in orbit, perhaps at space station Freedom.

Yet even the segments might be too massive for our current fleet of booster rockets. The space shuttle can lift only 50,000 pounds (22,680 kilograms) of payload. The old Saturn V boosters, which sent the Apollo missions to the Moon, can lift 250,000 pounds (113,400 kilograms), but that would still be a fraction of the weight of a Mars spacecraft.

Engineers, however, may have found a way to send segments of the Mars craft into orbit by modifying the old Saturn V rocket booster. The design would strap as many as six massive liquid-fuel booster rockets around the outside of a Saturn V engine, enabling it to lift as much as 600,000 pounds (272,160 kilograms) into orbit. It would be ironic to return to the technology of the 1960s to get the Mars mission off the ground. But in many ways, exploring Mars will require the same all-out effort that putting men on the Moon demanded. The payoff, however, would be an accomplishment that would dwarf the Apollo program.

NASA subjects new space-suit designs to rigorous testing for endurance and radiation resistance. The spacewear designated for the space station will have to withstand the most extreme conditions ever faced by astronauts.

DRESSING FOR SPACE

by Lillian D. Kozloski

You're an astronaut in the year 2010, assigned to work on the international space station Freedom. In addition to testing equipment for a planned lunar outpost, you will be helping to maintain the space station itself, performing work that will require long periods of time in open space repairing the station's outside hull. You will need a very special kind of suit to protect you from intense solar radiation, powerful cosmic rays, even micrometeoroids—small, sand-like and rocklike particles that whiz through space at speeds of 60 miles (96 kilometers) per second.

Luckily for you, National Aeronautics and Space Administration (NASA) researchers back in the late 20th century were already testing a new generation of space suits to meet your 21st-century needs. To appreciate their technological wizardry, let's start our story at the very beginning—flipping back to the very first pages in the family album of space suits.

THE NEED FOR PROTECTION

Early in the 19th century, novelists and daredevil pilots were already discovering the need for special suits and equipment for survival in the rarefied regions high above planet Earth. Science-fiction writer Jules Verne correctly predicted in his book *From the Earth to the Moon* that an astronaut would need "an encapsulating type of protection."

Scientists realized early on that an unprotected human would survive just 15 to 30 seconds in the almost-complete vacuum of space. They also knew that without the buffering effect of the atmosphere, a spacewalker would be exposed to extremes of cold and heat. In addition, he or she would not have any circulating air to carry off body heat. Other potentially deadly hazards for any unprotected spacewalker would include the occasional bulletlike micrometeoroid, radiation from solar winds and solar flares, and infrared and ultraviolet rays. Scientists, as well as science-fiction writers, correctly suspected that lunar explorers would face much the same hazards, since the Moon's thin atmosphere offers little protection.

PRECURSORS TO THE SPACE SUIT

To protect themselves from the elements, the first high-altitude pilots wore fleece-lined leather jackets, gauntlets, helmets, goggles, breeches, and boots. Due to the thinness of the upper atmosphere, they had difficulty breathing, and some even suffered hallucinations. To compensate, they breathed through tightly fitting face masks attached to oxygen tanks.

Air pressure at high altitudes presented another challenge. While a deep-sea diver experiences greater and greater atmospheric pressure as he or she descends into the ocean, a pilot experiences less and less pressure the higher his or her plane flies. Though the situations are reverse, the lifesaving principles are similar. What is crucial is to buffer the effects of a too-rapid drop in outside pressure. Early high-altitude pressure suits were based on the deep-sea-diving outfits designed to protect divers from too-rapid ascent out of the high pressures found at ocean depths.

In 1907 John Scott Haldane, an English respiratory physiologist, perfected an oxygen pressure suit for deep-sea divers. American balloonist Mark Ridge turned to Haldane for guidance in 1933. Ridge wanted to fly an open balloon gondola into the stratosphere, and knew he would need a pressurized garment to survive.

Haldane and his associate, Sir Robert Davis, did indeed create a suit that could have protected Ridge to an altitude of 50,000 feet (15,240 meters). But the British military ministry refused Ridge permission to flight-test the suit. Given international tensions at the time, officials feared that the balloon and the pressure suit might fall beyond Britain's borders and into the hands of an unfriendly country. So it was not until 1936 that British Flight Lieutenant F. R. D. Swain set a record for the Royal Air Force when he flew to almost 50,000 feet (15,240 meters) wearing a pressure suit based on the Haldane-Davis design.

American aviator Wiley Post was the next significant figure in the story of high-altitude flying. Post had already flown around the world several times. He identified the jet stream, a fast, westerly wind current that circles the Earth at altitudes from 10,000 to 15,000 feet (3,000 to 4,600 meters). Post knew he could not fly in the jet stream without special personal-survival equipment. He approached the B. F. Goodrich Company, which agreed to create the needed suit.

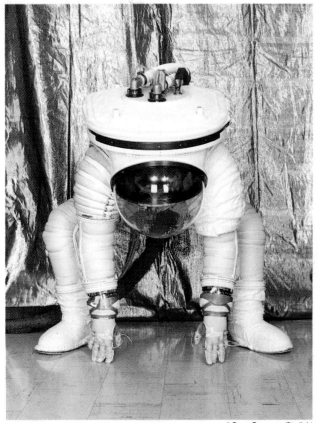

Shuttle astronauts have complained that their space suits lack flexibility. Makers of the Mark III suit (above) claim that its "soft" construction will allow space-station astronauts more freedom of movement.

The first B. F. Goodrich suit literally exploded like an overblown balloon the first time it was inflated. Goodrich engineer Russell Colley designed a second suit, but by the time it was completed, Post had gained weight. Post had to be extricated from the overly tight suit while standing in a refrigerated storage room.

Post's third—and finally successful—suit, completed only after careful measurement, produced compressed air from liquid oxygen, which also provided air for breathing. Wearing this suit, Post was able to fly into the upper atmosphere's jet stream in August 1934, upping his record-breaking speed from 180 to 280 miles (290 to 450 kilometers) per hour.

As pilots continued to fly at higher altitudes and faster speeds, they were increasingly plagued by pressure sickness, choking, gas pains, and other physiological problems. In the 1940s and 1950s, the military accelerated the development of pressure suits, the forerunners of today's space suit. During the 1940s pilots survived by forcing extra oxygen into their system, but this would take them no higher than 52,000 feet (15,850 meters). At 65,000 feet (19,800 meters), not even pressurized cabins were enough protection. In 1953 the development of dependable pressure suits enabled pilots to drop the first test hydrogen bomb from the edge of the stratosphere over the Bikini Atoll—an altitude high enough to protect them from the bomb's shock waves.

SUITS GEARED FOR THE MANNED SPACE PROGRAM

Mercury

The United States Manned Space Program began in 1958. Technology from the military pressure-suit program bore rich fruit. The high-altitude suits built by Goodrich soon evolved into the full-pressure space suit used by the Mercury astronauts. It was made of a neoprene-coated, nylon inner layer covered by a high-temperature-resistant aluminized nylon. According to plan, the astronauts would need the suits only if the Mercury spacecraft pressurization system should fail.

Gemini

The space suits that the Air Force developed for the next phase of manned spaceflight, the Gemini program, enabled astronauts to walk outside their spacecraft during extravehicular activities, or EVAs. The spacewalkers were connected to their craft by a life-supporting cord.

The basic Gemini suit was constructed of an inner bladder made from rubberized nylon, which was inflated during pressurization. To prevent this bladder from ballooning, designers used a Dacron fishnet layer, called Link-net. The outer protective layer was made of uncoated Nomex, a tem-

NASA evaluates the comfort and mobility of newly developed space suits in its Weightless Environment Training Facility, a 500,000-gallon water tank in which the water's buoyancy simulates the microgravity of space. Prevailing opinion seems to hold that neither the AX-5 "Michelin Man" all-metal suit (shown at right, about to be dunked) nor the Mark III "soft" suit (below) will be chosen for the job. Most likely, NASA will opt for a hybrid space suit that combines elements of the two main contenders.

Both photos: © Roger Ressmeyer/Starlight

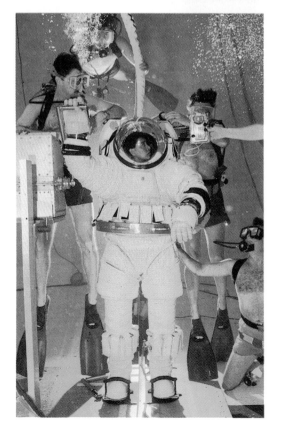

perature-resistant nylon material. During EVAs, astronauts would also don an additional cover, consisting of several layers of uncoated nylon over layers of Mylar and Dacron batten insulation, which in turn was covered with a layer of projectile-resistant nylon that would protect the astronauts from the impact of micrometeoroids.

Gemini astronauts, who were physically active during their EVAs, found several flaws in this suit. Without proper ventilation, they sweated profusely, and their helmet visors fogged. On one occasion in 1966, a small rip in a back zipper of astronaut Eugene Cernan's suit allowed the powerful solar radiation of space to overheat the suit and almost burn him.

Apollo

The Apollo astronauts and Moon walkers were not hampered by the umbilical cord that tethered the Gemini astronauts to their craft. Instead, the Apollo suits used a portable backpack life-support system. Apollo suits were also made to be ex-

THE EVOLUTION OF SPACE SUITS

Space-suit precursor: when pressurized, the outfit worn by aviator Wiley Post (above) during his 1934 high-altitude flights kept him from walking or standing.

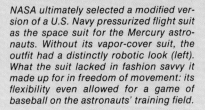

The evolution of space-suit technology has seen its share of reject designs. In the early 1960s, the man-in-a-can approach taken by the "tripod teepee" (above right) lost out, ostensibly because it required excessive space on already cramped spacecraft. In 1978, NASA turned down the "anthropomorphic rescue garment" (above), a space suit designed for the orbital transfer of an astronaut from a disabled shuttle.

NASA ultimately selected a modified version of a U.S. Navy pressurized flight suit as the space suit for the Mercury astronauts. Without its vapor-cover suit, the outfit had a distinctly robotic look (left). What the suit lacked in fashion savvy it made up for in freedom of movement: its flexibility even allowed for a game of baseball on the astronauts' training field.

tremely fire-resistant, in response to the tragic fire in 1967 that killed three astronauts—Virgil Grissom, Edward White, and Roger Chaffee—during ground training aboard *Apollo 1* at Cape Kennedy.

The basic Apollo space suit consisted of a five-layered torso-and-limb suit. It incorporated specially designed joints with convoluted bellows for greater joint mobility. A cable, block, and tackle assembly was needed to keep the suit arms bendable during pressurization while in Earth's atmosphere. The innermost layer resembled a pair of standard long johns, with an interlacing of small tubing stitched on nylon spandex fabric. This in turn was attached to a more comfortable liner made of nylon tricot. The tubed garment circulated cool water over the astronaut's body.

During lift-off and reentry, astronauts wore a three-layer cover for additional fire protection. Astronauts assigned to EVA

Top three photos: National Air & Space Museum; bottom left: Litton Industries; bottom right: UPI/Bettmann Newsphotos

When fully pressurized to five pounds per square inch, the space suit worn by Mercury astronaut M. Scott Carpenter (left) allowed limited mobility.

After years of research and development, the Mercury astronauts emerged in space suits that were aluminized to reflect heat.

Long before any Apollo missions had flown, work had begun on space suits that would protect astronauts on the upcoming lunar landings.

tasks wore a 14-layer protective garment over their basic space suits. This space garment was woven of Teflon-coated glass fibers and Beta cloth, which together resisted both extreme temperatures and the impact of tiny flecks of space debris.

For a workday spent inside the Apollo craft, astronauts changed into lightweight in-flight overalls, much like an automobile mechanic might wear, except that they were woven entirely of fireproof Teflon.

Skylab

Skylab, America's first space station, was launched in 1973. Three crews—all using modified Apollo suits—visited this orbital cluster, and stayed for increasingly longer periods of time. Their duties included extensive repair work on the outer hull of the craft, which had been damaged during launch. They also tended a number of scientific instruments mounted on the outside surface of their indoor workshop.

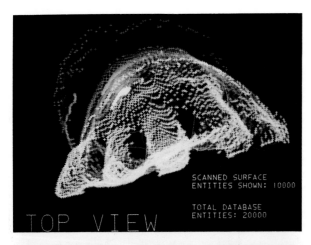
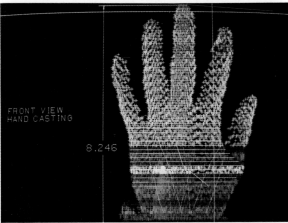

Custom-making an astronaut's gloves presents an engineering challenge to the manufacturers of space suits. At ILC Dover, lasers scan the astronaut's hand, producing accurate models based on 20,000 bits of data. Engineers then fit pieces of fabric to the models. The finished gloves will have ribbed convolutes—ring-like ridges—at the finger joints, and a complex structure of diamonds and squares in the palm.

During one EVA the astronauts installed a parasol over the "roof" of their workshop, thus lowering the inside temperature of the craft to more-bearable limits.

The Space Shuttle

When the space shuttle made its public debut, "shuttlenauts" wore modified U.S. Air Force high-altitude escape suits, similar to those worn by military pilots flying Blackbird reconnaissance aircraft. But shuttle astronauts stopped using such protective clothing after the fleet's first successful test flights.

In 1986 the explosion of the space shuttle *Challenger* forced officials to reassess the fleet's escape and survival systems. After two-and-a-half years of intensive study, the post-*Challenger* space fleet underwent significant modifications. One of the recommendations was to reinstitute emergency-rescue equipment. NASA therefore added to the orbiter a special crew-escape system that jettisons a hatch door if evacuation becomes necessary while airborne. A long telescoping pole would thrust through this open hatchway. Astronauts would hook onto the pole and quickly slide beyond the shuttle, to a point where their parachutes could safely open.

In addition, all crew members must now wear durable, temperature-resistant orange suits during launch and reentry. The new launch-and-entry suits are designed both to counter the low pressures of high altitudes and to protect the astronauts from exposure to temperature extremes. Crew members wearing the suits can survive at or below a 100,000-foot (30,500-meter) altitude for 30 minutes. Inflated bladders inside these suits cushion vital areas of the body. The suit can also be worn with a domed, full-pressure helmet for in-orbit emergency operations.

On some shuttle flights, designated astronauts also need true space suits for EVA assignments. The shuttle fleet's two-piece space suits were a vast improvement over the previous one-piece suits—with their long awkward body zippers. While it took Apollo astronauts, with assistance, close to an hour to suit up, shuttle astronauts can don their space suits in five minutes, unassisted. Shuttle space suits are also computerized to constantly monitor and adjust their own vital life-support systems. Astronauts are alerted when water or oxygen levels drop too low or when any system threatens to malfunction.

While the shuttle suit has proven its worth over the years, astronauts say it has its drawbacks. After several hours of wear, the suits can cause bruises on the shoulders and fingertips from pressure points between the suit and an astronaut's skin. As Kathryn Sullivan, the first American woman to walk in space, has noted, the contours of the shuttle space suit do not exactly match a human's shoulders and knees.

A SUIT FOR THE SPACE STATION

In looking ahead to its permanently manned space station, NASA is considering space suits with hard bodies. There are many advantages, the most important being that hard suits can maintain higher internal pressures in the vacuum of space than can conventional space suits. This reduces the "prebreathing," or adjustment time, now required before astronauts in soft suits are ready for the low pressures they will encounter during a space walk. At present, astronauts must prepare for low-pressure space walks by prebreathing concentrated oxygen to rid their system of nitrogen. Otherwise, they risk suffering from a potentially fatal condition known to deep-sea divers as "the bends" or caisson disease. This condition occurs when a rapid change in air pressure causes nitrogen dissolved in the body to form bubbles.

Several new space-suit designs are now being developed simultaneously, with modifications being made every step of the way. The top candidates are tested in the weightlessness of nose-diving jets (NASA's famed "Vomit Comets"), as well as at NASA's Weightless Environment Training Facility (WETF). The WETF, one of the world's largest "swimming" pools, is large enough to hold a full-size shuttle replica. It enables suit engineers and astronauts to test a new design's mobility in the microgravity of simulated space.

In the early 1990s, the two hottest space-suit prospects for the space station are the Mark III and the AX-5. Both are modular—that is, they are made of reusable, interchangeable parts that can be mixed and matched to fit different wearers. They are also made of enhanced materials and special layers to better protect astronauts during long periods in open space.

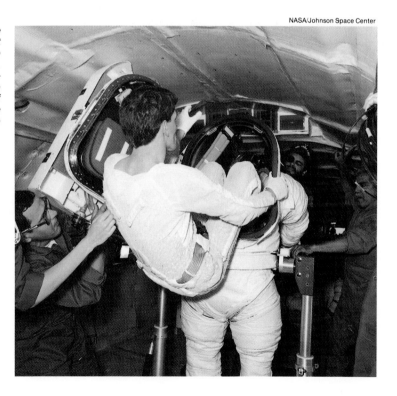

NASA/Johnson Space Center

NASA performs extensive tests to determine the amount of time it takes astronauts to get in and out of a space suit. To properly design the space-station airlocks, NASA also needs to know the volume of space behind and above the suit needed for an astronaut to don or doff the outfit.

SPACESUIT FOR SPACEWALKS

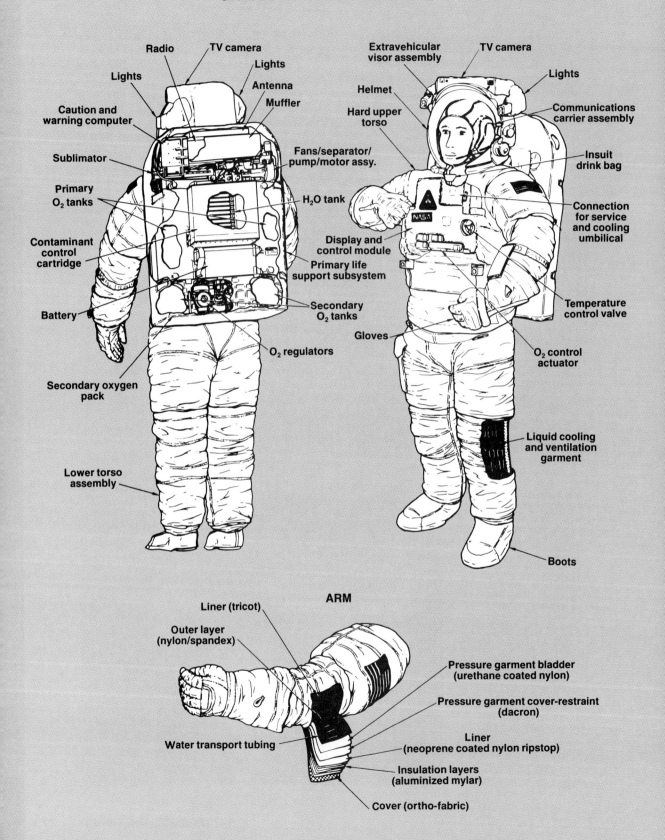

The Mark III Zero-PreBreathe Suit

Since 1979 NASA has been working on the development of a Zero-PreBreathe space suit (ZPS) for use by space-station astronauts. As its name implies, this suit will enable astronauts to make a quick transition from spacecraft to outer space, without the need to adjust to low pressures with prebreathing. Normally, it would be pressurized to about 8 psi (pounds per square inch). The current shuttle suits operate at only 4.3 psi; the shuttle cabin is 14.7 psi.

The Mark III ZPS, developed by NASA at the Johnson Space Center in Houston, combines new designs with pieces of the tried-and-true. Space aficionados will recognize the Apollo helmet and glove connectors and the shuttle suit's hard upper torso. However, astronauts "enter" this suit by climbing through a hatch in the upper back. As such, it is even faster to don than the shuttle space suit—an advantage that may prove crucial in case of emergency outdoor repairs.

The Mark III is worn with an enhanced thermo-micrometeoroid garment (TMG), an outer cover that provides extra protection from radiation, orbiting space debris, micrometeoroids, and the extreme temperature changes that can occur in the vacuum of space. In its entirety the Mark III system weighs about 150 pounds (68 kilograms), some 50 pounds (23 kilograms) more than the shuttle suit.

The AX-5

The AX-5, being developed at NASA's Ames Research Center in California, is the brainchild of engineer Vic C. Vykukal. This hard-bodied suit was quickly nicknamed the "Michelin Man" for its rounded, segmented joints and body parts. Its hard body covers not only the torso and limbs, but also the hands and feet. Unlike the Mark III, the aluminum AX-5 has no fabric parts, which may be an advantage in that there is no risk of torn stitches or worn inner bladders. Like the Mark III, the AX-5 is entered from the rear. Its hatch closes much like a refrigerator door, and must be latched shut by an assistant. It weighs in at about 185 pounds (84 kilograms).

The refinement of space-station suits is likely to continue for many years to come. In all likelihood the final design will be a hybrid of the best features of the AX-5 and Mark III, truly an apt "business" suit for the 21st century.

Facing page: Hamilton Standard/Courtesy of the National Air & Space Museum; below: NASA

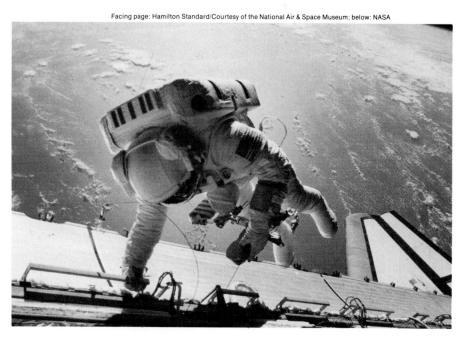

The space suits worn by the shuttle astronauts (diagram, facing page) need only withstand the rigors of comparatively brief periods of extravehicular activity (right). For the astronauts building the space station, the suits will have to endure hundreds of eight-hour days in the hostile space environment.

Before being catapulted into space, astronauts must prepare for the rigors of g-forces during lift-off and motion sickness during space travel.

ASTRONAUTS IN TRAINING

by Maura Mackowski

In 1959 no one knew if a human being could survive a flight in space. Could engineers design a vehicle to carry a traveler safely into orbit and back to Earth? Could a human body withstand the heat, vibration, noise, and radiation? Could the mind cope with total isolation in the void of space?

Neither the Soviets nor the Americans, competing to be first in space, knew the answers. Each decided to find out by launching test candidates chosen from the elite ranks of military pilots, proven to be fit, intelligent, adaptable, and able to handle danger.

Hundreds of applicants were screened on paper, and dozens were put through wearying days of medical and physical tests. They were whirled in centrifuges and flipped on tilting tables. They put miles on treadmills and stationary bicycles. They were probed with needles and X rays. Psychologists locked them in isolation rooms, where they couldn't see or hear anything, not even the sound of their own voices. In the end the U.S. chose seven Mercury astronauts, and the U.S.S.R. chose six Vostok cosmonauts as their respective countries' earliest pioneers in space.

U.S. SELECTION AND TRAINING PROGRAM

The centrifuge is gone now, as are most of the grueling tests that made astronaut selection seem like an obstacle course through a torture chamber. Today work experience, college education, and overall physical fitness are used to select America's astronauts.

Every two years the National Aeronautics and Space Administration (NASA) invites people to apply for its pilot and mission-specialist astronaut positions. In 1990–91, for example, 179 people applied as pilots and 2,245 as mission specialists. NASA's minimum requirements include an engineering or science degree and several years of related work experience.

There are no age limitations to being an astronaut, although the average candidate is 32 years old. There is no restriction on size, but weight must be in proportion to height. The current 98 astronauts range in size from 5 feet 1 inch to 6 feet 5 inches (155 to 196 centimeters). A candidate may be disqualified on medical grounds for a chronic condition that would require constant medication and monitoring, such as diabetes or epilepsy.

Of the 1,946 that passed this first hurdle of general good health in 1990–91, 363 were found by a 12-member NASA board to be "highly qualified." A subsequent screening looked at the type of college degree each candidate had, how recently it was earned, how relevant any work experience was to NASA's own programs, and what unique skills each person had to offer. Pilot candidates were also evaluated on their flying experience. The number was whittled to 106, and these candidates were invited to Johnson Space Flight Center in Houston, Texas, for a week of tests and interviews to determine who would fit best with the NASA team. Ultimately 23 people were chosen as astronaut candidates.

Boot Camp

Basic training is the first step for new astronaut candidates. It lasts one year. During this first year, astronaut candidates fly in the backseat of military trainers or ride special private jets modified to handle like the shuttle. The candidates undergo survival training, learning how to cope in any environment should the shuttle make a forced landing. They take scuba lessons and become accustomed to the strange sensation of weightlessness.

There is also plenty of classroom instruction. Astronaut candidates take courses in space physics, geology, meteorology, oceanography, star identification, and medicine. They learn observation techniques that will enable them to orient themselves over the Earth from the shuttle and to interpret what they see in the environment below. In basic training, they also learn how to operate the shuttle systems. They learn enough about electronics and computers so that they can help make repairs in space.

To understand how the shuttle and its experiments work, astronaut candidates travel to the aerospace companies that build them. They also visit many NASA control centers to familiarize themselves with the roles played by ground-support staff—from the crews that clean and repair the shuttle between flights to the mission controllers that monitor each flight. Candidates even take classes in public speaking so they can work with news reporters and give talks to schools, businesses, and other community organizations.

After graduation the candidate is ready to be assigned to a flight. Most crew selections are made a full year before the actual

Astronauts John Young (right) and Robert Crippen (left) are seated at the controls of the space shuttle Columbia while training for the very first shuttle mission in 1981.

NASA

Both photos: © Keith Meyers/NYT Pictures

A variety of escape systems have been incorporated into today's space shuttle. To prepare for emergency situations that might occur before lift-off, astronauts rehearse evacuating the shuttle using an aircraft-style inflatable slide (right). An astronaut must grow accustomed to a somewhat unwieldy space suit (left), which is equipped with a parachute, an oxygen supply, and a life raft in case of an emergency evacuation while in flight.

flight, and some may begin preparing for their assignment several years in advance. If the waiting period is long, NASA may assign the new astronauts to ongoing projects, such as a shuttle-orbiter-system upgrade or a long-term research project.

Usually, however, training becomes quite intense once an astronaut is assigned to his or her flight. Any given mission will typically involve a pilot, one or two mission specialists, and several payload specialists. The latter, who are in charge of the experimental equipment and testing procedures on board, may also include engineers and scientists from private industry and other non-NASA astronauts.

Two crew members, usually mission specialists, rehearse for extravehicular activities—EVAs, or "space walks"—in the neutral-buoyancy tank. Dressed in space suits and accompanied by NASA scuba divers, they work some 30 feet (9 meters) underwater, practicing shuttle repairs, satellite retrieval and deployment, and space-station construction.

When working as a team, the crews "fly" shuttle simulators that use the same software found on the real orbiter. They drill in preparation for dozens of anticipated problems, such as landing at emergency airfields, coming down in the water, putting out fires, dealing with pressure leaks, and any other potential emergencies NASA can dream up.

When any one shuttle is in orbit, you'll find the team training for the next flight serving as ground support. Some will work relaying messages from the crew in orbit to NASA officials and back.

And if the actual orbiting crew has to deal with a mechanical problem, the crew-in-waiting will simulate making the necessary repairs on Earth and fax written instructions and diagrams of the procedure up to the shuttle. If an astronaut in orbit has to take a space walk to make an unexpected repair, an astronaut on Earth will don a pressure suit and practice the task first in the neutral-buoyancy tank.

COSMONAUT TRAINING PROGRAMS

Information about cosmonaut selection and training was very difficult to obtain during the first 30 years of the Soviet program. Recently, however, the Soviets have tried to invigorate their economy by flying foreign experiments and foreign scientists aboard their space missions in exchange for hard cash. Because of this, it is becoming easier for space historians to learn how the Soviet system works.

Cosmonaut candidates are usually certified military jet pilots, 35 years old or younger. From time to time, however, civilian scientists, doctors, and engineers have been invited to join the Soviet program. Currently the ratio of military to civilian personnel is roughly 3 to 1. This may change as their *Buran* shuttle becomes operational and the next-generation space station, *Mir 2,* is placed in orbit. A change in the work that goes on in orbit will require more qualified engineers, scientists, and doctors than the military alone can provide.

Selection of cosmonauts is similar to the American process. Candidates must be interviewed, they undergo medical and psychological tests, and their references are checked. Typically, the entire process rules out 99 of every 100 candidates.

Once selected, cosmonauts begin a general education in space sciences, including aerodynamics, navigation, astronomy, computers, ballistics, and medicine. They learn about the systems that run the Soyuz capsules, the Progress supply ships, the Mir space station, and its Kvant and Kristall modules (areas that hold scientific equipment). New cosmonauts also study the launch complexes and the rockets that boost them into orbit. They observe flight-control and tracking operations.

An important part of cosmonaut training, particularly for missions lasting six to 12 months, is exercise. In addition to sports and survival training, the cosmonaut candidates fly jets, make parachute jumps, and undergo weightlessness training.

Basic training lasts roughly two years, but cosmonauts may wait years before being named to a particular team. Once they are assigned, however, they undergo intensive training tailored specifically to their particular mission. Like American astronauts, cosmonauts learn to respond to emergency situations and to operate the experiments they will tend in flight.

The experiments the cosmonauts will perform come from many different Soviet research agencies, including the Academy of Sciences, the Geophysical and Astrophysical institutes, the State Center of Nature, and the Ministries of Agriculture, Fishing, or Geology. Cosmonauts train with the scientists and engineers who designed each experiment to learn its purpose and how to operate necessary instruments. They must also learn how to care for any animals, insects, or plants involved in the studies, and how to accurately observe and record test results.

Most cosmonaut training occurs near Moscow, at the Gagarin Cosmonaut Training Center in Star City, where the cosmonauts live full-time. It includes a building with Soyuz, Salyut, and Mir mock-ups and a docking simulator that uses television, models, and realistic lighting under computer control. In the late 1980s, the Soviets added a training simulator for the Icarus, the "space motorcycle" that cosmonauts use to maneuver about during their EVAs.

Star City also has pressure chambers for testing spacecraft components, and soundproof rooms for sensory-deprivation tests. There is a planetarium where cosmonauts train to navigate by the stars. Another building contains a water tank, some 75 feet (23 meters) in diameter and 39 feet (12 meters) deep, where cosmonauts practice EVAs. Separate buildings house several centrifuges that can simulate the forces of blast-off and reentry.

Next door to Star City is Chkalov Air Force Base, where special aircraft take cosmonauts aloft for training flights in which they experience weightlessness.

The Soviet Union trains its cosmonauts for long stays on the permanently manned Mir space station (above). Soviet cosmonauts hold the record for space endurance—a full year spent in the weightlessness of outer space.

COMPARISON OF THE AMERICAN AND SOVIET PROGRAMS

While there are some similarities, Soviet and American training differ in several respects. Cosmonauts spend more time training for weightlessness aboard aircraft, while Americans put in more hours underwater in the neutral-buoyancy tanks.

Both space programs require guest payload specialists from other countries to be completely fluent in the host language, but the Soviets offer Russian-language training, while the American guests must already know English.

NASA has extensive written teaching materials, books, and how-to manuals for the various experiments and the shuttle itself. Cosmonaut trainees are given much less printed material and are expected to take copious notes.

The Soviet training program also includes extensive testing to match the personalities of its cosmonauts for the most compatible crews possible. To this end, cosmonauts are subjected to ground and inflight examinations of their pulse rates, brain waves, voice-stress levels, blood-chemistry levels, and even their facial expressions. Good matchups are essential on the Soviet program's many long-duration missions. Crew members may have only each other for companionship for months at a time.

Both the U.S. and the U.S.S.R. stress the importance of team building, but each country has different ways of familiarizing trainees with the philosophy, chain of command, and protocol of their space agencies.

Cosmonauts live together in Star City, an enclosed community of 4,000 people, and share after-hours activities. American astronauts usually live in Houston, a city of several million people, and generally team up only during working hours, much like workers at any large corporation.

GUEST CREW MEMBERS

In 1978 the Soviets began flying guests from its allied nations in a program known as Intercosmos. French and Indian guests also flew with the Soviets, and, in the 1990s, Japan, Austria, Spain, Germany, and the U.S. signed up to send astronauts to the Soviet space station Mir. The private Juno project bought training for four British cosmonauts and a flight for one in 1991. Tokyo Broadcasting Corporation purchased a seat for a reporter the same year, and a group of individuals in Texas bought another seat, which they plan to raffle off.

In all instances the Soviets required that guests pass physical exams and undergo six to 12 months of training. Three months of intensive Russian-language study are followed by general astronautic studies, lessons on how Mir functions, training in operating experiments, and finally, the instructions needed to return safely aboard a Soyuz craft. While U.S. shuttle astronauts ride much like airline passengers, guests aboard a tiny Soyuz literally block the reach of the ship's commander and must help by operating the radio, the atmospheric controls for the space suits, a navigational computer, the manual override for cabin oxygen, waste-water routing systems, and a TV system used for docking.

NASA has occasionally flown observers on spaceflights, including Senator Jake Garn of Utah and Saudi Prince Al-Saud. Researchers from private industry have bought seats on a shuttle, including McDonnell Douglas engineer Charles Walker, who has flown three times with an electrophoresis experiment. All these civilians had at least six months of full-time astronaut training. They learned how to eat, dress, maneuver, wash, and sleep in space, how to operate basic shuttle equipment, and how to respond in an emergency. They also received a rudimentary education in astronomy and aviation.

Several nations have space-science programs—but no launch vehicles. So they send astronauts to NASA for training and a shuttle flight. Canada, Germany, France, the Netherlands, and Mexico have already flown guest payload specialists. Japan, Italy, Switzerland, and Indonesia expect to send up astronauts in the 1990s. These astronauts are recruited in their own countries according to standards similar to NASA's. They undergo extensive training at home. Once assigned to a shuttle flight, they receive further training in the United States.

Soviet cosmonauts undergo training in a centrifuge. Its high-speed whirling motion creates punishing forces similar to those encountered during lift-off and reentry.

NASA

In 1960, NASA recruited 20 outstanding women pilots to undergo grueling physical and psychological tests to determine their suitability for space travel. Twelve women were selected to continue a second round of training. But in July 1961, NASA abruptly discontinued the female astronaut program. It would be 22 years before the first American woman, Sally Ride (above), would fly in space, serving as a mission specialist on the space shuttle Challenger in 1983.

WOMEN IN SPACE

Tass from Sovfoto

Soviet cosmonaut Valentina Tereshkova was the world's first woman in space. After her historic flight she toured the world, proclaiming, "on Earth, at sea, and in the sky, Soviet women are the equal of men." Some considered the flight a publicity stunt to extol the Soviet system, since it would be almost 20 years before another Soviet woman would fly in space.

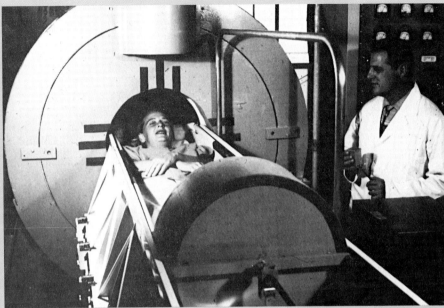

As one of the first female astronaut candidates, Jerrie Cobb (left) underwent extensive medical testing to ascertain that female physiology would be able to withstand the rigors of space travel.

Lovelace Medical Center

Early female astronaut candidates were subjected to ice-water injections in their ears and swallowing 3 feet of rubber tubing. Below, Myrtle Cagle undergoes pulmonary testing.

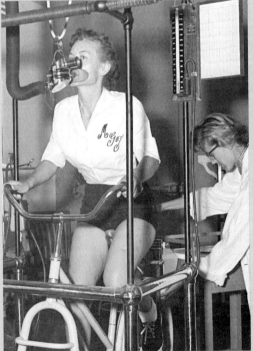

Lovelace Medical Center

NASA

Today NASA's female astronauts undergo the same rigorous preparation for space travel as their male colleagues. Here a female astronaut candidate enters a neutral-buoyancy tank to practice performing simple tasks in a weightless environment.

EUROPE RECRUITS ITS OWN ASTRONAUT CORPS

Citizens of Italy, France, the United Kingdom, Germany, and the Netherlands have already flown as guests of the U.S. and the U.S.S.R. Now the European Space Agency (ESA), a confederation of over a dozen nations, is developing its own habitable space vehicles and is recruiting an astronaut corps of its own. Construction has started on a new astronaut-training facility, located alongside an existing German Space Agency complex in Cologne.

Because ESA is an international consortium, training will be carried out at several sites, giving several countries the economic advantages of hosting a major technical center. For example, the $32 million Cologne facility will house mock-ups of the European *Hermes* shuttle, the Columbus space-station module, and a free-flying, human-tended module—plus offices, a neutral-buoyancy tank, classrooms, and computer centers. Other facilities, or Centers of Expertise, include pilot training in Brussels, Belgium; robotics in Nordwijk, the Netherlands; and underwater training in Marseilles, France.

Each ESA member nation can nominate three to five astronaut candidates. The goal is to have at least one from every member country, for a total of 40. The first group was screened and selected by the end of 1991; a second group will be chosen by the end of 1994, and the last of the 40 by 1996. The *Hermes* shuttle will require at least two three-person crews a year. Plans for the Columbus space station depend in part on how the U.S. proceeds with its proposed space station Freedom.

ESA will select astronauts for two distinct job classifications: space-plane specialists will be pilots for the *Hermes*; and lab specialists, college graduates with three years of scientific-research experience, will work on either *Hermes* or Columbus. They will have backgrounds in physics, astronomy, medicine, robotics, engineering, or similar disciplines. ESA's astronauts must be between 27 and 37 years of age. Their height must be between 5 feet and 6 feet 2 inches (152 and 188 centimeters). They also must be fluent in English.

Training is expected to last four years: one year of basic instruction, 18 months of specialized classes, then 18 months of practical work on a specific mission. ESA hopes to borrow from the best of the American and Soviet programs, and learn from those who have already worked in space.

WOMEN IN SPACE

At age twelve, Jerrie Cobb had her father fit an old biplane with cushions and blocks so she could learn to fly. By the age of 28, she was a professional pilot and an aircraft-company executive, having already logged over 7,000 flying hours and broken three world speed and altitude records.

To Dr. Randall Lovelace and Brigadier General Donald Flickinger, who had just completed the medical screening and selection of the first seven American astronauts, Cobb looked like the perfect candidate to test the idea of putting women into space. It was 1959, the Cold War was on, and word was out that the Soviets were already recruiting female cosmonauts.

Cobb reported to the Lovelace Clinic in Albuquerque, New Mexico, in February 1960 to begin what she hoped would be her trip into space. Over a period of five days, she underwent some 75 tests of health, strength, endurance, and intelligence.

As it became apparent that she was indeed highly qualified, Cobb was sent for a second round of tests in Oklahoma City. She was also asked to recommend other women for the female space corps. She recommended 31 women with the right credentials from among members of "The 99s," an organization founded by pioneer aviator Amelia Earhart and 98 other women fliers.

Twelve of these candidates passed their medical tests, and two more then reported to Oklahoma City to complete the next phase of testing. In 1961 Cobb passed the third and final battery of tests. Two other women who survived the first cut came to stay with her as they continued their grueling tests. Rhae Hurrle Allison and Wally Funk both proved highly qualified, with Funk's 10½ hour stay in the isolation tank setting a space-program record.

Then one day, Cobb got the devastating news from Dr. Lovelace. Without any explanation the Navy had canceled further tests. All the women candidates were sent home, some without the jobs they had quit to participate in the tests. Despite impressive test results, credentials, and the impassioned testimony of Cobb and several other women before Congress, their program was never revived.

It would be 22 years before Sally Ride, in 1983, would become the first American woman in space. Not until 1990 did Air Force Major Eileen Collins become the first American woman pilot astronaut.

Things were not on hold for Soviet women, however. In the early 1960s, several were evaluated, then trained. Ultimately, factory worker and amateur sky diver Valentina Tereshkova won a cosmonaut job and on June 16, 1963, became the first woman in space. But space historians agree that Tereshkova's flight was a publicity stunt initiated by Premier Nikita Khrushchev to demonstrate that women fared well under Communism and that space was not an elitist program in the U.S.S.R.

In 1982 Soviet engineer Svetlana Savitskaya became the second woman in space, aboard *Soyuz T-7*. With *Soyuz T-12* in 1984, Savitskaya became the first woman to fly a second space mission and the first to make a space walk. But Savitskaya's accomplishments are not backed up by the presence of other Soviet women on active cosmonaut duty, although Briton Helen Sharman flew to Mir as a paying customer in 1991.

It was the American space shuttle that really opened doors for women astronauts. For the first time, NASA recruited nonpilots—including physicist Sally Ride—to its ranks. Today the U.S. has more than a dozen women astronauts.

Women have also given their lives advancing the exploration of space. On January 28, 1986, Teacher Christa McAuliffe and specialist Judith Resnik were killed during the launch of *Challenger*.

Ten astronauts have given their lives advancing the exploration of space. The crew of the 25th space-shuttle mission (above), launched January 28, 1986, aboard the Challenger, became the first in-flight casualties of the U.S. space program.

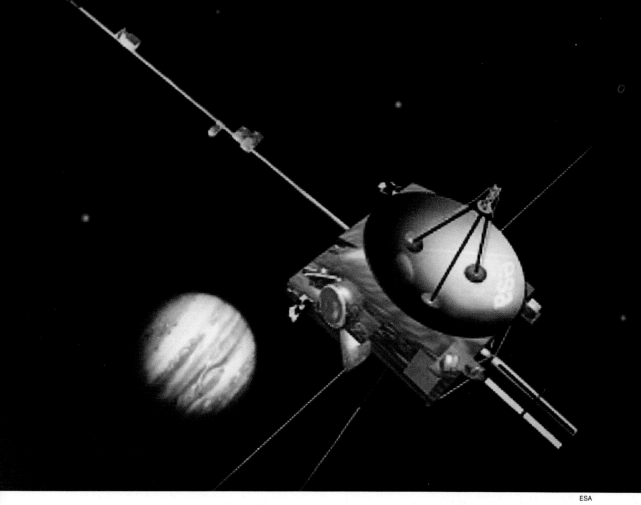

Astronomers use space probes like the Ulysses (above, nearing Jupiter) to explore the solar system.

SPACE PROBES

by Jeffrey Brune

The *Niña*, *Pinta*, and *Santa Maria* took Columbus and his crew across the Atlantic to the New World. Ships of another sort have taken us across the oceans of space to the next New World. They are called *space probes*, unmanned robots that we launch from Earth on missions to explore the Sun and the planets in our solar system. Loaded with instruments designed to add to our knowledge of the space atmosphere, as well as cameras that capture amazing images, these probes are sending home to Earth incredibly exciting data. For example, Voyager 2 sent us shots of the stormy blue world of Neptune. Magellan revealed rock formations on Venus that look like giant pancakes. Voyager 1 flew high enough above our solar system to snap the very first "family portrait" of the Sun, Earth, and most of our planetary neighbors.

Though we can't ride aboard space probes, we may use them as scouts for future human missions. For instance, Moon probes were used to determine the safest places to land on the Moon. The Viking probe provided enough preliminary information about Mars to set the stage for a manned mission in the not-too-distant future.

With each probe we launch, more secrets are revealed about our solar system.

And though we've learned a great deal about the worlds around us, there is still plenty to discover. As astronomer Carl Sagan says, "Somewhere, something incredible is waiting to be known." Let's take a closer look at certain space probes and their fantastic voyages. Prepared for launch?

PIONEERS

Even with the aid of the biggest and the best telescopes on Earth, the outer planets are just distant points of light. Those glimmers of light slowly began to come into focus in 1972 and 1973, when the United States sent twin probes, Pioneers 10 and 11, toward Jupiter for a close-up view. Scientists were most concerned about the probes' safety while traveling through the asteroid belt found between Mars and Jupiter. They knew there were lots of asteroids in the belt, but they didn't know how many. They also knew that a high-speed collision with a baseball-sized asteroid could easily turn a craft into a tin can. When the probes made it through the asteroid belt without crashing, scientists breathed a collective sigh of relief. That the Pioneer twins both crossed the belt without injury told scientists that there was a great deal of space between the asteroids. Crossing the belt in the future may be easier and safer than once thought.

Pioneer 10 became the first craft to fly by Jupiter and return pictures. It also charted Jupiter's brutally strong radiation belts, located the planet's magnetic field, and discovered that Jupiter is composed mostly of liquid hydrogen (see also pages 133–138).

After zipping past Jupiter, Pioneer 10 aimed to exit the solar system. On June 13, 1983, it became the first craft to cross the orbit of Neptune, which at the time was farther away than Pluto. Pioneer 10 continues to send us measurements of the *solar wind*, the stream of high-speed particles that boils off the Sun and shoots out far past the outermost planet. Ultimately, the probe is searching for the *heliopause*, the boundary area between our solar system and interstellar space. At this boundary, solar wind hits the cosmic rays produced by

In the early 1970s, NASA began sending space probes to destinations in the outer solar system. The first of these, Pioneer 10 (below), took close-up pictures of Jupiter and Neptune before heading out toward interstellar space.

NASA

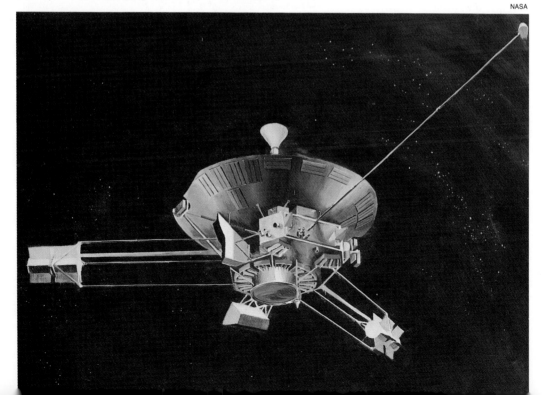

THE MOONS OF VOYAGER

Moons visited by Voyager ranged from tiny Iapetus (diameter: 900 miles), a moon of Saturn, to Jupiter's giant Ganymede (diameter: 3,270 miles). Earth' moon (diameter: 2,160 miles) is considered average size.

nearby stars and other astronomical bodies outside our solar system.

Pioneer 11 took a different path than its twin. It screamed past Jupiter toward Saturn at a dizzying 107,400 miles (172,903 kilometers) per hour, the fastest speed ever reached by a human-made object. The craft reached Saturn in 1979 and snapped the first close-up pictures of the planet. Pioneer 11 discovered two new moons and an additional ring. It also found Saturn's huge moon Titan to be too cold for life as we know it. Today, traveling in a path directly opposite to Pioneer 10, Pioneer 11 continues to send us information about the solar wind.

Both craft carry a message that intelligent aliens may be able to interpret. The message is a gold-plated plaque showing a man and a woman, where they live in the solar system, and where the solar system is located in the universe.

VOYAGERS

Two probes, Voyager 1 and 2, proved even more successful than the Pioneer twins. Thanks to the 100,000 or so pictures the Voyagers captured, we now have a more intimate knowledge of the outer planets. Launched in 1977, the Voyager duo embarked on a "grand tour" of the four giant planets—Jupiter, Saturn, Uranus, and Neptune—taking advantage of a rare planetary alignment that takes place about once every 176 years. When the probes reached Jupiter, they sent images of the planet's swirling red atmosphere and of active volcanoes on Jupiter's moon Io. Using Jupiter's gravity to gain speed, the two probes raced off to Saturn, where they snapped breathtaking pictures of the planet's complex ring system. At this point the duo took separate paths. Voyager 1 made a close flyby of Saturn's moon Titan, then swung out into interplanetary space. Voyager 2 went on to discover, among many other things, 10 new moons around the planet Uranus and an Earth-sized hurricane on the planet Neptune (see also page 139; page 147; and page 151).

Scientists never expected Voyager 2 to reach Neptune, let alone phone home with such revealing pictures. After all, Neptune was 2.8 billion miles away. And Voyager had become somewhat disabled—its main radio receiver had failed, one of its memory-storing computers had malfunctioned, and a jammed gearbox had made it difficult to swivel its instruments.

But the probe had been traveling for 12 years in space, so it was no wonder it was ailing. Space is no vacation spot. Temperatures can be a bone-chilling -400° F (-240° C). High-speed cosmic dust can create a vicious sort of sandstorm. Then there's radiation: high-speed protons, electrons, and other particles from the Sun and from exploding stars outside the solar system that can cause onboard computers to run amok.

To help the ailing Voyager 2, scientists radioed new computer programs that allowed the probe to work around its health problems. In August 1989, the probe's television cameras recorded images of Neptune and radioed the data to Earth. The radio signals were faint whispers across the lonely expanse of space. Back on Earth, scientists tuned in carefully with an array of 38 antennas that spanned four continents. The result was our first glimpse of an astounding world. Neptune appeared as a giant blue slushball with raging storms brewing in its methane atmosphere. On Neptune's largest moon, Triton, there was evidence of volcanoes spewing ice. Voyager 2's "last picture show" was truly spectacular.

Though both Voyager probes ventured past the orbit of Pluto, they never came close enough to that mysterious planet to take pictures. Pluto remains the one member of our solar system still shrouded in almost complete mystery.

The Voyagers are now racing toward distant stars. They have joined the Pioneer probes in searching for the heliopause. Scientists will be able to communicate with the Voyagers until about the year 2020. Around that time, their plutonium-based power generators will no longer be able to produce enough electricity to run the onboard computers, radios, and other systems. When the power fades, the Voyagers will con-

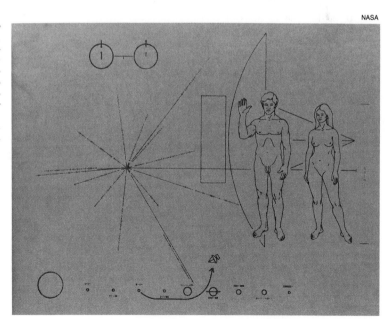

Should the Pioneer probes be intercepted by alien beings, plaques on board would provide the extraterrestrials with information on where the crafts originated and the type of beings that sent them. The Voyager probes visited many of the moons (facing page) in the outer solar system, but detected no trace of life as we know it.

tinue on their journeys, wandering through the Milky Way. Should intelligent aliens encounter either of the Voyager probes, they will find a videodisc aboard the craft, complete with greetings in 60 Earth languages, sounds and pictures from Earth, and even some interstellar rock and roll.

GALILEO

When the Pioneer and Voyager probes whisked past the outer planets, they gave us a glorious but brief glance. Scientists have since decided to launch new probes for a more detailed second look at each planet. Jupiter is the target of the space probe Galileo, named after the Italian scientist who used the first astronomical telescope to discover Jupiter's major moons.

The space probe Galileo was launched by the space shuttle in October 1989 after years of delay and a controversy over its onboard nuclear power system. Galileo contains a radioisotope thermoelectric generator (RTG) containing highly radioactive plutonium, which produces electricity for spacecraft instruments. While many spacecraft use solar panels to convert sunlight to electricity, Galileo would be too far from the Sun by the time it reached Jupiter to generate power this way.

The controversy centered around whether the RTG would remain intact if the shuttle were to explode, as the *Challenger* did in 1986. Some scientists were worried that, in such an event, the plutonium in the RTG would shower over parts of Florida, where the shuttle is launched. The U.S. National Aeronautics and Space Administration (NASA) assured the public that the RTGs were rigorously tested and would hold up under explosive conditions. In any case, the shuttle carrying Galileo roared into low-Earth orbit without a glitch and released the probe successfully from its payload bay. A booster rocket then sent the craft from low-Earth orbit on its journey to Jupiter.

© John T. Barr/Gamma-Liaison

Scientists closely track the progress of unmanned space probes through the solar system. Contact is maintained as long as the electrical power systems on the crafts function normally.

Originally, Galileo was designed for a 2.5-year direct flight to Jupiter. But changes in the launch system required engineers to devise a longer flight path, one that will take six years. Galileo will travel a course, shaped like a corkscrew, that will take it past Venus and back again. As the craft travels past Earth and Venus, it picks up speed by stealing some of each planet's gravitational energy. Galileo has already sent a wealth of data about Venus, Earth, and the Moon.

NASA hopes to have Galileo hurl an instrument-laden entry probe into Jupiter's violent atmosphere. A parachute will unfold, allowing the probe to descend slowly. The probe will radio back measurements of the chemicals in the atmosphere; it will determine whether superbolts of lightning exist; and it will clock wind speeds. After its 75-minute plunge for science, the probe will be crushed by the extreme pressures deep within the planet's atmosphere.

After the entry probe is released, Galileo will orbit Jupiter for a two-year photo-shoot. During this time, it will also determine the size and strength of Jupiter's radiation belt—useful information for designing future spacecraft to withstand the damaging particles trapped within the belt. The craft will also study the geology of some of Jupiter's moons, for example, to see if the volcanoes on Io are still active.

Cassini

Scientists plan to take a closer look at Saturn by launching the Cassini spacecraft in 1996. The craft, named after the 17th-century Italian astronomer who discovered four of Saturn's 19 moons, will orbit the planet for four years, beginning in 2002. The orbiter will study Saturn's icy moons, its 100,000 or so ringlets, and its atmosphere and magnetic field.

In past years the Voyager probes revealed that Saturn's largest moon, Titan, has a dense atmosphere, unlike any other moon in the solar system. Scientists want to know what Titan's atmosphere is made of and what its weather is like. To that end, Cassini will send a lander parachuting

NASA uses the space shuttle to deploy many of its unmanned probes. On October 18, 1989, the Galileo spacecraft was released from the shuttle Atlantis. It will take six years for Galileo to reach Jupiter.

through the moon's hazy covering. Scientists expect this lander to find methane snow and oceans of liquid ethane on the planet's surface.

Comet/Asteroid Probes

Comets are celestial chunks of rock, dust, and ice that speed through the solar system like huge, dirty snowballs. Scientists believe that comets contain the basic ingredients from which our Sun and planets were made. Naturally, scientists are curious to know more about them. Comets,

NASA plans to follow up the brief Voyager flybys with the proposed Cassini probe. Cassini will orbit Saturn for four years, making a detailed survey of the planet, its rings, and its moons.

however, rarely come close enough to Earth for study. The recent return (in 1985 and 1986) of Halley's comet—which swoops by Earth every 76 years—offered unique research opportunities.

Two Vega probes were launched by Soviet Proton rockets. These probes first studied Venus before flying to a rendezvous with Halley's comet in March 1986. The United States did not send a probe to Halley's comet, but opted to provide some instruments for the Vega probes, in cooperation with the Soviet Union and France. This was the first Soviet mission with an international payload.

There were other Halley probes. The European Space Agency (ESA) launched Giotto with a French Ariane booster in 1985. In March 1986, the probe approached to within 300 miles (500 kilometers) of the comet. Instruments aboard included a mass spectrometer, plasma-ion analyzer, and dust-impact instruments. Japan's Planet A probe also studied solar wind and included an ultraviolet imaging system.

The United States deployed its International Comet Explorer for rendezvous with the comet Giacobini-Zinner in September 1985. This comet approaches Earth about every six years. This was the first time in history that a space probe had passed through the tail of a comet. And NASA plans to do even better. The Comet Rendezvous/Asteroid Flyby (CRAF), scheduled for launch in 1995, will meet Comet Kopff in the year 2000 and fly in formation with it for three years as it travels toward the Sun. In 1998, en route to Comet Kopff, CRAF will fly past the asteroid Hamburga, one of many in a belt between Mars and Jupiter, to study its composition and photograph its surface.

Mars Probes

Of all the planets, Mars most closely mimics the life-sustaining conditions on Earth. That's why scientists believe life once may have evolved on Mars. Our first glimpse of the Red Planet came in 1965, when the NASA spacecraft Mariner 4, one of a series of probes that explored the inner solar system, sent home 21 fuzzy photographs of Mars. Mariners 6 and 7 flew past in 1969 and sent more pictures. Late in 1971 Mariner 9 reached the vicinity of Mars while the planet was being swept by a huge dust storm. Once the dust settled, the craft sent pictures of a great canyon, a towering volcano, and dried-up riverbeds.

Even more exciting were the U.S. Viking missions of 1976, which consisted of two orbiters and two landers. As the orbiters conducted a photographic survey of the planet, the landers, looking for signs of life, scooped up soil samples. Though the landers did not find life, the possibility still exists that some primitive life-form is hiding in a Martian crevice or that life once bloomed and later died out.

The Soviet Union has had only partial success with probes to Mars. After many early failures in the 1960s, Mars 5 managed to orbit the planet in 1973 and send back many pictures. That same year, Mars 6 and 7 sent back data from flybys, but their landers missed their targets. In 1988 the Soviets launched two Phobos probes to Mars. Contact with one was lost on its way to Mars. The second came within 100 miles (160 kilometers) of the planet, when its onboard computer went haywire. But before Phobos 2 failed, it sent valuable data, including pictures of Mars' potato-shaped moon Phobos and the first temperature map of the Martian surface.

NASA plans to launch the Mars Observer in 1992 to orbit the planet for a full Martian year (687 days). The orbiter will study Mars' weather and dust-storm activity, map the mineral content of its surface rocks, and try to determine the role water has played on its surface.

American, Soviet, and French scientists will team up in 1994 to explore the rugged Martian terrain with balloons made of Mylar. Once deployed by a Soviet orbiter, canisters will inflate the balloons with gas. During the Martian day, the Sun will heat and expand the gas, causing the balloons to drift along in the thin Martian atmosphere. During this time, cameras will snap panoramic views of the terrain and relay them to Earth via the Mars Observer. When night comes, the balloons will deflate and sink toward the surface, where they will drag instruments across the ground. The instruments will then search for water below the surface and map the terrain.

By the end of the century, both NASA and the Soviet Academy of Sciences plan to place rovers on Mars to scoop up rock and soil and then return home. On July 20, 1989, the 20th anniversary of man's first walk on the Moon, President George Bush proposed a manned mission to Mars for the early part of the 21st century.

Earth Probes

While scientists have worked hard to understand other planets of our solar system, they haven't forgotten our home planet. Since the 1960s, satellites have been launched to observe Earth from space and to take measurements of the atmosphere, oceans, and land. But with growing concern for the global environment, there is a pressing need to understand our planet better than we do at present.

In its "Mission to Planet Earth," NASA plans to launch a number of satellites equipped with instruments to monitor such things as temperature, ocean currents, emissions of "greenhouse gases," the extent of pollution, the water cycle, and the pace of rain-forest destruction. This information will allow us to assess the impact we have on the environment. It will also help us make accurate predictions about how, and to what degree, the environment will change in the future.

The mission will begin in the first half of the 1990s, when NASA launches a series of small satellites called Earth Probes. These small probes will be followed in the late 1990s by a series of 15-ton (14-metric-ton) platforms carrying a dozen or more remote-sensing instruments. The platforms, collectively called the Earth Observing System (EOS), will fly in space from pole to pole. This will allow the instruments aboard the platforms to monitor the entire planet as it rotates below. Japan and European countries plan to supplement EOS with additional space platforms.

Moon Probes

Exploration of the Moon was hastened by competition between the United States and the Soviet Union. The Soviets scored many early successes with their Luna program, which began in 1959. That year, Luna 1 flew past the Moon, Luna 2 slammed into the lunar surface, and Luna 3 went completely around the Moon. In its lunar orbit, Luna 3 photographed the side not visible from Earth, and transmitted the first photos of this area. Beginning with Luna 9, launched in 1966, several Luna probes have soft-landed on the Moon. These have transmitted pictures, density data, and other information. In 1970, 1972, and 1976, Luna probes scooped up Moon rock and dust samples and sent them to Earth by rocket, a feat the Soviets hope to repeat on Mars. In 1970 and 1973, Luna probes delivered unmanned wheeled vehicles that roamed the lunar surface, driven by remote control from Earth.

Though the United States got a later start in the race, it was the first—and, as yet, only—country to land men there. Unmanned probes helped pave the way. Rangers 7, 8, and 9 (1964–65) took more than 17,000 pictures of the Moon's surface before they crash-landed. A series of Surveyors soft-landed on the Moon and tested its surface. Lunar Orbiters photographed the Moon's surface, revealing suitable landing sites. On July 20, 1969, the *Apollo 11* spacecraft brought the first crew of Americans to the Moon's dusty surface.

In January 1990, Japan became the third nation to launch a probe to the Moon. MUSES-A rocketed from Earth aboard a booster built by Nissan, the car company. Once in orbit, the probe's transmission system failed. More Japanese probes will most likely follow.

It seems another race to the Moon has begun, this time an economic one. Several Japanese corporations plan to prospect the Moon for minerals, metals, and other natural resources. In 1992 a private U.S. research organization, Space Studies Institute, plans to launch its Lunar Prospector on a one-year mission to orbit the Moon. The Prospector will study the Moon's trace gases and search for frozen water and other resources.

NASA, too, is looking into the possibility of mining the Moon's resources. In

Space probes also study such heavenly phenomena as comets. In 1986, when Halley's comet made its much-awaited passage by Earth, it was intercepted by the Giotto probe (pictured below). The probe passed through the dust and gas surrounding the nucleus of the comet and transmitted data back to Earth about comet composition.

Space probes are manufactured to extremely precise specifications. Engineers assign especially high priority to the systems through which the spacecraft and ground stations will communicate. The Ulysses probe, for instance, has a 5.2-foot-diameter "high gain" antenna that continuously points toward Earth.

the late '90s, the agency plans to launch a Lunar Observer to scout the best place to build a human outpost. The probe will enter a polar orbit of the Moon for two years, surveying the lunar surface for frozen water, metals, and other resources. Metals could be used to construct a lunar base—a potential staging area for human trips to Mars, the asteroids, and beyond. Water brought to the Moon could be separated into oxygen and hydrogen and used as rocket fuel. A Moon base would also be ideal for telescopes, as the Moon is largely free of atmosphere, which blurs images, and completely free of the human-made space junk traveling through space, which threatens orbiting telescopes.

Venus Probes

Venus has long been called Earth's "twin" because the two planets have roughly the same size and density. But reports from more than 20 probes to Venus have reported an atmosphere that is more like Earth's worst nightmare.

In the world's first robotic flight to another planet, Mariner 2, launched by the U.S. in 1962, reported that Venus has a hellish surface temperature of 800° F (430° C), far too hot for water to exist in liquid form. Venus is victim of a runaway greenhouse effect; its thick cloak of carbon dioxide gas traps solar heat, which broils the surface (see also page 99).

In 1967 the U.S. sent Mariner 5 to Venus. From 1967 to 1970, the Soviets sent four Venera probes, which for the first time in history pierced Venus' clouds and landed on the planet's surface. From these early missions, we know that the air is so thick and heavy that it exerts a crushing pressure 90 times that felt on Earth's surface. In 1978 the American Pioneer-Venus 1 found swirling clouds of sulfuric acid amidst the carbon dioxide cover. En route to Halley's comet in 1985, two Soviet Vega probes released helium-filled balloons into Venus' atmosphere. The balloons were swept great distances by the 155-mile (250-kilometer)-per-hour winds.

Between 1972 and 1981, more Soviet Venera craft soft-landed on Venus, performed soil analysis, and returned color pictures of the rock-strewn surface sur-

Scientists have developed novel ways to control space probes from bases on Earth. An advanced system called "aerobraking" allows a spacecraft to change its orbit around a planet from elliptical (oval-shaped) to circular.

rounding the craft. But scientists wanted to have a global picture of Venus' topography. The planet's hazy atmosphere, however, kept the surface shrouded in mystery until the U.S. launched Pioneer-Venus 1 in 1978, and the Soviet Union launched Venera 15 and 16 in 1983. These three probes used radar to pierce the clouds and map many parts of the surface. They revealed, among other things, large craters, continent-sized highland areas, and huge mountains. But the images were fuzzy.

In September 1989, the NASA spacecraft Magellan started radar-mapping the Venusian surface with far better resolution. The craft sent sharp images of a bone-dry world strewn with bizarre patterns of cracks and fissures, channels cut by rivers of lava, mile-high "pancake domes," and giant craters from impacting meteorites. Such a landscape is far different than the plant-covered continents, blue oceans, and polar ice caps of Earth. Thanks to the data supplied by Magellan, scientists can now begin to figure out how Earth's "twin" planet turned out to be so different.

Mercury Probes

NASA's Mariner 10, launched in 1973, remains the only probe to have visited Mercury, the small but swift planet closest to the Sun. Through clever engineering, scientists managed to get *three* flybys of Mercury (and one of Venus) out of the craft. The Sun provided free sunlight, which the craft's solar panels collected to generate power for onboard instruments.

Mariner 10 began its trip with a flyby of Venus. During this time it took the first clear pictures of that planet's swirling atmosphere. Then the craft entered an orbit of the Sun, revolving like a miniplanet every 176 days, exactly twice the time it takes Mercury to orbit the Sun. This trajectory took Mariner 10 past Mercury on March and September 1974 and again on March 16, 1975, when it passed just 200 miles (322 kilometers) above Mercury's surface. After this third and final pass, the attitude-control gas used to stabilize the craft ran out and the craft slowly tumbled out of control and fell silent.

During its lifetime, Mariner 10 sent thousands of close-up photographs of Mercury, showing what no telescope on Earth could possibly see at the time. The craft revealed a gray planet that looked much like the Moon. Mercury's surface was strewn with impact craters—holes made when space rocks smash into the planet's surface. At first, scientists were going to name these craters after birds or famous cities. But they ended up choosing the names of famous artists, writers, and musicians. If you look at a map of Mercury, you'll see craters van Gogh, Mark Twain, and Beethoven, to name a few.

Sun Probes

From the late 1950s, many craft have been sent to orbit our star, the Sun, including the Soviet Luna 1, the U.S. Pioneers 6 through 9, and the German Helios 1 and 2. These probes studied the Sun itself, its powerful magnetic field, and the solar wind.

But because of the way these probes were launched, they viewed only the Sun's central area, or equator. Most probes, including the early solar probes, are launched in the same direction that the Earth orbits the Sun. That is because the probes can get a natural boost of 18 miles (30 kilometers) a second from the Earth's orbital speed. When these probes leave the Earth, they travel in the same plane that the Earth (and most other planets) travels around the Sun. If this plane, called the *ecliptic*, were solid, it would cut the Sun in half at its equator. Probes traveling on the ecliptic only can study solar activity coming from the Sun's equator. They can't measure the activity on the solar *poles*, the Sun's top and bottom.

Scientists have long desired a glimpse of the Sun's poles and the space above and below the ecliptic plane. (Imagine how limited our understanding of the Earth would be if we knew only the areas around our planet's equator!) To study the Sun's polar regions, scientists would have to send a probe perpendicular (at right angles) to the ecliptic plane. They couldn't use rockets to do this, because no rocket is powerful enough to overcome Earth's rapid motion around the Sun. Instead, scientists decided to enlist the help of the largest planet, Jupiter. Jupiter's gravity is strong enough to sling a craft out of the ecliptic plane and over the Sun's poles.

In October 1990, NASA and the European Space Agency launched the spacecraft Ulysses toward a rendezvous with Jupiter on February 1992. When Ulysses approaches Jupiter, the planet's gravity will pull the craft down and out of the ecliptic plane. Then Ulysses will come under the Sun's gravitational control and orbit the star in a way that brings the craft perpendicular to the ecliptic plane. By August 1994, Ulysses will whip around the Sun's south pole, and by July 1995, its north pole —the first craft to orbit the Sun's poles. It will send us more data about the star that holds all the planets together, that lights up our days, warms our bodies, and allows life to flourish on Earth.

NASA/JPL

When the Magellan probe (left) reached Venus (below), it used radar waves to produce detailed maps of the planet's surface.

NASA/JPL

SPACE SATELLITES

by Tom Waters

On October 4, 1957, when the Soviet Union launched the first artificial satellite, Sputnik 1, banner headlines blared around the world. "Russia Launches A Moon," proclaimed page one of the *London Daily Mail*. A *New York Times* editorial called the feat a "concrete symbol of man's coming liberation from the forces which have hitherto bound life to this tiny planet."

Sputnik was a metal sphere 23 inches (58 centimeters) in diameter and 185 pounds (84 kilograms) in mass, carried 560 miles (901 kilometers) above the Earth's surface by an SS-6 rocket. Moving at a speed of 17,896 miles (28,800 kilometers) per hour, Sputnik circled the Earth in 1 hour, 36.2 minutes. It was equipped with two radio transmitters, whose continuous broadcasts were powerful enough to be picked up by amateur radio operators around the world. Its purpose was to test Soviet technological ability and—perhaps more important—to proclaim that ability to

Satellites enhance worldwide communications, conduct scientific research, and provide illuminating photos of atmospheric conditions on Earth and on our neighboring planets.

the world, especially the Soviet Union's great rival, the United States.

U.S. space scientists responded by stepping up their own efforts. On December 6, 1957, the Navy attempted a satellite launch, but it failed. Finally, on February 1, 1958, the Army succeeded in getting Explorer 1, the first U.S. satellite, into orbit. The U.S. made several more launches in 1958, and the National Aeronautics and Space Administration (NASA) was founded on July 29 of that year.

Since these early stages of the space race, thousands of human-made objects have been placed in orbit by many countries. As the number of artificial satellites in orbit has increased, so has the number of satellite uses. Some satellites are used for radio, television, and telephone communications. Others have cameras used for observing the Earth, for weather forecasting, for scientific research, and for military purposes. Still others carry scientific experiments and instruments like the Hubble Space Telescope for observing the universe beyond the Earth. And in the 1980s, the United States began testing Strategic Defense Initiative (SDI, or "Star Wars") technology, satellites designed to seek out and destroy missiles or other satellites in time of war.

SATELLITE LAUNCHES

For the first 24 years of the space age, satellites were always launched on rockets. Rockets generally carry a single cargo, called a *payload*, into space. The rockets are then discarded and allowed to burn up as they fall back to Earth.

Since 1982 satellites have also been carried into orbit by U.S. space shuttles, reusable spacecraft that are launched by rockets but can glide back to Earth and land on a runway. When the space-shuttle program was first unveiled on September 17, 1976, its primary purpose was to launch most of the United States' satellites. But NASA has not been able to reach the flight-a-week schedule it first envisioned. The actual cost per launch has also proved much higher than planned. Since the explosion of the space shuttle *Challenger* in 1986, the shuttles have been used only for NASA's own satellites and a select few international scientific satellites.

The business of placing satellites into orbit has become very competitive. For years the U.S. Government launched almost all of the commercial satellites for the non-Communist world. Then, in 1984, the European Space Agency (ESA), a cooperative venture (originally eight countries, now 14) began launching commercial satellites on its own rocket model, Ariane. In 1990 China launched an American-made communications satellite for use by a group of 20 Asian countries. Japan is also developing a rocket known as H-2 for commercial use. And in 1988 NASA began allowing private companies in the United States to perform their own launches on single-use rockets. Today NASA launches no commercial satellites at all.

ORBITING THE EARTH

Once a rocket or shuttle has carried a satellite to the correct altitude, it must be placed in an effective orbit. That is, it must begin circling the Earth at the correct speed so that it does not immediately tumble back down to the ground. A guidance system turns the satellite so that it is pointing parallel to the Earth's surface. Then a rocket engine fires to send it speeding in this new direction. This may be either the same rocket engine that has carried it into space or a different one.

This motion parallel to the Earth's surface is what keeps the satellite in the sky. The force of gravity diminishes as one moves away from the Earth's surface, but it is still quite strong at typical satellite altitudes. As a result, an orbiting satellite is constantly being tugged back toward the Earth. Ideally, however, its motion keeps it from reentering the atmosphere and crashing to the ground because, as the satellite falls toward Earth, it also moves sideways so that it just misses the planet and flies off to the other side.

The most natural orbital path of a satellite revolving around a planet is an el-

The shuttle era heralded a new means for satellite deployment, formerly achieved by rocket launches. Here Australia's communications satellite, AUSSAT, is released from the shuttle Discovery.

lipse, with the center of that planet at one focus of this ellipse. This is also true of the Earth's orbit around the Sun, as well as the Moon's orbit around the Earth. Although the orbits of artificial satellites are generally elliptical, nearly circular orbits are often used as well.

Every elliptical orbit has an *apogee*, or point farthest from the Earth, and a *perigee*, or point nearest to the Earth. The apogee and the perigee vary widely with different satellites. Sputnik 1 had an apogee of 584 miles (940 kilometers) and a perigee of 145 miles (234 kilometers). The apogee of the U.S. Hubble Space Telescope is 385 miles (619 kilometers), and its perigee is 379 miles (610 kilometers). Sometimes the distance between the apogee and the perigee is very large. For example, the U.S.

craft Explorer 14 had an apogee of 60,835 miles (97,904 kilometers) and a perigee of 173 miles (278 kilometers).

Scientists must decide in advance of a flight what sort of orbit is needed and how high it should be above the Earth. The plane of an artificial satellite's orbit must always include the center of the Earth, but it may be directed in various ways. The actual orbit, worked out in advance, will depend upon the intended mission of the satellite.

The plane of the orbit may pass over both poles as well as the center of the Earth, in which case the orbit is said to be *polar*. An orbit just slightly off the poles can bring a satellite over a different part of the world each time it circles the globe, while eventually returning to pass over

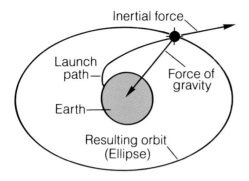

The force of a satellite launched at a speed of 5 to 7 miles per second, combined with the constant tug of Earth's gravity, pull a satellite into an elliptical orbit.

each spot at the same time of day. In this way a single satellite scans the entire planet under constant light conditions. This type of orbit is called *sun synchronous* and is well suited for reconnaissance satellites.

If the plane of a satellite's orbit lies along the equator, the orbit is said to be *equatorial*. The orbit may also be inclined by any number of degrees to the plane of the equator. For example, the inclination of the Explorer 1 was 33.6 degrees. If a satellite has a nearly circular, equatorial orbit 22,282 miles (35,860 kilometers) above the ground, then it has what is known as a *geostationary* orbit. It will take precisely one day to orbit the Earth. Since the Earth is also rotating once a day, the satellite will then remain fixed above one point on the Earth's surface. Geostationary orbits are especially useful for communications and weather satellites.

GUIDANCE IN SPACE

There are many opportunities for error and disaster in spaceflight. The smallest inaccuracy in calculation or the malfunction of a rocket can cause the space vehicle to veer far off target. This is why both rockets and satellites have guidance systems to keep them on course. There are various kinds of systems. For example, in the radio-command guidance system, changes in velocity and direction are broadcast by the spacecraft to a station on the ground. The data are fed into computers that make the calculations required to keep the craft on course. The necessary adjustments are then broadcast back to the craft, and control motors on the craft carry out the instructions. This procedure may be directed by ground personnel or it may be automatic.

In an inertial-guidance system, the velocity and direction of the satellite are measured by an onboard sensor system. This information is then fed to onboard computers and flight controls. And some spacecraft employ hybrid systems that combine some of the techniques from both radio and inertial systems.

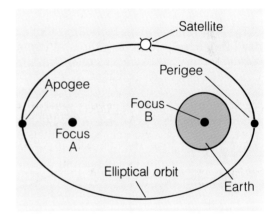

In an elliptical orbit the distance of a satellite's nearest point to Earth (perigee) and its farthest point (apogee) will depend on the satellite's mission.

Most communications satellites are placed in an equatorial orbit, while a polar or inclined orbit is well suited for many reconnaissance satellites.

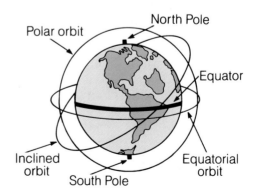

SPACE SATELLITES 305

POWER AND COMMUNICATIONS IN SPACE

All spacecraft require a power source to keep their instruments functioning and to provide the necessary radio transmission. Solar cells (waferlike silicon cells) convert sunlight into electric current. These cells must be exposed to the Sun in order to work, so whenever a satellite is in the Earth's shadow, it relies on current stored in a nickel-cadmium battery. This battery is, in turn, recharged by the solar cells. Engineers are now developing more-efficient silicon solar cells, as well as cells made from a gallium arsenide combination to provide lighter and more-durable power sources.

Other power sources are independent of the Sun. The SNAP generator, a device that produces electricity directly from atomic energy, proved successful in the U.S. Navy's Transit satellites in the early 1960s. The Soviet Union has used nuclear reactors to power its satellites since the 1970s. U.S. plans to use nuclear reactors in more satellites have come under criticism due to the danger posed by the radioactive fuel that may eventually fall back to Earth.

Almost all satellites are designed to transmit information to the Earth by means of a telemetering system. In this system, data collected by the scientific instruments on the craft are converted into radio signals, and these are transmitted to ground stations. At these stations, spools of magnetic tape record the signals, and computers decode them. Photographs can also be transmitted in this way, much as they are in television broadcasts.

THE END OF THE FLIGHT

The flight of an orbiting artificial satellite must come to an end eventually. Although the atmosphere is rarefied (much less dense) 124 miles (200 kilometers) from the Earth, no part of space is an absolute vacuum. Some molecules and other particles still exist that far up. When such particles strike an orbiting spacecraft, the individual collisions are insignificant, but the cumulative effect is damaging.

In time the craft will begin to lose velocity. It will not go so high at apogee, and, at its next perigee, it will be closer to the Earth. The spacecraft will eventually fail to balance the pull of gravity and will begin to penetrate deeper into the Earth's atmosphere. Unless special protective measures or devices are provided, a spacecraft reentering the atmosphere is doomed. Just like a meteorite that penetrates the atmosphere, it will be subjected to intense heat as it col-

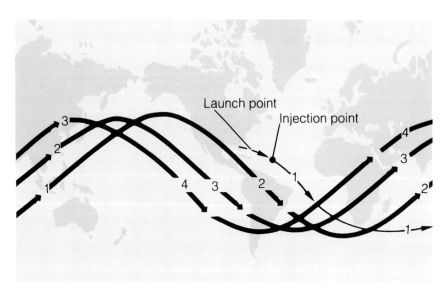

Satellites in polar and inclined orbits pass over different portions of the Earth each time they circle the globe. This is illustrated in these four successive passes of a satellite launched from a point in Florida that went into an inclined orbit at the injection point indicated.

COMSAT

Intelsat V is an advanced communications satellite that simultaneously provides 12,000 two-way telephone circuits and two television channels for communication between the United States and Europe. To keep these instruments functioning, the satellite is powered by solar cells that convert sunlight into electric current.

lides with the increasingly numerous molecules and ions, and it will burst into flames. The chances are very high that the satellite will be completely destroyed. Some fragments, however, may make their way down to Earth. When Skylab fell out of orbit in July 1979, many large pieces reached the ground. Fortunately, they landed in the Pacific Ocean, without injuring any people or property.

Satellites that escape from the Earth's gravitational field may be destroyed by way of collision. For example, they might crash on the Moon. But if they fail to collide with a planet, they may enter orbit around the Sun. After all, everything in our solar system is in the Sun's gravitational field. Once a satellite begins orbiting the Sun, it may continue circling indefinitely.

Today there are so many satellites roaming space that "space junk" has become a major concern at NASA and the rest of the world's space agencies. Old satellites, often broken into tiny fragments, now threaten to collide with and destroy new ones.

COMMUNICATIONS SATELLITES

Each day, tens of thousands of phone calls are made from the United States to Great Britain. Half of these are carried by a web of undersea fiber-optic cables that stretch across the Atlantic floor, but the other half are transmitted via satellite.

In satellite transatlantic telephone communications, the sound of a speaker's voice is converted to an electrical signal and carried over wires to a transmitting station. There it is translated into a powerful radio signal and aimed at a small, unmanned spacecraft orbiting 22,282 miles (35,860 kilometers) above the Earth's surface. An automated device on the craft receives the signal, then broadcasts it back down to a receiving station on the other

SPACE SATELLITES

The InfraRed Astronomical Satellite (IRAS), depicted here as it passes over Western Europe, was launched in January 1983. During its ten months of operation, its instruments collected data on the infrared radiation emitted from over 250,000 sources, including newborn stars, interstellar dust, distant galaxies, and quasars.

side of the Atlantic. The signal is then sent out over wires to the other speaker's home or office. There are now over 75 satellites, operated by more then 20 different companies, national governments, and international organizations, that provide this technologically advanced yet routine service.

Today's satellites are now used to carry all kinds of long-range signals. In addition to transmission of transatlantic calls, satellites direct many calls within the continental United States. Broadcast and cable television programs are relayed by satellite from network headquarters to local stations. Computers use satellites to convey enormous quantities of information to each other.

New uses for satellites are constantly opening up. People who live in remote areas without local TV stations sometimes use dish antennas up to 20 feet (6.1 meters) in diameter to pick up satellite TV signals that are meant primarily for rebroadcast by local stations. Japan and Germany now have satellites that broadcast TV programming directly to consumers, who need only small dishes to pick up these signals. Several companies are now preparing to offer a similar service in the United States.

Early Communications Satellites

The first satellite to relay messages from one Earth station to another was SCORE (Signal Communicating by Orbiting Relay Equipment), launched December 18, 1958. SCORE received and recorded messages up to four minutes long while flying over one station, then rebroadcast them while flying over another station. The experimental satellite lasted 35 days before its batteries ran down.

Early communications satellites did not amplify radio signals, but merely reflected them, and were thus called *passive* communications satellites. The first true passive communications satellite, Echo 1, was launched by the United States on August 12, 1960, less than three years after Sputnik 1 entered orbit. The Echo satellite was actually a plastic balloon 98 feet (30 meters) in diameter, covered with a thin coating of aluminum to reflect radio transmissions. A second, larger Echo began orbiting in 1964.

In Project West Ford in 1963, space scientists spread a thin ring of fine wire needles around the Earth. The wires were "tuned" to reflect signals at a certain frequency. Scientists eventually abandoned the program because the ring of wires threatened to disturb astronomical observations. In 1966 researchers tried another novel idea. An Echo-type balloon was covered with precisely spaced wire mesh. The balloon decomposed after a few hours in orbit, leaving a hollow metallic sphere that was five times more reflective than Echo. It was also much less affected by the Sun's radiation and was slowed down less by the atmosphere. These passive satellites were really not practical because of their large size and the powerful equipment required to send and receive messages.

Active Communications Satellites

Bell Telephone tested the commercial value of communications satellites in 1962, when Telstar 1 was placed in orbit. Telstar was an *active* satellite, both amplifying and retransmitting as many as 60 two-way telephone conversations at a time. Like other active satellites, Telstar obtained the electrical power that it needed to receive and transmit signals from banks of solar cells mounted on panels attached to the satellite. Ground stations were established in the United States, England, and France. A few months after Telstar 1, Relay 1 was launched for RCA (Radio Corporation of America), adding Italy and Brazil to the list of countries that received broadcasts from satellites in outer space.

In 1963, Syncom 2 became the first communications satellite to achieve geostationary orbit. It remained over the same position on the Earth's surface at all times and was in constant view of almost half the planet. Most communications satellites are now placed in orbits of this type.

From time to time, NASA has participated in programs to test experimental communications systems. NASA's Applications Technology Satellite, ATS-1, was launched on December 6, 1966, to test a variety of instruments, including a 600-channel repeater that could relay color-television programming, as well as radio transmissions between aircraft and ground stations. ATS-1 was also used to develop technology for weather satellites.

In addition to the United States, many countries have developed or purchased their own communications satellites, with numerous others planning to do likewise in the near future. The Soviet Union began building its orbiting system in 1965, and now has three series of communications satellites (called Molniy, Statsionar, and Gorizont) circling the Earth. These satellites bring telephone and television service to remote areas of the country, as well as to parts of Eastern Europe. Since the 1970s India has used a satellite called SITE to send educational-television programming to thousands of rural villages.

Satellite Networks

By far the most extensive satellite system is that of the International Telecommunications Satellite Consortium, or Intelsat. Intelsat is made up of 114 nations (39 nonmember nations also use its services) and operates over 800 ground stations worldwide. Beginning in 1965 with the Early Bird satellite, Intelsat has steadily expanded its capability, and now controls 15 satellites—with more planned in the years ahead. Spacecraft in the consortium's latest series, Intelsat VI, can handle 120,000 telephone circuits and three color television transmissions simultaneously. Intelsat VIIs, with improved television capabilities, are planned for launch in 1992.

Most of the members of Intelsat are national governments, but the American member of the consortium is the Communications Satellite Corporation (COMSAT), a privately owned company established by the U.S. Congress in 1962. COMSAT pays Intelsat for the use of its satellites, then sells services to companies

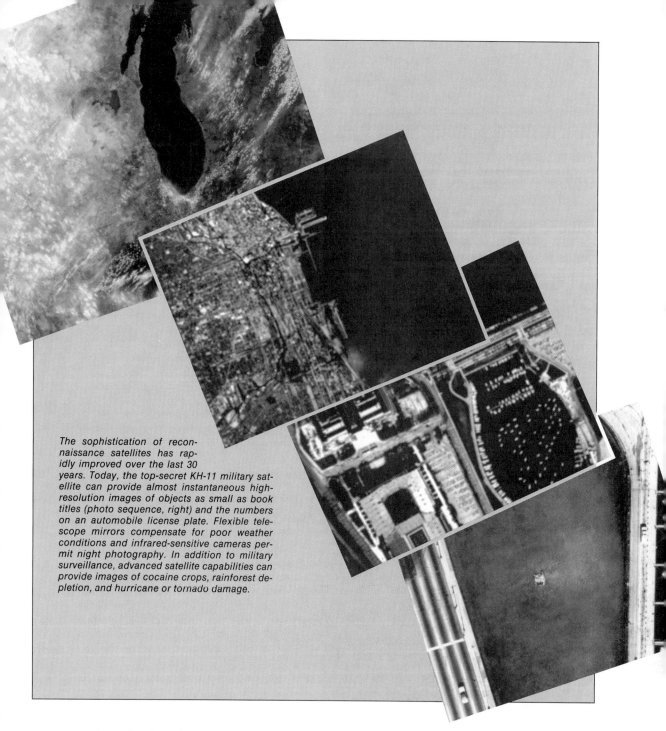

The sophistication of reconnaissance satellites has rapidly improved over the last 30 years. Today, the top-secret KH-11 military satellite can provide almost instantaneous high-resolution images of objects as small as book titles (photo sequence, right) and the numbers on an automobile license plate. Flexible telescope mirrors compensate for poor weather conditions and infrared-sensitive cameras permit night photography. In addition to military surveillance, advanced satellite capabilities can provide images of cocaine crops, rainforest depletion, and hurricane or tornado damage.

such as the American Telephone and Telegraph Corporation (AT&T). COMSAT also operates several of its own satellites. Several other American companies operate satellites, and many other countries have set up public or private satellite systems for domestic communications.

Other international networks operating communications satellites include Intersputnik, which serves the Soviet Union and the Eastern European countries; Arabsat, which serves 20 Arab nations; Eutelsat, which serves the communication needs of 26 European nations; and Inmarsat, which provides communication services to ships on the high seas. In addition, Indonesia operates a system of three satellites for use by a group of Pacific countries.

Communications Satellite Operation

Until recently, most communications satellites were launched by NASA. But now private companies have taken over most American launches. Europe, Japan, and other countries are also competing in the satellite-launching market. Intelsat has most of its satellites launched by the European Space Agency on Ariane rockets. The consortium also uses the launch services of General Dynamics and other American aerospace companies.

Once in its prescribed position in geosynchronous orbit over the Earth, a satellite is ready to go to work. Its ground station may belong to a country's government, a consortium like Intelsat, or even private individuals. The ground stations usually beam messages to the satellites at one frequency and receive messages at another by using an amplifying device called a transponder. This technology is very efficient because it allows a single ground station to send and receive signals at the same time.

Depending on a satellite's capability, it can be used for telephone service, encoding data, or television transmissions. Its various channels are usually leased to users in the form of half-circuits—two-way connections between the satellite and a ground station. Thus, a pair of half-circuits are required to complete an overseas call. These connections are normally operating round the clock.

Other channels may be used for television transmissions such as news stories and sports events. Finally, a portion of the circuits are held in reserve for emergencies, increased demand within the system, and other special needs.

The Future of Satellite Communications

Ironically, the major obstacles to designing new communications satellites no longer have to do with space technology. Instead, scientists are being stymied by the limited space in the electromagnetic spectrum. Any radio transmission takes place at a particular frequency, and there is a limit to how much can be transmitted on any one frequency. Almost every available frequency that can be used to transmit information is already being used, and the remaining frequencies are being held in reserve for specific purposes.

Engineers are now seeking ways to pack more information into each frequency channel, often by digitally processing signals. This approach has multiplied by five the number of phone calls that can be put through a single Intelsat channel, for example. But engineers believe that signals can be compressed even more in the future. Plans for new satellite-communications services—such as direct television broadcasts from satellites to consumers—depend on continued progress in these digital-signal-compression methods.

Satellite technology is invaluable to military commanders when planning an aerial attack. Here, in satellite photos taken during the Gulf War in 1991, a view of Baghdad, Iraq (above), is magnified many times (left) to capture an image of a bridge that was destroyed by the Allied forces during a bombing raid.

OBSERVATION SATELLITES

What better vantage point for observing the Earth could there be than a satellite orbiting high above the planet's surface? For many kinds of observations, apparently, there is none. In the past 30 years, satellite observations of the Earth have become routine in many areas.

Satellite images are used to predict the weather, for example, and to help explain it in televised weather reports. Military-reconnaissance satellites enable countries like the United States and the Soviet Union to judge the capabilities of their rivals in peacetime and to monitor the movements of their enemies in war.

As the availability of satellite images increases and the price falls, many more satellite uses keep turning up. An ecologist who wants to study certain tree species can locate large stands of them using commercially available satellite images. Satellites also monitor large-scale ecological changes, such as the clearing of rain forests in Brazil, Malaysia, and Indonesia. Since much of this tree cutting is done illegally or with little governmental supervision, there is no way to estimate the amount of land being cleared except by the use of satellite pictures.

The United States launched the first Earth-observation satellites: the Tiros 1, a weather satellite launched April 1, 1960, and Corona/Discoverer 14, a military-reconnaissance satellite launched August 18, 1960.

Tiros took pictures electronically and transmitted them to Earth on radio waves. But these pictures proved too blurry for military reconnaissance. Corona was developed to capture images with camera film. After exposing the film, the satellite ejected it in a capsule, which an Air Force plane then plucked from the sky. In Corona's pictures, objects as small as 1 foot (0.305 meter) long were clearly visible from space. Viewers could even identify the make of cars in Moscow's Red Square.

The Soviet Union entered the satellite-reconnaissance game in 1962, with its Cosmos 4 satellite, which also used the ejected-capsule method for getting its pictures back to Earth. China followed suit in 1970 with China 1.

Spies in the Sky

The technology of military reconnaissance by satellite has changed considerably over the years. During the 1960s and 1970s, the United States developed satellites that could use their own rocket engines and navigation systems to move from one orbit to another, or to swoop closer to the Earth for especially detailed observations.

The U.S. military continued to develop satellites that could relay their information to Earth on radio waves. But film-dropping satellites remained essential for high-resolution pictures until 1976. That year the first KH-11 went into orbit. KH-11s are U.S. satellites that use very large telescopes and video cameras to observe the Earth and continuously transmit pictures to ground stations. Their exact capabilities are a military secret—even their existence is not officially admitted. But it's rumored that KH-11s can pick out objects 6 inches (15.24 centimeters) long, and perhaps as little as 2 inches (5.08 centimeters) long. It may be possible to read automobile license-plate numbers with KH-11 pictures.

The United States stopped using film-return satellites in 1984, and has since developed the Advanced KH-11, with improved nighttime observing ability. While no one will explain how the images are relayed to Earth, the president or military officials can view images from KH-11s within an hour after they are taken.

The Soviet Union launched Cosmos 1426, its own answer to the KH-11, in 1982. But it still uses film-return satellites. It is believed that the Soviets do their military spying with less advanced, less expensive observation satellites, but they place more of them in orbit. Meanwhile, at any one time, the United States is said to use two KH-11s, two Advanced KH-11s, and a Lacrosse satellite, which uses radar to make observations at night or through thick cloud cover. All of these satellites are short-lived and must be replaced regularly. The United States and the Soviet Union also maintain separate networks of satellites that spy on radio signals around the world, including signals sent from other satellites.

In 1991 the United States, the Soviet Union, and China were still the only three countries with reconnaissance satellites, but this is rapidly changing. In 1988 Israel launched Offeq 1, an experimental satellite reportedly built with help from South Africa. Israel launched an Offeq 2 in 1990. France, Italy, and Spain are collaborating on the development of their own reconnaissance-satellite network. India and Germany each plan to develop military-reconnaissance satellites. Even Japan is reported to be working on a satellite that will make both military and scientific observations. Other countries are likely to undertake satellite-reconnaissance programs in the coming years.

Civilian Observation Satellites

Civilian satellite-observation technology has always lagged behind military capabilities. Civilian organizations lack the lavish funding available to the military.

Also, governments try to prevent civilian satellites from gaining capabilities that could affect national security. In 1978 the United States prohibited civilian satellites from producing pictures in which objects smaller than 33 feet (10 meters) can be seen. So civilian satellites had to be made 100 times less sensitive than military ones. In 1987, however, the Soviet Union started selling pictures with 16.5-foot (5-meter) resolution to anyone who could pay, thus rendering the U.S. ban ineffective.

Despite these limitations, civilian satellites are providing valuable information on meteorology, agriculture, forestry, geology, environmental science, and in other areas. The first weather satellite, Tiros, launched in 1960, was primarily an experimental craft, but meteorologists were able to get useful information from its images. More Tiros satellites went up through 1965 for combined experimental and operational use. In 1966 the Environmental Science Services Administration (ESSA) began launching a series of satellites based on the Tiros series to provide routine, daily satellite weather photography. ESSA has since been renamed the National Oceanic and Atmospheric Administration (NOAA), but it is still operating satellites that are advanced versions of Tiros. These satellites, now called NOAAs, fly in near-polar, sun-synchronous orbits, so that they scan the entire planet's surface, always passing over each spot at the same time of day.

NOAA also operates a series of satellites called Geostationary Operational Environmental Satellites (GOES), which fly in geostationary orbits so that they always seem to hover over the same spot on the Earth's surface. They can observe the same part of the Earth continuously, making it possible to see clouds and weather systems in motion. The GOES satellites also measure the heights of the cloud layers they see, making it possible to measure formations in three dimensions. The first GOES was launched on October 16, 1975.

The wavelengths of light reflected from the ocean are directly related to chlorophyll concentrations, which indicate biological activity. An image from the Coastal Zone Color Scanner satellite indicates that the highest concentrations of chlorophyll (red and yellow) hug the shallower coastlines, while the lowest levels (blue) occur in the deeper midocean areas.

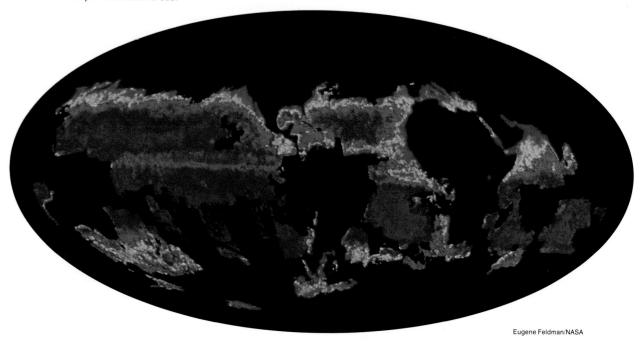

Eugene Feldman/NASA

Weather satellites often travel in a geostationary orbit, seeming to hover over the same area of the Earth's surface to make continuous observations of clouds and weather systems in motion. The course of a hurricane's swirling cloud formation can then be monitored.

Other countries operate satellites in both sun-synchronous and geostationary orbits. The United States participates in an international joint effort with Japan, the Soviet Union, and the European Space Agency to coordinate geostationary satellites that together scan the entire world. The United States launched Landsat 1 on July 23, 1972. It was designed to photograph the Earth's surface for geological, biological, and environmental research. It proved very successful, and the Landsat series continues today. The current satellite is Landsat 5, launched on March 1, 1984. It can resolve objects as small as 98 feet (30 meters) across and produce black-and-white or color pictures, as well as false-color images that reveal information from different regions of the normally invisible electromagnetic spectrum. Landsat 6, which will be able to resolve 49-foot (15-meter) objects, is scheduled to be launched in 1992. A satellite equipped to observe the planet's oceans, Seasat, was launched June 27, 1978.

Landsat was the first satellite whose photographs were sold to whomever would pay for them. Seventeen countries operate ground stations that receive Landsat information. On September 27, 1985, control of the Landsat system was transferred to a private company, EOSAT, which runs the satellite for profit. Since 1986, the Landsat satellites have faced stiff competition from a series of French satellites called SPOT. SPOT-2 was launched January 22, 1990, and can resolve 33-foot (10-meter) objects.

© Steven Hunt/The Image Bank

THE SEARCH FOR EXTRATERRESTRIAL LIFE

by David J. Fishman

To consider the Earth as the only populated world in infinite space is as absurd as to assert that in an entire field of millet, only one grain will grow.
—the Greek philosopher Metrodorus in the 4th century B.C.

Since the beginning of recorded history, people have speculated about humanity's place in the universe: are we alone, or do we share the incredible vastness with other intelligent forms of life? Today, for the first time in history, scientists are armed with knowledge and tools that can begin to answer this fundamental question. On Columbus Day, October 12, 1992, the National Aeronautics and Space Administration (NASA), the U.S. space agency, will begin sifting the sky for unusual radio signals as part of its 10-year Search for Extraterrestrial Intelligence (SETI) program.

Scientists intend to spend a great deal of time and money searching the heavens for intelligent extraterrestrial (not of this Earth) life. They claim that the conditions that led to the evolution of life on Earth are surely not unique in the universe. Our Sun is but one of 300 billion or 400 billion stars that form the Milky Way galaxy. The Milky Way is only one of 100 billion galaxies in the currently observable universe. And astronomers speculate that there may be many more galaxies beyond the observable range of our telescopes and sensing instruments. From the study of other nearby galaxies, astronomers now know that the conditions present in our galaxy and around our Sun are not significantly different from the many others in the universe.

EARLY FRUSTRATION

Twenty years ago, astronomers were not so optimistic. Scientists agreed it would be extremely unlikely for life to have evolved on or very close to stars themselves, as the high stellar temperatures would prevent the formation of the chemicals needed for life to evolve. Life as we define it, they reasoned, would have to evolve on planets far enough away from a star to allow the brewing of life-giving chemicals.

To the disappointment of many, searches of the visible universe during the 1970s and early 1980s yielded no absolute evidence of planets orbiting other stars. Astronomers concluded that planetary solar systems such as our own were extremely rare. And many astronomers concluded that life evolved on our planet and perhaps nowhere else in the universe.

One problem these astronomers faced in their search for hospitable planets, however, was that planets are very difficult to spot directly, even using high-powered telescopes. After all, planets don't emit light or other electromagnetic radiation in the same way that stars do.

Scientists now know that evidence that planets exist is generally indirect. One piece of evidence is derived from the movement of stars. Astronomers can calculate the effect that the gravity of orbiting planets has on a star, and then look to see if this characteristic pattern of movement appears among stars they observe over long periods.

Using recent advances in the tools of astronomy and physics, scientists are on the verge of proving that other planetary systems do exist. Many astronomers now believe that planets are not rare exceptions, but rather, a part of the star-formation process. Today NASA astronomers estimate that the number of planets in our own Milky Way galaxy alone might number in the hundreds of billions. Surely, they say, the conditions that helped bring about life on Earth must also be duplicated on one or more of these myriad planets.

Before we examine exactly how NASA intends to go about its search for intelligent extraterrestrial life, let's examine how life arose on this planet, and why many biologists feel that the evolution of life is an almost inevitable result of the chemistry of the universe.

SETTING THE STAGE FOR LIFE

Many cosmologists (scientists who study the origin and eventual fate of the universe) believe that about 15 billion years ago, a spectacularly energetic explosion created our universe. The explosion, often called the Big Bang, ejected seething matter at tremendous speeds in all directions. The universe has been expanding and cooling ever since.

According to this theory, about 5 billion to 10 billion years ago, large clouds of gas began to consolidate under the influence of gravity. As the force of gravity pulled more and more of the mass of gas together, large, ball-like clumps of gas began to form inside these large clouds.

Scanning the universe for signs of intelligent life might turn up unexpected results. Life-forms that evolved on planets with conditions dissimilar to those on Earth would likely appear bizarre to an Earthling's eyes.

© Kevin A. Ward and Alan Clark

When these clumps grew sufficiently large, nuclear fusion ignited them to create the bright, shining stars we see today.

The Big Bang would have produced only hydrogen and helium, the two lightest elements. None of the heavier elements, including those that make up much of our planet and our own bodies—elements such as carbon, oxygen, nitrogen, sulfur, and phosphorus—would have been present in the young universe. But these elements would slowly form by the fusion of lighter particles, a process that takes place inside massive stars.

Before stars run out of the nuclear fuel powering their bright burns, they often explode, creating huge supernovas. The clouds of an expanding supernova are filled with the heavy elements that have been cooked in the interior of the star. This interstellar dust is the potion from which life on Earth—and perhaps on other planets—arose.

Some of this ejected material is constantly undergoing chemical transformations in the cold space between the stars. By examining the light-scattering properties of such interstellar dust, astronomers have identified many of the molecules needed to start the evolution of life. Scientists call these chemical compounds "biogenic," for their potential ability to produce life.

Astronomers have found evidence of grains of silicate, carbon, ammonia, methane, and even water among the more than 70 types of molecules drifting in space. Some astronomers believe that comets and meteors transport these seeds of life from gas clouds to young planets.

Eventually the molecules making up interstellar gas clouds, including the biogenic compounds, collapse once again to form a new dense cloud of dust and gas, eventually forming a new generation of stars. In the case of our own solar system, one of the random gas clouds in the Milky Way collapsed to form our Sun about 4.5 billion years ago. Our Sun is an unremarkable star, similar in many ways to millions of others in the galaxy. At the time of our Sun's birth, a cooler disk of nearby interstellar material, called a protoplanetary nebula, coalesced and began orbiting the

Scientists have already determined that life does not exist on the other known planets in our solar system. People from planets beyond the solar system would need to travel many light-years to reach the Earth.

Sun. Ultimately the force of gravity sorted and separated this material into the nine planets we know today.

The InfraRed Astronomical Satellite (IRAS) has found evidence of protoplanetary nebulas around several nearby stars, a finding that strongly suggests that planets in our galaxy may indeed be more plentiful than previously realized.

Life's Chemistry Set

Fossil evidence suggests that Earth was without life for only a small fraction of its existence, perhaps less than 1 billion of its 4.5 billion years. For most of those early years, comets and meteorites fell continuously to Earth, contributing large amounts of biogenic compounds to the young planet. The comets and meteorites, together with volcanoes spewing lava and hot gases, helped create an atmosphere heavy in water vapor, but lacking free oxygen. As the Earth cooled, the water vapor eventually condensed and formed the oceans.

For millions of years, ultraviolet radiation from the Sun, cosmic rays from deep space, and frequent lightning brought large amounts of energy to the surface of the young Earth. Eventually this energy could have resulted in the synthesis of simple organic molecules. After many more millions of years, organic molecules formed more-complex systems. Somehow, perhaps with the help of clay surfaces, some organic molecules were able to use surrounding materials to replicate. And that replication was the hallmark of a living system.

Swiftly, the new forms of life began to have an enormous impact on the character of the immature Earth. A billion years after life began, blue-green algae evolved. They survived by tapping the energy of the Sun via photosynthesis. These algae pumped large amounts of oxygen into the atmosphere, setting the stage for more-advanced forms of life to evolve. About 1.7 billion

© Kevin A. Ward and Alan Clark

years ago, the first advanced cells appeared. These cells contained a nucleus and primitive structures inside their cell walls. About 1 billion years ago, multicellular organisms appeared, and life, as we know it, began in earnest.

Using the increased amount of oxygen in the atmosphere and dissolved in the

THE SEARCH FOR EXTRATERRESTRIAL LIFE 319

ocean, multicellular organisms proliferated rapidly. Fish appeared about 425 million years ago, and their relatives crawled out onto land about 325 million years ago. Modern humans arrived relatively recently, just 30,000 to 40,000 years ago.

Because humans had evolved appendages that can easily manipulate what we find in our world, our species developed technology—the first example of which was the creation and use of simple tools. Over the course of the past 10,000 years, humans have learned to build shelters, cultivate land for food, and to speak and write. Over the past 100 years, our species has developed the ability to communicate with each other at the speed of light, using electromagnetic radiation such as radio waves and microwaves.

THE MEANS TO COMMUNICATE

Communicating by electromagnetic radiation, or photons, is the fastest means allowed by the laws of physics. Photons can carry information, are easily generated and detected, are not seriously affected by the magnetic fields pervading outer space, and, at the correct frequencies, have a small probability of being scattered or absorbed. If intelligent life also evolved on other planets, scientists presume that these other beings, too, would have stumbled upon this ideal method for communicating across the heavens.

NASA's Search for Extraterrestrial Intelligence, or SETI, will focus exclusively on stars and planets outside of our own solar system. From observations made here on Earth, and with the many probes sent to investigate our eight neighboring planets, scientists have in effect ruled out the possibility of other forms of intelligent life orbiting our Sun. Still, many scientists are convinced that some of our neighboring planets contain, and are still producing, organic chemicals. They believe it is possible that primitive life-forms do now exist or once existed on other planets in our solar system.

Mercury, the planet nearest the Sun, is somewhat larger than our Moon. Like our Moon, Mercury has little or no atmosphere. Extremely high temperatures, especially on the daylight side of the planet, rule out the possibility that water or organic molecules could survive Mercury's surface. Venus, the second planet from the Sun, is very similar to the Earth in size and density. But it, too, is much hotter than the Earth. From a biological point of view, Venus is a grim place. The atmosphere is very unlike the Earth's; it consists mostly of carbon dioxide gas, with small amounts of carbon monoxide. Dense clouds of sulfuric acid droplets shroud Venus' surface. Temperatures range from 860° to 1,004° F (460° to 540° C), and atmospheric pressures can be 100 times greater than those at the surface of the Earth. The extreme heat and pressure make the existence of water and organic compounds extremely unlikely.

Mars, the fourth planet from the Sun, has a diameter half that of the Earth. Its atmospheric density is very low, similar to the Earth's atmosphere at an altitude of 31 miles (50 kilometers). Mars' atmosphere consists mainly of carbon dioxide with a trace of water. Because the air is so thin, ultraviolet radiation from space, which is harmful to most living things, can easily reach the Martian surface. At the Martian equator, the temperature may reach 75° F (24° C) during the day; at night it may drop to −99° F (−73° C).

But Mars does have many features that lead some astronomers to believe it could support life. Water may once have covered large parts of its surface. Researchers know that its polar ice caps are still composed of frozen water. The Viking landers examined soil samples that yielded water upon heating, as well as an unusual chemistry that seemed to mimic several lifelike reactions, including the breakdown of nutrient chemicals and the synthesis of organic compounds from gases. The debate about whether primitive life exists or once existed on Mars may have to wait for that eventual day when humans actually explore the Red Planet in person.

In the past, researchers did not seriously consider Jupiter, Saturn, Uranus, and Neptune as suitable homes for life. Be-

Judging from Viking 2's vantage point (above), the surface of Mars would seem inhospitable to even the most primitive forms of life. Alien life-forms, if they exist, would likely come from regions of the universe too distant for space probes to visit. Astronomers would, however, be able to pick up electronic messages or signals from faraway civilizations using today's powerful radio telescopes (right).

Top: NASA; above: National Radio Astronomy Observatory

cause of their great distance from the Sun, all are extremely cold. In spite of their immense sizes and masses, the outer planets are less dense than the Earth. Their solid cores are quite small when compared with their large volumes.

Nonetheless, astronomers now know that the ammonia-methane-hydrogen atmospheres typical of these giants can be breeding places for biochemicals. And surprisingly, the outer planets are not always as cold as astronomers once thought: new studies show that some of their cloud layers may be comparatively mild in temperature.

Jupiter is of particular interest to scientists. It may have an internal source of heat, making it warmer than it would be if heated by the Sun alone. And its many-colored cloud coat may contain a large variety of organic compounds.

Laboratory experiments show that organic and biological compounds can evolve in Jupiter-like conditions. The Pioneer and Voyager probes that examined Jupiter's clouds detected violent lightning flashes that might provide the spark of life for a primitive chemical soup. They have detected other hopeful signs that are precur-

THE SEARCH FOR EXTRATERRESTRIAL LIFE   321

EXTRATERRESTRIALS IN THE MOVIES

Science fiction has helped mold our perceptions of how the inhabitants of other planets would appear. Some of the first depictions of extraterrestrial beings appeared in science-fiction magazines of the early 20th century. The film industry's interest in science-fiction topics —including people from outer space—blossomed in the 1950s, perhaps as a result of public interest in rockets, space programs, and a growing number of UFO sightings.

Steven Spielberg's film E.T.: The Extraterrestrial *was the box-office hit of 1982 and the most popular science-fiction movie to date. E.T. (above) caught the hearts of moviegoers as he poignantly sought to return to his home planet.*

Not nearly so lovable were the extraterrestrials portrayed in the 1957 production Invasion of the Saucermen *(left). Though grotesque by Earth standards, any alien who devised a way to travel to Earth would likely have an oversized brain.*

Photos: Photofest

sors of primitive life on the moons of Jupiter and its close neighbor, Saturn.

Although very little is known about Pluto, this outermost planet of our solar system may be of little biological interest. We do know that it is a small, rocky body without an atmosphere. It appears to be too far from the Sun's warming energy for the chemical evolution of life.

WHAT ARE THE ODDS?

Knowing that we have to search outside of our own solar system to find intelligent life, how can we estimate our chances of success? American radio astronomer Frank Drake has developed a way to estimate the number of the universe's intelligent civilizations with which we might be able to communicate. The equation looks like this:

$$N = R_* f_g f_p n f_l f_i f_c L$$

The number of communicating civilizations is represented by the letter N. For all practical purposes, this number is limited to civilizations living within our own Milky Way. Even at the speed of light, a message would take more than 200,000 years to reach the next major galaxy outside of our Milky Way. Therefore, NASA will not bother trying to listen to other galaxies. The Drake

Television has also gotten aboard the science-fiction bandwagon. In the television show "Alien Nation," the extraterrestrials (as the one depicted at right above) have a distinctly human demeanor, though some obvious differences exist.

Most often, filmmakers allow their extraterrestrials an entirely human appearance; only their intelligence and behavior distinguish them from everyday Earthlings. Such was the case with Nyah (left), the title character in the 1955 sci-fi "classic" Devil-Girl from Mars.

Equation is solved by multiplying a number of factors, beginning with R_*, which stands for the rate of star births within the galaxy.

The fraction of stars suitable for life support is represented by f_g. The next factor is f_p, which represents the portion of stars that might have orbiting planets. The next factor, n, is the estimated percentage of Earth-like planets. All life on Earth is based upon the element carbon, an element that forms readily into complex molecular compounds. Highly complex molecules, such as DNA, incorporate long chains of carbon molecules. Biologists have yet to envision a chemistry for life that doesn't depend on carbon. Therefore, the n factor assumes that life elsewhere will similarly need carbon-based, Earth-like conditions to evolve.

The next f in the equation, f_l, corresponds to the number of habitable planets that could potentially produce life. The f_i represents the chances of such intelligent forms of life actually evolving. The final f_c is for the number of intelligent civilizations that might develop the ability and desire to communicate with others in the universe. The last factor, L, is the life span of a civilization. This factor is crucial in that scientists fear that any civilization smart enough to make radio contact might also have the ability for mass destruction and self-anni-

THE SEARCH FOR EXTRATERRESTRIAL LIFE

hilation. Intelligent societies may evolve all the time, some scientists believe, but because of the limited life spans of civilizations, they may transmit for only a few thousand years. If true, the chances of us listening for them at exactly the right time over the many-billion-year history of the universe might be quite small.

In 1961 an international conference on extraterrestrial life worked out the value of N to fall somewhere between 100 and 1 million civilizations. Even if our kind of intelligence is associated with just one star in every 100,000 of the same type as our Sun, there could be at least 400,000 technological civilizations in the Milky Way alone.

By and large, astronomers approach the concept of extraterrestrial life as an all-or-nothing proposition. Some calculate that in all probability the Earth is the only planet with intelligent life. Others still insist we are surrounded by countless inhabited worlds.

SEARCHING THE SKY

Since 1960 astronomers have attempted more than 50 radio searches for what might be interpreted as a "hello-there" signal. Though none of these searches have yielded evidence of extraterrestrial life, together they have covered only a tiny fraction of the space and possible radio frequencies that might contain such a message.

NASA's SETI program will scan 15 million frequency channels each second—more than 10,000 times the number previously tried. The NASA search, which will be led by the agency's California-based Ames Research Center and Jet Propulsion Laboratory (JPL), will also be at least 300 times more sensitive than any previous attempt. The complexity of the SETI search is possible only because of recent advances in computer-chip technology, which will allow signal processing at great speed.

In setting up their detection system, NASA researchers tried to think like space aliens. How could we recognize a signal as being of intelligent origin without knowing anything about the civilization sending the message? The first step was to choose a part of the electromagnetic spectrum (similar to a giant galactic radio dial) that would be the least cluttered by background noise such as cosmic static, dust-cloud emissions, and other natural interference. This rationale eliminated the frequencies on the lower end of the dial.

Unfortunately, frequencies on the high end of the dial are absorbed readily by the Earth's atmosphere, making them difficult to detect. This process of elimination led the NASA researchers to settle upon a quiet part of the spectrum known as the microwave region.

There are two nice side benefits of using microwave frequencies. First, very little energy is required to make signals at this frequency heard above the natural din of the universe. Second, and more important, hydrogen, the most abundant element in the universe, emits energy at a characteristic frequency of 1,420,405,752 times a second. Translated into radio frequencies (megahertz), this would set the radio dial right smack in the microwave region. NASA calls this the "magic frequency," and expects it would also be the choice of any intelligent, logical alien. The SETI program is therefore often called the Microwave Observing Project, or MOP.

As it begins its formal search, NASA will be using specially developed signal-detection equipment attached to existing radio-astronomy observatories. These will include the 1,000-foot (305-meter) Arecibo dish in Puerto Rico and the 230-foot (70-meter) and 111-foot (34-meter) dishes that form the agency's Deep Space Network (used to track the Voyager and Pioneer probes).

MAKING CONTACT

In order to recognize signals produced by technology as opposed to natural sources, NASA will again rely on logic. Natural signals tend to spread over a wide range of frequencies, and they are not "in phase," or well-tuned, so to speak. Artificial signals, produced by a transmitter and antenna, are often confined to a narrow

The universe could contain thousands of planets with all the conditions necessary to give rise to life. Astronomers hope to detect such planets in the near future.

range of frequencies, are highly polarized, and the peaks of the waves are "in phase," or "tuned in."

NASA will conduct the search in two steps. A *Targeted Search* will closely examine 800 stars just like our Sun, searching more than 2 billion channels in the microwave region. A second *Sky Survey* will search the entire sky, covering more than half a billion channels.

Scientists believe that if they ever detect a signal, it will most likely arrive in the universal language of mathematics. In order to confirm that a signal is indeed from another civilization, at least two observatories will have to be able to detect it. Once a signal is confirmed as being of extraterrestrial origin, an announcement of the discovery would be made as quickly and as widely as possible. A guidebook, called *The Declaration of Principles Concerning Activities Following the Detection of Extraterrestrial Intelligence*, was approved by five international space organizations, academic institutions, and several governments. It will dictate the researchers' actions after any discovery of extraterrestrial life-forms.

At present, NASA has no plans to reply if it does indeed receive a message. But officials expect that the nations of the Earth would join together to formulate any eventual response. NASA's 10-year search is certainly only the beginning of what will likely be a very long period of scrutinizing our galaxy.

Atmospheric phenomenon, optical illusion, or UFO? Sightings such as the one above often defy explanation.

UNIDENTIFIED FLYING OBJECTS

On June 24, 1947, a Boise, Idaho, businessman, Kenneth Arnold, was flying a private plane near Mount Rainier, Washington. Suddenly he was startled to see a group of strange-looking craft going through a series of amazing maneuvers. "They flew very close to the mountain tops," he said later, "flying . . . as if they were linked together. . . . I watched them for about three minutes—a chain of saucer-like things at least 5 miles [8 kilometers] long, swerving in and out of the high mountain peaks. They were flat like a pie pan, and so shiny they reflected the Sun like a mirror. I never saw anything so fast."

When Arnold reached his destination—Yakima, Washington—and reported what he had seen, he created a sensation. The "flying saucers" sighted by Arnold became an exciting topic of conversation. Within a few days, disklike flying craft were reported by observers in other parts of the country, and the flying-saucer scare was launched.

The scurrying disks that were now so frequently reported were merely the latest in a series of mysterious objects in the heavens that have startled humankind since time immemorial. Most of the earlier appearances had been definitively traced later to meteors, comets, atmospheric phenomena, and the like; a few have never been satisfactorily explained. Some aroused widespread interest for weeks or months at a time and then were forgotten.

The flying-saucer epidemic started by Arnold's report reached rather respectable proportions in 1947; but toward the end of the year, public interest seemed to be on the wane. It was powerfully revived by a tragedy that took place on January 7, 1948. Early in the afternoon of that day, observers at Godman Air Force Base in Kentucky saw a mysterious object flying overhead; it looked like "an ice-cream cone topped with red." Four pilots in National Guard F-51 planes were asked to investigate the strange aircraft. Captain Thomas F. Mantell, the flight leader, radioed to the control tower that he was "closing in to take a good look." After a time, he reported that the thing looked metallic and of tremendous size. "It's going up now and forward as fast as I am.... That's 360 miles [580 kilometers] per hour. I'm going up to 20,000 feet [6,100 meters] and if I am no closer, I'll abandon chase."

Following this last report, received at 3:15 P.M., there was no further radio contact with Mantell. Later that day, his body was found in the wreckage of his plane near Fort Knox. The official U.S. Air Force explanation was that Mantell had blacked out at 20,000 feet (6,100 meters) from lack of oxygen and had died of suffocation before the crash. The object that Mantell had pursued was at first identified as the planet Venus. But further probing showed that the planet had not appeared on that day in the quarter of the sky where the mysterious object had been sighted.

MANY SIGHTINGS REPORTED

There was now a new wave of flying-saucer sightings. Some of the mysterious craft were seen by planes, others by observers on the ground; a number of them were picked up on radar screens. Those making the reports included a variety of seasoned observers: trackers of guided missiles, radar operators, commercial air pilots, U.S. Air Force pilots, airport-traffic controllers, and weather observers. The saucers were sighted in many different parts of the United States, but particularly in the desert areas of the Southwest. There were also reports of sightings from Hawaii, Canada, Mexico, South America, Europe, the Far East, Australia, Africa, Greenland, and the Antarctic.

The U.S. Air Force, determined to get to the bottom of these mysterious appearances, launched a series of investigations. The investigators gave the official name of "unidentified flying objects" (UFOs) to the mysterious craft.

The UFOs reported by observers were of many different kinds. Flying saucers, or disks, predominated. There were large disks, up to 100 feet (30 meters) or so in diameter, medium-sized disks, and tiny disks with a diameter of only a few inches. These craft performed amazing maneuvers, sometimes hovering motionless in the air, then shooting skyward, making abrupt turns, and reversing their course with unbelievable suddenness. They attained the most fantastic speeds—up to thousands of miles per hour in some cases.

There were also rocket-shaped ships, ranging from 100 to 1,000 feet (30 to 300 meters) in length and also capable of tremendous speeds. In some cases, these rockets seemed to serve as mother ships for disk-shaped craft. In addition to disks and rockets, there were bright green fireballs moving silently through the heavens as swiftly as meteors.

Certain observers claimed to have seen members of the flying-saucers' crews. In a 1950 book called *Behind the Flying Saucers*, Frank Scully told of certain men from Venus whose dead bodies had been found after their saucer craft had crashed to Earth in New Mexico and Colorado. According to Mr. Scully's informant, a "Mr. Newton," the Venus men were "tiny creatures, from 40 to 45 inches [102 to 114 centimeters] long. They wore 1890 dress, of cloth that was not wool or cotton but that couldn't be torn." The Air Force, claimed Mr. Newton, had spirited the bodies away.

Not all UFO crew members were midgets, apparently. On September 12, 1952, a woman, three children, and a young National Guardsman espied a flying saucer near Sutton, West Virginia. Somewhat later they came upon a repulsive giant, apparently a member of the crew. He was 9 feet (2.7 meters) tall and had a red face and protruding eyes about 12 inches (30 centimeters) apart. When this monster started toward the startled observers with a hissing sound, they fled.

In addition to reported sightings, there were saucer photographs galore. Some of these were obvious hoaxes; in other cases, freak images had been produced in negatives because of faulty equipment or technique. A few could not be explained so easily, however.

Reports of UFO sightings increased sharply in the late 1940s and 1950s. Many people claimed the objects were disk-shaped, hence the term "flying saucer."

Flying-saucer appearances tapered off after 1953. Perhaps it would be more exact to say that fewer reports of unidentified flying objects appeared in newspapers and magazines, perhaps because editors felt that such reports were no longer particularly newsworthy. In the period 1965–67, there was a marked increase in reported sightings and then a tapering off again.

LOOKING FOR EXPLANATIONS

How are we to account for these mysterious, unidentified objects seen in the heavens? Are they really flying craft, incomparably more advanced than any previously known to humans? After all, it has been argued, humans have traveled to the Moon. Is it not possible that inhabitants of other star systems, further advanced than Earthlings, have perfected ships that can carry them to our solar system?

Soon after Kenneth Arnold sighted those "flying saucers" in 1947, the U.S. Air Force began keeping track of all reported UFO sightings in what came to be known as Project Blue Book. In 1966 it engaged a group of 36 scientists to conduct an independent investigation of these sightings. The study culminated in the publication of a voluminous report, called the Condon Report, which concluded that further UFO research would be of no value. In 1969 the Air Force ended Project Blue Book, although approximately 1,000 reports in their files remained unexplained.

Reports of UFO sightings dropped sharply, but the controversy over their cause didn't end. Some scientists continued to question the thoroughness of the Air Force's investigation. Dr. James McDonald, for example, noted that the authors of the Condon Report "conceded that about one-third of the 90 cases they investigated could not be explained. It is far from clear how this justified the conclusion that further study is not needed."

Then, in 1973, there was a sudden upsurge in UFO sightings. One of the more interesting accounts was related by two men from Mississippi, who said that they had been taken aboard a blue spaceship by creatures who had silvery, wrinkled skin and crab-claw hands. Astronomer J. Allen Hynek, an Air Force consultant on Project Blue Book, examined the men under hypnosis and felt that they were "telling the truth beyond a reasonable doubt."

Other sightings soon proved to have conventional explanations. For example, a number of UFOs were actually silvery weather balloons, which reflected sunlight at their 16-mile (26-kilometer) altitude. Others turned out to be meteorological phenomena such as low-hanging clouds or fog. Nonetheless, a survey of adult Americans showed that 51 percent believed that UFOs are real objects. And 11 percent said that they had seen UFOs.

In 1973 Dr. Hynek formed the Center for UFO Studies. "For a quarter century the UFO phenomenon has been the subject of gross misconceptions, misinformation, ridicule, buffoonery, and unscientific approach," he said. "The fact that reports persist—from many countries—presents a mystery that demands explanation." Shortly before Dr. Hynek died in 1986, he suggested that UFOs that appear briefly and then disappear without a trace may be compared to the duality of light, which acts either as a wave or a particle, depending on the particular situation.

Some Conventional Explanations

Many scientists, such as astronomer Donald Menzel of Harvard, maintain that many of the mysterious appearances are due to meteorological phenomena. In some cases, saucerlike appearances may be due to the reflection of light from ice-crystal formations in the atmosphere. If such crystals are falling or hovering in the air, reflected sunlight or moonlight may produce startling effects. For example, such reflections may cause a pair of concentric halos to appear around the Sun. Sometimes part of the halos may be almost as bright as the Sun itself and will form glowing mock suns, which are also known as sun dogs. At night, halos sometimes develop around the Moon, forming mock moons, or moon dogs. These might well resemble strange aircraft.

In other instances, flying-saucer appearances may have been due to the fact that air can act as a distorting lens, producing the effects known as mirages. Light rays are refracted, or bent, in various ways as they pass through air layers of different density. The effect is particularly noticeable when layers of sharply contrasting density are in close contact with one another.

If a hot layer of air lies close to the Earth, the image of the sky may be projected against the Earth, sometimes giving the illusion of ponds or lakes in the distance. If a cool layer is near the Earth (as in the desert at night), the image of the Earth may be projected against the sky. Distant lights, such as those of automobiles or a city, will seem to float in the air. If the air is turbulent, these lights will apparently dart hither and thither. Conditions that cause optical mirages can also produce radar mirages.

In some cases, people may have sighted flying craft, but not the exotic objects that they imagined. They may have seen ordinary airplanes flying at such great heights that only the reflection of the Sun from the fuselage was visible.

Other people have mistaken kites or weather balloons for UFOs. In 1951 physicist Urner Liddel asserted that most UFOs previously reported were really skyhook balloons—large plastic unmanned craft which are used to carry meteorological instruments aloft. These balloons reach great heights and often fly great distances; some have crossed the Atlantic. When viewed against the background of the sky, they often look quite disklike. Some authorities believe that it was while chasing a skyhook balloon that Captain Mantell crashed. The skyhook balloon was a secret device at the time, and information about it was not available to the public.

When this information was declassified a few years later, it solved a startling mystery. On March 17, 1950, the inhabitants of Farmington, New Mexico, were almost scared out of their wits as thousands of "flying saucers" soared over the town for an hour or so. Only later was it revealed that a skyhook balloon had burst over Farmington at an altitude of 11 miles (18 kilometers), scattering thousands of pieces of plastic in the air. These were the "flying saucers" that had invaded the area.

Products of the Mind?

Behavioral scientists also study UFO reports. Sociologist Robert Hall points out that "the sky, especially the night sky, is full of ambiguous stimuli, and people generally have a powerful need to reduce ambiguity . . . by explanations in terms of something familiar." Thus, some people, influenced by the general "system of belief" that has developed around the UFO phenomenon, transform naturally explainable stimuli into a ship from outer space. Other scientists link UFO sightings to the stresses of modern life. Under such stresses, people have illusions or delusions.

Dr. Carl Sagan of Cornell University has proposed that "certain psychological needs [are] met by belief in superior beings from other worlds." And today, he says, extraterrestrial visitors are a fashionable idea. Dr. Sagan, however, cautions against dismissing the extraterrestrial hypothesis. There is not enough evidence, he believes, to exclude the possibility that some UFOs are spaceships from advanced civilizations that live elsewhere in the universe.

MATHEMATICS

332–337	Introduction to Mathematics
338–345	Numerals
346–351	Arithmetic
352–365	Algebra
366–379	Plane Geometry
380–385	Solid Geometry
386–390	Trigonometry
391–400	Analytic Geometry
401–403	Non-Euclidean Geometry
404–413	Statistics
414–423	Probability
424–430	Game Theory
431–437	Calculus
438–445	Set Theory
446–452	Binary Numerals
453–460	Data Processing

Mathematics provides the symbols and operations through which many of the great theories of science have been developed.

INTRODUCTION TO MATHEMATICS

The word "mathematics" comes from the Greek *mathemata,* meaning "things that are learned." It may seem odd to apply this phrase to a single field of knowledge, but we should point out that for the ancient Greeks, mathematics included not only the study of numbers and space but also astronomy and music. Nowadays, of course, we do not think of astronomy and music as mathematical subjects; yet the scope of mathematics today is broader than ever.

Modern mathematics is a vast field of knowledge with many subdivisions. There is, first of all, the mathematics of numbers, or quantity. The branch of *arithmetic* deals with particular numbers, such as 3, or 10½, or 12.5. When we add, subtract, multiply, or divide such numbers or get their square roots or squares, we are engaging in arithmetical operations. Sometimes we wish to consider, not particular numbers, but relationships that will apply to whole groups of numbers. We study such relationships in *algebra,* another branch of the science of quantity. In algebra, a symbol, such as the letter a or b, stands for an entire class of numbers. For example, in the formula

$$(a + 2)^2 = a^2 + 4a + 4$$

the letter a represents any number. The relationship expressed in the formula remains the same whether a stands for 1, or 5, or 10, or any number.

Mathematics also studies shapes occurring in space, which may be thought of as a world of points, surfaces, and solids. We study the properties of different shapes and the relations between them, and we learn how to measure them. This space science is called *geometry. Plane geometry* is concerned with points, lines, and figures occurring in a single plane—a surface with only two dimensions (Figure 1). The study of the three-dimensional world is called *solid geometry* (Figure 2). *Trigonometry* ("triangle measurement") is an offshoot of

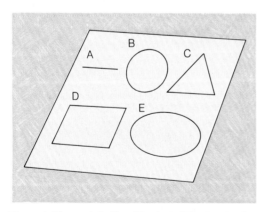

Figure 1. The straight line (A), circle (B), triangle (C), rectangle (D), and ellipse (E) occur in a single plane. Their study is a part of plane geometry.

Figure 2. The cube (A), prism (B), cylinder (C), and cone (D) are not bounded by a single plane. These figures are studied in solid geometry.

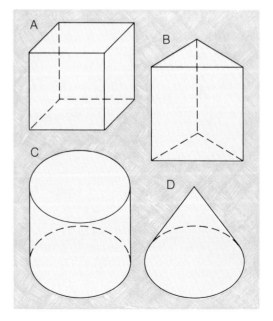

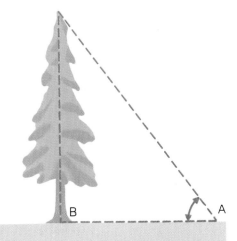

Figure 3. If we know the distance AB and the angle at A, we can calculate how high the tree is by using trigonometry, or triangle measurement, an offshoot of geometry.

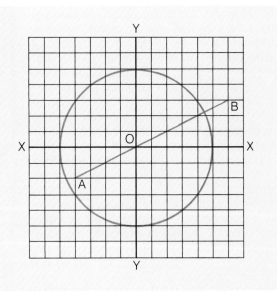

Figure 4. The x axis is at right angles to the y axis. The formula x = 2y indicates the position of AB with respect to the x and y axes. The formula $x^2 + y^2 = 25$ indicates the position of the circle. These relationships are derived from analytic geometry.

another and meet at the point O. To indicate the position of line AB with respect to the x and y axes, we use the formula $x = 2y$. To indicate the position of the circle in the diagram, we use the formula $x^2 + y^2 = 25$.

The branch of mathematics called *calculus* is based on the study of functions. If the value of a given quantity depends on the value we assign to a second quantity, we say that the first quantity is a function of the second. For example, we know that the circumference of a circle can always be found by the formula $C = \pi \times d$ (or $C = \pi d$), where C is the circumference, d is the diameter, and π (the Greek letter pi) represents a constant value approximately equal to 3.14159. In this case, C is a function of d since its value will depend on the value we assign to d. In *integral calculus,* we are interested in the limit of the different values of a variable function. In *differential calculus,* we determine the rate of change of a variable function.

Statistics, another branch of mathematics, involves the accumulation and tabulation of data expressed in quantities, and the setting up of general laws based on such data. The theory of *probability* is an important part of statistics. It enables one to calculate what will happen based on the chances of an event occurring. Commonly used to project outcomes of gambling games such as dice, poker, and lotteries, the theory of probability has countless applications in the social sciences, such as predicting the relative number of boys and girls that will be born in a particular place at a particular time.

These are only a few of the many subdivisions of mathematics. Besides being a most extensive field of knowledge in its own right, it represents a logical approach that can be applied to many different fields. It carefully defines the ideas that are to be discussed and clearly states the assumptions that can be made. Then, on the basis of both the definitions and the assumptions, it forges a chain of proofs, each link in the chain being as strong as any other. Mathematicians have displayed wonderful powers of imagination in determining what can

geometry. It is based on the fact that when certain parts of triangles are known, one can determine the remaining parts and solve many different problems (Figure 3).

Analytic geometry combines algebra and geometry—generalized numbers and space relationships. It locates geometrical figures in space. It explains circles, ellipses, and other figures in terms of algebraic formulas. In Figure 4, for example, the x and y axes are at right angles to one be proved and in constructing ingenious methods of proof.

It might seem rather far-fetched to think of mathematics as a search for beauty. Yet, to many workers in the field, mathematical patterns that are fitted together to form a harmonious whole can produce as pleasing effects as the color combinations of a painter or the word patterns of a poet. Bertrand Russell, an important 20th-century mathematician, wrote in his *Principles of Mathematics*, "Mathematics, rightly viewed, possesses a beauty cold and austere, like that of a sculpture, without any appeal to any part of our weaker nature, without the gorgeous trappings of painting or music, yet sublimely pure, and capable of a stern perfection such as only the greatest art can show."

Mathematics is also an endless source of entertainment. For many generations, mathematicians and others have prepared what are commonly known as mathematical recreations, ranging from simple problems and constructions to brain twisters that can be solved only by experts—and sometimes not even by experts. These recreations are a delightful challenge to one's wits. Sometimes, they bring us into a world of fantasy in which one "proves" that 2 = 1, or constructs a "magic ring" whose outside is its inside (Figure 5).

Mathematics and the Outer World

In the development of the different branches of mathematics, pioneers have often owed much to the observation of the world about them. It has been suggested, for example, that the concepts of "straight line," "circle," "sphere," "cylinder," and "angle," in geometry were derived from nature: "straight line," from a tall reed; "circle," from the disk of the sun or moon; "sphere," from a round solid object like a berry; "cylinder," from a fallen tree trunk; "angle," from the various positions of a bent arm or leg.

Pioneer mathematicians examined these shapes and the relations between them. At first, they applied the results of such studies to the solution of practical problems, such as the construction of canals or the dividing of land into lots for purposes of taxation. Later, they began to study the relationships between various geometrical forms in order to satisfy their intellectual curiosity and not to solve particular problems. In the course of time they built up a series of purely abstract concepts.

As certain mathematicians develop such concepts in geometry and calculus and other branches of mathematics, they are apt to turn their backs entirely on the

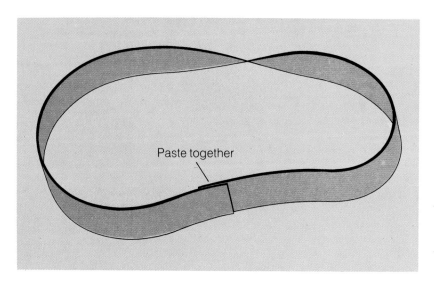

Figure 5. This "magic ring" is made by pasting together the two ends of a narrow strip of paper in the form of a ring, after giving one end a twist of 180°. If you then color one side of the ring, you will find that you have colored the whole ring, inside and out. The ring is called a Moebius Strip.

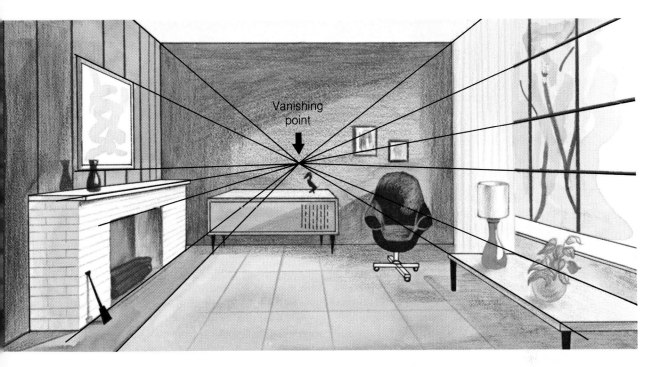

Figure 6. The realistic, three-dimensional effect in this drawing is obtained by having all the horizontal lines that recede from the observer meet at a vanishing point.

world of reality. They have no hesitation in working with equations involving four dimensions, ten dimensions, or any number of dimensions. The German mathematician Bernhard Riemann built up an entire geometry based on the *postulate,* or assumption, that no two lines are ever parallel. Another geometry (recognized independently by three great 19th-century mathematicians: Nikolai Lobachevsky, Karl F. Gauss, and János Bolyai) postulates that through a point not on a given line, there are at least two lines parallel to that line. The German Georg Cantor proposed a "theory of sets" that was highly ingenious and—at least, so it appeared at the time— utterly useless as far as any practical application was concerned.

It has been pointed out that the concern with pure abstraction is not without its dangers. The distinguished mathematician Morris Kline wrote in his *Mathematics and the Physical World,* "Mathematicians may like to rise into the clouds of abstract thought, but they should, and indeed they must, return to earth for . . . food or else die of mental starvation."

Yet even the most abstract speculations of mathematicians may find important applications, sometimes after many years have passed. Thus the Riemannian geometry was to prove invaluable to Albert Einstein when he developed his theory of relativity. Cantor's theory of sets has been applied to various fields, including higher algebra and statistics. As to multiple dimensions, they have been put to work in, of all things, the inspection of industrial products.

Applications

We do not have to give such extreme examples as Riemannian geometry and the theory of sets to show how mathematics has served mankind. Actually, it is deeply rooted in almost every kind of human activity, from the world of everyday affairs to the advanced researches of authorities in many different fields of science.

All of us are mathematicians to some extent. We use arithmetic every day in our lives: when we consult a watch or clock to find out what time it is; when we calculate the cost of purchases and the change that is due us; when we keep score in tennis, baseball, or football.

The accounting operations of business and industry are based on mathematics. Insurance is largely a matter of the compounding of interest and the application of the theory of probability. Certain manufacturers make use of calculus in order to be able to utilize raw materials most effectively. The pilot of a ship or plane uses geometry to plot a course. The surveyor's work is based largely on trigonometry, and the civil engineer uses arithmetic, algebra, calculus, and other branches of mathematics.

Mathematics also serves the branch of learning called the humanities, which includes painting and music. It is the basis of perspective—the system by which the artist represents on a flat surface objects and persons as they actually appear in a three-dimensional world. We may think of perspective as made up of a series of mathematical theorems. One theorem, for example, states that parallel horizontal lines that recede in the same plane from the observer converge at a point called the *vanishing point*. Figure 6 shows how this theorem is applied to the drawing of a room.

In music, too, mathematics plays an important part. The system of scales and the theories of harmony and counterpoint are basically mathematical; so is the analysis of the tonal qualities of different instruments. Mathematics has been essential in the design of pianos, organs, violins, and flutes, and also of such reproducing devices as phonographs and radio receivers.

Mathematics is so important in science and serves in so many of its branches that it has been called the "Queen and Servant of the Sciences" by the noted Scottish-American mathematician, Eric Temple Bell. Here are some examples:

Measurements and other mathematical techniques are vital in the work of the physicist. The physicist uses the mathematical device called the graph to give a clear picture of the relationship between different values—between temperature, for example, and the pressure of saturated water vapor in the atmosphere (Figure 7). The laws of physics are stated in the form of algebraic formulas. Thus to express the idea that the velocity of a body can be determined by dividing the distance covered by the time required to cover this distance, the physicist uses the formula $v = \frac{s}{t}$, when v is the velocity, s the space or distance covered, and t the time. The physicist employs geometry and trigonometry in the analysis of forces and in establishing the laws of optics, the science of light.

Like the physicist, the chemist continually uses arithmetical and algebraic operations and graphs. The chemist, too, presents laws in the form of algebraic formulas. The reactions are set down in the form of equations, which from certain viewpoints may be considered as mathematical equations. Chemists also use loga-

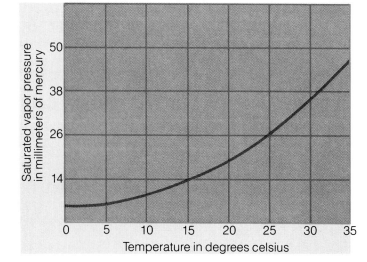

Figure 7. The graph at the right shows how saturated water-vapor pressure in the atmosphere varies with the temperature. The saturation point is reached when the atmosphere can hold no more water vapor at a given temperature.

The scope of mathematics today is broader than ever. Calculations are now performed by huge computers in minute fractions of a second. Sophisticated software is used to formulate and solve mathematical problems.

rithms, a mathematical technique, in calculating the degree of acidity of a substance—the so-called pH value. Plane and solid geometries are used in studying the ways in which atoms or ions (electrically charged atoms) are combined. Thus it can be shown that graphite atoms form a succession of hexagons (six-sided figures) in a series of planes set atop one another, and that the sodium and chloride ions that make up ordinary table salt (sodium chloride) are set at the corners of a series of cubes.

Mathematics has always been closely associated with astronomy. From the earliest days, astronomers measured angles and arcs and made a great many mathematical calculations as they followed the apparent motions of the sun, stars, moon, and planets in the heavens. Today such branches of mathematics as arithmetic, algebra, plane geometry, solid geometry, trigonometry, and calculus are just as useful to the astronomer, as the optical telescope, camera, radio telescope, and other devices that are used in this branch of science.

It would seem difficult to apply the formulas of mathematics to the infinitely varied world of living things. Yet mathematics serves even in biology, the science of life. It plays an extremely important part, for example, in genetics, which is concerned with heredity. To calculate the percentage of individuals with like and unlike traits in succeeding generations, the geneticist makes use of the theory of probability. Mathematics has also been applied to the comparison of related forms of life. Using the method called *dimensional analysis,* researchers have found that the frequency of the beating of a bird's wings can be summed up in a formula. It is rather amazing to see how often this formula applies to species of totally dissimilar birds. Dimensional analysis has also been used to analyze the growth patterns of certain animals, as well as to determine the ratio between the lifetime of a given animal and the time required for the animal to draw a single breath.

Such then, in brief, is the scope of mathematics. In the articles that follow, we shall present some of the more important fields of mathematics, including arithmetic, algebra, plane geometry, solid geometry, trigonometry, analytical geometry, and calculus. We shall tell you what these fields are about and the uses to which they are put by present-day mathematicians.

NUMERALS

The primitive cave man did not have to know much about counting or any other kind of mathematics to keep alive. Home was a cave; food could be gathered from native vegetation or hunted with primitive weapons. However, when people began to collect animals into herds, and particularly when one family entered upon social relations with others, it became necessary to decide how much belonged to each person. It probably sufficed, at the outset, to use such concepts as a little, some, or much. Later, when it became necessary to have a more definite means of determining "how much," people learned to count, and this was the beginning of mathematics.

At first, a person might count the number of animals in a herd by placing a pebble on the ground or tying a knot in a rope for each animal. Each pebble in the growing heap or each knot in the rope would stand for a single animal. Later, a man might use his ten fingers in his calculations. We may surmise that when the ten fingers had been counted, a little stone would be set aside to represent this first ten; the fingers would then be used to count another ten. Another stone would be added to the first one; the fingers would be used to count another ten, and so on. When the stones in the pile would equal the number of fingers, they would represent ten tens. The pile of ten stones would then be taken away, and a larger stone would be put in its place to indicate ten tens or one hundred (Figure 1). Thus three large stones, seven small stones, and eight sticks (standing for eight fingers), would represent three hundreds, seven tens, and eight units—in other words, 378 (Figure 2).

The ten fingers in this case would mark the halting place in a person's calculations; we would call it the *base*. Not all primitive people would use 10, or the number of fingers on both hands, as the base. Some would use only the two hands in counting (and not the fingers of the hands), and their halting place would be two. For others, the fingers of one hand would serve; their halting place would be five. Still others would combine the fingers of both hands and the toes of both feet; twenty would be their halting place.

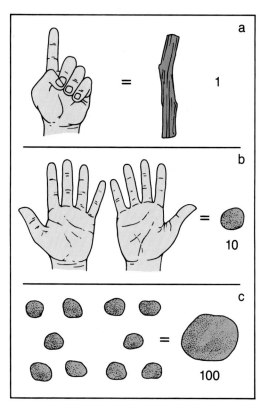

Figure 1. In the calculations of early times, the number 1 might have been represented by a finger or a stick (a). The number 10 could have been represented by the ten fingers of two hands or by a small stone (b). Ten small stones could then be equivalent to one large stone, which would represent 100 (c). In this system, the ten fingers marked the stopping place in counting; in other words, the system was based on 10.

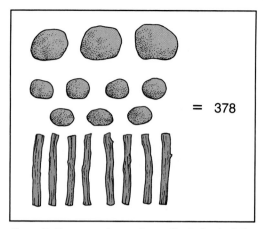

Figure 2. If we were to use the method of calculation illustrated in Figure 1, we would represent the number 378 by three large stones (representing 300), seven small stones (representing 70), and eight sticks (representing 8).

Since stones, pebbles, and sticks are awkward to handle, people created symbols to represent numbers as soon as they learned to write. The symbol that is used to write a number is called a *numeral*. About 5,000 years ago, the Egyptians developed a system for recording numerals on buildings on monuments. This system, called the *hieroglyphic system,* used pictures to represent numerals in the same way that pictures were used to represent words. The

Figure 3. With the hieroglyphic symbols shown below, the ancient Egyptians could write any number.

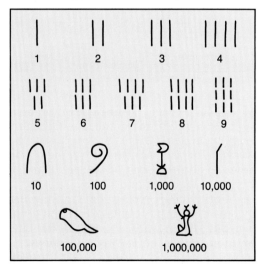

symbol for one was a small vertical stroke that might resemble a small stick. The symbol for 2 was two strokes, the symbol for 3 was 3 strokes, and so on until the symbol for 9 was written as 9 strokes. After the numeral 4, the strokes were grouped in rows of three or four to make the numeral easier to read. See Figure 3.

For the number 10, the base of the decimal system, the Egyptians used a new symbol that resembles an arch. The number 20 was written as two arches and so on until the number 90 which was written as 9 arches. For the number 100, which is 10 times 10, or 10^2, another new symbol resembling a link in a chain, was used. The number 900 was represented by 9 links. For the number 1,000, which is 10 times 100, or 10^3, another new symbol, a stylized lotus flower, was used. For 10,000, or 10^4, the symbol was a bent finger or possibly a bent stick; for 100,000, or 10^5, the symbol was a burbot fish that resembles a tadpole; and for 1,000,000, or 10^6, the symbol was a kneeling figure with arms outstretched.

To write large numbers the Egyptians repeated the symbols as needed. A system such as this is called an *additive system,* because it is necessary to add up the values of all the symbols to find the total number. Numbers of very large size appear in ancient Egyptian records. One record on a 5,000-year-old mace reports that 120,000 prisoners and 1,422,000 goats were captured in a major war. Figure 4 shows how both small and large numbers were written using Egyptian hieroglyphic symbols. The Egyptians wrote their language from right to left and their numbers were also written in this way.

For writing on papyrus, the Egyptians used another system of writing numerals that was not a simple additive system. This system, called the *hieratic system,* used special symbols for numerals instead of repeating the same symbol to show a larger number. The hieratic system of recording numerals is more efficient than the hieroglyphic system and makes it easier to do mathematical calculations. The hieratic numerals for 1 through 9 and for several larger numerals are shown in Figure 5.

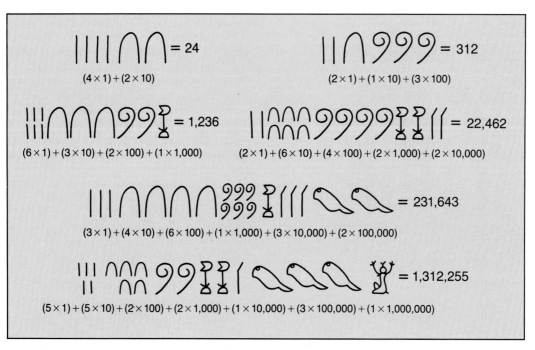

Figure 4. Hieroglyphic numbers were written with the symbols for the numbers of higher value at the right.

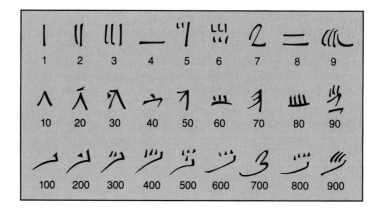

Figure 5. Egyptian hieratic numerals were used for cursive writing numbers on papyrus documents. The form of the numerals varied somewhat with the "handwriting" of each individual scribe.

The Babylonians recorded their language and their numerical records using a wedge-shaped tool to make impressions on wet clay tablets. The characters are called *cuneiform*, a word that means wedge shaped. The soft clay tablets were then baked to become a hard durable record. Many thousands of Babylonian cuneiform records have survived to this day.

The Babylonian numerals for the numbers 1 through 9 were written using groups of vertical wedges. The height of the individual wedges for the numbers 4 through 9 was made less than the height of the wedges for the numbers 1 through 3 in order to form groups of wedges that are easy to read. See Figure 6.

For numbers greater than 60, the Babylonians used positioning to create a numerical system with a base of 60. The position furthest to the right was used to record the number of ones from 1 through

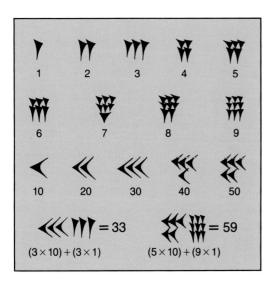

Figure 6. Babylonian cuneiform symbols were used to write numbers on soft clay with a wedge-shaped stick.

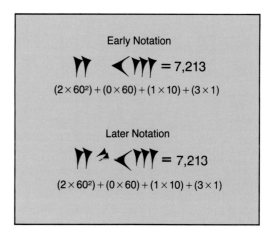

Figure 8. Later Babylonian numerals used a symbol to show an empty place in the middle of a number.

59 (the number of 60^0s). The next position to the left recorded the number of 60s (the number of 60^1s). In this second place the number of 60s was written using the same numerals used to record the number of ones. A small space was left between the symbols in the ones place and the symbols in the 60s place, to clarify the position occupied by a numeral in the second place.

The number 3,599 was written as 59 60s plus 59 ones. Beyond 3,599, numbers were written using the third place from the left to show the number of 60^2s (the number of 3,600s). The fourth space to the left was used to show the number of 60^3s (the number of 216,000s). The fifth space was used to record the number of 60^4s and so on. Several numbers written with Babylonian cuneiform numerals are shown in Figure 7.

In some numbers, such as 7,213, there are two 3,600s and 13 units but no 60s. In such a case an early system of writing numbers simply left a blank in the place for the 60s. Later a new symbol was added to show clearly that there were no sixties and to avoid confusion as to whether or not a blank had been left. This new symbol, Figure 8, was similar in function to our modern zero when the blank occurred in the middle of a number. However, the Babylonians did not use this symbol to show blank places at the end of a number. Thus the only way to know whether a number was 7,213 or 7,213,000 was from the context in which the number appeared.

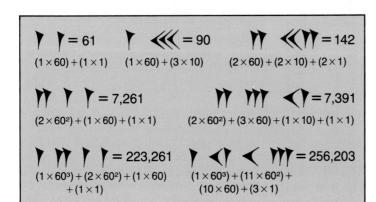

Figure 7. Babylonian numerals greater than 59 made use of positioning to make the writing of large numbers easier. Spaces between groups of symbols helped to identify positions.

NUMERALS 341

Figure 9. The ancient Greeks wrote numbers using the letters of their alphabet. They added three special characters not found in their alphabet for the numbers 6, 90, and 900.

Figure 10. The ancient Hebrews wrote numbers using the letters of their alphabet. In modern Israel, letters are used for numerals only when writing the Hebrew year and for special occasions.

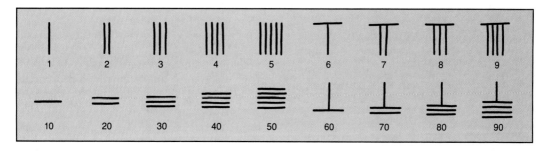

Figure 11. The ancient Chinese developed a system of numerals based on the way sticks were laid out on a table for calculating. This system was also used by early Japanese for writing and calculating.

The ancient Greeks developed several methods of writing numbers. In their most widely adopted version (Figure 9), they used all the letters of their alphabet plus three additional symbols. Each letter stood for a definite value. The first nine symbols represented the units from 1 to 9; the next nine, the tens from 10 to 90; the last nine, the hundreds from 100 to 900. They had no symbol for zero.

To represent thousands, the Greeks added a bar to the left of the first nine letters. Thus /Γ stood for 3,000; /Z for 7,000.

The total 4,627 was written /ΔXKZ, the horizontal base line indicating that these letters formed a numeral. The ancient Hebrews, also, used their alphabet in writing numbers (Figure 10).

A system of rodlike symbols was employed by the ancient Chinese to represent numbers. They had a *place system;* that is, a number symbol took on different values according to the place it occupied in the written number. As shown in Figure 11, the units from 1 through 5 were represented by vertical rods (one rod for each unit). For

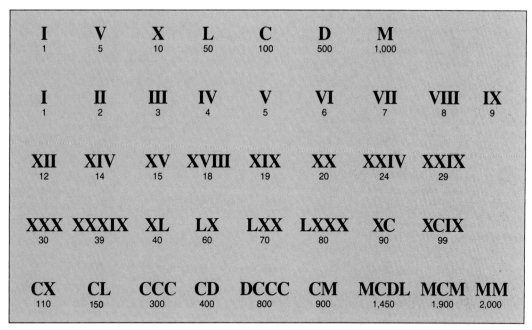

Figure 12. Roman numerals are written with 7 letters using both the additive and subtractive principles.

the units from 6 through 9, an upper horizontal rod denoted a value of five units (or 5), to be added to the vertical rods (each equal to 1). The tens from 10 through 50 were written as horizontal rods (one rod for each ten). For the tens from 60 through 90, an upper vertical rod denoted a value of five tens (or 50), to be added to the horizontal rods (each equal to 10).

The hundreds were written in the same way as the units. Thus the symbol ‖ would stand for either 2 or 200, depending upon its position in the number. The thousands were written in the same way as the tens, the ten thousands in the same way as the units, and so on. The number 7,684, therefore, would be written ⊥T≜‖‖‖‖ There was no symbol for zero, and a gap had to be left to indicate it. The number 7,004, for example, would be ⊥ ‖‖‖‖. If the gap were not recognized as such, the number might be read as 74 instead of 7,004.

The Romans probably derived their system of numbers from the Etruscans, earlier inhabitants of Italy. As shown in Figure 12, the Romans used only seven letters of the alphabet to record numbers. We still refer to numerals written with these seven letters as *Roman numerals*.

The early Roman numbers were written in accordance with the *additive principle*. Symbols for larger numbers were written first and followed by symbols for smaller numbers, all to be added together. Thus the number 4 was written IIII, the number 9 was VIIII, the number 90 was LXXXX, and the number 900 as DCCCC. Later the Romans developed the *subtractive principle* to avoid long strings of letters. In accordance with the subtractive principle, if the symbol of a smaller number precedes the symbol for a larger number, the smaller number is subtracted from the larger number. In this way the number 4 came to be written as IV and the number nine as IX, the number 90 as XC, and the number 900 as CM. See Figure 12.

Roman numerals are still used in English and other Western languages for certain specific purposes. They often indicate chapter or volume numbers of books. They are also used on the dials of some clocks as well as on commemorative monuments and tablets. To us, this number system may

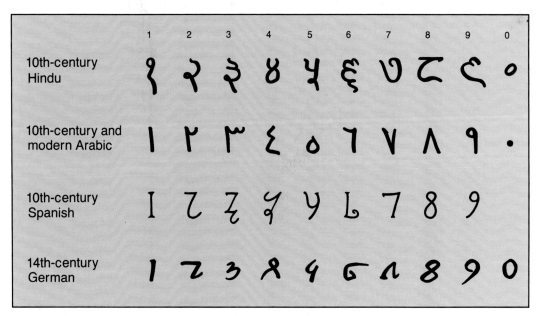

Figure 13. The evolution of our modern western numerals during the past one thousand years.

seem complicated and cumbersome. For example, in our number system, the year eighteen hundred eighty-eight can be expressed by four symbols: 1888. Written in Roman numerals, the same year is MDCCCLXXXVIII.

Our own system of numerals, the so-called Arabic numerals, should really be called the Hindu-Arabic numerals, since the system originated in India (not long before the Christian era) and was later adopted by the Arabs. The Arabs conquered a large part of Spain in the 8th century and in time introduced the Hindu-Arabic numerals in the conquered land. Figure 13 shows how these numerals looked in the 10th century. The system was gradually adopted by the other peoples of Europe. By the 15th century, the symbols of the system had acquired the form that is so familiar to us.

The base of our system is ten, and so it is called a *decimal system*. (*Decem* means "ten" in Latin.) It is a place system, in which the position of a symbol indicates its particular value. Moving one space to the left increases the value by a multiple of 10, as is shown in the following diagram:

10,000's	1,000's	100's	10's	1's
$(10 \times 1,000)$	(10×100)	(10×10)	(10×1)	1
10^4	10^3	10^2	10^1	10^0

In the number 4,962, the 4 stands for four thousands; the 9, for nine hundreds; the 6, for six tens; the 2, for two units. Therefore, 4,962 is really 4,000 + 900 + 60 + 2. Consider the number 7,004. The 7 represents seven thousands; the first 0, no hundreds; the second 0, no tens; the 4, four units. Because our number system includes a symbol for zero, there is no possibility of error in reading such a number.

With our ten symbols, we can write any number, no matter how large. Suppose we start with the number 1. If we put a zero to the right of it, the 1 is in the tens column and indicates ten. Continuing to put zeros to the right in this way, we make the value of 1 ten times greater for every zero that we write (10; 100; 1,000; and so on). We can also write extremely small numbers in our system by using a *decimal point*. The first numeral to the right of a decimal point indicates the number of tenths; the second, the number of hundredths; the third, the number of thousandths; and so on.

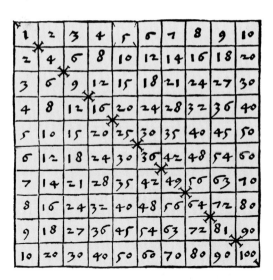

Figure 14. This multiplication table from Arithmetica Boetij *was printed in Germany in 1488. It shows the Hindu-Arabic numerals that were in use then. These numerals are very similar to those used today.*

$1,000,000.0 = 1$ million, 10^6
$100,000.0 = 1$ hundred thousand, 10^5
$10,000.0 =$ ten thousand, 10^4
$1,000.0 = 1$ thousand, 10^3
$100.00 = 1$ hundred, 10^2
$10.00 =$ ten, 10^1
$1.00 =$ one, 10^1
$0.1 =$ one tenth, 10^{-1}
$0.01 =$ one hundredth, 10^{-2}
$0.001 =$ one thousandth, 10^{-3}
$0.0001 =$ one ten-thousandth, 10^{-4}
$0.00001 =$ one hundred-thousandth, 10^{-5}
$0.000001 =$ one millionth, 10^{-6}

We can make the number even smaller by increasing the number of zeros between the decimal point and the 1. Extremely large or small numbers written in this way are awkward to write and use. In the "Algebra" article we show how to express such numbers conveniently.

Not every place system has ten as its base. The base-two place system is known as the *binary system*. It has only two symbols: 1 and 0. In the binary system, moving one space to the left multiplies the value by 2, the base of the system, as indicated in the following diagram:

32's	16's	8's	4's	2's	1's
(2×16)	(2×8)	(2×4)	(2×2)	(2×1)	1
2^5	2^4	2^3	2^2	2^1	2^0

Let us see how the number 101101 would fit into this scheme:

32's	16's	8's	4's	2's	1's
1	0	1	1	0	1

As the diagram shows, 101101 in the binary system represents: $(1 \times 32) + (0 \times 16) + (1 \times 8) + (1 \times 4) + (0 \times 2) + (1 \times 1) = 32 + 0 + 8 + 4 + 0 + 1 = 45$.

The binary system was ardently advocated by the great 17th-century German philosopher and mathematician Gottfried Wilhelm von Leibniz because of its simplicity and also because he thought that it mirrored creation. Unity (1), he thought, represented God: zero (0) stood for the void from which all things were created.

The binary system has certain practical uses in physics. It also serves in the calculating device called the electronic computer, or "electronic brain." This device is run by electricity; the current is either off or on. When it is off, the "0" in the binary system is indicated; when the current is on, the number "1" is indicated. The electronic computer can perform the most intricate calculations with the "0" and "1" of the binary system.

The Duodecimal Society of America has advocated the adoption of the base-twelve system known as the *duodecimal system*. In this, there are twelve symbols: 1, 2, 3, 4, 5, 6, 7, 8, 9, t (standing for ten), e (standing for eleven), and 0. Moving one space to the left, in the duodecimal system, increases the value twelvefold, as follows:

20,736's	1,728's	144's	12's	1's
$(12 \times 1,728)$	(12×144)	(12×12)	(12×1)	1
12^4	12^3	12^2	12^1	12^0

The numeral 6et4 in the duodecimal system represents: $(6 \times 1,728) + (11 \times 144) + (10 \times 12) + (4 \times 1)$. If we perform these multiplications and additions, we see that 6et4 is equal to the base-ten numeral 12,076.

For the ordinary purposes of calculation, our decimal system will not be replaced by a system with some other base.

Howard F. Fehr

ARITHMETIC

The computations carried out with the numbers of the decimal system make up the branch of mathematics called *arithmetic*. There are six fundamental operations in arithmetic: addition, subtraction, multiplication, division, involution, and evolution.

Addition

Addition represents the grouping together of numbers. If a primitive man wanted to find out how many skins he would have if he added three skins to two skins, he would lay three skins on the ground, put down two more, and then count the total number. After he had solved this particular problem again and again, he would no longer have to lay out the skins. He would recall that whenever he added three skins to two skins, the result would always be five skins, and he would do the problem $3 + 2 = 5$ in his head. Similarly, he would learn that $1 + 1 = 2$, $1 + 2 = 3$, $1 + 3 = 4$, and so on.

In the operation of addition, the numbers to be added are called *addends*; the answer is called the *sum:* In the problem $3 + 4 = 7$, 3 and 4 are the addends and 7 is the sum. If one has learned to add all the possible pairs of digits up to $9 + 9$, then one can solve any addition problem. Suppose the problem is $23 + 49$. Remember that 23 stands for 2 tens + 3 units and that 49 stands for 4 tens + 9 units. we can state our problem as follows:

$$\begin{array}{r} 2 \text{ tens} + 3 \text{ units} \\ + 4 \text{ tens} + 9 \text{ units} \end{array}$$

First, we would add the units together: $9 + 3 = 12$. (Remember, 12 is really equal to 1 ten + 2 units.) Keeping the 2 in the units column, we would add the 1 to the other numerals in the tens column, giving $1 + 2 + 4$ tens, or 7 tens. The answer then is 7 tens + 2 units, or 72.

Subtraction

In *subtraction,* we take one or more objects from another group of objects. Suppose our primitive mathematician had nine skins and wanted to find out how many skins he would have left if he took five skins away. He would lay out the nine skins on the ground, would take away five skins, and then would count the skins that remained. Later on, he would come to realize that five from nine always leaves four, and he would perform that subtraction in his head. He would then extend this type of calculation to other numbers. He would

All addition is based on the addition facts. An addition fact is made up of two numbers from 0 to 9, the addends, and their sum.

TABLE I

ONE HUNDRED ADDITION FACTS

0 +0 — 0	0 +1 — 1	0 +2 — 2	0 +3 — 3	0 +4 — 4	0 +5 — 5	0 +6 — 6	0 +7 — 7	0 +8 — 8	0 +9 — 9
1 +0 — 1	1 +1 — 2	1 +2 — 3	1 +3 — 4	1 +4 — 5	1 +5 — 6	1 +6 — 7	1 +7 — 8	1 +8 — 9	1 +9 — 10
2 +0 — 2	2 +1 — 3	2 +2 — 4	2 +3 — 5	2 +4 — 6	2 +5 — 7	2 +6 — 8	2 +7 — 9	2 +8 — 10	2 +9 — 11
3 +0 — 3	3 +1 — 4	3 +2 — 5	3 +3 — 6	3 +4 — 7	3 +5 — 8	3 +6 — 9	3 +7 — 10	3 +8 — 11	3 +9 — 12
4 +0 — 4	4 +1 — 5	4 +2 — 6	4 +3 — 7	4 +4 — 8	4 +5 — 9	4 +6 — 10	4 +7 — 11	4 +8 — 12	4 +9 — 13
5 +0 — 5	5 +1 — 6	5 +2 — 7	5 +3 — 8	5 +4 — 9	5 +5 — 10	5 +6 — 11	5 +7 — 12	5 +8 — 13	5 +9 — 14
6 +0 — 6	6 +1 — 7	6 +2 — 8	6 +3 — 9	6 +4 — 10	6 +5 — 11	6 +6 — 12	6 +7 — 13	6 +8 — 14	6 +9 — 15
7 +0 — 7	7 +1 — 8	7 +2 — 9	7 +3 — 10	7 +4 — 11	7 +5 — 12	7 +6 — 13	7 +7 — 14	7 +8 — 15	7 +9 — 16
8 +0 — 8	8 +1 — 9	8 +2 — 10	8 +3 — 11	8 +4 — 12	8 +5 — 13	8 +6 — 14	8 +7 — 15	8 +8 — 16	8 +9 — 17
9 +0 — 9	9 +1 — 10	9 +2 — 11	9 +3 — 12	9 +4 — 13	9 +5 — 14	9 +6 — 15	9 +7 — 16	9 +8 — 17	9 +9 — 18

soon come to know that 9 − 4 = 5 or that 7 − 6 = 1. He would need to know only a few such calculations to be able to subtract any number (the *subtrahend*) from any other number (the *minuend*) and get the correct answer (the *remainder*).

Multiplication

Multiplication is really a form of addition. If we did not know how to multiply, we could find the answer to the problem 5 × 7 by simple addition. For 5 × 7 means the same thing as five sevens, or 7 + 7 + 7 + 7 + 7. Adding the five sevens together, we would have 35. Sooner or later we would probably come to realize that when we solve the problem of adding together five sevens, the answer is always 35. We would memorize this particular operation and other similar ones such as three eights (3 × 8) and seven nines (7 × 9). These operations up to and including 12 × 12 are found in the familiar multiplication table. If we could get as far as 9 × 9, we could do any multiplication problem.

Of course, all this requires a certain amount of memorizing. One multiplication method that avoids the memorizing is called *duplation,* which involves only multiplication by two and addition. This is how duplation works: Suppose we wanted to find out the product of 24 and 18 (24 × 18). We would perform our calculations thus:

A	B	C	D
1	24		
2	48	2	48
4	96		
8	192		
16	384	16	384
			432

The 432, at the bottom of the last column, is the answer to our problem. All this looks complicated, but it is really quite simple.

As you see, we set up four columns (A, B, C, and D). The first number in column A is always 1, from which we start the process of multiplying by 2: 1 × 2 = 2. We write the answer 2 in column A and multiply it by 2: 2 × 2 = 4. We write 4 in column A and multiply it by 2: 4 × 2 = 8. We continue multiplying by 2 until we reach a

TABLE II
ONE HUNDRED MULTIPLICATION FACTS

0	0	0	0	0	0	0	0	0	0
×0	×1	×2	×3	×4	×5	×6	×7	×8	×9
0	0	0	0	0	0	0	0	0	0
1	1	1	1	1	1	1	1	1	1
×0	×1	×2	×3	×4	×5	×6	×7	×8	×9
0	1	2	3	4	5	6	7	8	9
2	2	2	2	2	2	2	2	2	2
×0	×1	×2	×3	×4	×5	×6	×7	×8	×9
0	2	4	6	8	10	12	14	16	18
3	3	3	3	3	3	3	3	3	3
×0	×1	×2	×3	×4	×5	×6	×7	×8	×9
0	3	6	9	12	15	18	21	24	27
4	4	4	4	4	4	4	4	4	4
×0	×1	×2	×3	×4	×5	×6	×7	×8	×9
0	4	8	12	16	20	24	28	32	36
5	5	5	5	5	5	5	5	5	5
×0	×1	×2	×3	×4	×5	×6	×7	×8	×9
0	5	10	15	20	25	30	35	40	45
6	6	6	6	6	6	6	6	6	6
×0	×1	×2	×3	×4	×5	×6	×7	×8	×9
0	6	12	18	24	30	36	42	48	54
7	7	7	7	7	7	7	7	7	7
×0	×1	×2	×3	×4	×5	×6	×7	×8	×9
0	7	14	21	28	35	42	49	56	63
8	8	8	8	8	8	8	8	8	8
×0	×1	×2	×3	×4	×5	×6	×7	×8	×9
0	8	16	24	32	40	48	56	64	72
9	9	9	9	9	9	9	9	9	9
×0	×1	×2	×3	×4	×5	×6	×7	×8	×9
0	9	18	27	36	45	54	63	72	81

All multiplication is based on the multiplication facts. A multiplication fact is made up of two numbers from 0 to 9, the multiplier *and the* multiplicand, *and their product.*

number that is equal to or just less than the *multiplier* in the original problem. (A multiplier is a number by which another is multiplied.) In the problem 24 × 18, the multiplier is 18 so in column A we stop at 16. The first number in column B is always the *multiplicand* (a number that is to be multiplied by another), which in this case is 24. Again, we begin multiplying by 2: 24 × 2 = 48. We continue the same doubling process, ending when column B has as many numbers as column A. For column C, we select from column A the numbers

Arithmetic is essential in every kind of work. In a small business, skill in the basic operations is needed to keep track of inventory, to set prices, and to keep track of the money coming in and going out.

that, added together, will equal the multiplier. Because the multiplier is 18, we take 2 and 16 and write them in column C directly across from their places in column A. In column D we write the numbers from column B that are directly across from those in column C. Finally, we add the numbers in column D for a sum, in this case, of 48 + 384 = 432. Therefore, through duplation, we have determined that the *product* (the answer to a multiplication problem) of 24 × 18 is 432. Of course, anyone knowing the multiplication table could multiply 24 by 18 in much less time than it would take to solve the problem by duplation.

Division

Division is a kind of subtraction. If we divide 12 (the *dividend*) by 4 (the *divisor*) we want to know how many times 4 goes into 12. We perform the following series of subtractions:

```
 12      8      4
 -4     -4     -4
 ──     ──     ──
  8      4      0
```

We have subtracted 4 from 12; then 4 from the remainder 8; then 4 from the remainder 4. Nothing remains. We used 4 as the subtracter three times. Hence the answer (or *quotient*) to the problem 12 divided by 4, or 12 ÷ 4, is 3.

Involution

In the process called *involution,* we raise a number to any desired *power*. The number that is to be raised to the power in question is called the *base*. To raise 2 to the third power, we repeat the base three times as it is multiplied by itself; thus: 2 × 2 × 2. We indicate the power by placing a small figure, called the *exponent,* to the right of the base and above it. For example, 2 to the third power, or 2 × 2 × 2, is written 2^3.

If the exponent is 1, it indicates that the number is not raised to a higher power but remains unchanged. Thus $2^1 = 2$; $5^1 = 5$. When we use the exponent zero, we show that the base is to be divided by itself. Any number with the exponent 0 is always equal to 1. For example, $4^0 = 4 \div 4 = 1$; $10^0 = 10 \div 10 = 1$. There are also *negative* exponents. (A negative number is one whose value is less than zero, such as −16 or −3.) A base with a negative exponent is equal to the *reciprocal* of the base with the corresponding positive exponent. (A reciprocal of a number is equal to 1 divided by the number; for example, the reciprocal of 4 is $\frac{1}{4}$.) Thus:

$$5^{-1} = \frac{1}{5^1} = \frac{1}{5}$$
$$3^{-2} = \frac{1}{3^2} = \frac{1}{9}$$

Evolution

In the process called *evolution* we are essentially doing the inverse (or opposite) of involution. Given a certain number (for example, 9), we try to find what other number, multiplied by itself a desired number of times (say, two times) will give us the first number. The number we are trying to find is called the *root*, and in this case, because we want a root that will be multiplied by itself two times to equal 9, we call it the *square root* of 9 and express it as $\sqrt{9}$. Because $3 \times 3 = 9$, we would say that the square root of 9 is 3, or $\sqrt{9} = 3$. We could also say that the two 3's are *factors* of 9. (Factors are numbers that can be multiplied together to equal a given product.)

A root that is multiplied by itself three times to give a certain number is called the *cube root* of that number. The cube root of 8, written as $\sqrt[3]{8}$, is 2 because $2 \times 2 \times 2 = 8$.

Fractions

Numbers such as 0, 1, 2, 3, 5, 10, 120, and 3,000 are called *whole numbers,* or *integers.* (Any whole number greater than zero is called a *positive integer.*) When one multiplies one integer by another, the answer is always an integer: $5 \times 6 = 30$; $7 \times 9 = 63$. However, it is not always possible to obtain an integer as an answer when one divides one integer by another. It is true that if we divide 8 apples into 4 equal shares, each share will consist of 2 apples. But if 8 apples are to be divided into 3 equal shares, the answer will not be an integer because $8 \div 3 = \frac{8}{3}$, which is a *fraction,* or "broken number." This does not mean that we have to divide each of our 8 apples into thirds and give each person 8 thirds. What we need to know is how many whole apples are in $\frac{8}{3}$ apples. By doing simple division we would find that $\frac{6}{3} = 2$, which means that 6 of the 8 thirds is equal to 2 and that 2 thirds will remain. In other words, the fraction $\frac{8}{3}$ can be *reduced* to a whole number and a remaining fraction:

$$\frac{8}{3} = \frac{6}{3} + \frac{2}{3} = 2 + \frac{2}{3} = 2\frac{2}{3}$$

So, to divide 8 apples equally among 3 people, each person would receive 2 whole apples plus $\frac{2}{3}$ apple, for a total of $2\frac{2}{3}$ apples. Because $2\frac{2}{3}$ represents the sum of an integer and a fraction, it is called a *mixed number.*

In a fraction such as $\frac{2}{3}$, the numeral above the line is called the *numerator,* and the numeral below the line is called the *denominator.* Although symbolized with two numerals, the fraction $\frac{2}{3}$ is actually one number. A fraction, like any other number, can be added, subtracted, multiplied, and divided. Operations with fractions follow the basic rules of arithmetic, but there are certain rules about fractions in particular that should be learned. Let us consider the following multiplication problem:

$$\frac{2}{3} \times \frac{4}{5} = \frac{8}{15}$$

Very simply, the product was obtained by multipling the numerators ($2 \times 4 = 8$) and then multiplying the denominators ($3 \times 5 = 15$). Division by a fraction takes an added step:

$$\frac{2}{5} \div \frac{7}{9} = \frac{2}{5} \times \frac{9}{7} = \frac{18}{35}$$

We notice from the example that the divisor $\left(\frac{7}{9}\right)$ was inverted $\left(\frac{9}{7}\right)$ and was then mul-

ARITHMETIC

Fractions are used every day in household weights and measures. If a pie serves 6, each portion is 1/6. This apple pie contains 3/4 cup sugar, 1/4 teaspoon of nutmeg, and 1/8 teaspoon of salt among its ingredients.

tiplied by the dividend $\left(\dfrac{2}{5}\right)$, following the rules for multiplying fractions.

Let us consider the following addition problem:

$$\frac{1}{8} + \frac{1}{8} + \frac{3}{8} = \frac{5}{8}$$

We can see that the three addends, as well as their sum, have the same denominator (8) and that only the numerators were added together (1 + 1 + 3 = 5). Adding fractions is very simple when all the denominators are the same, as in the example just shown, but when denominators differ it is necessary to find appropriate equivalents in order to "make" the denominators the same:

$$\frac{1}{8} + \frac{1}{4} = \frac{1}{8} + \left(\frac{2}{2} \times \frac{1}{4}\right) = \frac{1}{8} + \frac{2}{8} = \frac{3}{8}$$

As shown, $\dfrac{1}{4}$ "became" $\dfrac{2}{8}$ when it was multiplied by $\dfrac{2}{2}$. The value of $\dfrac{1}{4}$ did not change because $\dfrac{2}{2} = 1$ and multiplying any number by 1 will not change that number's value. The rule of like-denominators also applies to the subtraction of fractions:

$$\frac{1}{2} - \frac{5}{12} = \left(\frac{6}{6} \times \frac{1}{2}\right) - \frac{5}{12} = \frac{6}{12} - \frac{5}{12} = \frac{1}{12}$$

It is also helpful to remember that a fraction such as $\dfrac{17}{30}$ can be expressed as an operation of division. In other words, $\dfrac{17}{30} = 17 \div 30$, which means that a fraction's numerator is also a dividend and its denominator is also a divisor. The laws of fundamental mathematics prohibit the use of zero as a divisor (because zero multiplied by any number can result only in zero), and therefore, zero can never be the denominator of any fraction.

Negative Integers

If we return now to the integers, we note that when we subtract one positive integer from another, the answer is not always positive. It is true that 7 − 4 = 3. But suppose we want to subtract 7 from 4. To make the subtraction 4 − 7 possible, we invent a new kind of number called a *negative integer,* which is represented by an integer with a minus sign in front of it. The answer to the subtraction problem 4 − 7 is −3.

The mechanics of solving problems such as 12 − 6, 9 − 7, 6 − 13, and so on is simple enough. We subtract the smaller number from the larger one and we give the result the sign (plus or minus) of the larger number. (Numbers without signs are positive, and therefore "plus" signs are seldom necessary.):

$$12 - 6 = 6 \quad 4 - 9 = -5$$
$$9 - 7 = 2 \quad 3 - 7 = -4$$
$$4 - 1 = 3 \quad 6 - 13 = -7$$

There are as many possible negative integers as there are positive integers. Starting with 0, we can build up a list of negative integers to the left of 0 and a list of positive integers to the right of 0.

$$\ldots -3, -2, -1, 0, +1, +2, +3 \ldots$$

We could extend the list of positive and negative integers in this way indefinitely. There are negative fractions as well as negative integers. Every positive fraction has a negative counterpart, such as $\frac{3}{5}$ and $-\frac{3}{5}$.

The entire set of numbers we have been discussing (positive integers, negative integers, positive fractions, negative fractions, and zero) is called the *rational number system*. We can define a rational number as being zero, an integer (positive or negative), or a fraction (positive or negative) whose numerator and denominator are both integers. Therefore, 5 is a rational number, as are $\frac{1}{2}$ and $-\frac{3}{4}$.

Rules have been devised for the addition, subtraction, multiplication, and division of positive and negative rational numbers so that no illogical results will occur. One of these rules is: "The product of two like-signed numbers is positive." For example, $4 \times 4 = 16$; likewise $-4 \times -4 = 16$. Another of these rules is: "The product of two unlike-signed numbers is negative." For example, $4 \times -4 = -16$; $-7 \times 6 = -42$.

Irrational Numbers

The square root of a number, as we have seen, is one of its two equal factors. Thus $\sqrt{25} = 5$, since $5 \times 5 = 25$. But the square root of 2, or $\sqrt{2}$, cannot be expressed by any rational number. We know that $\sqrt{2}$ must be somewhere between 1.4 and 1.5, since $1.4^2 = 1.96$ and $1.5^2 = 2.25$. We could come closer to the square root of 2 by using more and more decimal places: $1.41^2 = 1.9881$, and $1.414^2 = 1.999396$. But no matter how many decimal places we add, we will never find a rational number whose square root is 2.

Yet numbers such as $\sqrt{2}$ result from many mathematical calculations, and, because they do not fit into the rational number system, we call them *irrational numbers*. Some examples of irrational numbers are $\sqrt{3}$, $\sqrt{5}$, and $\sqrt[3]{7}$. These are all *real numbers*, just as rational numbers are, even if we cannot express them by integers or fractions.

If we combine all the irrational numbers with all the rational numbers, we get a very large set of numbers called the *real number system*. All the numbers of the system can be represented by points on a straight line. Let one point represent the number 0. Then let points to the right represent positive integers and those to the left, negative integers, as follows:

If the space between the integers is subdivided into as many parts as possible, each of the subdivision points will represent a rational number. No matter how many divisions we may make in this way, however, gaps will always remain. If we fill in the gaps with points representing all possible irrational numbers, such as $\sqrt{2}$, $\sqrt{5}$, and so on, the line will be completely filled. This line is known as the real number *axis*, or *continuum*.

Imaginary Numbers

There are still other numbers besides the real numbers. Let us consider the number $\sqrt{-1}$. This seems to be a contradiction in terms, since a square is always the product of two equal numbers with like signs and is, therefore, always positive. Hence no number multiplied by itself can give a negative real number, and it would seem futile to try to get the square root of such a number. However, mathematicians have used the number $\sqrt{-1}$ to form the basis of a number system called *imaginary numbers*, or *complex numbers*. The imaginary number $\sqrt{1}$ is often indicated by the symbol *i*.

ALGEBRA

In the preceding article, "Arithmetic," we were concerned with particular numbers, which are expressed by symbols. "Sixty-seven" is a particular number. To do arithmetical problems in which sixty-seven plays a part, we use the symbols 6 and 7, combined as 67. We are now going to consider a branch of mathematics in which a symbol, such as the letter a, b, or c, stands not for a particular number, but for a whole class of numbers. This kind of mathematics is called *algebra*.

We can illustrate the difference between arithmetic and algebra by a very simple example:

Take the number 4.   4
Multiply by 5. $4 \times 5 = 20$
Add 4. $20 + 4 = 24$
Multiply by 2. $24 \times 2 = 48$
Subtract 8. $48 - 8 = 40$
Divide by original number (4). $40 \div 4 = 10$

In arriving at the final result, 10, we used the method of arithmetic, involving particular numbers, throughout.

Suppose now that we think of any number. Let us indicate "any number" by the symbol x, and let us go through the same operation as before:

Multiply by 5. $5 \times x = 5x$
Add 4. $5x + 4 = 5x + 4$
Multiply by 2. $2(5x + 4) = 10x + 8$
Subtract 8. $(10x + 8) - 8 = 10x$
Divide by original
  number (x). $10x \div x = 10$

Here we have been using the methods of algebra, because x can be replaced by any number. We could substitute for it 2, or 3, or 15, and the final result would always be 10.

When a generalized number, represented by a letter (such as a) is multiplied by a particular number (such as 5) or by another generalized number (such as b), we do not use multiplication signs, but indicate multiplication by putting these symbols close to one another. Thus $a \times b = ab$; $5 \times a = 5a$; $5 \times a \times b = 5ab$. We could not indicate the multiplication of two particular numbers in this way; 7×5 could not be given as 75, because 75 really stands for $70 + 5$.

Let us consider another example. In the equation $(2 + 3)^2 = 25$, we are dealing with the particular numbers 2 and 3, and the result is always 25. But suppose that instead of two particular numbers, we used the letters a and b to stand for any two numbers. We would then have the algebraic equation $(a + b)^2 = a^2 + 2ab + b^2$. (The derivation of this equation will be explained later.)

What is significant about $(a + b)^2 = a^2 + 2ab + b^2$ is that it indicates a general relationship that holds true for a great many particular numbers. If we substituted 3 for a and 2 for b, we could have $(3 + 2)^2 = 3^2 + (2 \times 3 \times 2) + 2^2 = 9 + 12 + 4 = 25$. Or we could substitute 5 for a and 6 for b, giving $(5 + 6)^2 = 5^2 + (2 \times 5 \times 6) + 6^2 = 25 + 60 + 36 = 121$.

Algebra, the mathematics of "any numbers," or *variables*, goes to the heart of the relationship between numbers. Generally speaking, it is concerned with particular numbers only insofar as they are applications of general principles. It is also used in the solution of certain specific problems in which we start out with one or more unknown quantities whose values are indicated by algebraic symbols.

An Ancient Discipline

The study of algebra goes back to antiquity. Recent discoveries have shown that the Babylonians solved problems in algebra, although they had no symbols for variables. They used only words to indicate such numbers, and for that reason their algebra has been referred to as *rhetorical algebra*. The Ahmes Papyrus, an Egyptian scroll going back to 1600 B.C., has a number of problems in algebra, in which the unknown is referred to as a *hau,* meaning "a heap."

Little further progress was made in algebra until we come to Diophantus, a 3rd-century A.D. Greek mathematician. He reduced problems to equations, representing the unknown quantity by a symbol suggesting the Greek letter Σ (sigma). He also introduced an interesting system of abbreviations, in which he used only the initial letters of words, after omitting all unnecessary words. If we were to use the method of Diophantus in presenting the problem "An unknown squared minus the unknown will give twenty," we would first state the problem as "Unknown squared minus unknown equals twenty." Then we would use initial letters for all the words except the last, for which we would use the numeral 20, as follows: "USMUE20."

In the 16th century, French mathematician François Viète, (or Vieta) used the vowels *a, e, i, o, u* to represent unknown numbers and the consonants *b, c, d, f, g,* and so on to stand for values that remained fixed throughout a given problem. The great 17th-century French philosopher René Descartes proposed the system of algebraic symbols now in use. In this system, *a, b, c,* and other letters near the beginning of the alphabet represent the fixed numbers. The last letters of the alphabet (*x, y, z,* and sometimes *w*) stand for the unknown numbers in a problem. As soon as this symbolism came into general use, algebra grew quite rapidly into a systematic set of rules and theorems that could be applied to all numbers.

The word "algebra" originated from the Arabic title of a work by a 9th-century Persian mathematician, Mohammed ibn Musa Al-Kwarizmi. The work was *Al-Jebr W'al Muqabala,* which means "restoration and reduction." By *al-jebr* or restoration, was meant the transposing of negative terms to the other side of an equation to make them positive. When the Arabs came to Spain, they brought this word with them. In the course of time, *al-jebr* was changed to "algebra," and the word came to be applied not to a single operation, but to all operations involved in modern algebra.

Three Fundamental Laws

Algebra generalizes—that is, expresses in general terms—certain basic laws that govern the addition, subtraction, multiplication, and division of all numbers.

(1) When we add or multiply two integers, the order in which we add or multiply them is immaterial. Thus $2 + 3$ is the same as $3 + 2$, and 4×3 is the same as 3×4. Since this is true for all integers, we set up the following algebraic formulas:

$$a + b = b + a$$
$$ab = ba$$

These are called the *commutative laws* of addition and multiplication.

(2) When more than two numbers are added or multiplied, we can group them in any order we choose, and the answer will always be the same. If 2 is added to (3 + 6), the result is the same as if we added (2 + 3) to 6. Similarly, 2 times the product of 3×6 is the same as 3 times the product of 6×2. These results are indicated in the following formulas:

$$a + (b + c) = (a + b) + c$$
$$a(bc) = (ab)c$$

These are the *associative laws* of addition and multiplication.

(3) If a multiplicand has two or more terms, a multiplier must operate upon each of these terms in turn. Suppose we wish to multiply (3 + 2) by 5, a problem we could set down as 5(3 + 2). We would first multiply 3 by 5 and then 2 by 5, giving 15 + 10, or 25. This rule is called the *distributive law of multiplication* and is given by the following formula:

$$a(b + c) = ab + ac$$

Suppose we want to multiply $a + b$ by $a + b$, which of course would be the same thing as $(a + b)^2$. We would set up the problem as follows:

$$a + b$$
$$\underline{a + b}$$

In accordance with the distributive law of multiplication, we (1) multiply the upper ($a + b$) by the *b* of the lower ($a + b$), (2) multiply the upper ($a + b$) by the *a* of the lower ($a + b$), and (3) add the results:

$$\begin{array}{r}a + b\\ a+b\\\hline ab+b^2 \quad (1)\\ a^2 + \ ab \quad\quad (2)\\\hline a^2 + 2ab + b^2 \quad (3)\end{array}$$

We use the same distributive law of multiplication in multiplying 25 by 25. Ordinarily, we would present our calculations as follows:

$$\begin{array}{r}25\\ 25\\\hline 125\\ 50\\\hline 625\end{array}$$

Because the 2 in 25 is really 20, the same problem can also be expressed by the following:

$$\begin{array}{r}20+5\\ 20+5\end{array}$$

Using the algebraic formula for the distributive law, having given a and b the values of 20 and 5, respectively, we can carry out the problem as follows:

$$\begin{array}{r}20+\ \ 5\\ 20+\ \ 5\\\hline 100+25\\ 400+100\\\hline 400+200+25 = 625\end{array}$$

The two methods shown for multiplying 25 by 25 are expressed differently but actually perform the same algebraic operations.

Formulas, Tables, and Graphs

There are different ways of showing how different quantities are related. We can use a formula, set up a table of values, or draw a graph.

Consider the rule: "The area of a square is equal to the square of the length of its side." This is a rather roundabout way of expressing the relationship in question. We could state it much more simply by using the symbol A to represent the number of square units in the area, and the symbol s to represent the number of units in the side. Therefore, the area of any square can be stated as $A = s^2$. This abbreviated rule is called a *formula*, from the Latin word meaning "little form."

If the length of the side of a square is 6, we can get A, the area, by substituting 6 for s in the formula:

$$A = s^2 = 6^2 = 36$$

We can also represent the relationship $A = s^2$ by setting up a *table of values*. Suppose the side of the square is equal to 1; A is then 1^2 or 1. If the side of the square is equal to 2, $A = 2^2$, or 4. Substituting for s the values from 1 through 10, in turn, we obtain the following table of values:

s	1	2	3	4	5	6	7	8	9	10
A	1	4	9	16	25	36	49	64	81	100

This table tells us that if $s = 1$, $A = 1$; if $s = 2$, $A = 4$; if $s = 3$, $A = 9$; and so on.

There is still another way of representing the area of a square. We could construct a *graph*, as in Figure 1. First, we draw two lines, called axes, which are perpendicular to each other. Along the horizontal axis, or base, we mark out a series of numbers at equal intervals, corresponding to the values of s in the table. We mark another series of numbers on the vertical axis that correspond to the values of A.

Within the framework of the axes we create a grid, made of lines perpendicular to both axes and originating from the designated values along those axes. A grid may consist of any number of perpendiculars (such as the ones in Figure 1 that fall between the designated values), which can indicate values not enumerated on the axes. For instance, the vertical line that lies between the s values of 3 and 4 indicates an s value of 3.5. You can see that there is no specific line on the grid for an s value of 3.1, for example, but that value *does* exist on the axis and would simply be estimated by eye to fall just to the right of the 3.

Because the purpose of the graph in Figure 1 is to show the relationship between the values of s and A, we want to translate the values from the table to a series of *points* on the grid. At each value of s (1, 2, 3, 4, and so on) we follow the vertical line upward until we reach the point at which a horizontal perpendicular would meet the corresponding value for A. For instance, the table tells us that when $s = 5$,

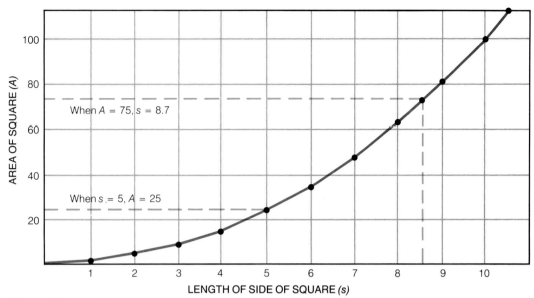

Figure 1. The graph of the area of a square, using horizontal and vertical axes, according to the method described in the text. The graph shows how the area of a square increases as the length of the side increases. It is one way of expressing the relation between the length of the side and the area.

$A = 25$. On our grid we find the perpendicular that would indicate $A = 25$ (halfway between the horizontal lines for 20 and 30), and where it meets the vertical line for $s = 5$ we draw, or *map,* a point. When we have mapped the points for each value of s from 1 through 10, we draw a line connecting all ten points, which in this case is an upward curve. This curve is called the graph of the area of a square. It shows how the area increases as the length of the side increases.

We can use the graph to find out the areas of squares with sides not given in the table of values. For example, if $s = 5.5$, we erect a perpendicular at this point, extending it until it meets the graph. From the point of meeting, we erect a perpendicular to the vertical axis. The point where this perpendicular meets the vertical axis will represent the area, which is about 30. (The correct value is 30.25.)

If we know the area of a square, we can find out the approximate length of the side by means of our graph. Suppose the area is 75. From the point corresponding to the number 75 on the vertical axis, we draw a horizontal line to the point where it meets the graph. From there we drop a perpendicular to meet the s axis, at about 8.7, a very accurate value for s. (The correct value, expressed to five decimal places, is 8.66025.)

Figure 2. Bar graph showing the populations of four countries in the mid-20th century.

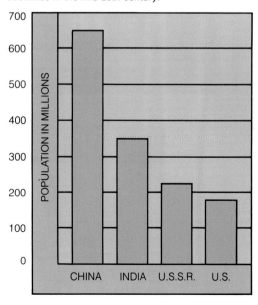

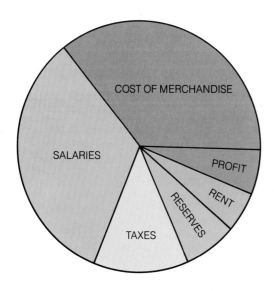

Figure 3. A circle graph indicating how each part of a dollar received by a department store is spent.

There are many different kinds of graphs. The most common are the *bar graph,* the *circle graph,* and the *line graph.* If, for example, we would like to show how the populations of China, India, the Soviet Union, and the United States compare, we could draw four bars (Figure 2). The lengths of the bars would be proportionate to the populations in question.

If we desired to compare parts of a whole quantity, we would use a circle graph. It could serve to indicate how each part of a dollar received in sales in a department store is spent by the store (Figure 3). Each sector of the circle, as compared to the whole circle, would show the proportion given to a particular service.

When a quantity is continuously changing, we would use a line graph. Hospital nurses often make graphs of their patients' temperatures. Figure 4 is an example of one such graph. It covers a 44-hour period, during which the patient's temperature was recorded (with a point on the grid) every 4 hours. As shown, each point is connected to the preceding point by a straight line. The doctor who consults the chart can see at a glance how the patient's temperature has been changing.

To repeat, then, we can show how quantities are related by a formula, a table of values, or a graph. It is the formula that is basic. If a table of values is worked up, a scientist tries to find the formula that will express the relationship in question. After plotting a graph to show the length of a steel cable under increasing tension, an engineer would work out an algebraic formula to sum up his or her findings.

Formulas are extremely important in many branches of pure and applied science. For example, to indicate the speed of a body having uniform rectilinear motion (that is, moving in a straight line at a con-

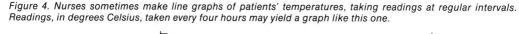
Figure 4. Nurses sometimes make line graphs of patients' temperatures, taking readings at regular intervals. Readings, in degrees Celsius, taken every four hours may yield a graph like this one.

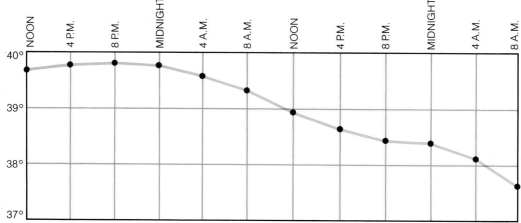

stant speed), the physicist uses the formula $v = \frac{s}{t}$, where v is the average speed of the body, s is the space or distance covered, and t is the time required to travel this distance. We can apply this formula to specific cases by substituting appropriate values for v, s, or t. For example, if a car takes 5 hours to travel 400 kilometers, we could find the average speed by substituting 5 for t and 400 for s. The average speed, then, would be $\frac{400}{5}$ kilometers, or 80 kilometers per hour.

The chemist often has occasion to use the law of J. A. C. Charles and J. L. Gay-Lussac, which states that if the pressure and the mass of a gas are constant, the volume is proportional to the *absolute temperature*. (Absolute temperature is based on the hypothetical lowest possible temperature, *absolute zero,* which is equivalent to $-273.16°$ Celsius.) This law can be stated very concisely by the formula $\frac{V_1}{T_1} = \frac{V_2}{T_2}$, in which V_1 is the volume of a gas at temperature T_1, while V_2 is the volume of a gas at temperature T_2.

Perhaps the most famous formula of all is the Einstein equation, $E = mc^2$. In this equation, E stands for the amount of energy, m for the amount of mass, and c for the speed of light (which is about 300,000 kilometers per second), measured in appropriate units. With this formula the great 20th-century physicist Albert Einstein indicated the amount of energy that appears when matter is transformed into energy.

How Equations are Employed

The equation plays an all-important part in algebra. It may be looked upon as a balance, with equal numerical values on each side of the "equal" sign (=). To show how an equation is applied, let us consider the formula for the perimeter (the outside boundary) of a rectangle: $P = 2l + 2w$, where P is the perimeter, l is the length and w is the width. Suppose we have 100 meters of wire with which to make a rectangular enclosure, which is to be 20 meters wide.

We wish to find out the length of this enclosure if we are to use the entire 100 meters of wire. We know that the perimeter will be 100 meters. We also know that the width is to be 20 meters. Substituting 100 for P and 20 for w in the formula, we have $100 = 2l + (2 \times 20) = 2l + 40$. In the equation $100 = 2l + 40$, the 100 on one side of the equal sign is exactly equal to $2l + 40$ on the other side. To find out what l is, we subtract 40 from each side of the equation, resulting in $60 = 2l$. If we divide both sides of the equation by 2, the result is $30 = l$. Therefore, the length of the enclosure will be 30 meters.

In solving the equation, we used the following rule: "If the same operation is performed on each member of an equality, then the results are equal." If we add the same quantity to each side of the equation or subtract the same quantity, the equality will be maintained. It will be maintained, too, if we multiply or divide each side of the equation by the same quantity. Of course, if we multiplied each side of the equation by zero, the result would be $0 = 0$, which would get us nowhere in the task of solving the equation. Division by zero is not permitted.

Equations may be used to solve problems in which no formula is involved but in which certain data are given. Here is a simple problem: "A man is 6 times as old as his son. In 20 years, the father will be only twice as old as his son. How old are the father and son at the present time?"

On the basis of the data, we can write an equation and solve the problem. First, we let x stand for the son's age. Since the father is 6 times as old as his son, his age can be given as 6 times x, or $6x$. In 20 years the son will be $x + 20$ years old. In 20 years, the age of the father will be $6x + 20$ years. At that time, the father's age will be twice that of the son, a relationship that can be expressed by the equation: $6x + 20 = 2(x + 20)$. Applying the distributive law to the right-hand side of the equation, we have $6x + 20 = 2x + 40$. We subtract $2x + 20$ from each side of the equation and get $4x = 20$. If $4x = 20$, $x = 5$. The son's age at the present time, therefore, is 5. Since the father's age at the present time is 6 times that of the son, or 6x, the father is 30 years old.

Not all problems are as simple as this one, in which the unknown is x. The equation may involve not only an unknown quantity, x, but also higher powers of x. If x^2 is the highest power occurring in an equation, it is called a *quadratic equation*. $x^2 + 6 = 5x$ is an example of such an equation. In various equations, the highest power of x may be x^3 or x^4, or there may be even higher powers.

Identities

When an equation is true for all the replacement values of the variables concerned, it is called an *identity*. A familiar example of an identity is $(a + b)^2 = a^2 + 2ab + b^2$. As we pointed out before, this equation holds true no matter what values we assign to a and b. It can be used as an aid in mental arithmetic. To square 22, we can think of this number as $(20 + 2)^2$, 20 being substituted for a and 2 for b. Mentally we square 20, giving 400; then we double 20×2, giving 80; then we square 2, giving 4. Finally, we have $400 + 80 + 4 = 484$. The answer, then, is 484.

Another identity is $(a - b)^2 = a^2 - 2ab + b^2$. You can verify this by performing the multiplication $(a - b)(a - b)$. We would have:

$$\begin{array}{r} a - b \\ a - b \\ \hline - ab + b^2 \\ a^2 - ab \\ \hline a^2 - 2ab + b^2 \end{array}$$

Still another identity is $(a + b)(a - b) = a^2 - b^2$. This also is useful in certain mental-arithmetic problems. If we wish to multiply 34 by 26 in our heads, we can change the problem to $(30 + 4) \times (30 - 4)$. Solving this in accordance with the identity $(a + b)(a - b) = a^2 - b^2$, we have $900 - 16 = 884$. Other well-known identities are:

$(a + b)^3 = (a^3 + 3a^2b + 3ab^2 + b^3)$
$(a - b)^3 = (a^3 - 3a^2b + 3ab^2 - b^3)$
$(a^3 - b^3) = (a - b)(a^2 + ab + b^2)$

Exponents

Exponents simplify the writing of algebraic expressions. Thus $aaabbbcc$, which is really a continuous multiplication (a times a times a times b times b times b times c times c), can be written $a^3b^3c^2$. The mathematician has derived a series of rules for combining exponents. The rules are stated in general terms: a^n stands for the base a raised to the nth *power*; a^m, for the same base raised to the mth power.

(1) $a^n a^m = a^{n+m}$. In multiplying powers, we add the exponents of like bases. Thus $2^2 \times 2^3 = 2^{2+3} = 2^5 = 32$.

(2) $a^n \div a^m = a^{n-m}$. In dividing powers, we subtract the exponents of like bases. This means that $2^5 \div 2^2 = 2^{5-2} = 2^3 = 8$.

(3) $(a^n)^m = a^{nm}$. To raise a given power by another power, we multiply the two exponents. For example, $(2^2)^3 = 2^{2 \times 3} = 2^6 = 64$.

(4) $(ab)^n = a^n b^n$. When a product is raised to a power, each member of the product is raised to that power. Thus $(4 \times 2)^2 = 4^2 \times 2^2 = 16 \times 4 = 64$.

(5) $\left(\dfrac{a}{b}\right)^n = \dfrac{a^n}{b^n}$. When a quotient (as a fraction) is raised to a given power, each member of the quotient must be raised to that power.

$$\left(\frac{2}{3}\right)^3 = \frac{2^3}{3^3} = \frac{8}{27}.$$

It should be noted here that a base with the exponent zero is equivalent to 1. Thus $10^0 = 1$; $3^0 = 1$; $1^0 = 1$. All the rules just stated for exponents apply to zero exponents. For example, $a^n a^0 = a^{n+0} = a^n$; $5^3 \times 5^0 = 5^{3+0} = 5^3 = 125$.

A base with a negative exponent is equal to the *reciprocal* of the base $\left(\dfrac{1}{\text{base}}\right)$ with the corresponding positive exponent. Thus $2^{-2} = \left(\dfrac{1}{2^2}\right)$. Negative exponents follow the rules for all exponents. Thus

$$a^{-5}a^3 = a^{-5+3} = a^{-2} = \frac{1}{a^2} \text{ and}$$

$$10^{-7} \times 10^5 = 10^{-7+5} = 10^{-2} = \frac{1}{10^2} = \frac{1}{100}.$$

Exponents can also occur in the form of fractions. Thus we have $a^{1/2}$, $a^{1/3}$, $a^{1/4}$, and so on: $a^{1/2}$ means the square root of a or $\sqrt{a}$; $a^{1/3}$ means the cube root of a, or $\sqrt[3]{a}$; $a^{1/4}$ means the fourth root of a, or $\sqrt[4]{a}$. The numerator in fractional exponents need not necessarily be 1. We frequently deal with exponents such as $\frac{2}{3}$ and $\frac{3}{5}$. In such cases, the numerator stands for the power of a base and the denominator for the root of a base. For example, $10^{2/3}$ is equal to $\sqrt[3]{10^2}$.

All fractional exponents, whether or not the numerator is 1, follow the rule for exponents. For example, $10^2 \times 10^{2/3} = 10^{2+(2/3)} = 10^{8/3} = \sqrt[3]{10^8}$

Expressing Very Large or Very Small Numbers

Exponents provide a convenient way of writing very large or very small numbers. We know that 1,000,000 is 10^6, the exponent 6 representing the number of zeros after 1. We could indicate 5,000,000 as $5 \times 1,000,000$, or 5×10^6. To write 5,270,000, we would multiply 1,000,000, or 10^6, by 5.27. The number would be written as 5.27×10^6. In other words, we can express a large number as the product of two numbers: the first a number between 1 and 10; the second, a power of 10.

The number 5,270,000 is not too formidable, and we can grasp it readily enough. But consider the problems that would arise if, in our calculations, we had to use a number such as 602,000,000,000,-000,000,000,000. It represents the number of molecules in 18 grams of water and is called Avogadro's number, after the early-19th-century Italian scientist Amadeo Avogadro, who worked out the value. It is used in a great many scientific calculations, but practically never in the long form. Instead, it is written as 6.02×10^{23}.

The 20th-century American mathematician Edward Kasner invented a new system of indicating extremely large numbers.

He coined the world "googol" to express the number 10^{100}, which would be equivalent to 1 followed by 100 zeros. He invented another term, the "googolplex," to stand for $10^{100^{100}}$, or the figure 1 followed by a googol of zeros, that is, 10,000 zeros.

Exponents can be used just as effectively to express very small numbers. Since a minus exponent indicates how many times the fraction $\frac{1}{\text{base}}$ is repeated as it is multiplied by itself, $10^{-3} = \frac{1}{10} \times \frac{1}{10} \times \frac{1}{10} = 0.001$. Note that the exponent 3, in 10^{-3}, represents the number of digits after the decimal point in the number 0.001. The number 0.005 could be written as 5×0.001, or 5×10^{-3}. Now consider a much smaller number. The wavelength of red light is 0.00000077 meters. We can write this number as 7.7×0.0000001 or 7.7×10^{-7} meters.

Writing a number as the product of a number between 1 and 10 and a power of 10 is called *scientific notation*. It is used widely by scientists and engineers.

Calculations with Logarithms

Exponents have also been put to work to simplify arithmetical calculations. Sup-

	TABLE I	
Number	Number expressed as power of 2	Exponent in preceding column
0.25	2^{-2}	-2
0.5	2^{-1}	-1
1	2^0	0
2	2^1	1
4	2^2	2
8	2^3	3
16	2^4	4
32	2^5	5
64	2^6	6
128	2^7	7
256	2^8	8
512	2^9	9
1,024	2^{10}	10

pose that we represent numbers as powers of 2. We know that $2^{-2} = \frac{1}{2^2} = \frac{1}{4}$; $2^{-1} = \frac{1}{2^1} = \frac{1}{2}$; $2^0 = 1$; $2^1 = 2$; $2^2 = 4$; $2^3 = 8$; $2^4 = 16$; and so on. Expressed as a power of 2, therefore, $\frac{1}{4}$, or .25, is 2^{-2}; $\frac{1}{2}$, or .5, is 2^{-1}; 1 is 2^0; 2 is 2^1; 4 is 2^2; 8 is 2^3; 16 is 2^4. Let us now make a table, (Table I), setting down (1) certain numbers, (2) these numbers expressed as powers of 2; (3) the exponents in question.

Consider the problem $0.25 \times 1{,}024$. The table shows that $0.25 = 2^{-2}$ and that $1{,}024 = 2^{10}$. The problem then becomes $2^{-2} \times 2^{10}$. Applying the first given law of exponents, we have $2^{-2} \times 2^{10} = 2^{-2+10} = 2^8$. Consulting the table, we find that $2^8 = 256$, which, then, is the answer to the problem $0.25 \times 1{,}024$. We have changed a problem in multiplication into a simple addition.

Let us take another problem: $1{,}024 \div 32$. Looking at the table, we see that 1,024 is 2^{10} and that 32 is 2^5. Applying the second law of exponents, we have $2^{10} \div 2^5 = 2^{10-5} = 2^5$. We now consult the table and find that 2^5 is equal to 32. This is the answer to $1{,}024 \div 32$. In this case, we have changed a problem in division into a simpler subtraction problem.

Our next problem is to raise 4 to the fifth power. In other words, we want to know what 4^5 would be. The table shows us that 4 is 2^2. From the third law of exponents, we know that $(2^2)^5 = 2^{2 \times 5} = 2^{10}$ which, according to the table, is 1,024. We have solved our problem by a single multiplication instead of multiplying $4 \times 4 \times 4 \times 4 \times 4$.

Suppose we wish to get the square root of 1,024. According to the table, 1,024 is 2^{10}. To get the square root of a given power, we divide the exponent indicating that power by 2. Hence the square root of $2^{10} = 2^{10} \div 2 = 2^5$. Dividing the exponent indicating the power by 2 is really in accordance with the third law of exponents. The square root of a number, as we have seen, is equivalent to the same number with the exponent 1/2. The square root of 2^{10}, therefore, can be expressed as $(2^{10})^{1/2}$. Remember that to multiply a number by 1/2 is the same thing as to divide it by 2. The table shows that $2^5 = 32$. So 32 is the square root of 1,024.

When a number is expressed as a power of a given base (in this case the base two), we call the exponent that indicates the power the *logarithm* of the number. All the exponents in the third column of the table are the logarithms, to the base two, of the numbers in the first column. When the base is two, -2 is the logarithm of 0.25. As a mathematician would put it, $\log_2 0.25 = -2$. Also when the base is two, the logarithm of 4 is 2 and the logarithm of 64 is 6. To multiply numbers, we first add their logarithms. To divide numbers, we first subtract their logarithms. To raise a number to a given power, we first multiply the logarithm of the number by the power in question. To obtain the root of a number, we first divide the logarithm of the number by the desired root. After we have added, subtracted, multiplied, or divided in this way, we find the number that corresponds to the resulting logarithm.

All the logarithms we have mentioned thus far are to the base two. Most tables of logarithms are given to the base ten. Let us now prepare another table (Table II), giving (1) a series of numbers; (2) the numbers expressed as powers of 10; and (3) the logarithms of the numbers, that is, the exponents when the numbers are expressed as powers of 10.

TABLE II		
Number	Number expressed as power of 10	Logarithm to the base ten (LOG_{10})
0.0001	10^{-4}	-4
0.001	10^{-3}	-3
0.01	10^{-2}	-2
0.1	10^{-1}	-1
1	10^0	0
10	10^1	1
100	10^2	2
1,000	10^3	3
10,000	10^4	4

TABLE III	
Number	Logarithm (base ten)
1	0
2	0.301
3	0.477
4	0.602
5	0.699
6	0.778
7	0.845
8	0.903
9	0.954
10	1

To solve the problem 0.0001×100, we consult the table and find the logarithms of 0.0001 and 100 (-4 and 2, respectively), add the logarithms ($-4 + 2 = -2$), and find the number corresponding to the logarithm -2. This number, as we see from the table, is 0.01. We can also do such problems as $10,000 \div 0.0001$; 10^4; and $\sqrt{10,000}$.

Of course, to be serviceable, a table of logarithms would have to include the logarithms of other numbers besides those in Table II. It would have to give, for example, not only the logarithms of 1 and 10, but also those of 2, 3, 4, 5, 6, 7, 8, and 9. We know that since $1 = 10^0$ and $10 = 10^1$, the logarithm of 2 would be between 0 and 1. Mathematicians have calculated that it is 0.301. This means that, expressed as a power of 10, the number 2 is $10^{0.301}$. The integer part of the logarithm (0 in this case) is called the *characteristic*. The decimal part (.301) is called the *mantissa*.

Here, we give only three decimal places for the sake of simplicity, but logarithms have been calculated to more than twenty places. Depending upon the accuracy desired, one would use a four-place table, or a five-place table, or a seven-place table, and so on.

The logarithms of the numbers from 1 through 10 have been displayed in Table III. Using the table, let us multiply 2 by 4. We add 0.301, the logarithm of 2, and 0.602, the logarithm of 4, and we get the logarithm 0.903. Consulting the table, we see that 0.903 is the logarithm of 8. A mathematician would say that 8 is the *antilogarithm*, or *antilog*, of 0.903. An antilogarithm is the number that corresponds to a given logarithm. Therefore, 8 is the answer to the problem 2×4. Let us now divide 9 by 3. The logarithm of 9, as we see from the table, is 0.954. The logarithm of 3 is 0.477. Subtracting 0.477 from 0.954, we get 0.477. The table shows that 0.477 is the logarithm of 3. Hence $9 \div 3 = 3$.

Our Table III gives the logarithms for only ten numbers, but mathematicians have prepared tables making it possible to find the logarithm of any number whatsoever. The tables give only the mantissas. We can determine the characteristic in each case by inspection. For example, the logarithm of the number 343 must be between 2 and 3, since $100 = 10^2$ and $1,000 = 10^3$. The logarithm, then, must be 2 and a fraction; so the characteristic must be 2. If we look up a five-place table in order to find the mantissa of 343, we observe that it is .53529. Putting together the characteristic 2 and the mantissa .53529, we have the logarithm 2.53529.

In calculations involving arithmetical problems, we can often save a tremendous amount of time by consulting a table of logarithms. Of course we would not use logarithms to get the answer to 4×5 or $72 \div 9$. But suppose we had to perform the various operations in the following:

$$\frac{-2.953 \times 5.913^5 \times \sqrt{5.973}}{49.743 \times 0.35947^3}$$

If the methods of arithmetic were used, this would be a most laborious task. It could be done in a few minutes if we employed logarithms.

Logarithms to the base ten are called *common logarithms*. In the so-called *natural logarithms*, the base is 2.71828..., generally indicated by the letter *e*. Natural logarithms serve widely in various types of higher analysis because they lead to comparatively simple formulas.

Algebraic Sequences and Series

Many events seem to recur in regular sequences. The sun "rises" every day. The planets revolve in their orbits around the sun so regularly that astronomers can calculate their positions years in advance. People have analyzed periodic happenings by means of algebraic *sequences* and *series*. A sequence is a succession of numbers. A series is a sum of numbers in a sequence. The results of these analyses are sometimes used to predict future happenings.

Arithmetic progression, also called *arithmetic series*, is a sequence in which each term, after the first one, is formed by adding a constant quantity to the preceding term. An example of such a sequence is 1, 3, 5, 7, 9, 11, 13, 15, in which 2 is added to each succeeding number of the sequence. If a stands for the first number, d for the constantly added number, and n for the total number of terms, we can represent the arithmetic sequence algebraically by this formula:

$$a, (a + d), (a + 2d), \ldots, [a + (n - 1)d]$$

Here, $[a + (n - 1)d]$ is the nth term.

If we add n terms together, the sum (s) of the terms is called a series and can be expressed by the following formula:

$$s = \frac{n}{2}[2a + (n - 1)d]$$

Let us apply this formula to the sum of the terms in the sequence 1, 3, 5, 7, 9, 11, 13, 15. There are 8 terms in all. The first term is 1. The quantity that is constantly added is 2. Substituting these values for n, a, and d, we get the following:

$$s = \frac{8}{2}\{(2 \times 1) + [(8 - 1) \times 2]\}$$

If we work out the arithmetic involved, we find that $s = 64$, which can be verified by adding the eight terms of the sequence.

Arithmetic progression is very useful in various types of calculations. It serves, among other things, in finding the total cost of an item bought on an installment plan. Suppose you buy a piano for $1,000. You pay $400 down and agree to pay the other $600 in 20 monthly installments of $30 each, plus the *interest* at 6 percent on the unpaid balance. Let us apply the arithmetic series to the problem in order to determine the total interest payments that will be required.

The 6 percent interest means "6 percent yearly" and is calculated by a factor of 0.06. The installment period is a month, or 1/12 of a year. The first of these interest payments is 1/12 × 0.06 × $600 (the unpaid balance), or $3.00. Each month the interest is less than in the preceding month since the unpaid balance is reduced by $30. You would pay 1/12 × 0.06 × $30, or $0.15 less interest than the month before. The consecutive interest payments, therefore, would be $3.00 (for the first month), $2.85 (for the second month), $2.70 (for the third month), and so on, until the 20 installments would be paid. Going back to the formula for the sum of an arithmetic sequence, we see that n (the number of terms) is 20; a (the first term) is $3.00; d (the constant addend) is $-$0.15. Making the appropriate substitutions in the formula, we get the following:

$$s = \frac{20}{2} \times \{(2 \times \$3.00) + [(20 - 1) \times -\$0.15]\}$$

The answer, representing the total interest paid, is $31.50.

Geometric progression is a sequence in which each term, after the first one, is formed by multiplying the preceding term by a fixed quantity. A typical geometric sequence is 1, 2, 4, 8, 16, in which the fixed multiplier is 2. Algebraically, geometric progression can be represented by

$$a, ar, ar^2, \ldots, ar^{n-1}$$

where a is the first term, r the constant multiplier, and n the number of terms. The sum of n terms of a goemetric sequence—a sum called a *geometric series*—is given by the following formula:

$$s = \frac{ar^n - a}{r - 1}$$

Applying it to the progression 1, 2, 4, 8, 16, the solution is as follows:

$$s = \frac{(1 \times 2^5) - 1}{2 - 1} = 31$$

If you were to add $1 + 2 + 4 + 8 + 16$, you would arrive at the same sum, 31. The formula, therefore, is a simple means by which to add the terms in any geometric sequence.

The geometric series plays an important part in the mathematics of finance. It is used, among other things, in figuring *compound interest*. Suppose that we put $100.00 (the *principal*) in a bank and that the *interest* is 3 percent, compounded annually. Interest is said to be compounded when it applies, not to the principal alone, but to the principal plus the interest that has been periodically added to the principal. To calculate the 3-percent interest on the $100.00 principal for the first year, we can multiply $100.00 by 0.03 to get $3.00. Our total, therefore, at the end of year one is $100.00 + $3.00 = $103.00. The two calculation steps just used are equivalent to this one step: $100.00 × 1.03 = $103.00. That is because "1.03" represents "100 percent + 3 percent."

For the second year we would receive 3-percent interest on our $103.00, which can be expressed as ($100.00 × 1.03) × 1.03, or $100.00 × 1.03^2. Following the same procedure, we would have $100.00 × 1.03^3 at the end of the third year, $100.00 × 1.03^4 at the end of the fourth year, and so on, representing the following geometric sequence in which 1.03 is the fixed multiplier:

$100.00, $103.00, $106.09, $109.27 . . .

A *binomial sequence* is the expansion of the power of a *binomial*. A binomial consists of two terms connected by a plus or minus sign. The expressions $a + b$, $2x + z$, and $x^2 - y^2$ are binomials. Using our knowledge of exponents, as well as the laws of multiplication, we know that $(a + b)^0 = 1$, $(a + b)^1 = (a + b)$, $(a + b)^2 = a^2 + 2ab + b^2$, and so on. If we continue to expand $(a + b)$ by a power of one, we can arrange the resulting sums in a triangular pattern, as shown in Figure 5.

Figure 6 is a modification of that triangle, and arranged within it are only the *coefficients* of the sums in Figure 5. (The coefficients are the numerals that precede the unknowns *a* and *b* as multipliers.) There is a distinct pattern of numerical progression within this arrangement of binomial coefficients. As you will notice, each coefficient is equal to the sum of the two coefficients immediately above it. For example, 15 (of the bottom row) is equal to 10 + 5 (of the next row up). There is actually no "bottom" to this "binomial triangle"; the power of $(a + b)$ can be

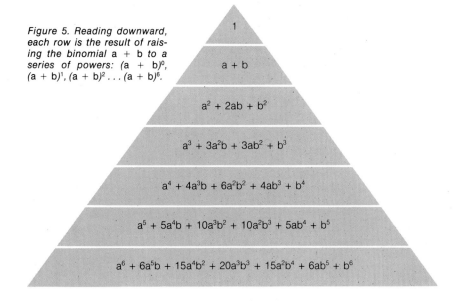

Figure 5. Reading downward, each row is the result of raising the binomial a + b to a series of powers: $(a + b)^0$, $(a + b)^1$, $(a + b)^2$ ... $(a + b)^6$.

expanded endlessly, as can the progressive pattern of the triangle.

Practical Applications

Let us now refer to Figure 7, where we have put a series of pegs in a shallow box, positioned to imitate the placement of numbers in the binomial triangle (Figure 6). The pegs are just far enough apart so that a small disk will be able to pass between them. We cut out a section of the box's top, as shown; and we built short walls to take the place of the bottom-row pegs, so as to form a series of compartments.

We keep the box in a tilted position so that when a disk is dropped through the gap at the top, it will make its way through the maze of pegs to one of the bottom compartments. Now we drop 64 disks one by one into the box through the gap. We can expect that half of the disks that hit a peg will fall to the left of it and the other half to the right. Hence 32 disks should drop to the left of peg A and 32 to the right of peg A. Of the 32 that fall to the left and strike peg B, 16 should fall to the left of B and should strike peg D; 16 should fall to the right of peg B and should hit peg E. Of the 32 disks that hit peg C, 16 should strike peg E and 16 should strike peg F. That means that 32 disks in all will hit peg E. We can indicate how the disks should fall on their way to the bottom compartments by the diagram in Figure 8.

As we have seen, 32 disks should strike peg B in the second row and 32 should strike peg C. The ratio (comparison of amount) of 32 and 32 is 1-1. In the third row, 16 should strike D; 32 should strike E, and 16 should strike F. The ratio of 16, 32, and 16, is 1-2-1. Going down the rows, the ratios of the disks striking the different pegs would be 1-3-3-1 in the fourth row, 1-4-6-4-1 in the fifth row, and 1-5-10-10-5-1 in the sixth. The disks in the seventh row would follow the distribution 1-6-15-20-15-6-1. Note that these ratios all correspond to the binomial coefficients of Figure 6.

In an actual experiment, the disks will not fall exactly as we have indicated. Some compartments will have one or two more than the number predicted. Others will have one or more less. Yet in every case, the result will be *nearly* that which was forecast. If the experiment is repeated over and over again, in a great number of trials, the number in each compartment will agree more and more closely with the expected number. Thus the coefficients of the binomial sequence provide an effective means for

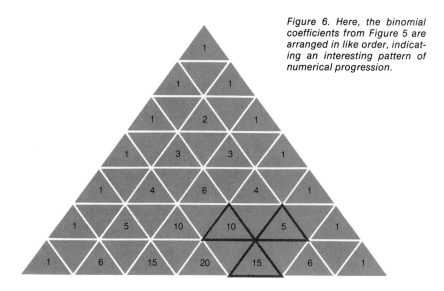

Figure 6. Here, the binomial coefficients from Figure 5 are arranged in like order, indicating an interesting pattern of numerical progression.

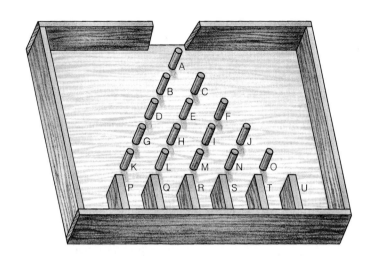

Figure 7. The pegs labeled A, B, C, D, E, and so on are positioned so as to imitate the placement of the numbers in Figure 6. The short walls replace the bottom-row pegs, so as to form compartments.

Figure 8. If we drop 64 disks, one by one, in the box shown in Figure 7, they will make their way down the pegs as indicated here. The ratios of the disks hitting the pegs in each row (right) reproduce the triangle in Figure 6.

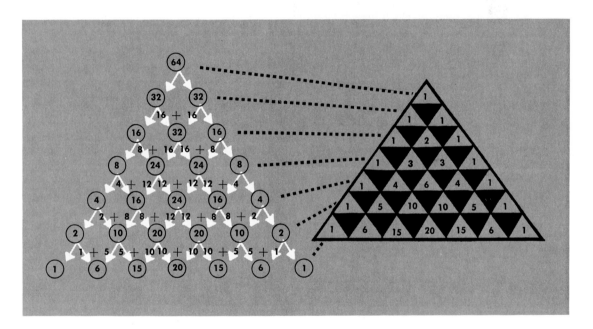

calculating *probabilities* or likelihoods, when the chances of an event occurring are even.

For determining probabilities when the chances are not even, other algebraic analyses have been made. The subject that deals with such analyses is called *statistics*, and its applications are both practical and far-reaching. An insurance company can, for example, use statistics to predict with fair accuracy life expectancies and how much in premiums and interest it will collect during the lifetime of an insured person. To make insurance work, the company must collect enough to be able to pay the insurance when the client dies. Probability and compound interest, as developed by algebra, are therefore the bases on which insurance is built. Thus algebra, which began by examining the relations of arithmetic operations, has become an interpreter of our experience and a guide for the future.

Howard F. Fehr

PLANE GEOMETRY

When we pass from arithmetic and algebra to geometry, we enter a world of shapes occurring in space—a world of points, lines, surfaces, and solids. We study the properties of these shapes and the relations between them. We learn to measure them. At the outset, geometry was used to solve specific problems, but in the course of its development it became a thoroughly abstruse subject. However, this abstruse branch of mathematics can often be put to practical use, as we shall see.

The beginnings of geometry go back far into prehistory. As the population of a given region grew, the natural dwelling places available did not suffice. It became necessary to build shelters, big enough to house families and strong enough to withstand winds, rain, and storms. To make a shelter the proper size, a person had to compare lengths. Thus the roof had to be higher above the ground than the top of the head of the tallest person.

The ancient Babylonians were pioneers in this branch of mathematics. The land between the Tigris and Euphrates rivers, where the Babylonians dwelt, was originally marshland. Canals were built to drain the marshes, and to catch the overflow of the rivers. For the purposes of canal construction, it was necessary to survey the land. In so doing, the Babylonians developed rules for finding areas. These rules were not exact, but the results they gave sufficed for canal construction.

In Egypt, the people who had farms along the banks of the Nile River were taxed according to their holdings. In the rainy season, the river would overflow its banks and spread over the land, washing away all landmarks. It became necessary, therefore, to remeasure the land so that each owner would have his rightful share. After the floods had subsided, specially trained men, called "rope-stretchers," would establish new landmarks. They would use ropes knotted at equal intervals so that they could measure out desired lengths and divide the land into triangles, rectangles, and trapezoids.

They devised practical rules for the areas of these figures. The rules were of the rough-and-ready variety and were often inexact. We know today, for example, that the area of any triangle is one-half the product of its *altitude* (height) and its base. The Egyptians erroneously gave this area as one-half the product of the base and a side. However, most of the triangles used in their surveying work were long and narrow (Figure 1), and in such triangles there is not too much difference in length between the long side and the altitude. Hence the results of the Egyptians' calculations served as a pretty fair basis for the allotting of land and the taxation of landowners.

Geometry Becomes a Discipline

The Greeks called the early Egyptian surveyors *geometers*, or "earth-measurers" (from the Greek *ge:* "earth," and *metria:* "measurement"). The geometers

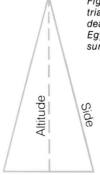

Figure 1. In this long and narrow triangle, each side is not a great deal longer than the altitude. The Egyptians used such triangles in surveying.

found out many facts about triangles, squares, rectangles, and even circles. These facts became a body of knowledge that the Greeks called *geometry,* or "the study of the measurement of the earth." Geometry today involves much more than it did at that early stage; yet it is still concerned with the sizes, shapes, and positions of things.

The Greeks made important advances in the field of geometry. They not only corrected many of the faulty rules of the Egyptians, but also studied the different geometrical figures in order to work out relationships. Thales, a Greek mathematician who lived 2,500 years ago, discovered that no matter what diameter is drawn in a circle, it always *bisects* the circle, that is, cuts it into two halves (Figure 2). He also noticed that if two straight lines cross each other, the opposite angles are always equal, no matter at what angle the lines cross (Figure 3). This was the beginning of the study of figures for the sake of discovering their properties rather than for practical use. The Greeks changed geometry from the study of land measurement to the study of the relations between different parts of the figures existing in space, which is what geometry means today.

After Thales, other Greek mathematicians discovered and proved facts about geometric figures. They also devised various instruments for drawing figures. By custom, the only instruments allowed in the formal study of geometry were an unmarked *straightedge,* (ruler) for drawing straight lines and a *compass* for drawing circles and transferring measurements (Figure 4).

The Greeks proposed various construction problems, to be solved with only the straightedge and the compass. Among these problems were the following: **(1)** *squaring the circle,* or constructing a square whose area exactly equals that of a given circle; **(2)** *duplicating the cube,* or constructing a cube whose volume will be exactly twice the volume of a given cube; and **(3)** *trisecting the angle,* or constructing an angle equal to exactly one-third of a given angle. For over 22 centuries, mathe-

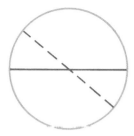

Figure 2. Each of the two diameters shown in the drawing cuts the circle into halves.

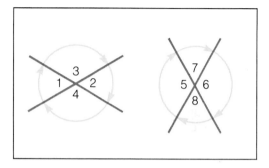

Figure 3. If two lines cross, the opposite angles are always equal. In the examples shown, angle 1 = angle 2; angle 3 = angle 4; angle 5 = angle 6; and angle 7 = angle 8.

Figure 4. By custom, the ancient Greeks used only an unmarked straightedge and compass in presenting and solving geometry problems.

maticians attempted to solve these problems, without success. Finally, in the 19th century, it was proved that it is impossible to square the circle, duplicate the cube, or trisect an angle if one uses only the straightedge and compass. It is possible, however, to make these three constructions with specially designed instruments. Such constructions fall in the domain of higher geometry.

Euclid: The Organizer of Geometry

By the 4th century B.C., there had grown up a vast body of facts concerning geometric figures, but for the most part these facts were unrelated. There were many theorems about triangles and circles, some about similar figures and areas, but no orderly arrangement. The learned Greek mathematician Euclid, who taught at the Museum of Alexandria, in Egypt, about 300 B.C., was the first man to apply a logical development to the mathematical knowledge of his time. He presented this development in his *Elements of Geometry*.

Euclid realized that it is not possible to prove every single thing we say and that we must take certain things for granted. He assumed that everybody knows and uses properly such words as "between," "on," "point," and "line"; hence it is not necessary to define them. He used these undefined terms to give definitions of various figures. Thus he defined a circle as "the set of all points that lie the same distance from a fixed point called the center." Again, Euclid noted that one cannot prove certain statements of relations between geometric figures; for example, "Only one line can be drawn between two points." (In geometry, the term "line" implies a "straight line that extends without end in both directions.") Euclid called such statements "common notions." Today we call them *postulates*.

Euclid used undefined terms, definitions, and postulates to prove *theorems* about geometric figures. A theorem is a statement that gives certain facts about a figure and that concludes from these facts that a certain other fact must be true. A typical theorem is "If two sides of a triangle are equal, the angles opposite these sides must be equal" (Figure 5). This theorem states two facts: There is a triangle and two sides of the triangle are equal. It then draws the conclusion that two of the angles of the triangle are equal. Once a theorem is proved, it can be used to prove other theorems.

Euclid built up a logical chain of theorems that introduced order in what had been a chaos of more or less unrelated facts. Besides organizing a vast body of knowledge about geometric figures, he introduced a method of treatment that became a model for the development of other branches of mathematics and pure science. This method is as valid today as ever.

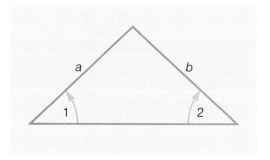

Figure 5. If two sides of a triangle are equal, the angles opposite these sides are equal; that is, if side a = side b, then angle 1 = angle 2.

Two Dimensions

The first branch of geometry we shall consider is *plane geometry*—the study of points, lines, and figures occurring in planes. Just what do we mean by these terms?

A *point* is the simplest element in geometry. It has no length, width, or thickness, which is another way of saying that it has no *dimensions*. We can represent a point by a dot, made with a pencil or a piece of chalk. Such a dot is not a geometric point but a physical point, since it has length, width, and thickness, however small these dimensions may be. In geometric constructions, we have to use physical points, such as pencil dots, to represent geometric points, because it would be impossible for us to set down on paper a point with no dimensions.

A *line*, like a point, is considered an undefined term because it can be represented (as in a drawing) but can be described only in relation to other geometric elements. If we consider a single point and ask how many lines can pass through it, the answer would be "an *infinite* (unlimited) number." But if we ask how many lines can pass through two distinct points, the answer will be, as Euclid determined, "exactly one line." The shortest distance *between* any two points is a *line segment*. A line segment is a portion of a line that is bounded by two points, or *endpoints*; it has only one dimension: *length*. It does not have width and thickness; hence, when we draw a line segment in constructing a geometric figure, we are again giving a physical representation of a geometric element.

If we were confined to a world having only one dimension, such as length, we would have a rather dull time of it. We would be points on a line, being able to move only forward and backward and always bumping into points ahead of us or behind us. In Figure 6, points A and B and segment CD are all parts of the line shown in the figure and must always stay within the line.

Suppose now that we selected a point P outside the line. Line segments drawn from point P to the original line create a series of figures existing in a *plane* (Figure 7). A plane is a surface having the two dimensions of length and width. The surface of a tabletop is a plane. A continuation of the surface would represent part of the same plane. If we were points in a two-dimensional world, we could move freely in any direction, except out of the plane. Our world would have other points like ourselves, and also lines. There would also be a great variety of figures made up of combinations of points and lines—figures such as triangles, squares, circles, and so on.

Angles in Plane Geometry

The name *ray* is given to the part of a line that starts at a given point. A plane figure formed by two rays having the same starting point is called an *angle*. In Figure 8, AB and BC are two rays with the same starting point, B. The angle formed by the two rays is ABC. You will note that in the expression "angle ABC," the "B" is between "A" and "C," indicating that "B" represents the point that the two rays share. That is how angles are always indicated.

If two lines meet so that all the angles formed are equal, the lines are said to be *perpendicular* and the angles are called *right angles*. In Figure 9, line AB is perpendicular to line CD and the four angles (AEC, BEC, AED, and BED) are all equal. If we draw a circle about point E, its length, called its *circumference,* can be divided into 360 equal units, called *degrees* and

Figure 6. Points A and B and segment CD are parts of this line and must always stay within the line.

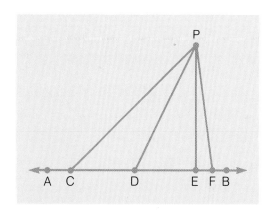

Figure 7. Lines drawn from point P to the line AB create a number of figures (such as PCD) all of which lie in a single plane.

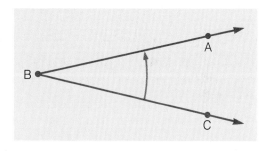

Figure 8. AB and BC are two rays having the same starting point, B. They form angle ABC.

written with the symbol °. The parts of the circle labeled IG, GH, HF, and FI are called *arcs*. (An arc is simply a portion of a circle's circumference.) Each of these arcs has 90° since the circumference of the circle is divided into four equal parts by lines AB and CD. The angle at the center of the circle has the same number of degrees as the arc it cuts off on the circle. Hence each of the four angles here has 90°; in other words, a right angle has a measurement of 90°.

If an angle is less than a right angle (that is, if it has less than 90°) it is called *acute*. It is *obtuse* if it is greater than a right angle (that is, if it has more than 90°). When

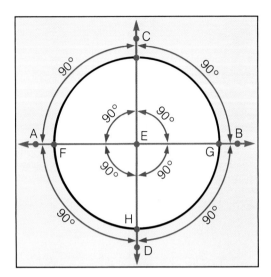

Figure 9. Lines AB and CD are perpendicular to each other; the angles at E are right angles, each having 90°. The arcs cut off on the circle by the intersection of AB and CD each have 90°.

The Study of Triangles

When three line segments connect three points in a plane, they form a *triangle*. Figure 12 shows examples of the three different kinds of triangles that can be made. There are literally thousands of theorems about the sides, angles, and lines in triangles.

One of the first theorems proved in plane geometry is "If three definite lengths are given, such that the sum of any two

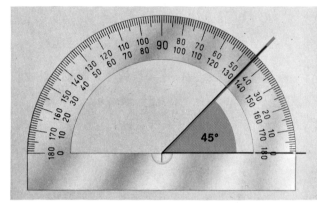

Figure 11. Angles are measured by a protractor.

the obtuse angle becomes so large that its sides form a straight line, it is a *straight angle* and has 180°. An angle larger than a straight angle (that is, more than 180°) is called a *reflex angle*. Figure 10 shows these different kinds of angles. Angles can be measured by the instrument called the *protractor*. It consists of a *semicircle* (half a circle) divided into 180 parts, each part representing one degree of angle at the center (Figure 11). As we shall see, angles play an all-important part in the study of geometry.

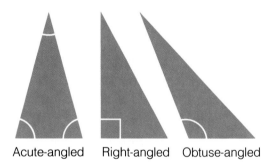

Acute-angled   Right-angled   Obtuse-angled

Figure 12. Three kinds of triangles.

Figure 13. If any two of the line segments 1, 2, and 3 are added, their sum will be greater than the third. Therefore, segments 1, 2, and 3 can be joined to form a distinct triangle.

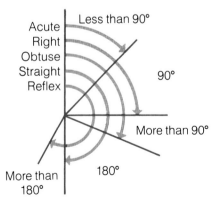

Figure 10. Angles are classified into groups and are named according to size, from acute to reflex.

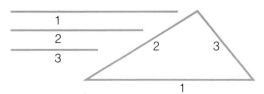

370   PLANE GEOMETRY

lengths is greater than the third length, it is possible to use the lengths in making a triangle that will have a definite size and shape" (Figure 13). Since the shape never varies, a construction built in the form of a triangle will be rigid and will not "give." Because of this property, interlocking triangles are used in bridge and building designs to prevent structural collapses. A figure of four sides, each of a definite length, could have many different shapes. As shown in Figure 14, such a construction would be collapsible. Hence it cannot be used in rigid construction unless it is braced by a *diagonal* (see Figure 15). Each diagonal makes two triangles of a four-sided figure, and each of these triangles is rigid.

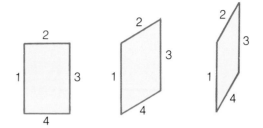

Figure 14. In these four-sided figures, the sides marked 1 are all equal; so are the sides marked 2, those marked 3, and those marked 4.

Figure 15. The four-sided construction is braced with diagonal BD, making two rigid triangles.

The most famous and perhaps the most important theorem in plane geometry is one dealing with a *right triangle* (a triangle with a right angle). It is called the *Pythagorean theorem*, after its discoverer, the Greek philosopher Pythagoras, who lived in the 6th century B.C. This theorem states that "in a right triangle, the sum of the squares of the *legs* (the two sides that form the right angle) is equal to the square of the *hypotenuse* (the side opposite the right angle)." In right triangle ABC in Figure 16, the sides a, b, and c are 3, 4, and 5 units, respectively, side c being the hypotenuse. According to the Pythagorean theorem, $a^2 + b^2 = c^2$; in this case, $3^2 + 4^2 = 5^2$, or $9 + 16 = 25$. We verify this by constructing, from each side of the triangle, an actual square, the sides of which are equivalent in unit length to the corresponding triangle side. Therefore, as shown in Figure 16, square ABED, for instance, consists of the number of square units represented by a^2. So the sum of square units in square ABED (9) added to those in square BCGF (16) should, and does, equal the number of square units in square ACHI (25).

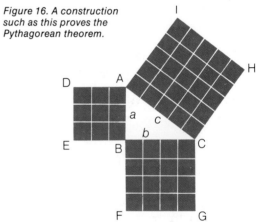

Figure 16. A construction such as this proves the Pythagorean theorem.

Figure 17 shows how the squares on the legs of a right triangle can be cut up so as to form the square on the hypotenuse. This is another confirmation of the Pythagorean theorem.

It follows from this theorem that if a triangle has sides such that the sum of the squares of the two shorter sides is equal to

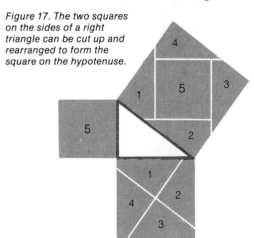

Figure 17. The two squares on the sides of a right triangle can be cut up and rearranged to form the square on the hypotenuse.

PLANE GEOMETRY

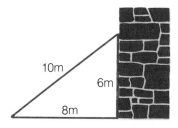

Figure 18. If the wall is perpendicular to the floor, the triangle will fit snugly.

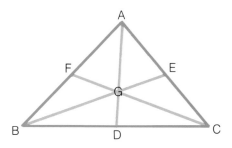

Figure 19. Here, D, E, and F are the midpoints of sides BC, AC, and AB. Note that the three lines CF, BE, and AD meet at G.

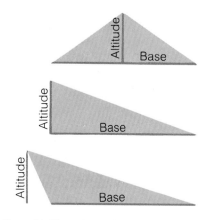

Figure 20. These three triangles have equal bases and altitudes. Hence the areas of the triangles are also equal.

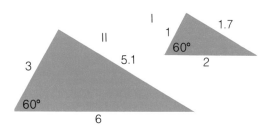

Figure 21. Triangles with equal corresponding angles are similar.

the square of the longest side, the angle opposite the longest side is a right angle. This theorem has various practical applications. The carpenter uses it to see whether a wall is perpendicular to the floor. If, for example, boards of 6, 8, and 10 meters are joined together, as shown in Figure 18, the angle formed by the 6- and 8-meter lengths must be a right angle, since $6^2 + 8^2 = 10^2$, or $36 + 64 = 100$. If the wall is truly perpendicular to the floor, the triangle will fit snugly.

The Pythagorean theorem is one of many that reveal an unexpected and important relationship. Here is another instance of such a theorem: In triangle ABC in Figure 19, D, E, and F are the *midpoints* of sides BC, AC, and AB, respectively. Points A, B, and C are the *vertexes* of the triangle. (A vertex is a point at which two sides of a plane figure intersect.) If we connect vertex A to the midpoint, D, of the opposite side, BC, the line segment AD is called a *median*. When we draw the two other medians, CF and BE, we learn that these three medians pass through the same point G, inside the triangle. By measuring, we would also learn that point G, on any of the medians, is two-thirds the distance from the vertex to the opposite side. Plane geometry offers proofs of these statements. Here is another interesting fact about G: If we cut out a triangle of cardboard and draw the three medians, as in Figure 19, we can balance the triangle on the blunt end of a pencil if we put this end directly under G. This point is called the *center of gravity*.

Equal and Similar Triangles

When triangles have the same size, or area, they are called *equal triangles*. All triangles with equal bases and altitudes are equal although they may have many different shapes (Figure 20). Some triangles have the same shape, but are different in size. They are known as *similar triangles*. The corresponding angles of similar triangles are equal and their corresponding sides are always in the same ratio. The two triangles in Figure 21 are similar because the corresponding angles are equal. Each side of triangle II is 3 times as great as the corresponding side of triangle I.

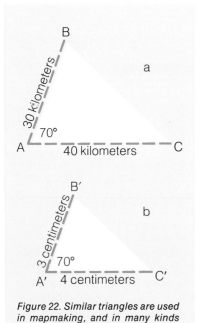

Figure 22. Similar triangles are used in mapmaking, and in many kinds of drawing to scale.

Figure 23. If we double the length of the sides of triangle X, producing triangle Y, the area of Y will be four times as great as the area of X. The area of triangle Z is nine times as great as that of X.

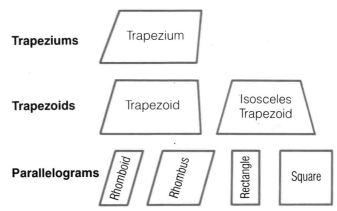

Figure 24. The three basic types of quadrilaterals, or four-sided figures.

Similar triangles are used in drawing to scale. In making a map, for example, we represent a large area of land on a small piece of paper. Suppose site A in Figure 22a is 30 kilometers from site B and 40 kilometers from site C and that the angle CAB is 70°. We are to show these sites on a map, where the scale is to be 1 centimeter = 10 kilometers. First, using a protractor, we draw a 70° angle. To reduce the 40-kilometer distance between A and C to our given scale, we divide 40 kilometers by 10 kilometers/centimeter to get a distance of 4 centimeters, which is the length we draw on our map from point A′, along the corresponding ray of our 70° angle, to endpoint C′. We follow the same procedure for the distance from A to B, which is reduced from 30 kilometers to 3 centimeters and is measured between A′ and B′ along the other ray of the 70° angle. Joining points B′ and C′, we have a proportional representation (Figure 22b) of the triangular area formed by the three sites.

Accurate maps, models, and photographs are similar to the original objects they represent. Hence the angles in such representations are exactly the same as in the originals, and all lines are changed in the same ratio. Their areas also will have a definite ratio, ratios that are the squares of the side-length ratios. For example, if we double each of the sides of a triangle, the area will be 4 times as great. If we triple each of the sides, the area will be 9 times as great. Figure 23 shows that this is so.

If photographic film 1-centimeter square is projected on a screen so that the picture on the screen is 40-centimeters square, the projection is 40^2, or 1,600, times as large as the original. The light used in projecting the film must cover 1,600 times as much area as the film; hence its intensity on the screen is only 1/1,600 as great as the film. This is a striking illustration of the manner in which the geometry of similar figures can be applied to the study of photographic phenomena.

Quadrilaterals

A figure with four sides is called a *quadrilateral*. As shown in Figure 24, there

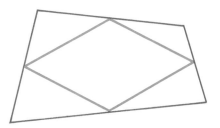

Figure 25. In any quadrilateral, the inner figure formed by joining the midpoints of the sides is a parallelogram.

Figure 27. Parallel rulers.

are three basic types of quadrilaterals: the *trapezium*, which has no parallel sides; the *trapezoid*, which has exactly one pair of parallel sides; and the *parallelogram*, with two pairs of parallel sides. The *isosceles trapezoid* is a trapezoid whose two nonparallel sides, or legs, are equal. A *rhomboid* is a parallelogram with no right angles and with only opposite sides equal. A *rhombus* is a parallelogram with four equal sides. A *rectangle* is a parallelogram with four right angles. A *square* is a rectangle with four equal sides (which means a square is also defined as a rhombus with four right angles). In any quadrilateral, the connected midpoints of the sides will create an inner quadrilateral that is a parallelogram (Figure 25).

In a parallelogram, the diagonals bisect each other, no matter how we distort the figure (Figure 26). This is a good example of an *invariant,* a property of a figure that remains true under all distortions. Another invariant is "The opposite sides of a parallelogram are equal."

The draftsman makes use of this invariant in the instrument called the *parallel rulers* (Figure 27). It consists of two straightedges that are joined by two rods of equal length (AC and BD). This device is flexible; hence AB can be at varying distances from CD. However, no matter how AB is moved, it always remains parallel to CD. Using this device, a parallel to a given line can be drawn at any accessible point in a plane.

The study of triangles and quadrilaterals forms much of the subject matter of geometry. All other *polygons* (closed figures having three or more angles and therefore sides) can be divided into triangles and quadrilaterals by drawing diagonals from the vertexes of the polygon. This then means that the basic rules for determining the perimeter and area of triangles and quadrilaterals can be applied to the study of other polygons.

Figure 26. The corresponding sides of these parallelograms are equal. The diagonals of both bisect each other.

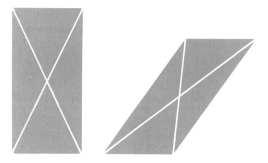

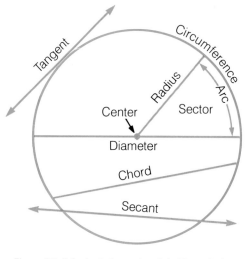

Figure 28. Principal elements related to a circle.

The Circle

A *circle* (see Figure 28) is defined as the set of points in a plane that are a given distance (the *radius*) from a given point (the *center*). A *chord* is a line segment that joins any two points on a circle. A *diameter* is a chord that passes through the circle's center and is actually two radii (the plural of "radius") that form a straight line. A *secant* is a line that *intersects* (or cuts off) a circle at two points. A tangent is a line that intersects a circle at exactly one point. A sector is a region bounded by an arc and the two radii that run to the arc's endpoints. To find the circumference of a circle, we use the formula $c = \pi d$, where c is the circumference, d is the diameter, and π represents a constant number approximately equal to 3.1416. To find a circle's area, we use $A = \pi r^2$, where A is the area and r is the radius.

Any diameter creates two semicircles. A simple and quite surprising theorem involving a semicircle is this: "If any point on a semicircle is joined to the ends of the diameter, an angle of 90° is formed at the point." In Figure 29, AB is the diameter and P is a point anywhere on the semicircle. It is easy to prove that angle APB is 90°: First connect P to the center of the circle (O). Line segments AO, PO, and OB are all radii of the circle and are, therefore, equal. In triangle APO, since AO = PO, the two angles marked $x°$ are equal because "if two sides of a triangle are equal, the opposite angles must also be equal." Likewise, in triangle OPB, PO and OB are equal, and so the angles marked $y°$ must be equal. Ignoring line segment PO, we have triangle APB, whose angles must total 180° because "the sum of the angles of a triangle is 180°." There are two $x°$ angles and two $y°$ angles in triangle APB; hence one $x°$ angle and one $y°$ angle must give half of 180°, or 90°. Since angle APB is composed of an $x°$ angle and a $y°$ angle, it must be equal to 90°; it must be a right angle.

There are various applications of this theorem. For example, a patternmaker can determine if the core box shown in Figure 30 is a true semicircle. He places a device called a "carpenter's square" in the box. If

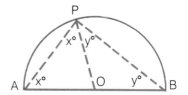

Figure 29. If point P on the semicircle is joined to the diameter at A and B, angle APB will be 90°.

Figure 30. Determining by means of a carpenter's square whether a core box gives a true semicircle.

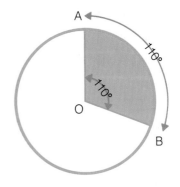

Figure 31. Both the angle at O and the arc that it intercepts (AB) are equal to 110°.

it makes firm contact at three points, as shown, he knows that the pattern will give a true semicircle.

An angle at the center of a circle has as many degrees as the arc that it intercepts on the circle: "A *central angle* is measured by its intercepted arc." In Figure 31, the obtuse angle at center point O is equal to 110°; the arc AB it intercepts is also equal to 110°.

An angle whose vertex is on the circumference of a circle and that intercepts an arc is called an *inscribed angle* (see

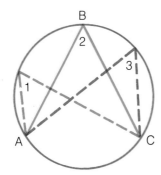

Figure 32. The angles 1, 2, and 3 intercept the same arc (AC) in this circle; therefore they are equal.

angle ABC in Figure 32). No matter where we place the vertex in arc AC's intercepting angle, the angle will remain the same size. In other words, "all inscribed angles intercepting the same arc are equal."

The Path of a Moving Point

It is often necessary in plane geometry to determine the path that a point describes in a plane when it moves according to a fixed rule. If, for example, a moving point must always remain 3 centimeters from a fixed point, it travels in a circle around the fixed point. The mathematician gives the name *locus* (Latin for "position") to the path described by a point. By studying the paths of moving points, we determine how machine parts move and how heavenly bodies appear to move.

We can illustrate the use of the locus by a very simple treasure-hunt problem, as illustrated in Figure 33. A treasure is reported to be buried in a spot *equidistant* (equally distant) from two intersecting roads and also 20 meters west of an oak tree. To find its location in relation to the tree, we think of the tree as the center of a circle with a 20-meter radius; the treasure lies somewhere on that circle's circumference. To find the treasure's location in relation to the two roads, we think of it as a moving point remaining at the same distance from each road. Because the roads intersect, forming an angle, we use the following theorem: "If a point moves so as to be equidistant from the rays of an angle, it will trace a line that bisects that angle." Hence the treasure must be on a line bisecting the angle made by the two roads. This line cuts the circle, as shown, at two places, T and G. The treasure must be at one of these two points. Of the two points, point T is the one west of the tree and must be the location of the treasure.

Machines that are designed to trace moving points are called *linkages* because they consist of linked bars. A compass, which is essentially a movable pair of hinged-together rigid legs, is the simplest linkage; the path it traces is a circle. Another linkage, called Peaucellier's Cell, changes circular motion into straight-line motion (Figure 34). As A in the figure moves around the circle, the point B moves up and down the straight line CD. There are other types of linkages that transform circular motion into linear motion. A study of linkage was necessary to help solve the problem of providing smooth motion in a

Figure 33. To find a treasure 20 meters from a tree and equally distant from two roads, use a locus.

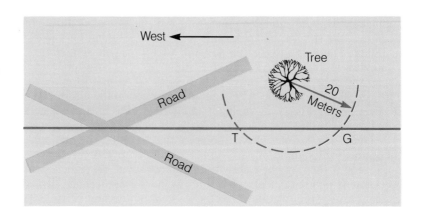

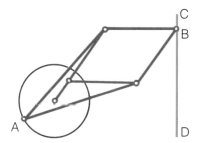

Figure 34. Peaucellier's Cell changes circular motion to straight-line motion. It is a type of linkage.

locomotive, where the straight-line motion of the drive shaft had to be converted into the circular motion of the wheels.

Conic Sections

The circle is the most common type of curve, but there are other kinds. The Greek geometers noticed very early that when a cone was cut by planes at different angles, the intersections gave different kinds of curves: circles, ellipses, parabolas, and hyperbolas collectively known as *conic sections* (Figure 35). The Greek mathematician Apollonius, who lived in the 3rd century B.C., wrote a treatise on the properties of these curves. In more recent times, it was discovered that they could also be defined as paths made in a plane by points moving according to certain rules. Such definitions are particularly meaningful when we put the curves to practical use.

An *ellipse* is the path traced by a point that moves so that the sum of its distances from two fixed points is always the same. The two fixed points are called *focuses,* or *foci*. It is easy to draw an ellipse, using the method illustrated in Figure 36. First we

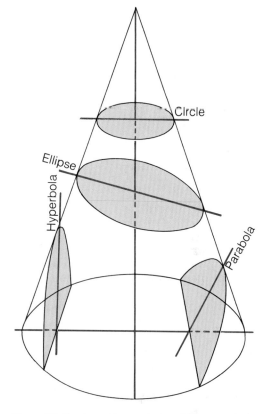

Figure 35. Conic sections—circle, ellipse, parabola, and hyperbola—produced as planes cut a cone.

insert thumbtacks at two fixed points, F_1 and F_2. We then take a piece of string that is larger than twice the distance between F_1 and F_2 and tie the ends together to make a loop. We attach the loop of string to the thumbtack at F_1 and to the thumbtack at F_2 (Figure 36). We draw the string taut with the point of a pencil (Figure 36), and as we move the pencil, its point will trace an

Figure 36. It is easy to draw an ellipse, using two thumbtacks and string, as shown here.

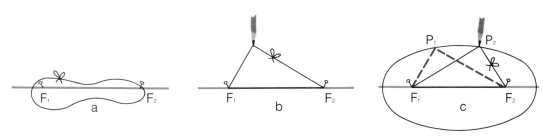

PLANE GEOMETRY 377

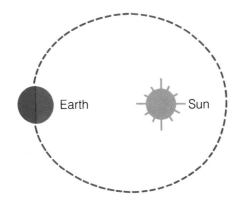

Figure 37. As the earth travels around the sun, its orbit is an ellipse, with the sun at one focus.

ellipse (Figure 36). The sum of the distances from the moving pencil point to the fixed points will remain constant: in Figure 36, $g + h = j + k$.

If a billiard table were elliptical in shape, any ball hit from one focus would rebound through the other focus. In an elliptical room, any sound issuing from one focus will be reflected by the wall to the other focus. This is the principle of the "whispering gallery." The elliptically shaped Mormon tabernacle in Salt Lake City, Utah, is an example. The foci in the tabernacle are clearly marked. A person standing at one focus can distinctly hear a whisper coming from a person at the other focus; those standing nearby hear nothing.

The ellipse has found many practical applications. Power punching machines use elliptical gears. At the narrow ends of the ellipse, the gears move faster, giving a quick return. At the flat parts, the gears move slower, exerting a greater force. Storage tanks and transportation tanks are made elliptical in cross section so as to lower the center of gravity and to lessen the danger of overturning.

The ellipse also serves to explain the movements of various heavenly bodies. All the planets move in elliptical orbits with the sun at one focus (Figure 38). A planet moves along the orbit so that the radius from the sun to the planet sweeps through equal areas in the same time. Knowing the elliptical orbit of any planet, astronomers can predict the position of the planet in its orbit at any time.

A *parabola* is the path of a point that moves so that its distance from a fixed line, called the *directrix,* always equals its distance from a fixed point, or focus. Thus, in Figure 38, as a point moves along the parabola, occupying positions P_1, P_2, P_3, and P_4 in turn, $a = b$, $c = d$, $e = f$, and $g = h$. A reflecting searchlight has a parabolic surface with the light source at the focus. All light beams emanating from the focus are reflected from the parabola in parallel rays (Figure 39). Sound detectors have parabolic surfaces. Sound waves are reflected

Figure 38. As a point moves along a parabola, its distance from the directrix equals its distance from the focus.

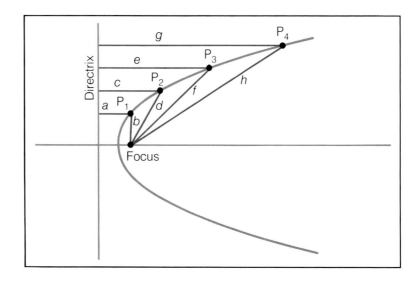

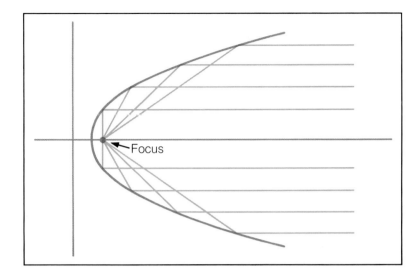

Figure 39. The light beams emanating from the focus of a searchlight are reflected from the parabolic surface in a series of parallel rays.

upon striking the surface and are concentrated at the focus. The mirror of a reflecting telescope is in the form of a parabola. Parallel rays of light from a distant heavenly body strike the parabolic surface and, reflected from it, meet at the focus within the telescope tube.

A *hyperbola* is the path of a point moving so that the distance to one fixed point minus the distance to another fixed point is always the same. The diagram in Figure 40 shows a hyperbola in which F_1 and F_2 are the fixed points, or foci, and P_1 and P_2 are points on the hyperbola. Therefore, $g - h = j - k$, which is equivalent to the distance between A_1 and A_2.

The hyperbola is applied to the Loran (*lo*ng *ra*nge *n*avigation) system by the use of radar. We can give a general explanation of Loran by referring again to Figure 40: Suppose there are two radar stations, F_1 and F_2, located on land 300 kilometers apart. Electric pulsations are sent out from each station. It takes longer for such pulsations to travel from F_1 to P_1 than from F_2 to P_1, and longer from F_1 to P_2 than from F_2 to P_2. The difference in time is a fixed constant for all points on the hyperbola. If the difference in time is greater or less, we have a different hyperbola.

A ship has a radar device that picks up the differences in time between the pulsations from the two stations. The navigator consults a map upon which are drawn the various hyperbolas corresponding to the various differences in time. He consults the map and locates his own ship on the hyperbola. The hyperbola passing through the "home port" is then picked out, and the difference in the time of the pulsations is noted. The course is changed, until the radar receiver indicates that the ship is on the home-port hyperbola. The difference in pulsations is kept constant and the ship sails home along the hyperbola.

Howard F. Fehr

Figure 40. Here, P_1 and P_2 are two positions of a point moving along a hyperbola. The distance to F_1 minus the distance to F_2 always equals the distance between A_1 and A_2.

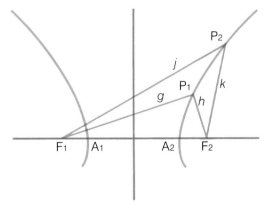

SOLID GEOMETRY

The two-dimensional world of plane geometry does not suffice to explain the world in which we live—a world of three dimensions. In it, there are many planes, which are boundless and extend in every conceivable direction. There are also many kinds of curved surfaces. We must consider not only north, south, east, and west but also up and down. To explain this three-dimensional world, the branch of mathematics called *solid geometry* has been developed. We use this kind of geometry in building machines, skyscrapers, airplanes, steamships, bridges, and automobiles and also in explaining the phenomena of the universe.

In solid geometry, there are many more possible relationships between geometric elements than in plane geometry. In a single plane, two lines are always either parallel or intersecting. In solid geometry, two lines may be parallel or intersecting, but they may also be *skew lines*. Skew lines are not in the same plane, are never parallel, and never intersect. Only one line, in a plane, can be drawn perpendicular to another line at a given point. In solid geometry, any number of such perpendicular lines can be drawn. For example, the spokes of a wheel are segments of lines, every one of which is perpendicular to the axle at the same point. In a plane, all points at a fixed distance from a fixed point are on a circle. In three-dimensional space, however, they are on a *sphere* containing an infinite number of circles passing through the center.

Angles in Solid Geometry

The simplest angle in solid geometry is called a *dihedral* ("two-faced") *angle*. It is formed by two intersecting planes. The size of this angle is measured by the *plane angle*. This is formed by two lines, one in each face, meeting the edge, or intersection of the two planes, at right angles (Figure 1). When an airplane *banks* (laterally tips), the angle of bank is a dihedral angle between the horizontal and tipped position of the wings (Figure 2). The dihedral angle is measured by an instrument in the airplane, and the size of this angle determines in part the speed with which the airplane will change its direction of travel.

When three planes meet at a point, they form a *trihedral angle* (Figure 3). Each of the angles making up a trihedral angle is called a *face angle*. In Figure 3, ADC, CDB, and ADB are all face angles. If more than three planes meet in a point, the angle is called a *polyhedral* (many-faced) *angle*. The sum of the face angles of a polyhedral angle must be less than 360°. As the sum of the face angles gets closer to 360°, the polyhedral angle becomes less pointed until, at 360°, it becomes a plane (Figure 4). The crystals of minerals show many kinds

Figure 1. Two intersecting planes form a dihedral angle. Angle ABC, between the planes, is called the plane angle.

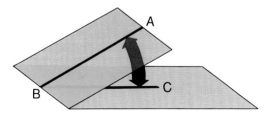

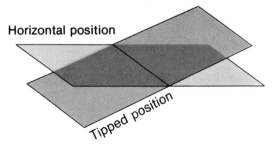

Figure 2. The dihedral angle between the horizontal and tipped positions of an aircraft's wing is the angle of bank.

of polyhedral angles. An analysis of these angles makes it possible to identify the various minerals.

Five Common Solids

The major part of the study of solid geometry is based on five common solids: the prism, the cylinder, the cone, the pyramid, and the sphere. (Much of the terminology used in solid geometry originates in the study of plane geometry; therefore, it may be helpful to review the article "Plane Geometry" before proceeding.)

The prism. In a prism, the bases, of which there are exactly two, are parallel and equal polygons. The number of sides to a prism's base is equal to the number of its side faces, or *lateral faces,* which are always parallelograms. When the lateral faces of any prism are perpendicular to the bases, that prism is called a *right prism* (Figure 5). A prism is generally named for the shape of its bases. For example, the bases of a *hexagonal prism* are two equal *hexagons* (six-sided figures) and, therefore, intersect with six lateral faces. A *square prism* has two square bases and four lateral faces. When the lateral faces of a square prism are also square, the prism is called a *cube* and all six faces are equal squares.

To find the area, or surface measure, of a prism, we must first calculate the area of each lateral face. Because any lateral face on a prism is a parallelogram, we use the formula for all parallelograms:

$$A = bh$$

where A is the area, b is the length of any base, and h is the height (or *altitude*) as measured perpendicular to b. The sum of the areas of the lateral faces is called the *lateral area,* and when it is added to the areas of the bases, the result is the *total area* of the prism. The interior of a room is often a rectangular prism. By calculating lateral and total areas, painters, wallpaperers, and remodelers can reasonably estimate the amounts of materials needed for any particular project.

The *volume,* or space-filling measure, of a prism is found by multiplying the area of the base by the altitude of the prism. This calculation is very important in the building of a house. Most builders estimate the construction cost as so much per *cubic unit* (the standard unit of volume). To estimate how much it will cost to build a particular house, we must first find the total volume of the prisms of which the house

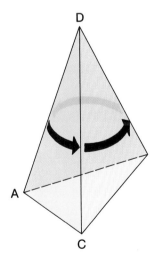

Figure 3. When three planes meet at a point, they form a trihedral angle.

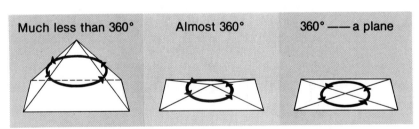

Figure 4. A series of polyhedral angles. As the sum of the face angles becomes greater, the polyhedral angle becomes less and less pointed. At 360° it is a plane.

Figure 5. The prisms shown here are known as right prisms because their bases are at right angles to their sides.

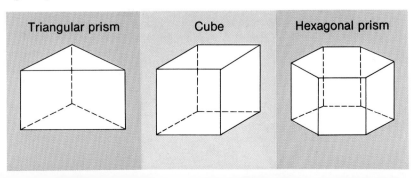

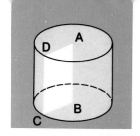

Figure 6. If rectangle ABCD is rotated about side AB, it will mark the boundaries of a cylinder, which will have AB as its axis.

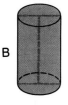
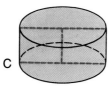

Figure 7. If the height of cylinder A is doubled and the base remains the same, as in B, the volume is doubled. If the height remains the same and the diameter is doubled, as in C, the volume of the cylinder is increased fourfold.

will consist. If the volume in question is 800 cubic meters and the builder gives an estimate of $60 per cubic meter, the cost will be approximately 800 × $60, or $48,000.

The cylinder. If one rotates a rectangle completely about one of its sides, as in Figure 6, it will *define* (mark the boundaries of) the solid called a cylinder. A cylinder is comprised of two flat, circular bases and one curved, lateral surface. The line segment that joins the centers of a cylinder's bases is called the *axis*. An ordinary soup can is a good example of a cylinder; it is specifically a *right cylinder,* in which the axis is perpendicular to the bases. The other classification of cylinders is the *oblique cylinder,* in which the axis is non-perpendicular to the bases.

The lateral area of a right cylinder is $2\pi rh$, in which π (pi) is approximately 3.1416, r is the radius of the base, and h is the height of the cylinder. Suppose that a canning company needs to know how much metal is required to make a particular size can. In other words, the company wants to know the total surface area of the can, which is determined by adding the lateral area to the areas of the bases (the area of a circle being πr^2).

The volume of a cylinder is found by multiplying the area of the base by the height of the cylinder, or $V = \pi r^2 h$. If the height of a cylinder is doubled and the diameter remains the same, its volume will also be doubled. You can see that this is so by placing one can on top of another just like it. If, however, the height of a cylinder remains the same and the diameter of the base is doubled, its volume will increase fourfold (Figure 7).

It is important to find the volume of a cylinder in computing the capacities of steel cans, gas-storage tanks, water reservoirs, and so on, and also in determining the rate of flow and pressure in pipes containing liquids. To see whether the economy size of a product sold in cans provides a real bargain, calculate the volume of the regular size and that of the economy size and then compare the two. Suppose the regular size can is 12 centimeters tall and has a base with a diameter of 8 centimeters. Suppose the economy-size can is also 12 centimeters tall and has a base with a diameter of 12 centimeters. The regular size costs $1.00, and the economy size $1.50. The problem is this: Will we save money if we buy the economy size? Knowing that $V = \pi r^2 h$, we find the volume of the regu-

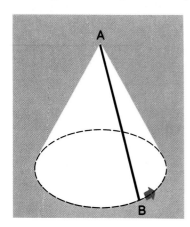

Figure 8. Point A is held fixed while point B follows a circular path. A cone is formed.

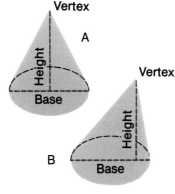

Figure 9. Cone A is a right cone. Cone B is an oblique cone.

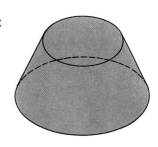

Figure 10. The frustum of a right cone.

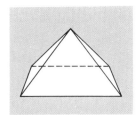

Figure 11. A pyramid with a square base is called a square pyramid.

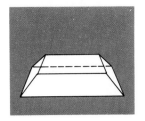

Figure 12. The frustum of a square pyramid.

lar can by substituting 4 for r (the radius is half the diameter) and 12 for h. Using 3.14 as the value of π, we get 3.14×16 square centimeters $\times 12$ centimeters $= 602.88$ cubic centimeters. The volume of the economy can is 3.14×36 square centimeters $\times 12$ centimeters $= 1{,}356.48$ cubic centimeters. By dividing 1,356.48 by 602.88 we learn that the economy can holds 2¼ times as much as the regular can. Since the economy can costs only 1½ times as much as the regular, it is, indeed, a bargain.

The cone. A cone is formed by holding one point of a line fixed and rotating the line, following a circular path (Figure 8). The line segment that joins the fixed point to the center of the circular base is the axis of the cone. If the axis is perpendicular to the base, the cone is a *right cone*; a nonperpendicular axis is part of an *oblique cone*. The height of any cone is the perpendicular distance from the vertex to the base. In a right cone, the height is equal to the length of the axis (Figure 9). The formula for the volume of a cone is $V = \frac{1}{3}\pi r^2 h$, where r is the radius of the base and h is the height of the cone.

When the top of a cone is cut off by a plane parallel to the base, the lower part is called a *frustum* (Figure 10). Many machine parts are in the form of cones or frustums of cones.

The pyramid. In the solid called the pyramid, the lateral faces are triangles whose vertices meet at a common point and whose bases form a polygon. Like a prism, a pyramid is generally named for the shape of its base (Figure 11). The Great Pyramid of Egypt is a *square pyramid*. So is the ancient Egyptian *obelisk* known as "Cleopatra's Needle." The formula for the volume of a pyramid is $V = \frac{1}{3}Bh$, where B is the area of the base and h is the height of the pyramid.

Like a cone, when the top of a pyramid is cut off by a plane parallel to the base, the lower part is called a frustum (Figure 12). Army squad tents and coal hoppers, among other things, have the shape of the frustum of a pyramid. To calculate the volume of a frustum, a rather complex formula is required. Yet there is evidence that the ancient Egyptians had an exact formula for making such a calculation. They used it in determining the amount of granite required to build sections of their pyramids.

The sphere. If a semicircle is rotated about a diameter (Figure 13), the solid defined is a sphere. When the sphere is cut by a plane, the intersection is a circle. Figure 14 shows various circles formed in this way. If the plane passes through the center of the sphere, the circle of intersection has the same radius as the radius of the sphere. Such a circle is called a *great circle*. All the other circles are *small circles*.

The earth may be considered a sphere in which the north and south poles are the ends of a diameter called the axis (Figure 15). The circles passing through both the north and south poles are great circles. They are known as *circles of longitude*. All the planes (except one) that cut the earth at right angles to the axis form small circles, called *circles of latitude*. There is just one plane that passes through the center of the earth and is at right angles to the axis. It forms the great circle called the *equator*.

Between any two points on the earth (not including the poles) only one great circle can be drawn. All other circles passing through the two points will be small circles. The shortest of all the arcs between the two points is the arc of the great circle (Figure 16). Pilots of aircraft, as far as possible, steer a course determined by the arc of a great circle between their starting points and destinations.

If we think of the earth as a rubber ball and cut this ball along one half of a circle of longitude, we can stretch the ball to form a flat rectangular sheet. The circles of longitude will then become parallel vertical lines and the circles of latitude parallel and equal horizontal lines (Figure 17). This sheet now represents a rectangular map of the world. A map of this type is called a *Mercator projection*, after the 16th-century

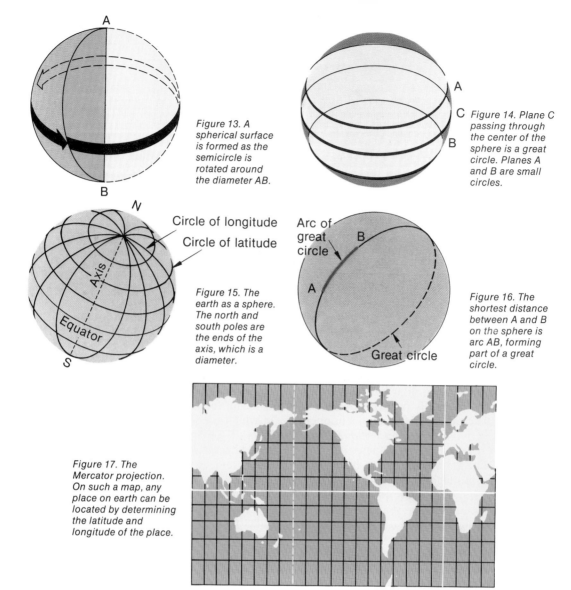

Figure 13. A spherical surface is formed as the semicircle is rotated around the diameter AB.

Figure 14. Plane C passing through the center of the sphere is a great circle. Planes A and B are small circles.

Figure 15. The earth as a sphere. The north and south poles are the ends of the axis, which is a diameter.

Figure 16. The shortest distance between A and B on the sphere is arc AB, forming part of a great circle.

Figure 17. The Mercator projection. On such a map, any place on earth can be located by determining the latitude and longitude of the place.

Flemish geographer Gerardus (or Gerhardus) Mercator, who developed it. The great longitudinal circle passing through Greenwich, England, is given the value of 0° longitude. The equator is given the value of 0° latitude. On such a map, any place on earth can be located by determining the longitude and latitude. Also on such a map, the farther we go from the equator, the more we find the original area stretched, so that land areas near the poles seem much larger on the map than they really are on the earth. Users of the map must take such distortions into account.

Figure 18. Three sets of similar solids. Here, the linear dimensions of the larger figures are exactly twice that of the smaller.

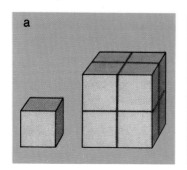
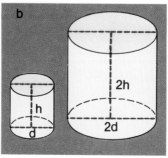
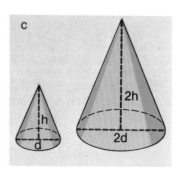

The surface area of a sphere is four times the area of a great circle; the formula is $4\pi r^2$. The earth's radius is approximately 6,380 kilometers; hence the total surface of the earth is $4 \times \pi \times (6{,}380 \text{ kilometers})^2$, or about 511,000,000 square kilometers.

The volume of a sphere can be expressed by the formula $(4/3)\pi r^3$. The volume of the earth, therefore, is $4/3 \times \pi \times (6{,}380 \text{ kilometers})^3$, or slightly over 1 trillion cubic kilometers.

Characteristics of Similar Solids

Solids that are of the same shape but of different sizes are said to be *similar*. The corresponding polyhedral angles of similar solids are equal, and the corresponding lines are in proportion.

The areas of similar solids have the same ratio as the squares of the corresponding linear parts. In each of the sets of similar solids in Figure 18, the surface area of the larger figure is 2^2, or 4, times the surface area of the smaller, because the linear parts are 2 times as large. If the linear parts were enlarged 3 times, the area would be 3^2, or 9, times as great.

The volume of similar solids has the same ratio as the cubes of the linear parts. In each of the sets of similar solids in Figure 18, the larger figure has 2^3, or 8, times the capacity of the smaller figure. In the case of the cubes in Figure 18a, you can count 8 small cubes in the larger cube.

The ratio of areas and volumes of similar solids has many practical applications. All spheres are similar. If oranges 8 centimeters in diameter sell for 30 cents a dozen, while oranges 10 centimeters in diameter sell for 50 cents a dozen, which would be the better buy? The volume ratio of a large orange to a small orange is as follows:

$$\left(\frac{10}{8}\right)^3 = \frac{1{,}000}{512} = 1.953$$

The large oranges are a better buy, then, because they have nearly twice the volume for less than twice the cost.

Use in Astronomy

Solid geometry has enabled astronomers to give a useful interpretation of the heavens and to calculate the distances and positions of the celestial bodies. The universe is conceived of as a huge *celestial sphere* with an infinitely great radius, which *appears* to revolve around the earth. Figure 19 gives a greatly simplified presentation of such a sphere as seen from the vantage point of a person stationed at latitude 50°. This person stands on a much smaller sphere, which of course is the earth. Directly overhead is the *zenith*; directly below is the *nadir*. The line where the sky seems to meet the earth is called the *horizon*. If the axis of the earth, which passes through the north and south poles, is extended, it will meet the outer bounds of our imaginary celestial sphere at the celestial poles—north and south. The line segment connecting the two celestial poles is the *celestial axis*. The plane of the earth's equator will cut the outer limits of the celestial sphere in a great circle called the *celestial equator*. A great circle passing through the poles and a star is the *hour circle* of the star. The *altitude circle* of the same star is a great circle passing through the zenith and the star.

These are but some of the features of the celestial sphere. They provide a frame of reference that enables the astronomer to trace the motions of celestial objects. This is one of the outstanding contributions of solid geometry to science.

Howard F. Fehr

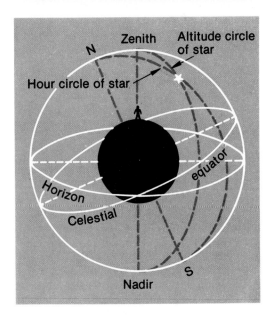

Figure 19. A simplified diagram showing the celestial sphere. Earth is the center of the sphere.

TRIGONOMETRY

An important offshoot of geometry is *trigonometry,* or triangle measurement. In trigonometry, when certain parts of triangles are known, one can determine the remaining parts and thus solve a great variety of problems.

The founder of trigonometry was the Greek astronomer Hipparchus, who lived in the 2nd century B.C. Hipparchus attempted to measure the size of the sun and moon and their distances from the earth. He felt the need for a type of mathematics that, by applying measurements made on the earth, would enable him to measure objects far out in space. He was thus led to the invention of trigonometry.

Boy Scouts have occasion to use trigonometry in their field world. A common problem that is put to them is to find the height of a tree. The scout first measures a distance, say 20 meters, from the base of the tree, as shown in Figure 1. This will be his *baseline.* At point A he measures the angle from the ground to the treetop by means of a *protractor.* Let us suppose that this angle is 35°. The scout now knows three facts about the triangle formed when he connects points B (the base of the tree), A (the end point of the line segment drawn from the tree), and C (the top of the tree): (1) side AB = 20 meters, (2) angle BAC = 35°, and (3) angle ABC = 90°. (We assume that the line segment representing the tree's height, BC, is perpendicular to the ground.)

On paper our scout now makes a triangle *similar* to the large one in the field. First he draws a line segment A′B′ ½ meter long, and at A′ he draws an angle of 35°— angle B′A′D′—with a protractor. Next he erects a perpendicular to segment A′B′ at B′. This line will intersect A′D′ at C′, and the angle A′B′C′ will be a right angle. The corresponding angles of the large and small triangles are equal: angle CAB = angle C′A′B′; angle ABC = A′B′C′; angle ACB = A′C′B′. Hence we have two similar triangles, and the corresponding sides will be proportionate.

The scout now measures side B′C′ and finds that it is 35 centimeters, or 0.35 meter. Since the sides of the similar triangles are in the same ratio, AB is to A′B′ as BC is to B′C′. We know all these quantities except BC, which we can call x. We now have the proportion 20 is to 0.5 as x is to 0.35, which we can write as 20 : 0.5 :: x : 0.35. In any proportion, the product of the *extremes* (the two outer terms) is equal to the product of the *means* (the two inner terms); hence:

$$0.5x = 20 \times 0.35$$
$$0.5x = 7$$
$$x = 14$$

Figure 1. The problem is to find the height of the tree when the angle at A (35°) and the distance AB (20 meters) are known. We show how to solve the problem by means of similar triangles ABC and A′B′C′.

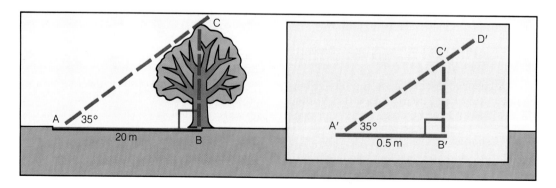

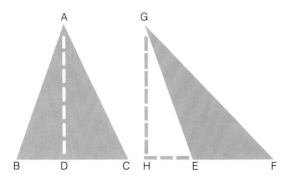

Figure 2. Any triangle can be converted into a right triangle by drawing a perpendicular from the vertex to the base.

The height of the tree, then, is 14 meters. By knowing one side and one acute angle of a right triangle, the scout was able to use basic trigonometry to solve the problem.

Trigonometric Functions

Trigonometry is based on the use of the right triangle. It can be applied to any triangle because by drawing an *altitude* (a perpendicular from the vertex to the base) we can always convert it into right triangles. In Figure 2, for example, the altitude AD divides the triangle ABC into the right triangles ADB and ADC; the altitude GH converts the triangle GEF into right triangles GHF and GHE.

Certain basic ratios or relationships between the sides of a right triangle are the very heart of the study of trigonometry. Among these ratios are the sine, cosine, tangent, and cotangent. To understand what these terms mean, let us draw a typical right triangle with angles X, Y, and Z and sides a, b and c (Figure 3). Angle Z is a right angle; the other two angles are acute angles. Side c, which is opposite the right angle, is the hypotenuse. Sides a and b are called legs. We can now define sine, cosine, tangent, and cotangent as follows:

The *sine* of either of the acute angles is the ratio of the opposite leg to the hypotenuse. The sine of angle X is a/c; the sine of angle Y is b/c.

The *cosine* of either of the acute angles is the ratio of the adjacent leg to the hypotenuse. The cosine of angle X is b/c; the cosine of angle Y is a/c.

The *tangent* of either of the acute angles is the ratio of the opposite leg to the adjacent leg. The tangent of angle X is a/b; the tangent of angle Y is b/a.

The *cotangent* of either of the acute angles is the ratio of the adjacent leg to the opposite leg. The cotangent of angle X is b/a; the cotangent of angle Y is a/b.

A sine, cosine, tangent, or cotangent of an angle is said to be a trigonometric function of that angle because its value depends upon the size of the angle. A given trigonometric function, such as the sine, is always the same for a given acute angle in a right triangle. In Figure 4, for example, angle BAC of right triangle ABC is 30°. This means that angle ABC must be 60° since angle ACB is 90° and the sum of the

Figure 4. Equal angles, such as BAC and B'A'C' shown here, have equal sines.

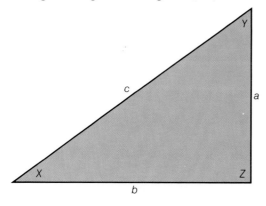

Figure 3. A typical right triangle with the hypotenuse labeled c, and legs a and b.

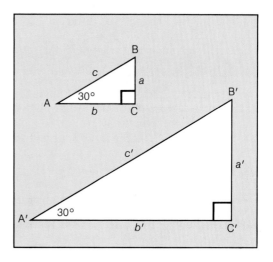

TRIGONOMETRY 387

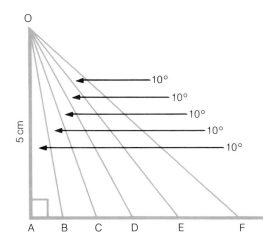

Figure 5. This diagram can be used to determine the tangents of certain angles, as discussed in the text.

interior angles of any triangle is 180°. Angle $B'A'C'$ of right triangle $A'B'C'$ is 30° and Angle $A'B'C'$ must be 60°. Hence the corresponding angles of the two triangles are equal, and the corresponding sides must be in the same ratio. Since this is so, a/c, (the sine of angle BAC) and a'/c' (the sine of angle $B'A'C'$) must be equal. Hence the sine of 30° is always the same no matter how large or how small the right triangle in which it occurs.

Problem Solving

Mathematicians have worked out the values of the trigonometric functions. By way of example, the sine of 40°, to five decimal places, is 0.64279; its cosine, 0.76604; its tangent, 0.83910; its cotangent, 1.1918. To give some idea of how such figures are derived, let us examine the procedure for finding out the tangents of different acute angles. You will recall that the tangent of an acute angle in a right triangle is the ratio of the opposite leg to the adjacent leg.

In Figure 5, side OA is equal to exactly 5 centimeters. Each of the small angles at O is equal to exactly 10°. Angle BOA, therefore, is 10°; angle COA, 20°; angle DOA, 30°; and so on. We now measure AB and find that it is about 0.9 centimeters. Since the tangent of angle BOA is $\frac{AB}{OA}$ and since OA = 5, the tangent of the angle is approximately $\frac{0.9}{5}$, or 0.18. This is the tangent of the angle 10°, whether the side OA is a centimeter, a meter, or a kilometer. Measuring AC, AD, AE, and so on in turn, we can find the tangents of 20°, 30°, 40°, and so on.

The values of the trigonometric functions can be found in almost any book of mathematical tables and are now one-key operations of many hand-held calculators. Knowing such values, we can work out a variety of measurements with great ease. Let us return to the Boy Scout's problem in Figure 1. We know that AB = 20 meters and that angle CAB = 35°. Because BC (the height of the tree) is our unknown, we will call it x. Therefore, the tangent of angle CAB = $x/20$ meters. Using a table or a calculator, we find that the tangent of 35° = 0.7; so, 0.7 = $x/20$ meters. Multiplying both sides of the equation by 20 meters, we find that x = 14 meters.

In the foregoing problem, the unknown quantity was a part of the tangent ratio. In other trigonometric problems, the cotangent, sine, or cosine might be involved. In still other cases, the unknown quantity might be an angle, as in the following problem.

The cable car in Figure 6 is going up a uniform slope, rising 12 meters in a horizontal distance of 30 meters. What is the angle (x) of the slope? Solving for the tan-

Figure 6. The cable car rises 12 meters in a horizontal distance of 30 meters. What is the angle (x) of the slope?

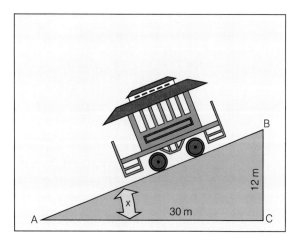

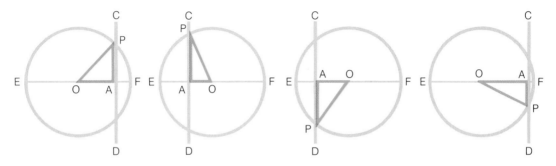

Figure 7. As radius OP goes around the circle in a counterclockwise direction, it makes a series of triangles with the diameter of the circle and with the rod CD kept in a vertical position at the end of the radius.

gent of *x*, we get $\frac{12}{30} = 0.4$. Consulting the table of trigonometric functions, we learn that 0.4 is the tangent of a 22° angle (to the nearest degree); therefore, the angle of the slope is approximately 22°.

The surveyor makes extensive use of trigonometry. First, the surveyor tries to get a fixed line that has no obstruction so that it can be measured fairly accurately. This becomes the baseline, from which numerous calculations can be made. Other measurements, as far as possible, are measurements of angles. Trigonometry is also of vital importance in engineering, navigation, mapping, and astronomy.

Studying Periodic Phenomena

Trigonometry is used in still other ways. For one thing, it serves in the study of various *periodic phenomena.* Any phenomenon that repeats itself in regular intervals of time is called periodic. The tides, for example, are periodic since they rise and fall in regular sequence. The motion of a pendulum bob is also periodic. Let us show how we can describe all periodic phenomena in terms of the sine of an angle.

We know that the spoke of a moving wheel sweeps through 360° as it makes a complete turn. It repeats the same sweep in the second complete turn, in the third complete turn, and so on. Obviously such rotation is periodic. The spoke of a wheel is really the radius of a circle. We can analyze the motion of the radius around the center of the circle by examining the diagrams in Figure 7. We are to suppose that a rod, CD, is kept in a vertical position at the end of the radius as the latter moves around the circle. You will note that a series of right angles is formed as the rod maintains its vertical position. The line segment AP, joining the points where CD meets the circumference of the circle and the diameter EF, grows longer and then shorter. Angle POA, which is called the *angle of rotation,* also changes as the radius goes around the center of the circle. The hypotenuse of right triangle APO is the radius OP and, therefore, never changes length. If we give OP the value unity (that is, 1), AP will represent the sine of the angle of rotation (POA) since the sine of POA = $\frac{AP}{OP}$.

Let us now analyze the different sine values of the angle of rotation as the radius sweeps around the circle. Sine values above the diameter are expressed as positive values. Values below the diameter are negative values. In Figure 8a, the sine is zero, corresponding to a zero angle of rotation. The sine continues to grow as the radius revolves around the center (b) until it reaches the value 1 in c (when the radius OP forms a 90° angle with the horizontal diameter, the sine = 1). The sine becomes smaller (d) until it reaches zero again in e. The radius continues to rotate below the diameter, where the sines are of negative values. The sine decreases (f) until it reaches the value -1 in g (when the radius OP forms a 270° angle with the horizontal diameter, the sine = -1). The sine becomes greater (h) until, after a complete 360° rotation (i), it becomes zero again.

We can show the variation in the sine by the line graph in Figure 9. Within the circle, we label the different values of the sine (PA) as P_1A_1, P_2A_2, and so on. When the positive and negative sine values are "removed" from the circle and plotted separately, the result is a *sine curve,* which will

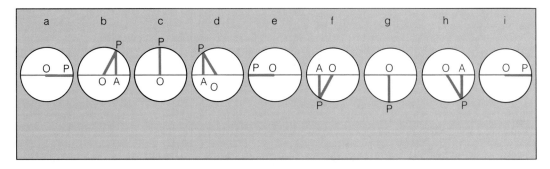

Figure 8. As the radius OP sweeps around the circle, the sine values vary along with the angles of rotation.

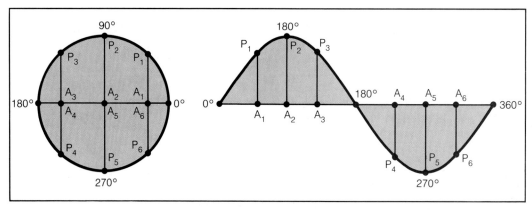

Figure 9. Variations in sine value are shown here by means of a graph. The different values of the sine PA are given here as P_1A_1, P_2A_2, P_3A_3, and so on. The curved line at the right of the circle is known as a sine curve.

repeat itself every 360°. Each fluctuation between the minimum and maximum values of a sine curve is called an *oscillation*.

The sine curve can be applied, among other things, to the periodic phenomena of sound. For example, when a tuning fork is struck, its vibration results in sound waves being sent out. If, immediately after being struck, a tuning fork that gives a tone of middle C is drawn very rapidly over a sheet of paper covered with soot (Figure 10), the vibration will describe a very specific repeating sine curve: In one second, 264 complete oscillations will be produced; middle C always corresponds to 264 vibrations per second, which can also be expressed by saying that middle C has a *frequency* of 264 hertz. Suppose we strike a tuning fork that gives a tone one *octave* (an 8-degree interval between tones) higher than middle C and draw it over the paper as before. In this case, 528 vibrations will occur in one second, which means that the frequency is exactly twice that of middle C.

Sine curves of various amplitudes and frequencies are used to explain phenomena of electricity, light (both polarized and plane), and force, as well as those of sound. Thus trigonometry helps to explain and control are physical environment.

Figure 10. Sine waves made by a vibrating tuning fork as it was passed rapidly over a sheet of paper covered with soot.

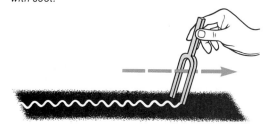

ANALYTIC GEOMETRY

Analytic geometry explains geometric figures in terms of algebraic formulas. From the earliest times, algebra and geometry had been studied as separate subjects. No one had seriously considered before the 11th century that numbers could be used to represent a point or line, or that a geometric figure could serve to represent the value of a number.

One of the first to try to combine algebra and geometry was the 12th-century Persian poet and mathematician Omar Khayyám. He wrote a work on algebra that was clearly influenced by earlier Arab and Greek writings. In this work, Omar showed how to solve algebraic equations by the use of squares, rectangles, and cubes. For example, for a number multiplied by itself, as in $x \times x$, he would use a square, each side of which had a length equal to the value of x. If x were equal to 5, each side of the square would be 5 units long, and the square would be made up of 25 units. If a number were to be a factor three times, as in $x \times x \times x$, Omar would use a cube, each side of which had a length equal to x. See Figure 1.

It was because of this method of solving equations that a number multiplied by itself, or "raised to the second power," came to be known more commonly as the *square of the number*. For the same reason, "x^3" represents "x to the third power" but is read more commonly as "x cubed."

Omar Khayyám made no known effort to solve an equation of the fourth degree—that is, an equation containing a term to the fourth power—by the geometrical method. It is likely that few, if any, mathematicians in his day had even an inkling of a fourth dimension. Today, however, the fourth dimension is an accepted and vital concept in modern physics and mathematics.

The pioneering efforts of Omar Khayyám to break down the barriers between algebra and geometry amounted to very little in his lifetime. Five centuries passed before analytic geometry was developed, through which the mathematical relationship between algebra and geometry finally gained the recognition it deserved.

Cartesian Coordinate System

The fundamental idea of analytic geometry was worked out by the great 17th-century French philosopher and mathematician René Descartes, who claimed that the idea came to him in a dream. He presented this new approach to mathematics in his *Discourse on Method* in 1637. In it he introduced a *system of coordinates* (called the "Cartesian" system, after Descartes),

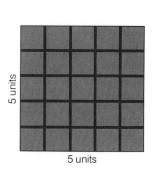
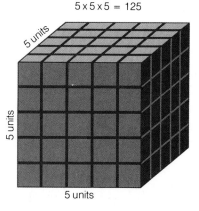

Figure 1. To express a number multiplied by itself, for example 5 × 5, Omar Khayyám used a square, each side of which was equal to that number of units. To express a number used as a factor three times, as in 5 × 5 × 5, he used a cube.

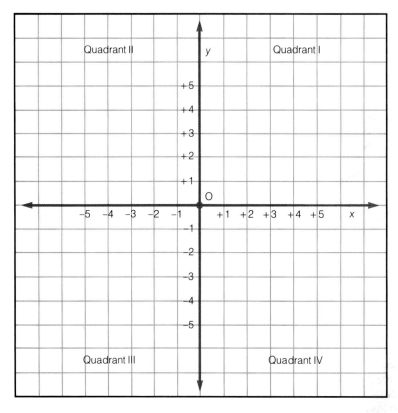

Figure 2. Axes labeled x and y are used to divide a plane into four quadrants. Positive and negative units of distance extend along each axis from the point of intersection, or origin (point O).

by which the location of any point in a plane can be described. Tremendous advances have been made in the study of analytic geometry (also called *coordinate geometry*) since 1637, and the original Cartesian system has been further developed along the way, but the system used today is essentially that of Descartes. Advancements notwithstanding, the coordinate system is really quite simple and can be explained as follows:

Consider two perpendicular lines, x and y, as shown in Figure 2. The two lines x and y are called the *x-axis* and *y-axis*, respectively. The axes divide the plane into four *quadrants*, as labeled in Figure 2. The point of intersection, point O, is called the *origin* and is considered to have a distance value of zero. Units of distance that run along the x-axis are positive to the right of the origin and negative to the left of the origin. Units along the y-axis are positive above the origin and negative below. Thus, by using this coordinate system, any point in the plane can be described in reference to the x- and y-axes.

Point A in Figure 3, for instance, is located by counting +4 units along the x-axis and +3 units up, parallel to the y-axis. The x-distance 4 is called the *abscissa* (or *x-coordinate*), and the y-distance 3 is called the *ordinate* (or *y-coordinate*). Because the

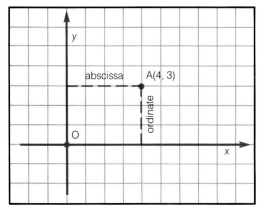

Figure 3. The coordinates of point A are (4, 3) because it is located +4 units along the x-axis and +3 units up, parallel to the y-axis.

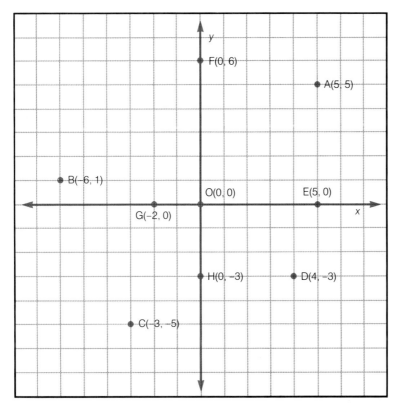

Figure 4. An x and y coordinate in a Cartesian plane can be positive, negative, or zero, depending on its reference to the x- and y-axes.

abscissa is always expressed before the ordinate, point A can be described simply by the ordered pair of numbers, or coordinates, (4, 3).

When Descartes developed his coordinate system, he described points only in the first quadrant. Today we define points in all four quadrants. For any point in a Carte-

Figure 5. For both K and L the value of y is twice that of x. The equation of the line is thus y = 2x.

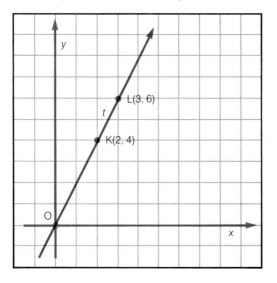

Figure 6. Any point whose coordinates satisfy the equation y = 2x can be found on line t.

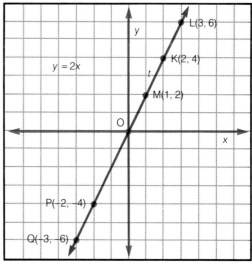

ANALYTIC GEOMETRY 393

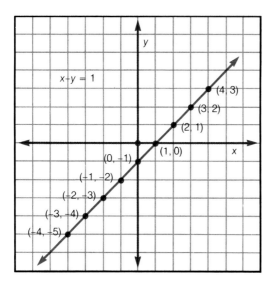

Figure 7. Plotting points from a table based on a line's equation is one way of constructing a graph of the line.

sian plane, the value of each coordinate can be described in exactly one of three ways: positive, negative, and zero. Figure 4 shows examples of each, and from such, certain statements regarding the coordinate system can be made:

1. The abscissa of any point within quadrants I or IV is positive (see points A and D).
2. The abscissa of any point within quadrants II or III is negative (see points B and C).
3. The ordinate of any point within quadrants I or II is positive (see points A and B).
4. The ordinate of any point within quadrants III or IV is negative (see points C and D).
5. The abscissa of any point on the y-axis is zero (see points F, O, and H).
6. The ordinate of any point on the x-axis is zero (see points G, O, and E).
   Note that the origin is always represented by the coordinates (0,0).

The Cartesian coordinate system is particularly valuable because it enables us to analyze a geometric figure described by a *variable* point. ("Variable" here means "able to assume any given value.") If we think of a point as being able to "move" and thereby "trace out" a figure, we can assign different point coordinates for specific point positions. Under certain conditions, we can write an equation that will hold true for all possible positions of such a point. The equation can then be used in place of the geometric figure that consists of all the positioned points. There are, for example, equations for straight lines, circles, ellipses, parabolas, and hyperbolas.

Graph of a Line

Let us first consider the equation for a straight line, using line t in Figure 5 as an example. To find the equation that defines line t, we determine the coordinates of any two points on the line (excluding any point of intersection with the x- or y-axis). As shown in Figure 5, the coordinates for points K and L are (2, 4) and (3, 6), respectively. You will note that in each instance, the y-coordinate is twice the value of the x-coordinate. Hence the equation of the line t is $y = 2x$.

Knowing this equation, we can use coordinates to locate any point on line t. For example, when $x = 1$ we know that $y = 2 \times 1$, or 2. Translated to coordinates, the point (1, 2) must lie on line t. The values of x and y can be negative as well, as shown in Figure 6. The coordinates of point P, for instance, fit into line t's equation because $2 \times -2 = -4$. Line t, like any line, is a collection of points that continues indefinitely in either direction; the path we draw to represent the line is called its *graph*.

Suppose that the equation of a straight line is given as $x - y = 1$ and that we are required to draw a graph of the line. First, we could set up a table of values for x and y, based on the given equation.

x	4	3	2	1	0	−1	−2	−3	−4
y	3	2	1	0	−1	−2	−3	−4	−5

Once we have set down the x and y values, we can locate the points (4, 3), (3, 2), (2, 1) and so on, as indicated in Figure 7, and we can draw the line connecting the points. This, then, is the straight line indicated by the equation $x - y = 1$.

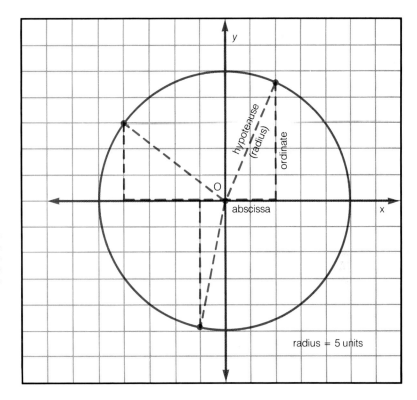

Figure 8. By applying the Pythagorean theorem, as explained in the text, we can determine that the equation of this circle is $x^2 + y^2 = 25$.

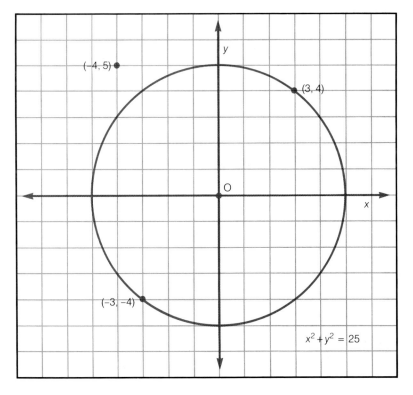

Figure 9. Points (3, 4) and (−3, −4) are both on the circle because their coordinates satisfy the circle's equation: $x^2 + y^2 = 25$. Point (−4, 5) does not satisfy the equation and is, therefore, not on the circle.

ANALYTIC GEOMETRY

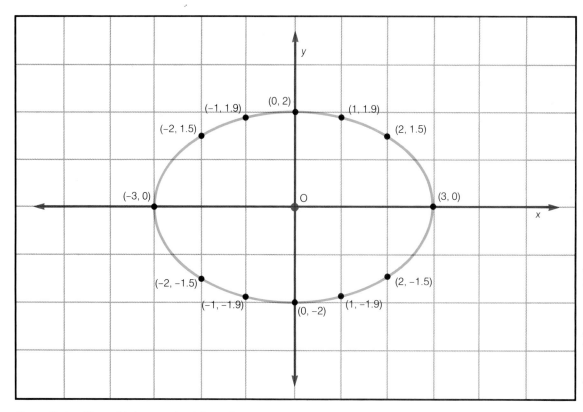

Figure 10. An ellipse with the equation $4x^2 + 9y^2 = 36$.

Graph of a Circle

We can derive the equation of a circle if we apply the Pythagorean theorem (defined in detail in the article "Plane Geometry") to the system of Cartesian coordinates. The Pythagorean theorem states that in a right triangle (a triangle with a right angle), the square of the hypotenuse (the side opposite the right angle) is equal to the sum of the squares of the other two sides.

Let us draw a circle with its center at the origin and with a radius of 5 units (Figure 8). No matter what point on the circle we select, the abscissa along the x-axis, the ordinate, and the radius will form a right triangle, with the radius as the hypotenuse. According to the Pythagorean theorem, in each of the three triangles shown in Figure 8, the square of the abscissa plus the square of the ordinate will be equal to the square of the hypotenuse, that is, 5^2 or 25. In other words, $x^2 + y^2 = 25$ is the equation of the circle in Figure 8.

Suppose we want to see whether a given point is on the circumference of this particular circle. If the sum of the squares of the point's coordinates is equal to 25, the point is on the circle. Consider the point (3, 4), as shown in Figure 9. The sum of 3^2 and 4^2 is $9 + 16 = 25$; hence (3, 4) is on the circumference of the circle. Also on the circumference is $(-3, -4)$, since the sum of -3^2 and -4^2 is also $9 + 16 = 25$. Point $(-4, 5)$, however, is not on the circle, since the sum of -4^2 and 5^2 is $16 + 25 = 41$.

Graph of an Ellipse

Suppose that we are given the information that $4x^2 + 9y^2 = 36$ is the equation of a geometric figure. Let us prepare a graph and see what sort of a figure it is. First, we set up a table of values:

x	0	±1	±2	±3
y	±2	±1.9	±1.5	0

The symbol ± in the table stands for "plus

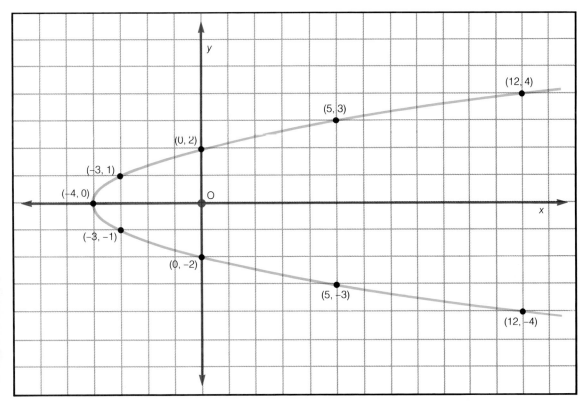

Figure 11. A parabola with the equation $x = y^2 - 4$. It is an open curve.

or minus." For example, 4 is the square of $+2$; it is also the square of -2. Hence a square root of 4 is either $+2$ or -2, or ± 2.

The values in the table are derived as follows. Applying the value $x = 0$ to the equation $4x^2 + 9y^2 = 36$, we have $0 + 9y^2 = 36$; $9y^2 = 36$; $y^2 = 4$; $y = \sqrt{4} = \pm 2$. If x is 1, $4x^2$ in the equation $4x^2 + 9y^2 = 36$ is equal to 4. Then $4 + 9y^2 = 36$; $9y^2 = 32$; $y^2 = 3.6$; $y = \sqrt{3.6} = \pm 1.9$. The other values in the table are derived in the same way.

Let us now locate the points (0, 2), (0, -2), (1, 1.9), (-1, 1.9), (1, -1.9), (-1, -1.9), and the others given in the table, and let us draw a smooth curve to connect these points. As Figure 10 shows, the geometric figure whose equation is $4x^2 + 9y^2 = 36$ turns out to be an ellipse.

Graphs of Open Curves

A typical equation for a parabola is $x = y^2 - 4$, for which we prepare a table of values, as follows:

x	-4	-3	0	5	12
y	0	± 1	± 2	± 3	± 4

If we plot the values $(-4, 0)$, $(-3, 1)$, $(-3, -1)$, $(0, 2)$, $(0, -2)$, and the others given in the table, we have the parabola shown in Figure 11. Since the greater the values of x, the greater the corresponding *absolute values* of y, the parabola is an open curve extending without limit. (An absolute value indicates the magnitude, or "size," of a number without regard for its positive or negative sign.)

A typical hyperbola is $x^2 - y^2 = 4$. To draw a graph of the figure, we first prepare a table of x and y values:

x	± 2	± 3	± 4	± 5	± 6
y	0	± 2.2	± 3.5	± 4.6	± 5.7

When we plot the values given in the table, we obtain a curve with two branches, as shown in Figure 12. The branches extend indefinitely in both directions.

ANALYTIC GEOMETRY

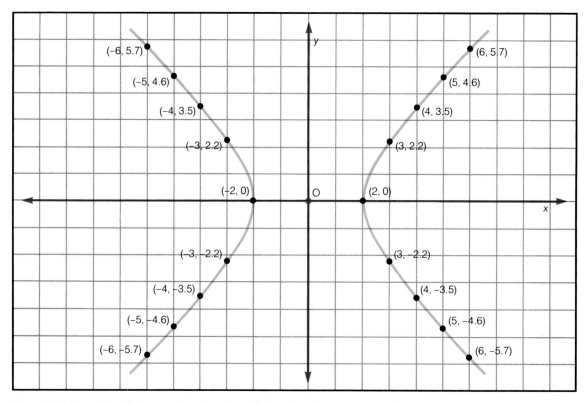

Figure 12. A hyperbola with the equation $x^2 - y^2 = 4$. Its two branches are open curves.

Using the Equations

Equations of geometric figures can be added and subtracted, and many other operations can be performed with them. The equations resulting from such operations can be interpreted by means of graphs drawn up on the basis of x and y values. In this way, algebra can be used to discover geometric relationships. American mathematician Eric T. Bell once said, "Henceforth algebra and analysis [that is to say, analytic geometry] are to be our pilots in the uncharted seas of space and its geometry."

Analytic geometry is often put to practical use. For example, in many towns, gas and water mains can be located by the equations of straight lines using the center of the town as the origin (Figure 13). *Concentric circles* (circles that share the same center) can be used to give zone distances from the center of the town. The intersections of circles and lines help describe the location of breaks in the mains.

SOLID ANALYTIC GEOMETRY

To locate a point in three-dimensional space, we must give the distance above or below the plane formed by the x- and y-axes. We add a third axis: the *z-axis*, which extends straight up and down (Figure 14). Suppose we wish to locate point A, shown in the figure, in a three-dimensional system. Point A is 2.5 kilometers east of the origin (x-distance), 2 kilometers south of the origin (y-distance), and 0.5 kilometers up (z-distance). We would locate it by giving the x-, y-, and z-coordinates (2.5, 2, 0.5). To locate a point in solid analytic geometry, therefore, we need three ordered numbers as our coordinates. In the three-dimensional world, if some point is taken as the origin, any other point in space can be represented by an ordered triplet of numbers and can be definitely located.

In solid analytic geometry, since there are three dimensions, there will be three unknowns to be related in our equations. The equation of a plane is $ax + by + cz = k$, where a, b, c, and k are given numbers. If the coordinates of a point are such that they satisfy the equation of a given plane, the point will be located on the plane. The coordinates of points that are located outside the plane will not satisfy the equation.

The equation of a sphere is $x^2 + y^2 + z^2 = r^2$, r being the radius. It is the extension of the circle equation of plane analytic geometry: $x^2 + y^2 = r^2$. The equation $x^2 + y^2 + z^2 = 25$ represents a sphere whose center is at the origin and whose radius is 5 units long. All the common solids can be represented by an equation or a series of equations with the three unknowns x, y, and z. Solid analytic geometry uses such equations to study the points, lines, surfaces, and solids that exist in space.

HIGHER GEOMETRY

The mathematician is not confined to the analysis of the three-dimensional world, in which an ordered triplet of numbers represents a point. The mathematician

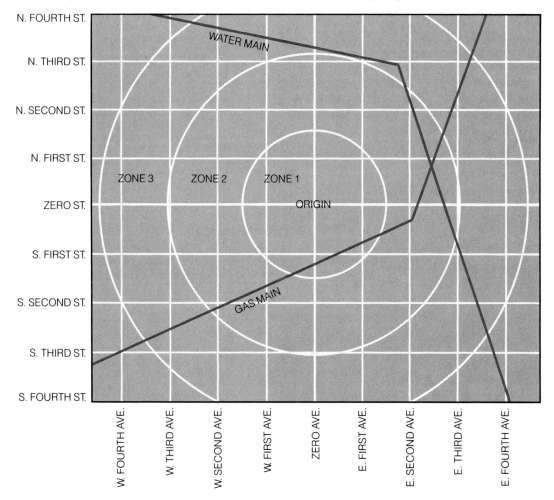

Figure 13. The streets of the town represented here are laid out to form square blocks. Zero St. and Zero Ave. are considered the x- and y-axes; the center of town is the origin. Concentric circles form zones. The geometric relationships between the circles and lines can be used to locate points along the gas and water mains.

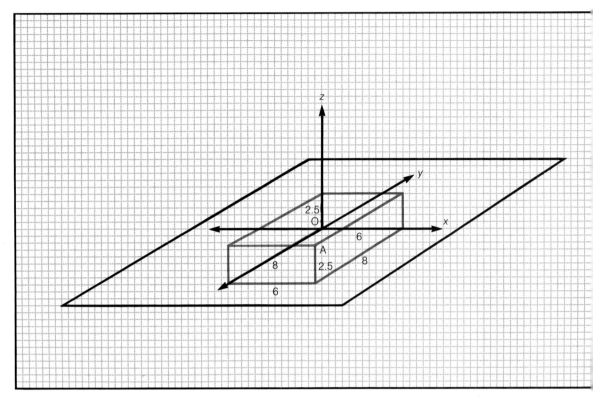

Figure 14. In solid analytic geometry, the three axes x, y, and z, correspond to the three dimensions. Point A, shown here, is 6 kilometers east of the origin, 8 kilometers south, and 2.5 kilometers up.

might ask, "What would an ordered *quadruplet* (foursome) of numbers represent?" One possible answer could be that the coordinates (3, 4, 2, 5), for instance, locate a point in four-dimensional space. The equation $x + y + z + w = 3$ would represent the equation of what our mathematician would call a *hyperplane*; (3, 4, 2, 5) would be a point on this plane. What sort of thing would this hyperplane be? Mathematicians do not know, because they have never seen one. That, however, does not prevent them from working out and using the equation of such a plane and from discovering the properties of the plane by various algebraic operations.

How can one maintain that a four-dimensional world does not exist? Suppose we were two-dimensional creatures living in a two-dimensional world. We would be quite unaware of a three-dimensional world, of which our own limited world would form a part. Yet this three-dimensional world does exist, as we all know. Is it not entirely possible that creatures like ourselves, confined to the best of our knowledge to a world of three dimensions, are really living out our lives in a four-dimensional world, which our senses cannot perceive? It may be, too, that this four-dimensional world is embedded, in turn, in a five-dimensional world, and the five-dimensional world in a six-dimensional world, and so forth. The mathematician finds such speculation fascinating, whether or not it is based on reality.

As a matter of fact, the equations of these higher-dimensional worlds have certain practical applications. For example, the equation of a *hypersphere* (a sphere with more than three dimensions) has been applied in the manufacture of television cathode-ray tubes.

by Howard F. Fehr

NON-EUCLIDEAN GEOMETRY

The geometry presented in the articles "Plane Geometry" and "Solid Geometry" and that which is commonly taught in secondary schools is called *Euclidean geometry*. It is so named because it is based on the system established by the Greek mathematician Euclid and taught by him in Alexandria, Egypt, about 300 B.C. He established a logical series of *theorems,* or statements, which were so arranged that each one depended for its proof on (1) the theorems that preceded it and on (2) certain assumptions, or *postulates*. Euclid called these assumptions common notions and he accepted all of them without proof.

In order to prove some of the theorems in the first part of his system, Euclid found it necessary to assume that through a point outside a given line, only one line can be drawn parallel to the given line. He tried hard to prove that this was so, but failed. Finally he had to consider the statement about parallel lines as a common notion, or postulate, which was to be accepted because it was self-evident. The entire system of Euclid depends upon the validity of this particular common notion, which has been called *Euclid's postulate* (Figure 1).

In the centuries that followed, many mathematicians tried to prove that through a point outside a given line, only one line can be drawn parallel to the given line. They were no more successful than Euclid had been. They had to accept Euclid's postulate as self-evident but not proven.

In modern times, not all mathematicians have conceded that the postulate is self-evident. They have felt justified in making different assumptions, and they have built up entire geometries based upon these assumptions. The 19th-century German mathematician Bernhard Riemann based his geometry, called *Riemannian geometry,* on the postulate that no two lines are ever parallel. Another type of geometry was created in the 19th century by the Russian Nikolai Lobachevsky and, independently, by the Hungarian János Bolyai. It was based on the assumption that at least two lines can be drawn through a given point parallel to a given line. The geometries of Riemann and of Lobachevsky-Bolyai are known, therefore, as *non-Euclidean*.

Different Assumptions

To understand the difference between Euclidean and non-Euclidean geometries, let us consider the shortest distance between two points in (1) a *plane,* or flat surface; (2) a *sphere,* or solid circular figure; and (3) a *pseudosphere,* the surface of which suggests two wastepaper baskets joined together at their tops (Figure 2). The shortest distance between two points on any kind of surface is called a *geodesic*.

On a plane, a hemisphere (half of a sphere), and a hemipseudosphere (half of a pseudosphere), we can measure the same distance AB, as shown in Figures 3, 4, and 5. On a plane (Figure 3), the geodesic of

Figure 1. According to Euclid's postulate, one line and only one line, DE, can be drawn parallel to the line AB through point C.

Figure 2. A pseudosphere.

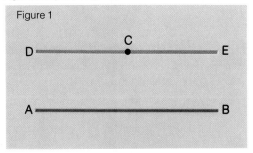

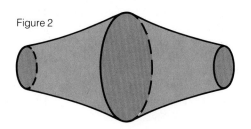

points A and B is a straight line. On a hemisphere (Figure 4), when points A and B lie on the *great circle* (or largest circle of the hemisphere; the center of a great circle is the center of the sphere), the geodesic AB is an arc of the great circle. On a hemipseudosphere (Figure 5), when points A and B lie on the "waist" (also, the largest circle) of the pseudosphere, the geodesic AB is an arc of the waist.

On the plane, hemisphere, and hemipseudosphere (Figures 3, 4, and 5), we draw geodesic AC, perpendicular to AB at A, and geodesic BD, equal to AC in length and perpendicular to AB at B. Finally, we draw the geodesic CD. On the plane, the geodesic CD is a straight line; on the hemisphere, it is an arc of the great circle; on the hemipseudosphere, it is an arc of a lesser circle.

How large are angles 1 and 2 in Figures 3, 4, and 5? The answer to this question will provide an insight into the differences between Euclidean geometry and the non-Euclidean geometries of Riemann and Lobachevsky-Bolyai.

In the case of the plane in Figure 3, it would seem to be obvious that angles 1 and 2 are *right* angles (90°). But we cannot prove this as a geometric theorem unless we first agree that only one line passing through C (that is, the line CD) is parallel to line AB. If we make this assumption, we are accepting Euclid's postulate.

On the hemisphere (Figure 4), angles 1 and 2 are apparently *obtuse* angles (greater than 90°). Can we prove this? Not unless we first assume that every geodesic through C will meet line AB at two points. If we accept this, we can prove that angles 1 and 2 are greater than 90°.

Riemann assumed that even in a plane, any line (such as CD in Figure 3) drawn through an external point (such as C) will meet any other line (AB) at two points. Hence there are no parallel lines in his geometry. He developed a perfectly logical set of theorems based on this assumption.

"But," you will say, "anyone can see that line CD in Figure 3 is parallel to line AB and that the two lines will never meet."

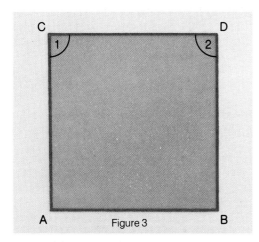

Figures 3, 4, and 5. The figure ABCD is shown on a plane (Figure 3), a hemisphere (Figure 4), and a hemipseudosphere (Figure 5). In each case, geodesics AC and BD are equal to each other and perpendicular to line AB.

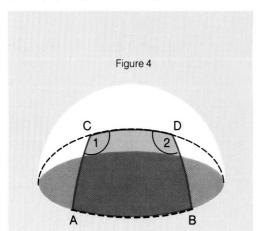
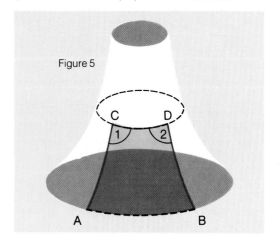

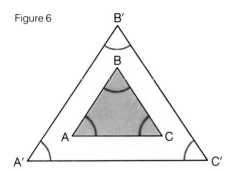

Figure 6. In the geometry of Euclid, the sum of the angles in a triangle, such as ABC, is 180°. If the area of such a triangle is increased so that we have the triangle A'B'C', the sum of the angles is still equal to 180°.

Figure 7. According to the geometry of Riemann, the sum of the angles of a triangle, such as ABC, is always greater than 180°. It increases as the area of the triangle increases, as in triangle A'B'C'.

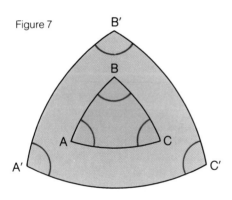

But as long as you cannot prove that this is so, you cannot deny that it is perfectly logical to develop a new kind of geometry based on another assumption.

Let us now examine angles 1 and 2 in the hemipseudosphere (Figure 5). These angles are apparently *acute* (less than 90°). Again, we cannot prove that this is so unless we make an assumption—in this case, that there are two geodesics through C that never meet AB. Lobachevsky and Bolyai made this assumption and applied it to all kinds of surfaces, including those of spheres and planes. Their geometry describes a world quite different from that of Euclid and Riemann—a world in which through every external point there are two lines parallel to a given line.

Of course, three geometries based on such different assumptions are bound to show many striking points of difference. Consider, for example, the sum of the angles of any triangle. In the Euclidean geometry, the sum is 180°. It always remains the same no matter how much the size of a given triangle increases (Figure 6). In the Riemannian geometry, the sum of the angles of a triangle is always greater than 180°. As the area of the triangle increases, the sum of the angles increases (Figure 7). In the geometry of Lobachevsky-Bolyai, the sum of the angles of a triangle is always less than 180° and decreases as the area of the triangle increases (Figure 8).

One cannot say that one of these geometries is correct and that the others are incorrect. Rather, they are different explanations of space, based on different assumptions. We accept the one that offers the most satisfactory interpretation of a particular phenomenon with which we are concerned. For example, the Riemannian geometry has given a better explanation of certain astronomical aspects of Einstein's relativity theory than either the Euclidean or Lobachevsky-Bolyai geometry.

Howard F. Fehr

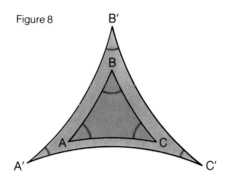

Figure 8. In the geometry of Lobachevsky-Bolyai, the sum of the angles of a triangle, such as ABC, is always less than 180°. It decreases as the area increases, as in triangle A'B'C'.

STATISTICS

Have you ever noticed how much of the information you receive comes to you through numbers? Particularly since the coming of age of electronic computers, more and more kinds of information are being coded, processed, and presented numerically. On any given day, we can expect to see numerical presentations of weather information, the stock market, political polls, business transactions, census data, government operations, and many other types of data.

In most instances, the numerical information in its original form would be difficult to interpret. For this reason, the information is usually organized, summarized, and presented to us in a form that can be more readily interpreted. Frequently this is accomplished by reducing the numerical data to a table or graph or by reporting one number, such as the average, to represent an entire set of numbers. The process by which numerical data are collected and eventually presented in a usable and understandable form is an important part of the mathematical science of *statistics*.

Statistics is important not only for communication; it also provides a basis for decision making. The government makes extensive use of statistics in estimating its budget needs and setting its tax rates. Statistics enables manufacturers to compare production processes when they seek to improve their products or increase their profits. Store managers may rely upon statistical analyses to determine which items they should stock. Scientists employ statistics in comparing the effects of critical variables upon their experiments. Insurance companies rise and fall on the accuracy of their statistical predictions. Engineers base the design of highways and bridges upon statistical studies of materials and traffic. School officials may modify their curricula on the basis of statistical analyses of student achievements and needs. The list of such decision-making uses of statistics is almost endless.

Collection of Data

The science of statistics involves a variety of tasks. Even the seemingly simple business of collecting numerical data requires careful study. Obviously the conclusions of a statistical study can be no more reliable than the figures upon which they are based. The statistician must be sure that the data collected are accurate, relevant to the problem being studied, and representative of the problem. Invalid conclusions drawn from statistical evidence are often due to inadequacies in the data collected. The matter of data collection will not be discussed in detail, but its importance is so great that we should be aware of some of the problems involved.

First, the *population* to be studied must be well defined. What do we mean by "population" here? To the statistician, it may consist of a set of cities, automobiles, books, or even scientific experiments. In fact, the population for a statistical study might be any set of objects having a common characteristic to which a number might be assigned. Of course, the population selected must supply the appropriate numerical data for the problem being studied. If the population of a statistical study is not well defined or is not representative of the problem being studied, the results of the study will be difficult to interpret or apply. For example, surveys of the voting preferences of high-school students would be of questionable value in predicting the results of a national election, since few high-school students can vote.

Once the population has been identified, the particular characteristics to be studied must be represented numerically. Sometimes the numerical data are already available in recorded form. For example, if you wanted to study the rainfall in your city over the past year, you could probably obtain the needed data from your local weather bureau. In this case, the population might be defined as the set of days in the year.

Sometimes the numerical data may be obtained by a simple counting process. You might be interested, for example, in a study of the books in a school library. This project would involve counting the volumes devoted to each of several subjects.

More often, the data for a statistical study are obtained by measuring some common characteristic of the population being studied. If the population were a set of scientific experiments, the scientist might be concerned with such characteristics as time, temperature, volume, and mass. In each case, the scientist would need to use a suitable measuring instrument to assign a number to the characteristic.

In some cases, no measuring instrument is available and the investigator must create a measuring device. A teacher, for example, creates examinations. Each classroom test is an instrument by which student achievement can be measured. Test grades are the numbers assigned to that measurement. As with any other measuring instrument, accuracy is a prime consideration. The investigator would need to know how well the number assigned represents the true value of the characteristic being measured—in this case, student achievement.

The ultimate value of a statistical study depends to a large extent upon the quality of the measuring instrument that is employed. For this reason, the construction and evaluation of such an instrument is often a critical task in a statistical study

Sometimes it is possible to obtain numerical data about each member of a population that is being studied. When this is true, the data are completely representative of the population, and the task of the statistician is to describe the numerical data obtained. This branch of statistics is called *descriptive statistics*.

Sampling a Population

Often it is necessary or practical to collect data only for a *sample* of the population and to make *statistical inferences* about the population itself. An inference is a conclusion about the unknown based upon something that is known. A statistical inference, therefore, is one based upon statistical data. When data are available only for a sample, the sample represents the known and the population the unknown. Any subject of a population would constitute a sample, but statistical inferences are valid only when the sample is representative of the population. Many techniques are employed by statisticians to insure that the samples they select are representative.

When each member of the population has an equal chance of being chosen, we have what is called a *random sample*. This is usually assumed to be representative of the population. In some special problems, the statistician uses a *stratified sample,* which insures that specific segments of the study population are represented in the sample. The process of identifying a representative sample is a critical task in many statistical studies. In the examples cited in this article, we will assume that the samples used are representative of the populations from which they have been selected.

To summarize the discussion of collecting data, let us consider the following example: Suppose that the population being studied is the set of students in a given grade. If each of these students was assigned to an English class by a random process, the English class would constitute a representative sample of the population. If you were to measure the height, weight, or age of each student in the class, or record each student's score on a particular test, or count the number of people in his or her family, you would obtain a set of numbers. These numbers could then become the data for a statistical study. The data could be used to describe the English class (the sample). They could also be used to make *estimates* or inferences about the total set of students in the grade (the population).

Organizing the Data

Once data have been collected, they must be arranged in some systematic order before a useful interpretation can be made or conclusions drawn. Sometimes a simple table or graph can be quite helpful as a first step toward the statistical analysis of numerical data.

TABLE I

Vocabulary Test Scores in Sixth-Grade Class 6-1

Name	Score	Name	Score
Alex	14	John	16
Barry	19	Joseph	18
Carolyn	16	Karen	14
Catherine	13	Larry	14
Charles	19	Loretta	15
Cynthia	16	Michael	15
David	18	Monica	17
Diane	17	Rebekah	18
Doris	20	Sandra	13
Drew	17	Sean	16
Elaine	16	Seth	17
Elizabeth	16	Sharon	15
Eric	15	Sheila	15
Gabriel	15	Stephanie	14
George	17	Stephen	16
Greg	16	Virginia	13
Janice	12	Zachary	12
Jennifer	16		

TABLE II

Frequency Distribution of Vocabulary Test Scores in Sixth-Grade Class 6-1

Score	Frequency
0	0
1	0
2	0
3	0
4	0
5	0
6	0
7	0
8	0
9	0
10	0
11	0
12	2
13	3
14	4
15	6
16	9
17	5
18	3
19	2
20	1
21	0
22	0
23	0
24	0
25	0

$n = 35$
n = total number of scores

The numerical data presented in Table I are scores obtained by 35 students from one sixth-grade class, Class 6-1, in a vocabulary test. Each score represents the number of test questions that have been answered correctly by a given student. In this case, the word "score" is used in its usual sense. However, regardless of the nature of the numerical data, statisticians often use the term *raw score* to indicate the individual numbers obtained as a basis for a statistical study.

It is difficult to make any useful interpretation of the data in Table I. The simplest way of organizing these data would be to arrange the scores in numerical order. It is common to record only the different raw scores (in this case, 16, 18, 14, and so on) and to note the *frequency* with which each score occurs. Table II is a frequency table prepared from the data in Table I.

Even a cursory examination of Table II permits some elementary interpretation of the data. We can easily observe the highest and lowest scores (20 and 12) and the most frequent scores (15, 16, and 17). We can even begin to have some feeling for the way the scores seem to cluster about a central point—in this case, the score 16.

Further clarification of the data may be obtained by translating Table II into a graphical form. Figure 1 is a common type of *frequency graph* used to present frequency distributions. The numbers below the horizontal line represent scores; the numbers at the left represent the frequency distribution—that is, the number of times each score occurs.

The *frequency polygon* shown in Figure 2 is based on the same idea. In this

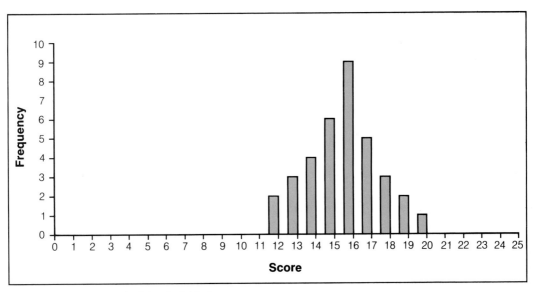

Figure 1. Frequency graph of data in Table II.

case, we imagine lines perpendicular to the scores on the bottom line and lines perpendicular to the frequencies at the left. A point indicates the intersection of a score line and a frequency line. Thus we have a point where the score 12 and the frequency number 2 meet, and a point where the score 16 and the frequency number 9 meet. The points are connected by straight lines. Note that to "complete the appearance" of the polygon, a zero-frequency-point is assigned to the value one unit lower than the lowest score and to the value one unit higher than the highest score.

An examination of these two graphs will reveal exactly the same information

Figure 2. Graphs, such as this frequency polygon, are often easier to interpret than tables.

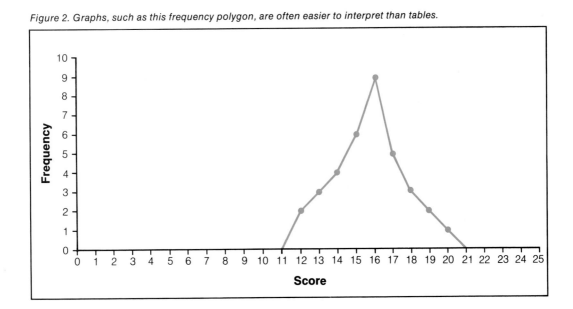

TABLE III

Frequency Distribution of Vocabulary Test Scores In All Sixth-Grade Classes

Score	Frequency
0	0
1	0
2	0
3	1
4	2
5	2
6	3
7	6
8	9
9	7
10	7
11	12
12	20
13	10
14	12
15	24
16	28
17	22
18	13
19	11
20	9
21	5
22	4
23	0
24	0
25	0
	$n = 207$

n = total number of scores

TABLE IV

Grouped Frequency Distribution of Vocabulary Test Scores In All Sixth-Grade Classes

Score	Frequency
2-4	3
5-7	11
8-10	23
11-13	42
14-16	64
17-19	46
20-22	18
	$n = 207$

n = total number of scores

available in Table II. The graphical form often is easier to interpret than the tabular form. In particular, the tendency of scores to cluster around a central point becomes more apparent when the data are presented pictorially.

Grouping Raw Scores

The frequency distribution for raw scores made by all 207 students from all sixth grade classes in the school is shown in Table III. In many statistical studies, particularly when the range of scores is great, it becomes cumbersome to work with all the individual scores. In these cases, it is common to condense the data by grouping the raw scores into class intervals. In studying the scores on the vocabulary test, instead of considering each score individually, we combine a number of adjacent scores to form an interval. Thus we would combine the individual scores 20, 21, and 22 to form the interval 20–22. Intervals are treated in much the same way as we would treat individual raw scores.

A grouped frequency table, based on intervals, is shown in Table IV. It presents scores made by 207 students from all the sixth-grade classes.

When the number of possible individual scores is large, the grouping of data into appropriate intervals enables the investigator to work with a manageable number. However, this grouping method has the disadvantage of obscuring individual scores. Thus all eighteen individual scores in the interval 20–22 might be 22 instead of 9 at 20, 5 at 21, and 4 at 22. We ignore this consideration when working with class intervals. Once the data are compressed by grouping, in all subsequent analysis and computation, we treat individual scores as if they were evenly distributed throughout the interval to which they belong.

Graphical representations of grouped frequency tables are very similar to those presented earlier for ungrouped data. One special kind of graph, the *frequency histogram,* is worthy of note. To construct a histogram from Table IV, as in Figure 3, the frequency of scores in each interval is represented by a rectangle with its center at

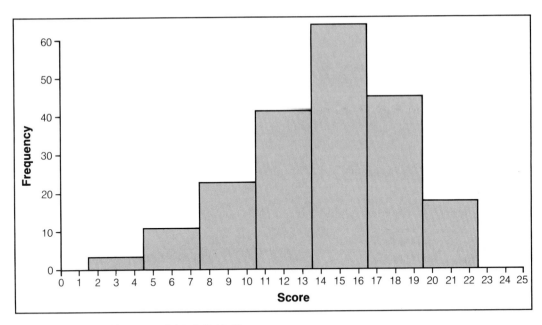

Figure 3. Frequency histogram of data in Table III.

the midpoint of the interval, its height equal to the frequency, and its width equal to the width of the interval's *limits*.

Here it is necessary to define and distinguish between two types of limits. When an interval is identified as 20–22, the limits reported are called *score limits*. They identify the lowest score and highest score that belong to the same interval. For purposes of mathematical treatment and graphical representation, it is common to use the *real limits* 19.5–22.5 to identify this same interval. The interval is represented as extending halfway to the scores immediately preceding and following. Such an interpretation is consistent with the way we usually report measurements. For example, if we were measuring to the nearer meter, any measurement between 19.5 meters and 20.5 meters would be reported as 20 meters.

Calculating Averages

Tables and graphs can help us obtain considerable understanding of a set of scores. However, for many purposes, it is more desirable to try to represent the set of scores by a single number. When selecting a single number to represent a whole set of numbers, the first thing we usually think of is the average. As we have noted earlier, the scores we have been examining seem to cluster around a central point. It is this point of central tendency that statisticians identify when they report an average score. In statistics, there are several types of averages. Three are common in statistical analysis—the *mode,* the *median,* and the *mean*. Each is called a measure of central tendency. For the data with which we have been working in this article, the mode, median, and mean are close in value. This is not always the case. They may be appreciably different.

Mode. The mode is quite easily identified from a frequency table or frequency graph. It is the score that occurs most frequently—in a sense, the most popular score. From the data presented in Table III, we can determine readily that the most frequent score is 16. Thus 16 is the mode of this set of scores. In the grouped data, the *crude mode* would be identified as the midpoint of the interval with the highest frequency. From either Table IV or Figure 3, we can see that the interval of highest frequency is 14–16. Hence 15 would be the crude mode of this distribution since it is the midpoint of the interval. Since individual scores have been obscured, we cannot be sure that it is actually the most frequent individual score, but it is our best estimate of the most frequent score.

Median. The median is the middle score in a set of scores. The data in Table III contains 207 scores. The median is, then, the value of the 104th score. We find that the 104th score is among the 24 scores that all have a value of 15, and the median is therefore 15. If there had been an even number of scores, the median would have been reported as a figure that is halfway between the two middle scores if the two are different.

Since the set of data, presented in Table IV and Figure 3, has been condensed by grouping, we cannot work with individual scores and must find a new procedure for identifying the median. From Table IV, we find that the middle score, the 104th score, must fall in the interval 14–16. Since 79 scores out of the total number of 207 scores fall below this interval, we know that the median will be the 25th score in the interval (79 + 25 = 104). Since for grouped data we must assume even distribution within an interval, we will assume that the median lies 25/64 of the width of the interval above its lowest boundary. We then multiply 25/64 by 3 (the number of scores in the interval) for a product of 25/64 × 3 = 1.2 (rounded to the nearest tenth). We now add 1.2 to the least-value endpoint of the interval 14–16. (We noted previously that this endpoint is 13.5.) Hence we have 13.5 + 1.2 = 15.7. The median, then, as calculated from the grouped data is 15.7.

Mean. The mean is the most commonly used measure of central tendency, and it is the average most of us think of first. It is found by dividing the sum of all the individual scores by the number of scores in the set. The calculation of the sum of the scores can be shortened when a frequency table is available if we multiply each score by its frequency and then find the sum.

The mean for the data in Table III is calculated as follows:

Score		f		
3	×	1	=	3
4	×	2	=	8
5	×	2	=	10
6	×	3	=	18
7	×	6	=	42
8	×	9	=	72
9	×	7	=	63
10	×	7	=	70
11	×	12	=	132
12	×	20	=	240
13	×	10	=	130
14	×	12	=	168
15	×	24	=	360
16	×	28	=	448
17	×	22	=	374
18	×	13	=	234
19	×	11	=	209
20	×	9	=	180
21	×	5	=	105
22	×	4	=	88
		207		2,954

2,954 ÷ 207 = 14.27 = 14.3 = the mean

When computing the mean for grouped data, we assume even distribution of scores within each interval. We multiply the value of the midpoint of each interval by the frequency and divide the sum of the resulting number by the total number of scores.

The mean for the data in Table IV is calculated as follows:

Score		f		
3	×	3	=	9
6	×	11	=	66
9	×	23	=	207
12	×	42	=	504
15	×	64	=	960
18	×	46	=	828
21	×	18	=	378
		207		2,952

2,952 ÷ 207 = 14.26 = 14.3, the mean

Selecting Averages

The mode is used less frequently than either the median or the mean. It is useful only when we want to identify the number occurring most frequently in a set of numbers. As a matter of fact, if the mode is to be truly meaningful, one number in a set must occur quite a bit more frequently than any other number in the set. The advantage of the mode is that, like the median, it is easy to identify and understand. But the term "mode" is sometimes ambiguous because there may be more than one score with the "highest frequency" (see Figure 4). Also, the mode is not reliable as an indication of central tendency because the most popular score is not always near the center of a given distribution.

The chief advantage of the median is that it is not affected by extreme scores. The "average" income in a community, for example, is often more accurately reflected by the median than the mean because the value of the median is not influenced by a few very high or very low incomes. The idea of the median is closely related to the concept of *percentiles*—a type of score students receive on certain standard tests in school. The median corresponds to the 50th percentile. Other percentiles can also be used in connection with tests. They are valuable as a basis for comparing individual scores with other scores in a distribution.

For most purposes, the mean is the best measure of central tendency. It is the only one of the three measures that depends upon the numerical value of each score in a distribution. It is a reliable indicator of central tendency because it always identifies the "balancing point" or "center of gravity" in the distribution. Since the mean lends itself better to mathematical computation, it is more suitable for deriving other statistical measures. For example, the means of two sets of data can be used to compute a mean for the combined set of data. This cannot be done with the mode or the median. However, the mean can give us information only about the central point in a distribution. To understand a set of scores more fully, we also need to know how the scores spread out around this central point. For this reason, statisticians develop measures of *dispersion* or *variability*.

The simplest measure of distribution is the *range*, which is defined as the difference between the highest and lowest scores in a distribution. The range of scores reported in Table III is easily identified as being 19. We simply subtract the lowest score (3) from the highest score (22). Since the range is sensitive only to the two extreme scores in a distribution, it is not considered a very satisfactory measure of dispersion. Its weakness is dramatized in the two distributions whose graphs are presented in Figures 4 and 5. In both distributions the mean is 12 and the range of scores is 12. Yet the distributions are obviously different.

As with the median and the mode, the weakness of the range is that it does not take into account the numerical value of each score. A natural measure of dispersion involving every score is the *average difference* between the individual scores and their mean. As we have seen, the mean serves as a "balancing point" for a distribution. Hence we can measure the deviations from the mean in terms of positive and negative differences. Deviations from the mean in one direction would be positive; in the other, negative.

The sum of the deviations from the mean is always zero. This is because the magnitude of negative differences always equals that of the positive differences. To avoid using negatives in the measurement of deviations, we use their *absolute values*. (The absolute value of a number disregards its positive or negative sign; the absolute value of any number x is represented as $|x|$.)

The average of the absolute values of the differences between individual scores and the mean is called the *mean deviation* and is a simple and accurate measure of dispersion. The calculation of the mean deviation for the distribution in Figure 4 is as follows:

$$
\begin{array}{rl}
f & \text{Deviation from mean} \\
5 \times & |6 - 12| = 30 \\
10 \times & |9 - 12| = 30 \\
20 \times & |12 - 12| = 0 \\
10 \times & |15 - 12| = 30 \\
\underline{5} \times & |18 - 12| = \underline{30} \\
50 & 120
\end{array}
$$

$120 \div 50 = 2.4$, the mean deviation

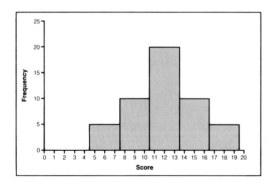

Figure 4, above, and Figure 5, below, show two different distributions of test scores. Both distributions have the same mean, 12, and the same range of scores, 14. In both distributions the lowest score is 5 and the highest is 19. However, the distributions of scores are very different. The mean and range are not sufficient to describe the distributions. A measurement of the deviation of scores from the mean is also needed.

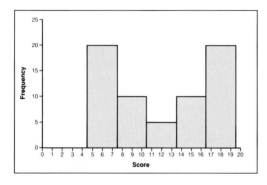

A similar calculation for the distribution of test scores in Figure 5 would yield a mean deviation of 4.6. Therefore greater dispersion is indicated for the distribution in Figure 5.

The use of absolute values presents mathematical difficulties that can be avoided by using another measure. The positive and negative signs that led to the introduction of absolute values could also have been eliminated by squaring the deviations from the mean, since the square of a positive or negative number is always positive. Such a procedure preserves the descriptive qualities of the mean deviation while providing a measure that is easier to handle mathematically. Hence statisticians prefer to use the *standard deviation*, indicated by the symbol "σ", as a measure of dispersion. The standard deviation is defined as the square root of the average squared deviation from the mean. Thus to calculate the standard deviation, we first find the average of the squares of the deviations. This number is called the *variance*. Suppose we wish to find the variance and standard deviation (σ) for the data in Table II.

Scores in Sixth Grade Class 6-1. The number of scores in this class is 35, and the mean is 15.7. Calculation of the variance and the standard deviation, σ, is as follows:

Frequency	Deviation from Mean Squared	
2 ×	$(12 - 15.7)^2$	= 27.38
3 ×	$(13 - 15.7)^2$	= 21.87
4 ×	$(14 - 15.7)^2$	= 11.56
6 ×	$(15 - 15.7)^2$	= 2.94
9 ×	$(16 - 15.7)^2$	= 0.81
5 ×	$(17 - 15.7)^2$	= 8.45
3 ×	$(18 - 15.7)^2$	= 15.87
2 ×	$(19 - 15.7)^2$	= 21.78
1 ×	$(20 - 15.7)^2$	= 18.49
35		129.15

$129.15 \div 35 = 3.69$, the variance

$\sqrt{3.69} = 1.92$, the standard deviation, σ

Similar calculations would yield a standard deviation of 4.1 for the data in Table IV, Scores in All Sixth-Grade Classes. If we compare these two measures of dispersion, we see that the scores in Table II the scores of Class 6-1, (the sample) are not so "spread out" as the scores in Table IV, the scores of all sixth grade classes (the population).

Interpreting the Data

Together, the mean and the standard deviation give us a reasonably clear picture of a distribution because they describe both its central tendency and its dispersion. Sometimes, if we know the general nature of the distribution, we need only these two numbers to reconstruct the distribution. For example, many sets of measurements have the *normal* distribution shown in Figure 6.

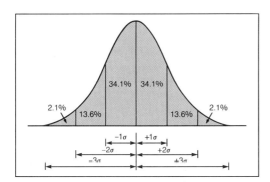

Figure 6. Human intelligence as measured by intelligence tests is believed to be distributed at random. Scores on intelligence tests form a normal distribution. Many other random variables form normal curves.

When a set of numbers "fits" such a standard distribution, we can determine approximately how many of the numbers fall within a given distance of the mean. For the normal distribution, 68.2 percent of the scores fall within one standard deviation of the mean. Thus, given the mean of 14.3 and the standard deviation of 4.1 for the set of scores in Table III, we could "predict" that 68.2 percent of the scores would be between 10.2 and 18.4. Since the distribution is actually given in Table III, we see that 152 of the 207 scores, or 68.1 percent, actually do fall in this interval. The prediction is accurate because the number of scores is relatively large and they do fit the normal distribution.

The knowledge of such general models for distribution coupled with the laws of probability form the basis of *predictive statistics*. Both statistics and probability have to do with distributions, and it is upon this common focus that we capitalize in predictive statistics. In probability, the sample space (population) is known and we predict the composition of a set of outcomes (sample). In statistical inference, the sample (set of outcomes) is known, and we infer the composition of the population (sample space). Thus predictive statistics, also called statistical inference, can be thought of as an application of the laws of probability in reverse.

If we could be certain that the distribution of scores in a sample reflected exactly the distribution of scores in the population from which it was chosen, statistical inferences would be exact and simple to make. But even when a population is known, probability theory tells us that samples will not always be the same. The best we can hope for is that, if the sample is large enough and is carefully chosen, the sample characteristics will closely approximate those of the parent population.

Suppose we had only the data recorded in Table II for Class 6-1 and wished to estimate the mean score for all the English classes. By calculation, we have determined that the scores for Class 6-1 have a mean of 15.7 and a standard deviation of 1.92. Our best estimate of the mean for all classes would be equal to the mean of the sample, Class 6-1. But knowing that samples vary, we would hedge on this estimate. We would give a *confidence interval* within which we would expect the true mean of the population to fall. By assuming the total set of scores to be normally distributed and applying basic laws of probability, we could determine that there is a 95 percent chance that the population mean falls within 1.96 standard deviations or 3.69 score points (1.96×1.92). Thus there is a 95 percent chance that the population mean will be 15.7 plus or minus 1.8, or will be between 13.9 and 17.5. The 95 percent is a measure of the confidence or reliability we can place in our estimate. It means that 95 of every 100 populations from which a sample with the given characteristics might be chosen would have a mean within the determined interval. Because of the uncertainties involved in sampling, such an interval estimate, accompanied by a statement of the degree of confidence we can place in the estimate, is preferable to a single number approximation.

The process of establishing confidence intervals permits us to test hypotheses about a population. Modifications of this process permit us to make statistical comparisons of two samples drawn from the same population, to compare a sample to a known population, or to infer other characteristics of an unknown population.

F. Joe Crosswhite

PROBABILITY

How often should we expect all three of the children in a family to be boys?

What are your chances of winning a sweepstakes contest or the door prize at a party?

How likely are you to land on "Boardwalk" in your next turn in a game of Monopoly?

If a baseball player averages 3 hits in 10 times at bat, what are his chances of getting 4 consecutive hits?

Have you ever wondered about the answers to such questions? Scientists often have occasion to study questions of this general type (though much more complicated than the four given above). In seeking the answers, they apply the mathematical theory of *probability*.

Probability is a mathematician's way of describing the *likelihood* that a certain event will take place. It is used to predict the outcome of an experiment when this outcome is governed by the laws of chance. Probability theory enables us to determine probable characteristics of a sample drawn from a population whose characteristics are known. It also provides the basis for part of the related science of *statistics*. Statistical inference is an extension of probability theory. In it, the outcome of an experiment is used to estimate the conditions governing the experiment. We would be drawing a statistical inference if we were to make an educated guess about some population by examining the characteristics of a sample taken from that population.

Let us now consider some examples. Suppose we know that there are 400 boys and 200 girls in a certain school and we conduct the experiment of choosing 1 student from this school at random. In a random selection, each of the 600 students would have exactly the same chance of being chosen. In this experiment, we would say that the probability of choosing a boy is 400 ÷ 600, or ⅔, and the probability of choosing a girl is 200 ÷ 600, or ⅓. The numbers ⅔ and ⅓ indicate that on the average we should expect to choose a boy "two times out of three" and a girl "one time out of three."

If we repeated this experiment 15 times, our best estimate would be that the sample chosen should contain ⅔ boys and ⅓ girls. This is an example of how probability is used to predict the outcome of an experiment. The probable composition of the sample was determined from the known composition of the population from which it was chosen.

Suppose that in a second school we do not know the proportion of boys and girls but know only that the total number of students is 600. To estimate the ratio of boys to girls, we might conduct an experiment consisting of selecting 20 students at random. If the sample chosen in this experiment contains 10 boys and 10 girls, our best estimate would be that there are equal numbers of boys and girls in the school—that there are 300 boys and 300 girls. In this case, the composition of the sample permits us to make a statistical inference about the composition of the population from which it was drawn.

As you see from these examples, probability and statistical inference are very closely related. The latter may be thought of as an application of probability in which the reasoning process is reversed. Statistical inference is only one aspect of the field of statistics, which also involves the collection, presentation, and analysis of data.

The examples just given illustrate the basic idea of probability and suggest how it may be applied, but they also may be misleading. They were oversimplified in order to avoid some of the more difficult problems that arise in the study of chance phenomena. In the first example, if we repeated the experiment a very large number of times, we might expect to choose a boy at random instead of a girl just about two times out of three. But the chances of selecting exactly 10 boys in a given sample of 15 are really quite small. This would happen only about two times in any ten trials. Most of the samples of 15 would be close to the theoretical ratio of ⅔ (11 boys and 4 girls or 9 boys and 6 girls, for example), but some would not be close at all. We might even choose a random sample of 15 made up entirely of boys. In other words, the "most probable" outcome may not be very probable at all.

In the second example, the fact that our sample contained an equal number of boys and girls could lead us to a very poor estimate of the proportion of boys and girls in the school. We might even have drawn such a sample from the first school, in which there were twice as many boys as girls. In that case, we would have estimated the probability of choosing a boy as ½ when it was actually ⅔.

Many interesting and useful problems in probability are complicated because there are so many *possible* outcomes. If the number of possible outcomes is infinite, we must use calculus in our analysis. In this short introduction to probability theory, we shall deal with simple experiments producing a relatively small number of outcomes. However, the basic principles of probability are the same in simple problems as in more complicated ones. As you read, try to understand these principles, so that you can apply them to more complex situations. It will help you to grasp the subject if you work out all the situations given in the text as well as the special problems.

Probability and Games of Chance

Games of chance involving coins, cards, and dice provide us with simple experiments producing a small number of outcomes. The use of such experiments to illustrate the principles of probability is historically appropriate. Historians tell us that a 17th-century French gambler, the Chevalier de Méré, was interested in the odds involved in a game of chance played with dice. He decided to get in touch with a famous mathematician and scientist, Blaise Pascal, so that the latter might help him with his calculations. Pascal became intrigued with the questions that arose in his study of de Méré's problem. He began a correspondence with other mathematicians concerning the matter, and this led to the development of probability theory.

Consider the simple experiment of tossing coins. There are two possible outcomes when we toss a single coin: heads or tails. (We would ignore any toss in which the coin came to rest on its edge.) If the coin is in fair condition and if we toss it vigorously, it seems reasonable to say that heads and tails are equally likely outcomes. A mathematician would say that the probability of heads is ½—that is, that the coin would land heads one time in two on the average.

Suppose we complicate this experiment just a little by tossing the coin twice or, what comes to about the same thing, tossing two coins. We can see that three things might happen in this experiment. We could get two heads, or one head and one tail, or two tails. The diagram in Figure 1 illustrates the possible outcomes in this experiment.

If we examine the diagram carefully, letting H stand for "heads" and T for "tails," we see that there are really four individual outcomes, or elementary events, possible—HH, HT, TH, and TT—when we toss two coins. Note that HT is not considered the same thing as TH. We might, for example, use a nickel and a dime as our two coins.

Since each of the four individual outcomes is equally likely to occur, we would expect to obtain HH one time in four, or, as a mathematician would indicate it, $P(HH) = ¼$. Similarly, we would say $P(HT) = ¼$; $P(TH) = ¼$; $P(TT) = ¼$. Because what happens to the first coin has no effect upon what happens to the second coin, we say that the two tosses are *independent*. When two events are independent, we can use their individual probabilities to compute the probability that both will happen in a single trial of an experiment. In this case, we could have used the probabilities associated with tossing a single coin (that is, $P(H) = ½$; $P(T) = ½$) in order to compute the probabilities

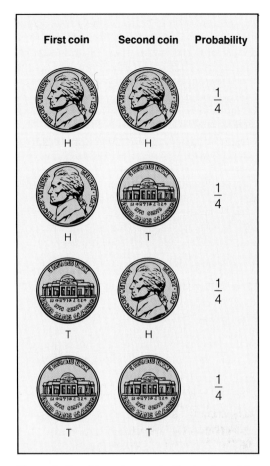

Figure 1. Possible outcomes in the tossing of two coins. An analysis of the likelihood of each of the four possible outcomes appears in the text.

associated with tossing two coins. For example, $P(HT) = P(H) \times P(T) = \frac{1}{2} \times \frac{1}{2} = \frac{1}{4}$; $P(HH) = P(H) \times P(H) = \frac{1}{2} \times \frac{1}{2} = \frac{1}{4}$. Since the event "one head and one tail" could occur in two ways, either HT or TH, we would expect to obtain "one head and one tail" two times in four on the average and would assign probability to this event.

We could also use the individual probabilities assigned to the outcomes HT and TH to compute the probability that *either* HT *or* TH will occur. When two events cannot both occur in a single trial of an experiment, the probability that one or the other will occur is the sum of their individual probabilities. Thus $P(HT \text{ or } TH) = P(HT) + P(TH) = \frac{1}{4} + \frac{1}{4} = \frac{1}{2}$. In these examples, we obtain the probability of an event either by counting outcomes of an experiment or by using previously determined probabilities.

We can apply the same principles in other experiments. Consider the question "How often should we expect all three of the children in a family to be boys?" The diagram in Figure 2 illustrates the possible outcomes. If we assumed that equal numbers of boys and girls are born (this is not quite true), we would say that the probability of a boy, $P(B)$, is ½ and that the probability of a girl, $P(G)$, is also ½. The possible outcomes for this experiment would be equal in number to the outcomes for tossing three coins—BBB, BBG, BGB, BGG, GBB, GBG, GGB, and GGG. Since BBB occurs in only one of the eight possible outcomes, we would say $P(BBB) = \frac{1}{8}$. We could also have used the rule for computing the probability that all of several independent events will occur: $P(BBB) = P(B) \times P(B) \times P(B) = \frac{1}{2} \times \frac{1}{2} \times \frac{1}{2} = \frac{1}{8}$. Either way, we would conclude that we should expect all three children in a family to be boys about one time in eight.

Sample Spaces

To determine the probability of a particular outcome of an experiment, we must be able to identify all the possible outcomes. Mathematicians call the set of possible outcomes of an experiment a *sample space* for the experiment. In the simple examples, we shall usually find it convenient to list the sample spaces.

When we considered the experiment of tossing a single coin, our sample space consisted of the two possible outcomes H and T (heads and tails). When we extended the experiment to the tossing of two coins (or one coin twice), we used a sample space of HH, HT, TH, TT. In the experiment of choosing a single student from a known school population, the sample space involved was the set of all students in the school. Since we were concerned only with whether we chose a boy or girl and not with which boy or girl, we could have used a sample space of only two elements—boy and girl. In this case, B could be used to

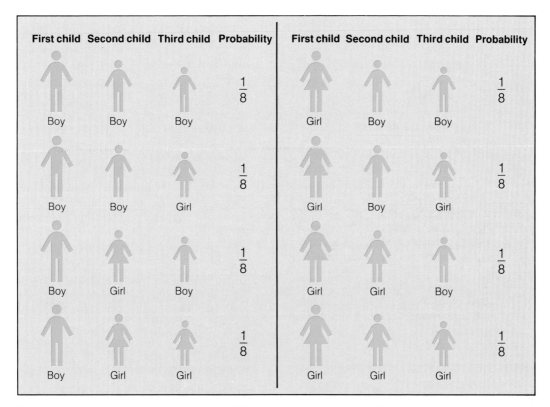

Figure 2. The diagram shows that the probability of having either three girls or three boys in a family of three children is 1/8. The probability of the next child being a boy is 1/2, the same as the probability of the first child being a boy.

stand for the set of boys and G for the set of girls. Each individual outcome of a sample space is called a *point* or an *elementary event*.

Although a single experiment will produce only one set of individual outcomes, we may be able to use any one of several different sample spaces, depending on what we are investigating. For example, consider the experiment of drawing one card from a well-shuffled deck of cards. If we are concerned with the individual card drawn, we could use a sample space consisting of 52 elements—every single card in the deck. In the same experiment, we might be concerned only with the face value of the card shown. In that case, our sample space would consist of only 13 elements (2, 3, 4, 5, 6, 7, 8, 9, 10, jack, queen, king, ace). Or we might be interested only in the suit drawn. In that case, we would use the sample space of clubs, diamonds, hearts, spades. If we were concerned only with color, we could use a sample space of only two elements (red, black). In most experiments, we use the sample space identifying the characteristics on which we wish to concentrate in the experiment. We can use the sample space of elementary events to build up other sample spaces by collecting the elementary events with like characteristics.

You have probably played board games, such as Monopoly and Parcheesi, in which your moves were determined by the throw of a die or a pair of dice. A sample space for the experiment of tossing a single die would be the set {1, 2, 3, 4, 5, 6}. It is rather more difficult to generate a sample space for the tossing of a pair of dice. To help us keep things straight, let us suppose that one die is red and the other one green. An outcome of 4 on the red die and 3 on the green die would be different from an outcome of 3 on the red die and 4 on the green die. If we agreed to write the outcome on the red die first and the outcome on the green die second, we could show this difference by setting down the pairs (4,3) and (3,4).

Figure 3 shows the sample space for the experiment of tossing a pair of dice. Suppose that you tossed the dice and that the red die came up 4 and the green die came up 3. Moving horizontally from 4, at the extreme left of the table, until we came to the column headed by 3, we would find the pair (4,3). We shall have occasion to refer to this sample space several times in the course of this article.

See if you can list sample spaces for the following experiments:

1. Toss a coin and a die. [Hint: two points in this sample space could be (H,3) and (T,2).]

2. Toss a nickel, dime, and quarter. If the nickel comes up heads, the dime tails, and the quarter heads, we can indicate this by (H,T,H). You should find 8 points for this sample space.

Probability of an Event

An *event* is by definition any subset of a sample space. In other words, if each member of set A is also a member of set B, we say that A is a subset of B. If set A = {1, 2, 3} and set B = {1, 2, 3, 4, 5}, then set A is a subset of set B. Thus an event is a set of individual outcomes of an experiment. An elementary event, as we have indicated, is a single individual outcome. In order to assign probabilities to an event—that is, to a set of individual outcomes—we must first be able to assign probabilities to individual outcomes of the experiment.

If we consider, for example, the experiment of drawing a single card from a deck of cards, there are 52 elementary events or individual outcomes possible. If we make a random draw, each card has exactly the same chance of being drawn. Thus to each of the 52 possible outcomes we would assign the same probability: $\frac{1}{52}$. Note that the sum of the probabilities assigned to the elementary events is 1. The probability of an event that is certain is also 1.

The same basic principle may be used to answer the question "What are your chances of winning a sweepstakes contest or the door prize at a party?" In each case,

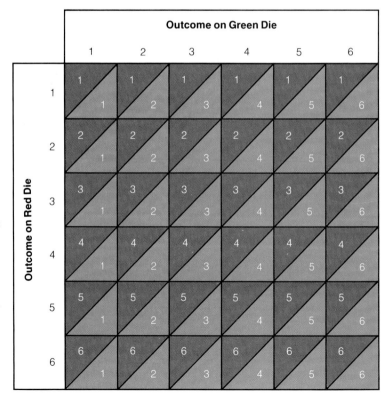

Figure 3. At the right is a sample space for the tossing of a pair of dice. It is assumed that one of the dice is red and the other green; also that the outcome of the red die is given first in each of the pairs shown. Note that in the text the two outcomes, separated by a comma, are given in parentheses. Thus we would refer to (2,5), (4,3), (6,1) and so on.

if the total number of tickets is *n* and you hold only 1 ticket, then you would have only 1 chance in *n* of winning. Thus the probability of your winning would be $1/n$.

In the experiment of drawing a card from a deck, we may be concerned only with whether the card is an ace. Then the event with which we are concerned consists of four elementary events: the ace of clubs, the ace of diamonds, the ace of hearts, and the ace of spades. We still would have to draw at random from the entire deck of cards, numbering 52. Since 4 out of the 52 cards belong to the event just described (4 aces), we would say that the probability that the outcome is an ace is $4/52$. We could use the symbols $P(\text{ace}) = 4/52$ to represent this statement.

Suppose that in the preceding experiment, we were concerned only with the color of the card chosen. Do you see why $P(\text{red}) = 26/52$? We can use the probabilities assigned to elementary events to assign probabilities to the points in other sample spaces. For example, if we were using the sample space clubs, diamonds, hearts, spades, we could assign the probability of $1/4$ to each point in the sample space since $P(\text{clubs}) = 13/52 = 1/4$.

Rule 1.

If an experiment can result in *n* different but equally likely outcomes and if *m* of these outcomes correspond to event X, then the probability of the event is $P(X) = m/n$.

Applying this rule, let us find the probabilities of some events in the experiment of tossing a pair of dice. Figure 3 lists 36 possible outcomes for this experiment. Thus we could assign a probability of $1/36$ to each of the elementary events. Now consider the event "The sum of the numbers shown is 7." If we let *r* stand for the number on the red die and *g* for the number on the green die, we can represent the event as $r + g = 7$. How many of the elementary events correspond to this amount? In other words, how many of the pairs in the table add up to 7? Consult the table to find the answer. You will see that $P(r + g = 7) = 6/36$. If we considered the event "The same number appears on both dice," we could call the event "$r = g$" or, as is common in many games played with dice, "double." There are six such pairs: (1,1), (2,2), (3,3), (4,4), (5,5), and (6,6). Thus $P(r = g) = P(\text{double}) = 6/36$. What is $P(r + g > 7)$? (The symbol $>$ stands for "is greater than.") Are you able to find 15 points in the sample space that correspond to this event? You should be able to, for $P(r + g > 7) = 15/36$.

These examples should suggest a way to answer the question "How likely are you to land on 'Boardwalk' in your next turn in a game of Monopoly?" Suppose, for example, that you are located on "Pennsylvania Avenue," which is 5 spaces from "Boardwalk." To determine the probability of your landing on "Boardwalk," carry out the experiment of tossing two dice. You would land on "Boardwalk" if the numbers shown on the dice totaled 5. What is $P(r + g = 5)$?

Try the following problems:

1. If a box contains 4 red marbles and 5 white marbles, what is the probability of drawing a red marble on the first try?
2. What is the probability that you will draw a face card (jack, queen, king) from a deck of cards?
3. What is the probability that you will get a 5 when you toss a single die?
4. What is the probability that you will not get a 5 when you toss a single die?

Complementary Events

There is a relationship that frequently simplifies the computing of a probability. Consider the example pictured in Figure 4. There are 10 points in the sample space S. The event A contains 4 of these points and the remaining 6 points are not in A. The set of points in sample space S that are not in a given set A is called the *complement* of A and is usually indicated by the symbol $\overline{A}$. If the elementary events in the sample space S are equally likely, then $P(A) = 4/10$ and $P(\overline{A}) = 6/10$.

In a sample space of n equally likely events, if an event A occurs in m of the outcomes, then the event $\overline{A}$ will occur in $n-m$ outcomes. Thus if $P(A) = m/n$, then $P(\overline{A}) = (n-m)/n$. Now $(n-m)/n = 1 - m/n$. We obtain this result by dividing both the numerator and the denominator by n, as follows:

$$\frac{\frac{n}{n} - \frac{m}{n}}{\frac{n}{n}} = \frac{1 - \frac{m}{n}}{1} = 1 - \frac{m}{n}$$

We now can give the following general rule:

Rule 2.

$$P(\overline{A}) = 1 - P(A)$$

Sometimes it is easier to compute the probability of one of two complementary events than it is to compute the probability of the other. In such cases, we compute the easier probability. We then use the relationship indicated in Rule 2 to derive the probability of the complementary event. For example, in some games played with dice, there is either a premium or a penalty associated with throwing doubles. We could compute the probability of not throwing a double by counting the sample points in Figure 3 that are not doubles. (There are 30 of them.) Or we could compute the probability of throwing a double ($r = g$) and use the relationship in Rule 2 to derive the probability of not throwing a double ($r \neq g$). (The symbol $\neq$ stands for "is not equal to.")

Probability of "A or B"

A situation that often arises is the finding of the probability of an event that might be expressed as "either event A or event B." By the event (A or B) we mean the set of outcomes that correspond to either event A or event B, or possibly both A and B.

Consider the example shown in Figure 5. There are 10 points in the sample space S. In this sample space, $P(A) = 3/10$ and $P(B) = 2/10$. Since the event (A or B) contains the 3 elements of A and the 2 elements

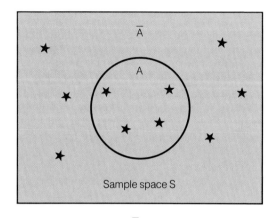

Figure 4. Subsets A and $\overline{A}$ are complementary events in sample space S.

of B, then $P(A \text{ or } B) = P(A) + P(B) = 5/10$. In this instance, there are no elements that are in both A and B; so A and B here are *mutually exclusive* events, meaning that they cannot both occur in a single trial of an experiment. When events A and B are mutually exclusive, we can find the probability of (A or B) simply by adding the probability of A to the probability of B as follows:

Rule 3.

If events A and B are mutually exclusive, then
$$P(A \text{ or } B) = P(A) + P(B).$$

For example, suppose we were required to find the probability of throwing either a 7 or a double in a toss of two dice. Since 7 is an odd number, it is not possible for a single toss of two dice to produce both a 7 and a double. Thus the event ($r + g = 7$) and the event ($r = g$) are mutually exclusive. We have already seen that $P(r + g = 7) = 6/36$ and that $P(r = g) = 6/36$. Hence, applying Rule 3, $P(r + g = 7 \text{ or } r = g) = 6/36 + 6/36 = 12/36$. Can you verify this by counting the appropriate points in the sample space shown in Figure 3?

When both events are not mutually exclusive—that is, when they *can* both occur in a single outcome—we cannot use the addition principle of Rule 3. Consider the situation presented in Figure 6. In this example, the sample space S consists of 10 equally likely elementary events; 5 outcomes correspond to event A, and 4 outcomes to event B. Hence $P(A) = 5/10$ and $P(B) = 4/10$. If we were to apply Rule 3 here, we would have $P(A \text{ or } B) = P(A) + P(B) = 5/10 + 4/10 = 9/10$. This answer would be incorrect. The reason is that events A and B are not mutually exclusive, since 2 outcomes belong to both A and B. To obtain a correct answer in this case for $P(A \text{ or } B)$, we need to apply the following rule:

Rule 4.

If events A and B are not mutually exclusive, then
$P(A \text{ or } B) = P(A) + P(B) - P(A \text{ and } B)$.

Therefore, for Figure 6, $P(A \text{ or } B) = 5/10 + 4/10 - 2/10 = 7/10$.

Let us consider another example in which this rule could be used. Suppose we wanted to find the probability of drawing either a face card or a spade from a deck of cards. $P(\text{spade}) = 13/52$ and $P(\text{face card}) = 12/52$. (Remember that each of the 4 suits has 3 face cards: jack, queen, and king.) There are 3 cards that are both face cards and spades. Therefore $P(\text{spade and face card}) = 3/52$. Applying Rule 4, we would conclude that $P(\text{spade or face card}) = 13/52 + 12/52 - 3/52 = 22/52$. Could you list the 22 cards that would belong to the event (spade or face card)?

Here are two rather simple problems for you to try:

1. What is the probability that you could throw either a 7 or an 11 on a single throw of 2 dice?
2. What is the probability that you would throw either a double or a total of more than 9 on a single toss of 2 dice?

Probability of "A and B"

When two events, A and B, are *independent*, we can compute the probability of the event (A and B) by using the following rule, involving the multiplication principle:

Rule 5.

If events A and B are independent, then
$P(A \text{ and } B) = P(A) \times P(B)$.

Intuitively, we would expect two events to be independent when they have nothing to do with each other. Although this is usually a reliable rule of thumb, it must be used with care. For example, when two events are mutually exclusive, our first thought might be that they have nothing to do with each other. But when one of two mutually exclusive events occurs, the other cannot possibly occur. Thus the occurrence of one of these events certainly affects the probability that the other also occurs. Mutually exclusive events, therefore, are never independent of one another. For two events to be independent, the occurrence of one must not affect the probability that the other occurs at the same time.

Here is a problem involving two independent events, in which Rule 5 could be applied: Suppose that we toss two dice, one red and one green. What is the probability that the number 1 would come up on the red die and that the number 3 would come up on the green die? In this case, the outcome on the green die has nothing to do with the outcome on the red die. We therefore have two independent events. The probability of any one of the numbers 1, 2,

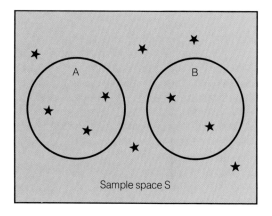

Figure 5. Subsets A and B are mutually exclusive events in sample space S.

3, 4, 5, and 6 coming up on the toss of a single die is ⅙. The probability of the outcome (1,3) on a toss of two dice would be calculated as follows, in accordance with Rule 5: $P(1 \text{ and } 3) = P(1) \times P(3) = 1/6 \times 1/6 = 1/36$.

By using the multiplication principle, we can provide an answer of sorts to the question "If a baseball player averages 3 hits in 10 times at bat, what are his chances of getting 4 consecutive hits?" If we assume that the times at bat are independent trials, and if we let H stand for each hit, we could compute the possibility of 4 consecutive hits in this way: $P(HHHH) = P(H) \times P(H) \times P(H) \times P(H) = 3/10 \times 3/10 \times 3/10 \times 3/10 = 81/10{,}000$. We would expect the batter to get 4 hits in a row about 81 times in every 10,000 sequences of 4 times at bat. This probability would hold if we were right in assuming that the 4 times at bat would be independent trials. The assumption might not be a sound one in this case. For one thing, the batter would probably hit better against certain pitchers than against others. Also, getting a hit or not getting a hit might have an effect on his chances the next time at bat.

Consider the event (double and even) in the experiment of tossing two dice. By "double," of course, we mean that the same number would come up on both dice; by "even," that the sum of the two numbers would be an even number. In this case, $P(\text{double}) = 6/36$ and $P(\text{even}) = 18/36$. If we applied the multiplication principle for independent events, as stated in Rule 5, we would have $P(\text{double and even}) = 6/36 \times 18/36 = 108/1{,}296 = 3/36$. But if we examine Figure 3, we will find 6 points that correspond to the event (double and even). Therefore the true value of $P(\text{double and even})$ is $6/36$ and not $3/36$, as our calculation had seemed to indicate. The reason is that in this case the two events are not independent. If we throw a double, we are certain to throw an even number. This means that if we have thrown a double, the probability that we have also thrown an even number is 1. Again, if we know that we have thrown an even number, the probability that we have also thrown a double is ⅓. This is

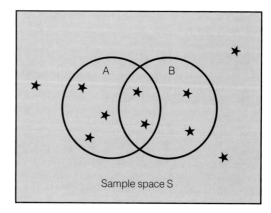

Figure 6. Subsets A and B are nonmutually exclusive events in sample space S.

determined by counting the number of ways an even number can occur (18 in all) and the number of these occurrences that are doubles (6).

In order to calculate $P(A \text{ and } B)$ when the events are not independent, the multiplication principle for independent events (Rule 5) can be generalized as follows:

Rule 6.

If events A and B are not independent, then
$P(A \text{ and } B) = P(A) \times P(B/A)$.

The term "B/A" would be read as "B given A." When we know that the event A occurs, we can compute $P(B/A)$ by thinking of the event A as a reduced sample space. Consider Figure 7. The sample space S consists of 10 equally likely events, of which 5 are in A, 3 are in B, and 2 are in both A and B. Given that A occurs, then any outcome obtained is one of the 5 in A. If we think of these 5 points as a new sample space, then 2 of the equally likely outcomes belong to the event B. Thus $P(B/A) = 2/5$. Applying Rule 6, we can compute $P(A \text{ and } B) = P(A) \times P(B/A) = 5/10 \times 2/5 = 10/50 = 2/10$. This, as you can see, agrees with what we would obtain by a direct count of the points that belong to both A and B in Figure 7.

If events A and B are independent, then the probability that B occurs will not be affected by the fact that A occurs. Thus for independent events, $P(B/A) = P(B)$. Suppose we were to select 1 card from a deck of 52 cards. What is the probability that the card selected will be both a red card (R) and a face card (F)? Since half the cards are red, we know that $P(R) = 26/52$. Of the 26 red cards, 6 are face cards. Hence $P(F/R) = 6/26$. Since $P(F) = 12/52 = 6/26$ and $P(F/R)$ is also $6/26$, the events F and R are independent and $P(R \text{ and } F) = P(R) \times P(F)$. Thus the multiplication rule for independent events (Rule 5) is actually just a special case of the more general Rule 6: $P(A \text{ and } B) = P(A) \times P(B/A)$.

Here are some problems involving the probability of event (A and B). (Assume that cards are drawn from a regular 52-card deck.)

1. A coin is tossed and a die is thrown. What is the probability of obtaining heads and a 3?

2. A card is drawn and a die is tossed. What is the probability of getting two 6's?

3. A card is drawn. What is the probability that it is niether a face card nor a red card?

Degree of Confidence

In this brief introduction to probability, we have touched on only a few of the basic principles involved. The scope of probability theory is much broader than our discussion would suggest. For example, we simplified matters a good deal by dealing only with situations in which the individual outcomes of an experiment were equally likely. In many situations this is not the case. Calculations then become more complex, and new dimensions are added to the study of probability.

In our simplified presentation, we merely pointed out that predictions based on probability are uncertain. Scientists accept this but want to know *how* uncertain. They recognize that in trying to predict the outcome of an experiment subject to the laws of chance, they may often be wrong. They need to know how often and how wrong. They want to establish the degree of confidence they can place in their predictions. Probability theory provides the basis for establishing this degree of confidence.

Many Uses

If you go at all deeply into the study of mathematics, you will find occasion to work with some of the more complicated and intriguing aspects of probability theory. You will also obtain a clearer idea of the many ways in which this theory serves man: It enables scientists to fix a limit within which the deviations from a given physical law must fall if these deviations are not to count against the law. It has been used to calculate the positions and velocities of electrons orbiting around the nuclei of atoms. The fluctuations in density of a given volume of gas have been analyzed by applying probability theory. It has played an important part in genetics; among other things, it has made it possible to calculate the percentage of individuals with like and unlike traits in successive generations. Manufacturers use probability theory to predict the quality of items coming off mass-production lines. Insurance experts make extensive use of the theory; It enables them, for example, to calculate life expectancies so that they may set appropriate life-insurance rates. Finally, because of probability theory, electronic computers can be programmed to predict the outcome of elections on the basis of comparatively few returns and with what is generally a surprising degree of accuracy.

F. Joe Crosswhite

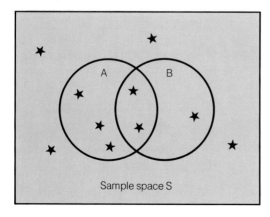

Figure 7. Subsets A and B are nonindependent events in sample space S.

GAME THEORY

Five criminals huddle in a building after committing a crime. Outside waits a lone policeman, determined to capture the leader. The criminals do not know he is out there. Nonetheless, they plan to leave randomly, one at a time, to avoid attracting attention. All the policeman knows is that the leader is the tallest of the criminals. He cannot capture more than one of the men, for he would be overpowered. He has no way of getting help. As the men come out, which should the policeman arrest? The first? The second? The last?

If he made the arrest on a random basis, the policeman would have only a 20 percent chance of getting his man. Game theorists say he can do better by following a certain strategy, or plan of action: Let the first two men go, and arrest the next man to come out who is taller than the first two men. If he does this, the policeman stands a 40 percent chance of capturing the leader.

This is a simple problem in game theory, a branch of mathematics that deals with risk in conflict situations. It gets its name from the fact that so many conflict situations in the real world are similar in basic structure to bridge, poker, checkers, chess, even tic-tac-toe. Both in games and in everyday events, people compete with one another. There are rules by which the game must be played; there are outcomes, or payoffs—such as win, lose, and draw—that result from the opponent's different moves, or strategies; and there is generally some information available to one or more of the players involved.

In game theory it is assumed that the objective of each player is to maximize gains or minimize losses. It is also assumed that the player is faced with a rational opponent whose objective is to do the same. Of course, the cop-and-robbers situation is really a *one-person game*. The robbers

Two boys playing chess, a zero-sum game involving a pair of players following set rules, with a clear objective—simply to win.

Mimi Forsyth, Monkmeyer

424 GAME THEORY

The complicated game of football involves two teams pitted against each other to win a game by attaining the higher numerical score. There are rules, but all kinds of strategies and tactics can be worked out. Left, *the coach explains a play to members of the team.* Right, *a play in action.*

don't know they are competing with the policeman. If they did, they would try a strategy of their own. They might assume that the policeman is a game theorist and therefore send their leader out first. On the other hand, the policeman might realize that they would think this, and therefore arrest the first man to come out. In short, a game can be complicated.

Aside from one-person games, which are apt to have little value and be easy to solve, perhaps the simplest games are games of two individuals or teams where the losses of one are the gains of the other. This kind of game is called a *two-person zero-sum game*. In a zero-sum game, what one side gains, the other side loses—and vice versa. The total value of the game doesn't increase or decrease.

It is easy to imagine games where this is not the case. Two advertisers competing for a market, for example, might not only increase or decrease their shares of the market through their advertising, but might actually increase the total market. To best illustrate game theory, however, we will stick to the two-person zero-sum game.

Case of the Complicated Collection

Most games are finite. That is, there are only a certain number of strategies, or alternatives, that each player can follow, and the game is "solved" in a certain number of plays, or moves. Some games are also games of *perfect information,* that is, we know, or can see, each move our opponent makes. Checkers, chess, and tic-tac-toe are examples of finite games of perfect information. Chess tends to be interesting because so many different moves are possible. A game such as tic-tac-toe tends to be less interesting because there are so few moves and it can be solved so easily. In fact, in tic-tac-toe the first player can be assured of at least a draw, or tie, if the first mark is put in one of the corner squares. The second player can guarantee a draw by making a mark in the center square.

To illustrate game theory, let's consider a finite, two-person zero-sum game of *imperfect information.* Let's assume that Abe owes you $60 and Bill owes you $40. Abe and Bill get paid on Friday, and the owed money will be available from each only on payday at 5 P.M. at their respective places of employment. Because you cannot be in two places at once, you will be able to collect from either Abe or Bill, not from both. The situation is further complicated by the fact that these two men owe the same amounts of money to a fellow we will call your opponent: Oscar.

Whoever is at Abe's or Bill's place of employment at 5 P.M. gets the money owed there. If both of you show up at one place, you split the money. But if neither of you shows up, Oscar gets the money because he is a special friend of both Abe and Bill. Both you and Oscar have this information. Thus you must determine where you should go to maximize your gain: to Abe's or Bill's?

This is a problem that can be solved by using game theory. It is a conflict situation: You want to get as much money as you can, while Oscar, who could get all the money if you didn't go to either place, wants to minimize his losses. It is a zero-sum game because the total amount to be paid out ($100) stays the same whatever strategies are used.

Choice of Strategies

The game you and Oscar play is called a "2 × 2 game"; there are 2 strategies available to you and 2 available to Oscar: Each of you can go to see either Abe (strategy A) or Bill (strategy B). Some two-person games have many more strategies available to the players. For example, in a 2 × 3 game, your opponent has 3 strategies to choose from while you have only 2. In a 4 × 3 game, you have 4 strategies to your opponent's 3. In a 3 × m game, m means that your opponent has many strategies from which to choose.

Large games can usually be reduced to more manageable proportions because some of the strategies will not make sense in terms of the game. For example, if all the payoffs for a particular strategy are zero, there is no reason to choose this strategy. It may as well be eliminated from consideration.

We can best illustrate a game by a *payoff matrix*. This indicates what happens when players select their different strategies. Figure 1 shows a payoff matrix for the game between you and Oscar. It is conventional to have a matrix show payoffs for the player whose strategies are listed on the left side of the matrix. This is the player who wants to maximize his or her gains. The payoffs shown are gains, in a zero-sum game, for the player whose strategies are listed at the left. The payoffs are losses for

Paul Conklin, Monkmeyer

A shopper studying brands of products in a grocery. Manufacturers and advertisers apply game theory to capture a share of the food market.

the person who is named at the top of the matrix.

The matrix in Figure 1 shows that if both you and Oscar choose strategy A (going to see Abe), your payoff will be $30. If you choose A and Oscar chooses B (going to see Bill), your payoff will be $60. If you choose B and Oscar chooses A, your payoff will be $40. If you both choose strategy B, your payoff will be $20.

Some additional figures are shown outside the squares in Figure 1. These help us "solve" the problem following game-theory procedure. The particular theory we will follow was first proposed by the mathematician John Von Neumann in 1929 and extensively developed in a book he wrote with Oskar Morgenstern, published in 1944, titled *Theory of Games and Economic Behavior.*

The figures outside the matrix are called *rim figures*. The column at the right gives the *row minima:* the minimum amount in the row opposite which it appears. The row at the bottom of the matrix gives the *column maxima:* the maximum amount in the column under which it appears. Game theory says that you should choose the strategy that provides the "maximum of your minimum" gains. This is called your *maximin*. Your opponent rationally should choose the strategy that provides the "minimum of his or her maximum" losses. This is called the opponent's *minimax*. In Figure 1, your maximin and Oscar's minimax are circled. By choosing strategy A you guarantee yourself a gain of at least $30; if your opponent should choose strategy B, you will get the entire $60 from Abe. Oscar, however, will likely choose strategy A because it guarantees him a maximum loss of $40. Either one of you could, of course, try to outsmart the other. But then you would be *gambling*, and this is what game theory avoids. It is, instead, a mathematical technique for guaranteeing a minimum gain (or maximum loss) from a conflict situation.

Mixed Strategies and Saddle Points

We assumed that your game with Oscar would be played only once. That is,

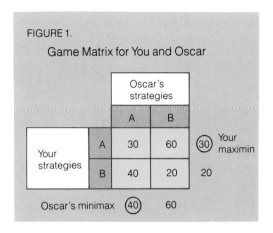

FIGURE 1.

Game Matrix for You and Oscar

each player was to make only one move. But most games, both in real life and among parlor games, involve a number of moves. Matching pennies is a good example. In this game, if you choose a "pure" strategy, such as playing heads all the time, you could wind up losing your shirt (or at least all your pennies). Frequently, such a situation calls for a "mixed" strategy, such as playing heads and tails each 50 percent of the time on a random basis. Tossing the coin before each play would achieve this aim. In fact, even those who know nothing about game theory usually play the game this way.

Now lets look at your game with Oscar again and assume that it is to be played many times. Does a pure strategy then make sense? Not quite. We've determined that Oscar will choose strategy A because, if he doesn't, you might get $60 instead of $30. But let's fool him. Let's choose strategy B, getting $40 instead of $30. How long will Oscar allow this? If you choose B, then he also will choose B, thus reducing your winnings to $20. You can switch back to strategy A and gain $60. But he also will switch back to strategy A. Before long, both of you are employing mixed strategies, trying to best each other. Now the question is what kind of mixed strategy should you employ?

Before we answer this question, let's consider whether or not a mixed strategy should be used in a particular game. Game theory says that a mixed strategy should

GAME THEORY

not be used if the game has a *saddle point*. A game is said to have a saddle point if the maximin equals the minimax; that is, if the maximum of your minimum gains equals the minimum of your opponent's maximum losses. The game with Oscar does not have a saddle point, as you can see by comparing the two circled numbers in Figure 1. Therefore, the game calls for a mixed strategy. If the two circled numbers were equal, then the game would have a saddle point, and each player should choose the pure strategy dictated by his or her maximin or minimax.

To illustrate this, consider Figure 2, in which we have changed the payoffs (or *payouts,* as they are sometimes called) to create a saddle point for the game. Your maximin equals Oscar's minimax. Under these conditions it doesn't make sense for you to choose strategy B. You would only lose. Oscar, who assumes that you are rational, cannot do other than choose strategy A, too. If he chooses strategy B he stands to lose $100 instead of $50. So both of you will continue to choose strategy A. Such a game is said to be *strictly determined.*

Value of the Game

Every game has a certain value. The value of the strictly determined game in Figure 2 is $50. The value of the original game (in Figure 1), played on a one-time basis, is $30. (Both these values are, of course, yours, and we will continue to

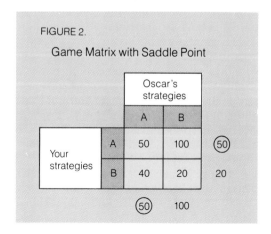

FIGURE 2.
Game Matrix with Saddle Point

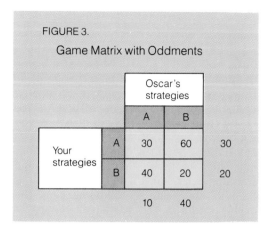

FIGURE 3.
Game Matrix with Oddments

speak of the value of the game in your terms.)

If the original game is to be played on a continuing basis, both you and Oscar will begin mixing your strategies, each trying to get the better of the other. What the two of you are really fighting over is the $10 spread between your guaranteed minimum gain of $30 and his guaranteed maximum loss of $40—if each of you plays a pure strategy. By mixing strategies each of you hopes to get as much of this $10 as possible. This suggests that the value of the game lies somewhere between $30 and $40. How can Oscar assure himself that you get no more than that amount? The answer lies in mixing strategies in a prescribed way.

After determining that the game does not have a saddle point, you may erase the rim figures. Now, determine the absolute differences between the payoffs in each row and column and write these figures in the margins. These figures are called *oddments,* or part of the odds in using each of the strategies. We've done this for you in Figure 3. You can see that the differences in the row payoffs are $30 and $20, while the differences in the column payoffs are $10 and $40. These figures are the ratios with which the different strategies should be played (3 to 2 for you, 4 to 1 for Oscar). Each figure, however, applies to the opposite strategy. In other words, your mixed strategy should be to play strategy B three times and strategy A two times out of every five plays; Oscar's mixed strategy should be to play strategy A four times and strategy B one time in the same number of plays.

Even though Oscar can easily figure out your overall strategy, it is best to keep him guessing about each individual move. Otherwise he might outfox you. You should choose each move on a random basis. For example, you might throw three red cards, representing strategy B, into a hat together with two black cards representing strategy A. Before each move mix the cards and withdraw one. Let this card determine the strategy you choose. Throw the card back into the hat and repeat the process for the next move. In the long run you will play your strategies in a ratio of 3 to 2, and the probability is high that you will achieve the *value* of the game. The value of the game is easily determined. For each row, multiply the oddment times each payoff in the other row. Add the products together and divide by the sum of the oddments. Average your answers for the rows. This gives the value of the game if you play your prescribed mixed strategy. Do the same for the columns. This gives the value of the game if your opponent plays his prescribed mixed strategy. These two values should be the same. Over the long haul, the value of the game will be achieved if both of you play your prescribed mixed strategies. If one player deviates, chances are that player will not achieve the value of the game.

The arithmetic involved in determining the value of the game is illustrated below:

rows:
$$\frac{20 \times 30 + 20 \times 60}{20 + 30} = 36$$
$$\frac{30 \times 40 + 30 \times 20}{20 + 30} = 36$$
$$\text{Value of the game} = \frac{36 + 36}{2} = 36$$

columns:
$$\frac{40 \times 30 + 40 \times 40}{40 + 10} = 56$$
$$\frac{10 \times 60 + 10 \times 20}{40 + 10} = 16$$
$$\text{Value of the game} = \frac{56 + 16}{2} = 36$$

The value of the game, if both players play rationally (i.e., use their prescribed mixed strategies), is $36.

Non-Zero-Sum and N-Person Games

When we move away from the two-person, zero-sum game, problems become a bit more complicated. For one thing, psychology, negotiation, and communication may become factors in the problems.

A non-zero-sum game, you will recall, is one in which the losses of one player are not necessarily the gains of the other. That is, the total value of the game does not necessarily remain the same throughout the play. An example of a two-person, non-zero-sum game is "the prisoners' dilemma":

Imagine that two criminals suspected of a bank robbery are arrested and placed in separate jail cells so they cannot communicate. If one confesses and turns state's evidence, he will go free. The other will be sentenced to 20 years in prison. If both criminals confess and throw themselves on the mercy of the court, they will each receive a 5-year sentence. If neither confesses, they will each get 1 year for carrying concealed weapons. If you were one of the prisoners, what would you do?

This conflict situation is pictured in Figure 4. The strategies are confession (C) and no confession (NC). Because this is a non-zero-sum game, the payoffs for both players are listed in each square, with player A's payoff to the left of the comma

FIGURE 4.

Game Matrix for Prisoners' Dilemma

		Prisoner B	
		C	NC
Prisoner A	C	5,5	0,20
	NC	20,0	1,1

and player B's payoff to the right. Assume you are player A, and study only your own payoff. As a good game theorist, you should choose the strategy that provides your minimax, the minimum of your maximum losses. Thus you should confess. The worst that could happen would be a 5-year sentence.

Your opponent, prisoner B, should choose the same strategy. But this may be oversimplifying the game. Suppose you believe that your opponent is the kind of person who would never confess. What would you do then? It depends on what type of person you are.

Suppose you two were cellmates and could discuss the situation. Would you both agree not to confess? That would make sense. But what about the danger of a double-cross? As you can see, for many (if not most) games, mathematics does not have all the answers.

Games involving more than two people (*n*-person games) can be even more complex. Imagine a situation involving three people: A, B, and C. If A cooperates with B they will split $6. If A cooperates with C they will split $8. If B cooperates with C they will split $10. This is illustrated in the triangle in Figure 5. Who should cooperate with whom to divide the money?

"Look," A says to B, "cooperate with me and I'll give you $4 and keep only $2."

"Don't be silly," says B to A. "I can probably get $5 cooperating with C."

"Nothing doing," says C. "I think I can make a better deal with A. How about it, A? Will you take $2 and leave me $6?"

"Okay," says A.

Then B panics. "Look, A," he says, "I'm willing to split the $6 fifty-fifty."

"Don't lose your head," says C to B. "I'll give you $4.

"But that's what *I* offered you," says A to B.

Round and round it goes. The division of the money will depend on the negotiating talents of those involved. All game theory can tell us is that, by cooperating with one of her opponents, the most A can hope to receive is $2; the most B can hope to get is $4; and the most C can hope for is $6.

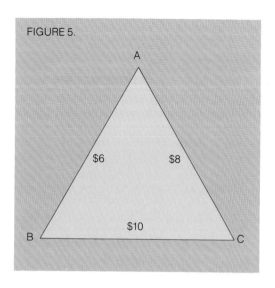

FIGURE 5.

But let's change the rules. Suppose A, B, and C could gain a total of $18 if all three of them cooperated. How should the money be split? Using the previous resolution, we can determine what ratio of the total each player should receive. First we add the maximum amounts receivable by each player, according to the original rules: $2 + $4 + $6 = $12. Player A's ratio, therefore, is $2/12$, or $1/6$; B's is $4/12$, or $1/3$; C's is $6/12$, or $1/2$. If we apply these ratios to the new rules, in which the receivable total is $18, A should receive $1/6$ of $18, or $3; B should receive $1/3$, or $6; C should receive $1/2$, or $9.

As you can see, these simple games give only a hint of the complexity of larger, non-zero-sum games. Game theory has had its greatest success to date with only the most elementary of games, or conflict situations. Undoubtedly it will come to have increasing importance in the future, for so much of life has to do with conflict and competition. As John D. Williams has written in his book *The Compleat Strategyst*, "The concept of a strategy, the distinctions among players, the role of chance events, the notion of matrix representations of the payoffs, the concepts of pure and mixed strategies, and so on give valuable orientation to persons who must think about complicated conflict situations."

Joseph G. Cowley

Nearly rectangular windows fill out the area under the roof of M.I.T.'s Kresge Auditorium. In calculus we use rectangles to find the area under a curve.

CALCULUS

One of the greatest contributions to modern mathematics, science, and engineering was the invention of *calculus* (or as it is sometimes called, *the calculus*) near the end of the 17th century. It is safe to say that without this fundamental branch of mathematics, many technological accomplishments, such as the landing of men on the moon, would have been very difficult, or impossible, to achieve.

The word "calculus" comes from the Latin word for pebble. This name probably originated because pebbles were used thousands of years ago for counting and doing problems in arithmetic. Similar words that we often use are *calculate, calculation,* and *calculator.*

Two people who lived during the 17th century are credited with the invention of the calculus: Sir Isaac Newton of England and Baron Gottfried Wilhelm von Leibniz of Germany. The basic ideas of calculus were developed independently by them within a few years of each other.

Newton, who was one of the greatest physicists of all time, applied the calculus to his theories of motion and gravitation. These theories, often referred to as *Newton's laws,* enabled him to describe mathematically the motion of all objects in the universe—from the tossing of a ball into the air to the revolution of the earth around the sun.

Before Newton and Leibniz, the mathematics used for solving problems was the kind commonly taught in modern secondary schools. This involved subjects such as arithmetic, algebra, geometry, and trigonometry. The basic principles of these subjects were known at least 1,500 years before Newton and Leibniz. Although the mathematical principles studied in these subjects were useful in solving certain kinds of problems, they were not at all suited to solving problems dealing with *changing,* or *varying, quantities.* It was for the purpose of working with such quantities in everyday life that the calculus was invented. We therefore can say that calculus is the "mathematics of change."

Changing, or Varying, Quantities

If a person travels in an automobile at a velocity of 50 kilometers per hour, we know that in 2 hours that person will travel a distance of 100 kilometers. In practice, of course, a driver rarely travels at the same velocity of 50 kilometers per hour for 2 hours. The driver will sometimes stop, sometimes travel at a velocity of 80 kilometers per hour, and sometimes at 40 kilometers per hour.

Scientists use calculus to calculate a rocket's escape velocity as well as its velocity and position at any given time.

NASA

The velocity of the automobile is usually a changing, or varying, quantity. When the velocity is increasing, as when the auto goes from 30 to 40 kilometers per hour, we say that the automobile is being *accelerated,* or that it is undergoing *acceleration.* Similarly, when the velocity is decreasing, as when the auto goes from 40 to 20 kilometers per hour, we say that the automobile is being *decelerated,* or undergoing *deceleration.* If the automobile maintains the same velocity, then we say that it is traveling at a *constant,* or *uniform, velocity.*

Another example of a changing quantity involves a ball that has been dropped or thrown. Suppose we drop a ball from a building. At the instant we drop it, the velocity is zero. Gradually, the velocity increases; that is, the ball accelerates. Finally, when the ball hits the ground, it is traveling at its greatest velocity.

Similarly, if we throw a ball up into the air, it at first travels fast—that is, with a large velocity. Gradually it slows down, or decelerates, until its velocity is zero and it stops for an instant. At this point the ball has reached its maximum height and starts to come down. As it comes down, its velocity increases until the ball hits the ground. The change in the velocity of the ball—or any object that is thrown—is due to the attraction of the earth, which is called the *force of gravity,* or the *gravitational force.*

The same ideas apply to a rocket launched from the earth's surface. By using calculus, it is possible to find the velocity the rocket must have in order to escape the earth's gravity—that is, not return to the earth. This velocity, often called the *escape velocity,* turns out to be about 11 kilometers per second, or about 40,000 kilometers per hour. By means of calculus, we can calculate the time it would take for the rocket to get to the moon and how much fuel would be needed.

There are many other examples of changing, or varying, quantities. When a raindrop or snowflake falls, its size gradually increases. The population of a country changes each year, or even each day. The cost of living changes. The amount of

This car's velocity is hardly ever constant. The car accelerates at the start of the race, slows down before each curve, and accelerates again on straightaways.

a radioactive substance, such as radium, changes.

What is a Variable?

Any quantity that is changing, or varying, is referred to in mathematics as a *variable*. In the examples of the automobile, ball, and rocket, the velocity is a variable. Also, the distance of the ball or rocket from the earth's surface is a variable. Even the time, which is changing, is a variable. The population of a country or the cost of living is a variable. When we put air in our tires, the air pressure changes and is a variable. The outdoor temperature and humidity are variables.

In mathematics we often represent variables by symbols, such as letters of the alphabet. Thus we can let v represent velocity, t represent time, p represent pressure in a tire, and so on. These letters stand for numbers. Thus when the velocity is 50 kilometers per hour we can say that $v = 50$. When it is 25 kilometers per hour we can say that $v = 25$, and so on.

In the case of time, we usually measure it from some specific instant, which is often called the *time origin* or *zero time*. For example, the instant at which we drop a ball from a building is zero time, or $t = 0$. After 3 seconds have elapsed, we say that $t = 3$; after 5 seconds, $t = 5$; and so on.

Very often, in practice, we find that one variable depends in some way on another variable. For example, the distance a rocket or ball travels depends on the time of travel. In this instance, we call distance the *dependent variable* and time the *independent variable*.

CALCULUS 433

In general, time is considered as an independent variable, while any variable that depends on it is a dependent variable. Thus the cost of living, which depends on time, is a dependent variable. The outdoor temperature is also a depenent variable.

Many independent variables besides time can occur. For example, the area of a circle depends on the radius. We can call the area A the dependent variable, and the radius r the independent variable.

What is a Function?

If one variable depends on another, mathematicians say that the first variable is a *function* of the second variable. The distance a rocket or ball travels is a function of the time of travel. The area of a circle is a function of the radius of the circle.

Suppose we designate by the letter x any independent variable such as time, radius of a circle, and so on. Suppose further that we designate by the letter y any dependent variable that depends on x, such as the distance traveled by a rocket or the area of a circle. We then can make the statement that "y depends on x" or "y is a function of x," often abbreviated as $y = f(x)$. This is read "y equals f of x."

In terms of this functional notation, we could abbreviate the statement that the cost of living C is a function of the time t by writing $C = f(t)$. Similarly, we could express the fact that the area A of a circle is a function of the radius r by writing $A = f(r)$. A function always indicates that there is a relationship between the dependent and independent variables. One of the important problems in mathematics and its applications to other fields is to determine the nature of these relationships.

In some cases, the relationship between variables is simple. For example, we know from geometry that the area A of a circle is given in terms of its radius r by means of the relationship $A = \pi r^2$, where π is a constant whose value is approximately 3.14159. In other cases, the relationship between variables can be very difficult to obtain. For example, the relationship between the cost of living and time is not known, although we can have some idea about such a relationship based on past experience. However, even though we cannot find such a relationship, it does not mean that there is none.

Rates of Change

Suppose that at 10:00 A.M. an automobile driver is 30 kilometers from a certain town, and at noon the driver is 160 kilometers away from that town. The change in distance in 2 hours, then, is 160 − 30, or 130 kilometers. On dividing this change in distance by the change in time—that is 130 kilometers divided by 2 hours—we obtain 65 kilometers per hour. This is called the driver's *average velocity*. It should be noted that, during the 2 hours, the driver may have been traveling at 80 kilometers per hour some of the time, and 50 kilometers per hour at other times. On the average, however, the driver traveled at 65 kilometers per hour.

We see that the average velocity is the "time rate of change in distance." The actual velocity of the driver at a particular instant is called the *instantaneous velocity*. We can get a good idea of the instantaneous velocity of a driver by finding the average velocity over a very brief interval of time. Thus suppose we know that at 10:00 A.M. the driver is 50 kilometers away from a certain town and that at 10:05 A.M. the driver is 55 kilometers away. The driver, then, has traveled 5 kilometers in 5 minutes, or about 1 kilometer a minute; that is, 60 kilometers per hour. This is the driver's *average velocity,* but it is also a very good approximation to the *instantaneous velocity,* since if the driver did stop during this time or was traveling much slower or faster than 60 kilometers per hour, it certainly couldn't have been for long.

The average velocity is, as we have seen, the time rate of change in distance. The instantaneous velocity is the "instantaneous time rate of change of distance." The problem of finding instantaneous time rates of change of distance, or of other quantities, is one of the most important parts of calculus and has many applications. We shall try to see how such instantaneous rates of change can be found.

Process of Differentiation

Let us consider a ball being dropped from a tall building, such as the Leaning Tower of Pisa, shown in Figure 1. If we let s be the distance (in meters) that the ball falls in a time t (in seconds), then s will depend on t; that is, s will be a function of t, or $s = f(t)$. A formula that reveals the relationship between s and t is

$$s = 4.9t^2 \qquad [1]$$

From this formula we see that at $t = 0$ we have $s = 0$, which is not surprising, since the ball falls zero distance in zero time. After 1 second (that is, $t = 1$) we see from equation [1] that $s = 4.9$, which means that in 1 second the ball has fallen 4.9 meters. Similarly, after 2 seconds (that is, $t = 2$) we see that $s = 19.6$, so that in 2 seconds the ball has fallen 19.6 meters.

Suppose now we want to find the velocity of the ball at the time t (that is, the instantaneous velocity at time t). We increase the time t by a small amount, which we denote by dt. We can think of dt as a "little bit of t."

We now try to find out the distance the ball will travel in the time $t + dt$ (that is, the time t plus the extra bit of time dt). This distance will be the original distance s plus an extra little bit of distance, which we call ds. Since the distance traveled in time $t + dt$ is $s + ds$, we can modify equation [1] to

$$s + ds = 4.9(t + dt)^2 \qquad [2]$$

which can be expanded by multiplication to

$$s + ds = 4.9t^2 + 9.8t(dt) + 4.9(dt)^2 \qquad [3]$$

The extra little bit of distance ds that the ball travels can now be obtained by subtracting s from the left side of equation [3] and subtracting its equal value $4.9t^2$ (see equation [1]) from the right side.

$$ds = 9.8t(dt) + 4.9(dt)^2 \qquad [4]$$

Now since dt represents a small number (less than 1), $(dt)^2$ is a very small number. It is so small, in fact, that the last term on the right of equation [4] can for all practical purposes be removed. We thus have

$$ds = 9.8t(dt) \qquad [5]$$

On dividing both sides by dt, we obtain

$$\frac{ds}{dt} = 9.8t \qquad [6]$$

The quantity on the left of equation [6] which is the little bit of distance ds divided by the little bit of time dt, is the instantaneous velocity of the ball at time t. If we let this instantaneous velocity be v, we have

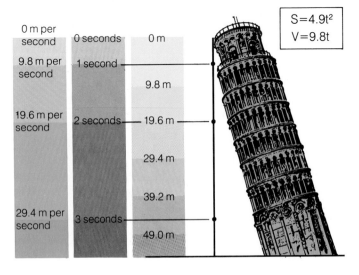

Figure 1. Suppose a ball is dropped from the top of the Leaning Tower of Pisa. The diagram at left shows the velocity of the ball and the distance it has fallen at given times. Using calculus, one can determine the velocity and distance traveled at any given time.

$$v = 9.8t \qquad [7]$$

If we put $t = 3$ in equation [7] we find $v = 29.4$. This means that, after 3 seconds, the ball is traveling 29.4 meters per second, which is its instantaneous velocity.

The process that we used to obtain the result in equation [6] is known in the calculus as *differentiation*. That part of the calculus that deals with such processes is called *differential calculus*. The quantity ds/dt is called the "*derivative* of s with respect to t," or simply the "derivative of s."

If we like, we can use equation [7] to find the derivative of v with respect to t. To do this, we use exactly the same procedure given for the derivative of s. We increase the time t by dt so that the total time is $t + dt$. Then the velocity v increases by a little bit, which we call dv, so that the new velocity is $v + dv$. From equation [7] we then see that

$$v + dv = 9.8(t + dt) \qquad [8]$$
or
$$v + dv = 9.8t + 9.8dt \qquad [9]$$

Subtracting v from the left side of equation [9] and subtracting its equal value 9.8t (see equation [7]) from the right side, we have

$$dv = 9.8dt \qquad [10]$$
or
$$\frac{dv}{dt} = 9.8 \qquad [11]$$

The left side of equation [11] is the instantaneous time rate of change of the velocity, which is called the *instantaneous acceleration*. From equation [11] we see that the instantaneous acceleration is a constant. The significance is that when the ball falls toward the earth it increases its velocity by 9.8 meters per second in each second.

Since v is the derivative of s, we see that dv/dt is the derivative of the derivative of s, which is called the *second derivative* of s. It is written as

$$\frac{d^2s}{dt^2} = 9.8 \qquad [12]$$

Minima and Maxima

As we have already mentioned, the process of differentiation—that is, finding derivatives—is studied in that part of calculus known as differential calculus. There are many applications of differential calculus besides those involving velocity and acceleration. One such application is illustrated by the following problem. Let us suppose that we are in the business of making metal cans for a soup company. We are asked by the company to make the cans cylindrical in shape, with the requirement that the cans have a capacity of 1 liter.

We could make the cans tall and narrow or short and wide (Figure 2). How shall we decide what to do? We know that the metal out of which the cans are made will cost money. It is therefore natural to ask ourselves whether we can make the required can by using the least amount—that is, the minimum amount—of metal for the total surface of the can. By the method of differential calculus we can determine the exact measurements (diameter and height) that the can must have in order to contain a given volume and at the same time have the least possible surface area and therefore the least cost. Since we are interested in the least, or minimum, surface area, this is called a problem in finding *minima*.

As another illustration, suppose we have a rectangular piece of cardboard that measures 3 meters by 5 meters. We wish to make an open box from it by cutting out equal squares from the corners and then bending up the sides. The question is, What

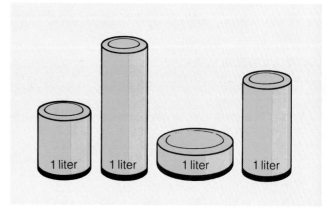

Figure 2. A manufacturing company wants to make 1-liter containers. Calculus can be used to determine the minimal surface area needed to contain the required volume.

size squares should be cut out so that the box will contain the greatest, or maximum, volume? This problem, which is one in finding *maxima,* can also be solved by differential calculus.

The Process of Integration

We have already seen how, if we are given $s = 4.9t^2$, we can find $ds = 9.8t(dt)$. The process amounts to starting with the total distance s and finding the little bit of distance ds traveled in time dt. We gave the name differentiation to this process.

We now ask, If we are given the equation $ds = 9.8t(dt)$, can we get back to $s = 4.9t^2$? This is the reverse, or inverse, of differentiation. Since ds represents a little bit of distance, it is natural that to find the total distance s we must add up all the little bits of distance ds. The process of such adding is represented mathematically by

$$\int ds = s \qquad [13]$$

The symbol $\int$ is called an *integral* sign, and equation [13] is read "the integral of ds is s." The process of finding integrals is called *integration* and is studied in a part of calculus that is known as *integral calculus.* By methods of integral calculus, we can find, for example, that

$$s = \int ds = \int 9.8t(dt) = 4.9t^2 \qquad [14]$$

so that we have recovered the formula $s = 4.9t^2$. It follows that integration is the reverse, or inverse, of differentiation.

Applications

Just as there are many applications of differential calculus, there are also many applications of integral calculus. One important application is that of finding areas bounded by complicated closed curves or of finding volumes bounded by complicated closed surfaces. Another application is that of finding the total length of a complicated curve or the total area of a complicated surface (Figure 3). The idea involved in such cases is that of *summation,* or addition, of little bits of area or little bits of volume and so involves integration.

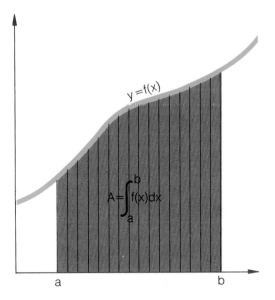

Figure 3. What is the area under y from a to b? We can divide the area into nearly rectangular regions whose areas can be approximated. The sum of these areas is very close to what we want. Integration gives the "best possible approximation," which is the actual area under the curve.

We have seen how the process of integration enables us to go from equation [6] back to equation [1]. Equation [6] involves an instantaneous rate of change. It is a derivative. Equations involving derivatives of quantities that we want to determine are called *differential equations.* Such equations often arise in the mathematical statements of problems in science and engineering, because it is in many cases easier to arrive at a relationship between derivatives of quantities rather than between the quantities themselves. The process of solving differential equations is often also known as "integrating the differential equation."

In the preceding paragraphs we have been able to provide only a glimpse into some of the many important ideas of calculus. In order to appreciate the power of the calculus in solving the many important problems of mathematics, science, and engineering, the student should consult some of the many books on calculus that are available.

Murray Spiegel

SET THEORY

A herd of cows, a flock of birds, a school of fish—each of the words "herd," "flock," and "school" could be replaced by the word "set." A *set* is simply a collection of objects or ideas.

The concept of sets was developed into a new branch of mathematics in the late 19th century by a German mathematician, Georg Cantor. Since the beginning of the 20th century, set theory has developed rapidly, and today it has important applications in nearly every branch of mathematics. In fact, most of our mathematics can be derived from set theory.

Two kinds of notations for sets are in common use. One is the *enumeration notation,* in which, for example, we write $\{1, 3\}$ for the set consisting of the numbers 1 and 3 and {Roberts, Roye} for the set consisting of the first two presidents of Liberia.

The second notation is the *set-builder notation,* in which we write $\{x|x$ is a whole number$\}$ for the set of whole numbers, and $\{y|y$ is one of the first two presidents of Liberia$\}$ for the set {Roberts, Roye}. We read $\{x|x \ldots\}$ as "the set of all x such that $x \ldots$" Thus $\{x|x$ is a whole number$\}$ is read "the set of all x such that x is a whole number."

Of the two notations, the set-builder notation is the more frequently used in mathematics. Note that the letter, or symbol, used in the set-builder notation is not significant. Thus, for example, $\{x|x$ is a whole number$\}$ is the same as $\{y|y$ is a whole number$\}$.

Capital letters are usually used as symbols for sets. We write, for example, $A = \{x|x$ is a whole number$\}$.

The *members,* or *elements,* of a set are simply those things that make up the set. Thus the members of the set $\{2, 3\}$ are the numbers 2 and 3. The members of the set $\{x|x$ is a United States citizen$\}$ are the citizens of the United States. When we say that two sets A and B are equal, we mean that every member of A is a member of B and, conversely, that every member of B is a member of A. For example, $\{5, 7\} = \{7, 5\}$. Note that the order of listing the members is not significant. If x is a member of a set A, we write $x \in A$. If x is not a member of A, we write $x \notin A$. Thus $2 \in \{2, 3\}$, but $4 \notin \{2, 3\}$.

Subsets of Sets

Every member of the set $\{a, b\}$ is a member of the set $\{a, b, c\}$. We say that $\{a, b\}$ is a *subset* of $\{a, b, c\}$ and write $\{a, b\} \subseteq \{a, b, c\}$. Every member of the set $A = \{x|x$ is an even number$\}$ is a member of the set $B = \{x|x$ is a whole number$\}$. We say that A is a subset of B and write $A \subseteq B$.

Our examples illustrate the "natural" use of the word "subset" to indicate a part of a whole (as, for example, "subtotal" is a part of the total). It is useful, however, to consider that $\{a, b\} \subseteq \{a, b\}$. In general, every set is a subset of itself: $A \subseteq A$ for all sets A.

If A and B are sets, then $A \subseteq B$ means that whenever $x \in A$, then $x \in B$. If A is a subset of B but $A \neq B$, we write $A \subset B$ and say that A is a *proper* subset of B. Thus, for example, $\{a, b\}$ is a subset of $\{a, b\}$ but is not a proper subset of $\{a, b\}$, whereas $\{a, b\}$ is both a subset and a proper subset of $\{a, b, c\}$.

Union of Sets

Just as two numbers can be combined by addition or multiplication to yield a third number, so can two sets be combined in various ways to yield a third set. In particular, given two sets A and B, we can form their *union,* $A \cup B$. This is the set whose members are members of A or B (or both). Thus, for example, $\{a, b\} \cup \{c, d\} = \{a, b, c, d\}$ and $\{x|x$ is an odd number$\} \cup \{x|x$ is an even number$\} = \{x|x$ is a whole number$\}$. In "set-theory language," if A and B are sets, then $A \cup B = \{x|x \in A$ or $x \in B$ (or both)$\}$.

This concept of union is used in many parts of mathematics. In elementary school, for example, the idea of addition of

Figure 1.

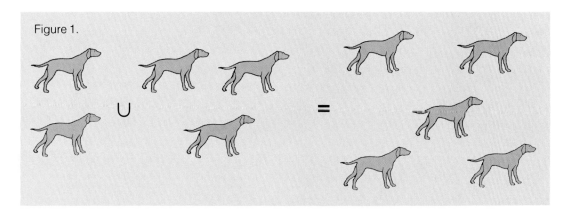

whole numbers is made real to children by the use of set union. Thus the union of the two sets shown in Figure 1 corresponds to the addition fact 2 + 3 = 5. In geometry, the concept of union is used to define a triangle as the union of three line segments, each of which has a common end point with each of the other two, as shown in Figure 2.

Suppose we consider

$$\{a, b\} \cup \{a, c\} = \{a, b, c\}.$$

Can we also write

$$\{a, b\} \cup \{a, c\} = \{a, a, b, c\}?$$

Yes, we can. It is true that {a, b, c} = {a, a, b, c}, but we do not "add" to the set {a, b} by repeating the symbol "a". For example, consider the set {you, reader of this article} = {you} = {reader of this article}. You can't become two persons by naming yourself twice. So we agree, in listing the members of a set, not to list an element more than once.

Intersection of Sets

Similar to the concept of the union of two sets is the concept of the *intersection* of two sets. If A and B are sets, then the intersection of A and B, A ∩ B, is

$$A \cap B = \{x | x \in A \text{ and } x \in B\}.$$

Thus, for example, {a, b, c} ∩ {a, e, f} = {a} and {$x|x$ is a brown-eyed human being} ∩ {$x|x$ is a woman} = {$x|x$ is a brown-eyed woman}.

Like the concept of union of sets, the concept of intersection of sets finds wide application in mathematics. For example, we can find the set of common divisors of 12 and 18 thus:

$$\{x|x \text{ divides } 12\} \cap \{x|x \text{ divides } 18\} =$$
$$\{1, 2, 3, 4, 6, 12\} \cap \{1, 2, 3, 6, 9, 18\}$$
$$= \{1, 2, 3, 6\}$$

In geometry, we can symbolize the fact that two lines L_1 and L_2 intersect in a point P:

$$L_1 \cap L_2 = P$$

What about {$x|x$ is an even number} ∩ {$x|x$ is an odd number}? Since no number is both even and odd, it may seem as if no answer is possible. To handle this and many similar situations, however, we introduce the concept of the *empty* set. Also called the *null* set, the empty set is symbolized by { }, or ∅. Thus, {$x|x$ is an even number} ∩ {$x|x$ is an odd number} = ∅. Likewise, if the lines L_1 and L_2 are parallel, $L_1 \cap L_2 = \emptyset$.

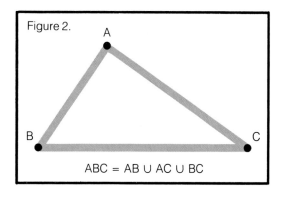

Figure 2.

ABC = AB ∪ AC ∪ BC

It is often useful to consider a fixed set, called the *universal* set, such that all the sets under consideration at a particular time are subsets of this universal set. For example, if we are discussing subsets of the whole numbers, such as the set of natural numbers {1, 2, 3, 4, . . .} and the set of even numbers {0, 2, 4, . . .}, we could take our universal set as the set of whole numbers. Similarly, we would take our universal set as the set of all points in the plane if we were discussing subsets of the set of points in a plane, such as the sets of points forming circles and sets of points forming triangles.

Complements of Sets

Suppose our universal set (U) is the set of all students at City High School, and A is the set of all female students at City High. Then the set of all male students at City High is called the *complement* of set A, and is indicated by A'. The complement A' of a set A consists of all elements in the universal set that are not members of A. In symbols, if $A \subseteq U$, then $A' = \{x | x \in U$ and $x \notin A\}$. Thus if our universal set is the set of real numbers and A is the set of rational numbers, then A' is the set of irrational numbers.

Venn Diagrams

It is often helpful to picture sets and set relations by means of what are commonly called *Venn diagrams*. In Figure 3 the region enclosed by each rectangle represents the universal set, and regions inside the rectangle that are enclosed by circles (or other curves) represent subsets of the universal set. Figure 4 shows Venn diagrams that illustrate some concepts of set theory.

Venn diagrams can be used to make plausible—although not actually prove—various statements of equality between sets, such as the two equalities that, in algebra, are called the *distributive properties*:

$$A \cup (B \cap C) = (A \cup B) \cap (A \cup C)$$

and

$$A \cap (B \cup C) = (A \cap B) \cup (A \cap C)$$

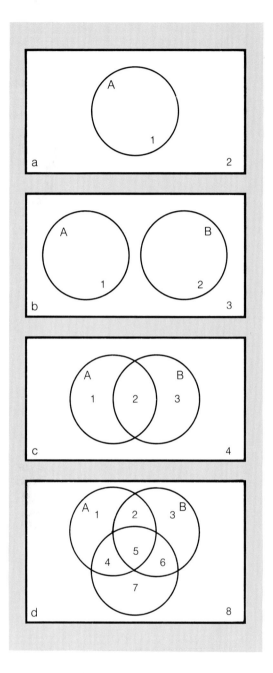

Figure 3. In each Venn diagram, the region enclosed by the rectangle represents the universal set. Subsets are represented by circles (labeled A, B, and C), the placement of which divides the universal sets into distinct regions (labeled 1, 2, 3, . . . , 7). The regions shared by intersecting sets contain the members, or elements, common to both, or all, intersecting sets. Sets A and B in diagram b are called "disjoint" sets because they share no region.

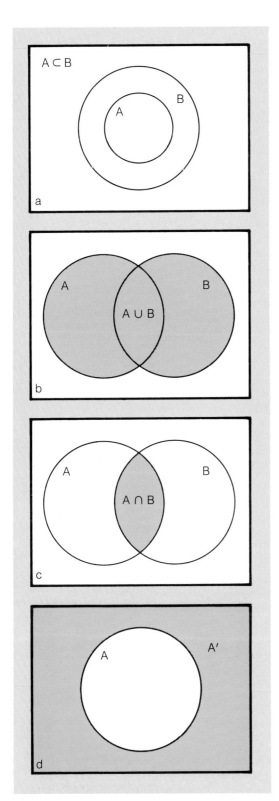

Figure 4. Each Venn diagram represents a particular set relationship: a) a proper subset; b) a union; c) an intersection; d) a complement.

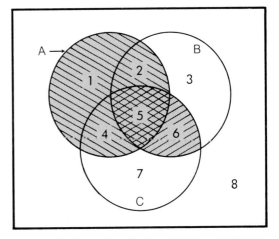

Figure 5. Set A consists of regions 1, 2, 4, and 5. Set B ∩ C consists of regions 5 and 6. Of what regions, then, does set A ∪ (B ∩ C) consist?

[note their similarities to the more familiar algebraic equation a(b + c) = ab + bc]. Thus the entire shaded region in Figure 5 illustrates A ∪ (B ∩ C) while the doubly shaded region in Figure 6 illustrates (A ∪ B) ∩ (A ∪ C). In both cases we obtain the same region. The corresponding situations for A ∩ (B ∪ C) and (A ∩ B) ∪ (A ∩ C) are shown in Figure 7.

Algebra of Sets

There are many analogies between the operations of union and intersection on sets and the operations of addition and multiplication on numbers. It is easy to see, for example, that just as $a + b = b + a$ and $a \times b = b \times a$ for all numbers a and b, it is also true that, for all sets A and B, A ∪ B = B ∪ A and A ∩ B = B ∩ A. Likewise, just as 0 has the property that $a + 0 = a$ and $a \times 0 = 0$ for all numbers a, so does A ∪ ∅ = A and A ∩ ∅ = ∅ for all sets A. Furthermore, just as $a \times 1 = a$ for all numbers a, so A ∩ U = A for all sets A ⊆ U (the universal set).

The analogies are not complete, however. There are properties of our number system for which there are no analogous properties for the algebra of sets, and there are properties of the algebra of sets for which there are no analogous properties for our number system. For example, given a nonzero number a, there exists a number b (the multiplicative inverse of a) such that a

SET THEORY 441

× b = 1. Now U is analogous to 1 in the sense that A ∩ U = A for all sets A, and yet, unless A = U, there is no set B such that A ∩ B = U.

On the other hand, the fact that A ∪ A = A ∩ A = A for all sets A has no analog in arithmetic. We have stated that A ∩ (B ∪ C) = (A ∩ B) ∪ (A ∩ C) for all sets A, B, and C, and this does have an analog in a(b + c) = ab + ac. We have also stated, however, that A ∪ (B ∩ C) = (A ∪ B) ∩ (A ∪ C) for all sets A, B, and C, and this corresponds to a + bc = (a + b)(a + c) for all numbers a, b, and c. The latter statement, however, is not true. (Try $a = 1$, $b = 2$, and $c = 3$, for example.)

Table I lists a number of identities that hold for all sets A, B, and C that are subsets of some universal set U. Note the parallel-

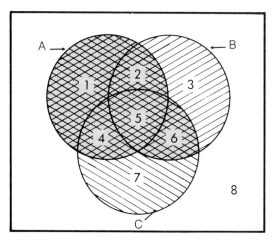

Figure 6. The sets A ∪ B and A ∪ C are indicated here by slanting lines. Their intersection is the region where the lines crisscross. It is the same set as that in Figure 5.

Table I:	
A ∪ A = A	A ∩ A = A
A ∪ B = B ∪ A	A ∩ B = B ∩ A
A ∪ (B ∪ C) = (A ∪ B) ∪ C	A ∩ (B ∩ C) = (A ∩ B) ∩ C
A ∪ (B ∩ C) = (A ∪ B) ∩ (A ∪ C)	A ∩ (B ∪ C) = (A ∩ B) ∪ (A ∩ C)
A ∪ ∅ = A	A ∩ U = A
A ∪ A' = U	A ∩ A' = ∅
U' = ∅	∅' = U
A ∪ U = U	A ∩ ∅ = ∅
	(A')' = A

Figure 7. The shaded area of diagram a represents the set A ∩ (B ∪ C), while the shaded area of diagram b represents the set (A ∩ B) ∪ (A ∩ C). Both sets are the same, confirming the identity.

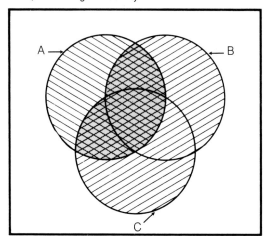
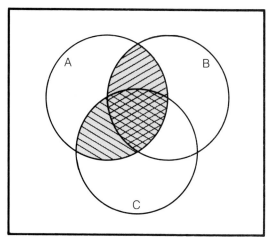

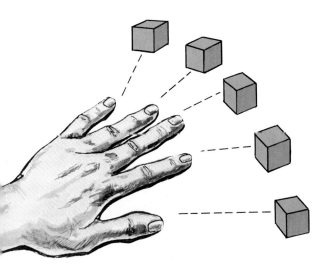

ism between the corresponding identities in the left- and right-hand columns. All these identities are rather easily seen to hold as a consequence of the definitions of union, intersection, empty set, universal set, and complement, and can be made plausible, as we have seen by use of Venn diagrams.

Infinite and Finite

What is meant by the word "infinite"? Many people think of infinity as being a very large or incalculable number. When mathematicians speak of an infinite set, however, they do not think of a set with a large number of elements, such as the set of grains of sand on the seashore. Indeed, they find it worthwhile to avoid entirely any direct reference to counting in describing the difference between finite and infinite sets.

To do this, the mathematician first considers the concept of *matching* sets, or, to use more technical language, sets that are in *one-to-one* (1–1) *correspondence*. This is a very natural concept that is used even in the very early stages of instruction in mathematics when a child learns to manually pair off blocks.

A formal definition is as follows: A 1–1 correspondence between two sets A and B is a pairing of the elements of A with the elements of B such that each element of A is paired with precisely one element of B and each element of B is paired with precisely one element of A. If such a 1–1 correspondence exists between the two nonempty sets A and B, we say that the two sets are *equivalent* and write A ~ B. (We also say that $\emptyset \sim \emptyset$, although we will not use this fact here.)

For finite sets we normally do not bother to attempt a pairing to see if the two sets are equivalent; we simply count the elements in each set. Thus we observe that each of the two sets shown in Figure 8 has five elements, so that we "know" that they are equivalent. Actually, however, the concept of 1–1 correspondence precedes the idea of counting. "Five," for example, is simply the number name that we attach to sets that are in 1–1 correspondence with the fingers of one hand. Primitive man, long before he knew how to count, could keep track of the size of his flock by matching each animal with a pebble. If he dropped a pebble into a container as each animal left the enclosure in the morning, and then removed a pebble as each animal returned in the evening, he would know whether or not the same number of animals returned as had left, without counting.

Now if we have any finite set A, we sense intuitively that no proper subset of A is in 1–1 correspondence with A. Try, for example, matching the set of fingers minus the thumb on one hand with the entire set of fingers on the other hand. But what about infinite sets? Consider the set N = {1, 2, 3, . . .} of natural numbers and the set E = {2, 4, 6, . . .} of even natural numbers. Clearly E is a proper subset of N, and yet E ~ N as shown by the 1–1 correspondence in Figure 9.

We generalize these two examples to make a formal definition: A finite set is a set that has no proper subset equivalent to itself; an infinite set is a set that has at least one proper subset equivalent to itself.

Figure 9.				
Members of N	1	2	3 . . .	n . . .
	↕	↕	↕	↕
Members of E	2	4	6 . . .	2n . . .

Cardinal Number of an Infinite Set

If finite sets are equivalent, they have the same number of elements. That is, they are *equinumerous*. What about infinite sets? For example, is the set E of even numbers equinumerous with the set N of natural numbers? From one point of view it is certainly natural to argue that E and N are not equinumerous, since we take something away from the set N = {1, 2, 3, . . .} to get the set E = {2, 4, 6, . . .}. If we take something away, say 1, from the set {1, 2, 3}, the set we obtain, {2, 3}, is not equinumerous with the set {1, 2, 3}.

Nevertheless, it turns out to be useful to agree that E and N are equinumerous and, in general, to make the following formal definition: Two sets A and B (finite or infinite) are said to be equinumerous, or to have the same *cardinal number* of elements, if and only if A ~ B. We conclude from this definition that the finite sets of Figure 8 have the same cardinal number and that the sets E and N of Figure 9 have the same cardinal number.

Do all infinite sets have the same cardinal number? Consider, for example, the set Q of all positive rational numbers—the set of all natural numbers together with all the positive fractions: ½, ¹⁷⁄₁₉, 1⅞, ¹⁄₂₅₂, 0.12, and so on. Are Q and N equinumerous? Again, we may feel intuitively that the answer should be "no," since Q contains many (indeed, an infinite quantity of) numbers not in set N. We can, however, show that Q ~ N, so that, according to our definition, Q and N are equinumerous.

A simple argument that Q ~ N can be based on the diagram shown in Figure 10. It shows the positive rational numbers in an array extending indefinitely to the right and down. (Ignore the arrows for the moment.)

Now suppose we are given a number $n \in$ N and wish to describe a rule for associating with n a definite rational number $r \in$ Q. We simply follow the arrows in Figure 10, starting with ⅟₁ in the top left-hand corner.

As we follow this line we count "one, two, three, . . . , n" as we go through each fraction—except that we don't count repetitions of fractions. For example, suppose

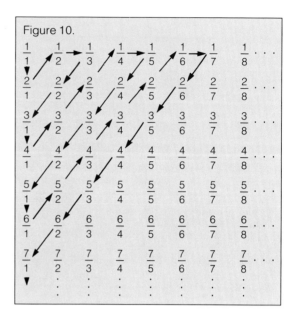

Figure 10.

we have $n = 5$. Then, following the arrows, we count off the five fractions ⅟₁, ²⁄₁, ½, ⅓, and ³⁄₁ (omitting ²⁄₂ = ⅟₁) and conclude that corresponding to $5 \in$ N we have ³⁄₁ $\in$ Q. Similarly, if $n = 13$ we obtain ⅟₁, ²⁄₁, ½, ⅓, ³⁄₁, ⁴⁄₁, ³⁄₂, ⅔, ¼, ⅕, ⁵⁄₁, ⁶⁄₁, and ⁵⁄₂ (omitting ²⁄₂ = ⅟₁, ²⁄₄ = ½, ³⁄₃ = ⅟₁, and ⁴⁄₂ = ²⁄₁). Thus to $13 \in$ N corresponds ⁵⁄₂ $\in$ Q.

Conversely, given any $r \in$ Q, we take its representation as a fraction in lowest terms (such as ½ for ³⁄₆) and count "how far" it is from ⅟₁ along the line (again omitting repetitions). Thus if we have given ⁴⁄₆ $\in$ Q, we observe that ⁴⁄₆ = ⅔ and count ⅟₁, ²⁄₁, ½, ⅓, ³⁄₁, ⁴⁄₁, ³⁄₂, ⅔; so that corresponding to ⅔ $\in$ Q we have $8 \in$ N.

You may now suppose that all infinite sets have the same cardinal number as N. Certainly this is a plausible conjecture at this point. It turns out, however, that if we add to the set Q of positive rational numbers all the positive irrational numbers, such as $\sqrt{2}$, $\sqrt[3]{2}$, π, $\sqrt[5]{1 + \sqrt{2}}$, and so on, we obtain a set with a cardinal number different from N. Even if we consider only the set Z of rational and irrational numbers between 0 and 1, we obtain, as Georg Cantor showed in 1874, a "larger" set which cannot be matched with N. The union of the set of rational numbers and the set of irrational numbers is the set of real numbers.

The proof that Z is not equinumerous with N rests upon the fact that every real number between 0 and 1 can be written as an infinite decimal. For example, $\frac{1}{2} = 0.5000\ldots$, $\frac{1}{3} = 0.3333\ldots$, $\frac{1}{7} = 0.142857142857\ldots$, $\sqrt{2}/2 = 0.7071\ldots$, where the rational numbers (such as $\frac{1}{2}$, $\frac{1}{3}$, and $\frac{1}{7}$) are expressed as repeating decimals, and the irrational numbers (such as $\sqrt{2}/2$) are expressed as nonrepeating decimals.

Now suppose we claim that we have a 1–1 correspondence between the elements of Z and the elements of N, such as:

```
      N             Z
      ↓
      1    ↔    0.183478412001... = r₁
      2    ↔    0.369715400000... = r₂
      3    ↔    0.579321715432... = r₃
      4    ↔    0.481762314000... = r₄
      5    ↔    0.673216732167... = r₅
      6    ↔    0.591416789143... = r₆
      7    ↔    0.001326841841... = r₇
      .    ↔    . . . . . . .    = ..
```

The digits of r_1, r_2, and so on are chosen arbitrarily so that $r_1 \neq r_2 \neq \ldots$.

Now we form another number in Z as follows: the first digit to the right of the decimal point is chosen as any digit different from the first digit of r_1. The second digit is chosen as any digit different from the second digit of r_2, and so on. Thus the resulting number could be $0.2783254\ldots$ or $0.3723689\ldots$. None of these numbers so chosen, however, can be in our list, since $0.2783254\ldots \neq r_1$ because it differs from r_1 in at least the first decimal place; $0.2783254\ldots \neq r_2$ because it differs from r_2 in at least the second decimal place; and so on. Hence $0.2783254\ldots$ was not present in the (supposedly complete) list, and so our correspondence is not 1–1 as alleged. No matter what listing we propose, the same process will show the listing is incomplete, and we conclude that no 1–1 correspondence between Z and N is possible.

Mathematicians denote the cardinal number of N by $\aleph_0$ (read "aleph-null"—aleph is the first letter of the Hebrew alphabet). They denote the cardinal number of the real numbers by c. A famous question in mathematics is, Does there exist an infinite set of numbers whose cardinal number lies between $\aleph_0$ and c? That is, is there a subset S of the real number R such that the cardinal number of S is neither $\aleph_0$ nor c? The *continuum hypothesis*, as formulated by Cantor, was the conjecture that no such set S exists.

The question of the truth or falsity of the continuum hypothesis claimed the attention of many first-rate mathematicians after it was first formulated by Cantor, but was not settled until 1963. However, it was not settled by a simple "yes" or "no." What was finally shown by the work of Kurt Gödel and Paul Cohen was that within the framework of the set theory accepted by most mathematicians today, either the acceptance of the continuum hypothesis or its rejection yields equally valid systems of mathematics. This still leaves open the question of whether there exists another equally "useful" formulation of set theory in which the continuum hypothesis can be proved true or false. In a sense, then, the continuum hypothesis has been settled in only a relative fashion, and research still continues on this and related problems.

Paradoxes of Set Theory

There are many paradoxes associated with sets that still concern mathematicians today. One very famous one was formulated by the noted mathematician and philosopher Bertrand Russell (1872–1969). It is known as "Russell's paradox." Imagine a barber in a village. The barber, said Russell, shaves only the men who do not shave themselves. The paradox is, Who shaves the barber? The contradictory answer is, If the barber does not shave himself, then he shaves himself; but if he shaves himself, he cannot shave himself. When stated mathematically, in terms of classes of sets, Russell's paradox challenged the very foundations of set theory.

The complete resolution of this and other paradoxes of set theory is still a matter of active concern today. In the meantime, however, most mathematicians continue to use the very useful concept of a set without waiting until all the foundations of set theory are made solid.

Roy Dubisch

BINARY NUMERALS

The *binary numeral system* offers an interesting glimpse into the world of mathematics. It is the simplest numeration system in which addition and multiplication can be performed, for there are only 4 addition facts and 4 multiplication facts to learn. By contrast, in our more familiar *decimal system* there are 100 addition facts and 100 multiplication facts to learn.

The significance and power of the binary system in modern mathematics, however, lie in its practical application in computer technology. Most computers today are binary computers; that is, they use binary numerals in computing and in processing data. With binary numerals, modern computers can perform well over 1 billion computations per second.

Before going into the binary numeral system, we shall take a closer look at the Hindu-Arabic numeral system and the concept of place value in it. An understanding of place value is vital to an understanding of the binary numeral system.

Hindu-Arabic Numeral System

The numeration system most widely used throughout the civilized world today is called the Hindu-Arabic system, probably because it originated with the Hindus and was carried to the Western world by the Arabs. It is the only numeration system most of us have ever known, and it fits most of our personal, commercial and technical needs very well.

The Hindu-Arabic system is called a decimal, or *base-ten*, system because it needs only 10 symbols to represent any number. These symbols, which are called *digits*, are 1, 2, 3, 4, 5, 6, 7, 8, 9, and 0. These 10 symbols stand for the numbers one, two, three, four, five, six, seven, eight, nine, and zero, respectively.

The Hindu-Arabic system is also called a *positional decimal system,* because the number each digit represents depends on its "position," or "place," in the numeral. The far-right place in any numeral is the unit, or one, position. The place value of the next position to the left is ten. In general, the place values from right to left, in any decimal numeral, are ones, tens, hundreds, thousands, and so on. For example, the place values represented by 4,081 are $(4 \times 1,000) + (0 \times 100) + (8 \times 10) + (1 \times 1)$.

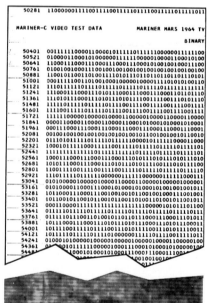

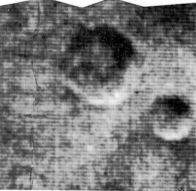

The binary numeral system is used to transmit photos from space to receiving stations on earth. The photos are transmitted as radio signals, representing the zeros and ones of the binary system (upper photo right). The digits are then converted by computer to an image consisting of a series of dots. The photo of the Martian surface at right was obtained this way.

NASA

Another way of looking at a positional decimal system is through the idea of grouping. Suppose we have a set of 13 dots marked on a sheet of paper. We draw a ring around 10 of these dots. There will be 3 dots remaining. We have 1 set of 10 dots plus 3 remaining single dots. This fact may be expressed as $13 = (1 \times 10) + (3 \times 1)$.

Now suppose we have 37 dots, and that we "ring" them in groups of 10. We will have 3 sets of 10 dots, plus 7 single dots. We could write this as $37 = (3 \times 10) + (7 \times 1)$. Similarly, if we have 128 dots we could have 1 set of 1 hundred or 10 tens, 2 sets of 10, and 8 dots remaining. We could write $128 = (1 \times 100) + (2 \times 10) + (8 \times 1)$. All this emphasis on "tens" seems quite natural to us. After all, we do have 10 fingers and 10 toes. But how do you suppose we would be counting today if our anatomy were not structured as it is?

Binary Numeral System

Imagine all of us with one finger on each of two hands and one toe on each of two feet. Suppose further that all the numbers we use could be expressed with the digits 0 and 1. Since we are going to use only two symbols to write any number, we can call this a binary, or *base-two*, system. In a binary system the value of any place in a numeral is twice as large as the place to its right. Thus the place values—from right to left—in a binary system are ones, twos, fours, eights, and so on.

The number 1 expressed in the binary system would be written as 1_{two}. Thus the number represented by 1_{ten} and 1_{two} is the same. (The subscript *two* indicates that we are expressing numbers in the binary system; the subscript *ten* indicates the decimal system.) How would we indicate 2_{ten} as a binary numeral?

Suppose we have a set of 2 stars. If we draw a ring around them, we have 1 set of 2 stars, and no single stars remaining. We would write $2_{ten} = 10_{two}$, which is read as "1 two plus 0 ones." If we begin with a set of 3 stars and draw a ring around 2 of them, we would have 1 set of 2 stars plus 1 set of 1 star. This would be indicated as $3_{ten} = 11_{two}$, read as "1 two plus 1 one."

How would 4_{ten} be expressed? Notice that we have been making pairs of equivalent sets wherever possible. Thus if we had a set of 4 stars, we could first form 2 sets of 2 stars each. Now draw a ring around these sets. This approach suggests that 4_{ten} may be thought of as 1 four, no twos, and no ones, and written as 100_{two}. In similar fashion, $5_{ten} = 101_{two}$, which is "1 four, no twos, and 1 one."

Note how 6_{ten} would then be treated. We would first have 3 sets of 2 stars in each set. Then we could pair up 2 of these sets, and wind up with 1 set of four, 1 set of two, and no ones. We write 6_{ten} as 110_{two}.

Using the same development, it is clear that $7_{ten} = 111_{two}$, interpreted as 1 set of four, 1 set of two, and 1 one. We can analyze 8_{ten} in the same way. First we have 4 sets of 2 stars in each set. Then we have 2 sets with (2×2) or 4 stars in each. Finally, we have 1 set with $(2 \times 2 \times 2)$ or 8 stars in it. This would be written as 1000_{two} and interpreted to mean 1 eight, 0 fours, 0 twos, and 0 ones. We now summarize what we have learned so far about binary numeration in Table I, and see if we can express our ideas.

TABLE I
Place-Value Chart

Binary				Decimal	
8 eights	4 fours	2 twos	1 ones	10 tens	1 ones
			0		0
			1		1
		1	0		2
		1	1		3
	1	0	0		4
	1	0	1		5
	1	1	0		6
	1	1	1		7
1	0	0	0		8
1	0	0	1		9
1	0	1	0	1	0
1	0	1	1	1	1
1	1	0	0	1	2
1	1	0	1	1	3
1	1	1	0	1	4
1	1	1	1	1	5

How would we express 106_{ten} as a base-two numeral? Remember, the place values in binary numeration are ones, twos, fours, eights, sixteens, thirty-twos, sixty-fours, one hundred twenty-eights, and so on. The largest place value contained in 106 is sixty-four, so we can figure the binary equivalent as follows:

There is 1 sixty-four in 106.
$106 - 64 = 42$

There is 1 thirty-two in 42.
$42 - 32 = 10$

There are 0 sixteens in 10.
$10 - 0 = 10$

There is 1 eight in 10.
$10 - 8 = 2$

There are 0 fours in 2.
$2 - 0 = 2$

There is 1 two in 2.
$2 - 2 = 0$

There are 0 ones in 0.
$0 - 0 = 0$

We conclude that $106_{ten} = 1101010_{two}$. Our answer is correct, but the method used is tedious. There is a simpler process that we could use to derive the same conclusion:

To convert any decimal numeral into its binary form, divide it successively by 2. For example if we divide 106_{ten} successively by 2, we obtain the following seven-step process:

1. $2\overline{)106}$ $\underline{106}$ R0 quotient 53	2. $2\overline{)53}$ $\underline{52}$ R1 quotient 26	3. $2\overline{)26}$ $\underline{26}$ R0 quotient 13	4. $2\overline{)13}$ $\underline{12}$ R1 quotient 6
	5. $2\overline{)6}$ $\underline{6}$ R0 quotient 3	6. $2\overline{)3}$ $\underline{2}$ R1 quotient 1	7. $2\overline{)1}$ $\underline{0}$ R1 quotient 0

The remainders (R), in inverse order, make up the binary equivalent of 106_{ten}. Reading from step 7 back to step 1, the remainders are 1, 1, 0, 1, 0, 1, 0. So $106_{ten} = 1101010_{two}$, which checks with the result we obtained using the longer process.

Binary Addition

Suppose we wish to add 11_{two} and 110_{two}. (In the examples that follow, we shall discard the subscript "two" to avoid cluttering up our notation.) As in base ten, addition in base two is always possible; the sum of any two numbers is unique; addition is commutative and associative; and 0 names the identity element. With these structural properties in mind, it is relatively easy to develop addition in base two.

We begin by expressing the addition facts we know:

$$1 + 0 = 0 + 1 = 1$$

and

$$0 + 0 = 0$$

The only remaining fact we need is $1 + 1$. But we have already found out that $1 + 1 = 10$ in the binary system. We therefore write:

$$1 + 1 = 10$$

We can summarize our addition facts in the form of a matrix for easy reference.

+	0	1
0	0	1
1	1	10

Now let's add $110 + 11$:

$$\begin{array}{r} 1\leftarrow \text{carry over} \\ 110 \\ + 11 \\ \hline 1001 \end{array}$$

Discussion: $0 + 1 = 1$, so we write 1 in the ones' place in the sum. Then $1 + 1 = 10$. Write 0 in the twos' place in the sum, and carry over the 1 to the fours column. Again, $1 + 1 = 10$, so write 0 in the sum's fours' place and 1 in the sum's eighths' place. (Note: The binary sum $110 + 11 = 1001$ represents the same number values as the decimal sum $6 + 3 = 9$.)

Here are two more examples. See if your sums agree with these.

```
  110101         1101
+1100011        +1001
10011000        10110
```

Binary Subtraction

Subtraction in the binary system is treated essentially the same as in the decimal system. What is needed is some additional flexibility with binary notation.

From 110 let us subtract 101:

```
 110
-101
   1
```

Discussion: We cannot subtract 1 one from 0 ones, so we take 1 unit from the next base position (twos) and think of it as '10' ones. Because, in the binary system, 1 one from 10 ones leaves 1 one, write 1 in the ones' place in the difference. Now, 0 twos subtracted from 0 twos leaves 0 twos, and 1 four subtracted from 1 four leave 0 fours; therefore, 1 is the correct answer.

Here is another example:

```
1001
-  11
 110
```

Discussion: Because 1 one from 1 one leaves 0 ones, write 0 in the ones' place in the difference. We can't subtract 1 two from 0 twos. The base position to the left of twos is fours, but there are no fours either. So we go to the eights' place. We change 1 eight to '10' fours. We take one unit from the 10 fours, which leaves 1 four, and change that to '10' twos. Now, 1 two from 10 twos leaves 1 two, so we write 1 in the twos' place in the difference. We still have 1 four remaining, which we show by writing 1 in the fours' place in the difference.

Let us do a more difficult example:

```
10110
-  111
 1111
```

Discussion: Because 1 is greater than 0, change 1 two into '10' ones. Now, 1 from 10 is 1. Change 1 four into '10' twos; 1 from 10 is 1. Change the 1 sixteen into '10' eights. Take 1 of these eights, which leaves 1 eight, and change it to '10' fours. Now we complete the subtraction: 1 from 10 is 1; 0 from 1 is 1. The difference is 1111.

Here are two more examples. See if your differences agree with these.

```
10011100      10101101
-  110101     - 1111001
  1100111       110100
```

Binary Multiplication

Suppose we wish to multiply 10_{two} and 11_{two}. As in base ten, multiplication in base two is always possible; the product of any two numbers is unique, multiplication is commutative and associative; and 1_{two} names the identity element. Also we assume that the distributive laws hold. That is, multiplication is distributive over addition. Keeping these assumptions in mind, let us develop multiplication in the binary system. We begin by expressing the multiplication facts we know:

$$1 \times 0 = 0 \times 1 = 0$$

and

$$1 \times 1 = 1$$

Our multiplication matrix would then be quite simple.

×	0	1
0	0	0
1	0	1

Now let's multiply 11 × 10:

```
   10
 × 11
   10
  10
  110
```

Discussion: Because 1 × 0 = 0, write 0 in the product. Because 1 × 1 = 1, write 1 to the left of 0 in the first partial product. Now, 1 × 0 = 0, so write 0 under 1, in the second partial product; $1 \times 1 = 1$, so write 1 to the left of the 0 just written. Now add.

Now we are ready to examine a more difficult multiplication example.

$$\begin{array}{r} 1101 \\ \times\ 101 \\ \hline 1101 \\ 0000 \\ 1101 \\ \hline 1000001 \end{array}$$

Discussion: For the first partial product, $1 \times 1 = 1$, $1 \times 0 = 0$, $1 \times 1 = 1$, and $1 \times 1 = 1$. Every entry in the second partial product is zero. The entries in the third partial product are the same as in the first.

Here are two more examples. See if your products agree with these.

$$\begin{array}{r} 10111 \\ \times\ 1101 \\ \hline 10111 \\ 10111 \\ 10111 \\ \hline 100101011 \end{array} \qquad \begin{array}{r} 11100 \\ \times\ 111 \\ \hline 11100 \\ 11100 \\ 11100 \\ \hline 11000100 \end{array}$$

Binary Division

To be able to divide with binary numerals, you must be very familiar with binary multiplication and subtraction. The first few examples that we shall work will be done by repeated subtraction. Then we shall resort to the more conventional algorithm.

Let's divide 11011_{two} by 11_{two}:

$$\begin{array}{r} 1001 \\ 11\overline{)11011} \\ \underline{11} \\ 0011 \\ \underline{11} \\ 0 \end{array}$$

Discussion: The divisor 11 is contained in 11 just 1 time; $1 \times 11 = 11$. Subtract and get 0 for the difference; bring down the 0. The 11 is contained 0 times in 0; write 0 in the quotient and bring down the 1. The 11 is contained 0 times in 1; write 0 in the quotient and bring down the 1. The 11 is contained 1 time in 11. The remainder is 0.

Let's look at another example:

$$\begin{array}{r} 110 \\ 101\overline{)11110} \\ \underline{101} \\ 101 \\ \underline{101} \\ 00 \end{array}$$

Discussion: The divisor 101 is contained in 111 just 1 time; $1 \times 101 = 101$. Subtract and get 10; bring down the 1. The 101 is contained in 101 also 1 time; $1 \times 101 = 101$. Subtract and get 0; bring down the 0. The 101 is contained 0 times in 0.

Here are two more examples. See if your quotients agree with these. (You may check your work by multiplying the divisor by the quotient to get the dividend.)

$$\begin{array}{r} 1110 \\ 10011\overline{)100001010} \\ \underline{10011} \\ 11100 \\ \underline{10011} \\ 10011 \\ \underline{10011} \\ 0 \end{array}$$

$$\begin{array}{r} 1101 \\ 1101\overline{)10101001} \\ \underline{1101} \\ 10000 \\ \underline{1101} \\ 1101 \\ \underline{1101} \\ 0 \end{array}$$

Binary Fractions to Decimals

Converting fractional numbers in binary notation to equivalent decimal numerals is not particularly difficult. The position to the right of the "binary point" has a value of ½; the next position has a value of ¼; the next a value of ⅛; and so on. In each position, the numerator is 1, and the denominator is a power of 2 that depends on the position. For example, the third place to the right of the binary point is ½³, or ⅛. The following example illustrates the conversion process.

$$1.1011_{two} = \underline{?}_{ten}$$

NIM: A game based on binary numerals

An interesting game for two persons, called Nim, is based on binary numeration. In this game the players take turns drawing chips from three stacks before them. A player may draw as many chips as he chooses from any stack in a single move. In his next move he may take chips from the same stack or any other stack as he wishes. The player who takes the last chip left from the three stacks is the loser.

If this game is played by an experienced player and a novice, the beginner will rarely win. The secret of the game, as an experienced player will know, is to select the proper number of chips from the correct stack so that your arrangement will be "binary even." This then forces the other player to be "binary odd."

Here is what is meant by "binary even" and "binary odd." Suppose the stacks of chips, designated as A, B, and C, contain nine, eleven, and fifteen chips respectively. Represent nine, eleven, and fifteen in binary form: nine = 1001_{two}, eleven = 1011_{two}, and fifteen = 1111_{two}.

$$\begin{array}{c|c|c|c|c} A \rightarrow & 1 & 0 & 0 & 1 \\ B \rightarrow & 1 & 0 & 1 & 1 \\ C \rightarrow & 1 & 1 & 1 & 1 \end{array}$$

After this is done, focus your attention on the columns of numbers instead of on the original row representations. If the sum (using decimal system addition) of each column is 0 or 2, then your position is "binary even." If not, your position is "binary odd." When it is your turn to move, hope that the position you are in is "odd," because if it is even, any move you make will force you into an odd position and may cause you to lose—if your opponent knows the game.

Suppose you are faced with the illustrated situation of the nine, eleven, and fifteen chips. Your position is clearly "odd," so you must make a move that will leave your position "even." With a little practice you will see that you should remove thirteen of the chips from stack C. If you do, your position will look like this:

$$\begin{array}{c|c|c|c|c} A \rightarrow & 1 & 0 & 0 & 1 \\ B \rightarrow & 1 & 0 & 1 & 1 \\ C \rightarrow & & & 1 & 0 \end{array}$$

Quite clearly, the resulting sum of every column is either 2 or 0. In other words, you are binary even. Once you reach a safe (even) position, no matter what your opponent does, you will win, provided you make an "even" move every time.

Solution:
$$1.1011 = (1 \times 1) + (1 \times \tfrac{1}{2}) + (0 \times \tfrac{1}{4})$$
$$+ (1 \times \tfrac{1}{8}) + (1 \times \tfrac{1}{16})$$
$$= 1 + \tfrac{1}{2} + 0 + \tfrac{1}{8} + \tfrac{1}{16}$$
$$= 1\tfrac{11}{16}_{ten} \text{ or } 1.6875_{ten}$$

Decimals to Binary Form

To convert a decimal fraction to its binary equivalent, follow these steps: First, write down the fraction. Then, directly below it, write twice the original amount. (In other words, multiply the fraction by 2, using base-ten multiplication.) Then remove the integer to the left of the decimal point and record it to the right (having written the resulting fraction directly below the previous one). For example, if the decimal fraction we wish to convert to binary is 0.721, we write:

```
0.721
1.442        (0.721 × 2)
0.442    1   (the "1" from 1.442)
```

To complete the conversion, we continue by repeating the last two steps (multiplying the most recent amount by 2 and then removing the integer to the left of the decimal point). For 0.721, then, we have (from the beginning):

```
0.721
1.442    1
0.442
0.884    0
0.884
1.768    1
0.768
1.536    1
0.536
1.072    1
0.072
```

The process is repeated until the decimal fraction is reduced to zero, or until the desired number of places in the binary fraction is obtained. The binary equivalent is obtained by reading the recorded digits from top to bottom. Thus, to five places, $0.721_{ten} = 0.10111_{two}$.

Binary Numerals to Octals

A number expressed as a binary numeral is readily expressed as an equivalent in base eight. For a natural number the technique is simple. Since $2^3 = 8$, we begin with the smallest unit and group the "digits" in clusters of three "digits" each.

Example: Convert 10110111_{two} to the equivalent base-eight numeral.

Solution:
$$(10)\ (110)\ (111)_{two} = 267_{eight}$$
$$\uparrow\ \uparrow\ \uparrow$$
$$2_{eight}\ 6_{eight}\ 7_{eight}$$

Example: Convert 0.1101101_{two} to base eight.

Solution:
$$0.(110)\ (110)\ (100)_{two} = 0.664_{eight}$$

Summary

The binary numeral system offers an interesting glimpse into the world of mathematics. Addition and multiplication, at first glance, appear to be very simple operations in the binary system because there are so few facts to learn. A more careful study reveals that numbers of any sizable magnitude require an almost endless sequence of 0's and 1's for their representations. Operations very quickly become bogged down in these long and labored numerals. But computers, because of their tremendous speed, can utilize binary notation to great advantage. Many of the digital computers of earliest design employed decimal notation, but it was soon apparent that binary notation has many advantages over other numeration systems for computer circuit design.

There are many interesting phenomena that lend themselves to binary explanation. If a question can be answered by "yes" or "no"; if a circuit is either "closed" or "open"; if a light bulb is either "on" or "off"; if a choice involves either "male" or "female"—these are the kinds of situations to which mathematicians apply binary notation and find it particularly convenient. The "new mathematics" programs that are currently taught in some elementary and secondary schools treat the subject of binary numeration in some detail.

Irwin K. Feinstein

Modern microcomputers are powerful enough to perform complex data-processing tasks. In a small drycleaning business a computer keeps track of the work being done and assists in preparing the necessary financial records.

© Texas Instruments

DATA PROCESSING

by Elias M. Awad

What are the most profitable brands of soup? Which are money-losers? How does the manager of a supermarket determine the answers to questions such as these?

First he must gather *data*—facts and figures—on soup sales in his store. He must find out how many cans of each brand of soup are sold during a given period of time; how much he pays for a can; how much the customer pays for the same can; and so on. This is called *originating*, or bringing out, data. It must be done carefully. The data should be checked to be certain that it is accurate. Then the manager must put this data into an understandable, finished form, such as a chart or a graph. The data, called the *input*, is now in such shape that it can easily be handled, or *manipulated*, by anyone.

From this input, the manager wishes to draw meaningful information on the profitability of different soup brands. This information is the *output* of his manipulated data input. The process *originating-input-manipulating-output* is the cornerstone of all data processing.

The manager may manipulate the data input in several ways. Paper and pencil may

DATA PROCESSING 453

In a grocery store a scanner reads information encoded on an item. The data is fed into a central computer which tabulates the price and maintains inventory records of all items in stock.

be enough. Or he may need a calculating machine. If the input is very large, he may have a computerized system.

The output from his manipulation of the data may be a written statement describing the profitability of different soup brands. Or the output may be in the form of a mathematical equation that relates profits to costs.

The procedure described above is an example of data processing. The manager (1) *originated* the data, (2) arranged it for *input*, (3) *manipulated* it and (4) produced an *output* of orderly information.

We use data processing all the time in our daily lives without even realizing it. For example, whenever we see, hear, smell, taste, or feel anything, data in the form of nerve impulses races to the brain. The brain manipulates this data to obtain the information we need to think or behave in a certain way. In this case, the information and the reactions it produces may be considered the output of the brain's activity.

In the broadest sense, data is anything that produces information. Information is data that has been refined, summarized, and arranged in a logical fashion.

There are many ways to communicate, or relay, data. Speech and writing are probably the most familiar ways. Shorthand, electrical pulses, and magnetic patterns are some other methods.

DATA PROCESSING BY MACHINE

For many centuries people have used speech, written language, and numbers to process data. Most of this data arises from human activities such as historical events, government and business dealings, social life, and various personal affairs.

Until about a century ago, speech and writing were sufficient to process such data. But faster methods were needed as populations increased and institutions became more complex. The amount of information increased dramatically, and man turned to machines for help.

Like human beings, many machines— typewriters, electronic calculators, cash registers, and certain parts of computers— make use of ordinary letters, words, and numbers. But other devices use a "language" of their own: electric pulses, magnetic patterns on tapes or disks, and so on.

Beginning about 1870, data-processing machines began to be adopted on a wide scale. These early devices were strictly mechanical. They did not use electricity or electronics. Some of them—such as cash registers and adding machines—are still in use today, although they are no longer manufactured in the United States.

These mechanical processors greatly speed up data handling. Operations that used to take weeks and months can be done more accurately in a few hours or days.

But government and business institutions became still larger and more complex. Even more speed in data processing became necessary. To meet the need, electricity was pressed into service. Electricity made typewriters faster. It made calculators both faster and capable of greater precision. New business machines were developed. Calculations and other data operations were cut from days to hours and even minutes. Until about 1940, these so-called *electromechanical processors* served the needs of government and industry well.

But since 1940, the *electronic computer* has revolutionized data processing. The first serious application of computers was their use in solving complex military problems during World War II. Computers were used to make complicated ballistics calculations. Since then, the use of computers has spread into business, industry, government, education, engineering, and other fields.

There is hardly any part of our lives that has not been touched by computers. Computers and computerlike devices often take care of our bills, checks, loans, credit, and other financial transactions. They are used as teaching aids in school. A nation's economic needs and future may be determined by using computers. Such computer applications process and produce mountains of data and information.

The computer has cut the time needed for calculations enormously. Millions of calculations can be finished in one second. Also, data processing that could never before be tried, because it was so terribly difficult, can now be handled quickly on computers.

Unlike mechanical and electromechanical data processors, the heart of a computer has no moving parts. Computer input and output are, however, often handled mechanically or electromechanically. But even these phases have been speeded up recently.

PROCESSING A PAYROLL

Let us say you hold a job in some business or industrial firm. You work a certain number of hours a day, so many days per week and so many weeks in a year. You get a certain number of days off a year: weekends, legal holidays and paid vacations. In addition, you may sometimes be unable to work a full day or may be out sick.

The firm must keep an exact record of the time you work and your time off. It must do this, not only to rate your work performance, but to calculate your pay. You receive a definite wage, based on an hourly rate of compensation. From this, the company must deduct certain employee expenses: taxes, pension, insurance, and social security. At the end of each week or every two weeks or monthly, you get a paycheck.

These *payroll operations* can be complex and difficult, especially in companies that employ hundreds or thousands of people. In the following description of a typical payroll operation, we assume a computer is not being used. That is, the work is done by hand or on ordinary, electromechanical office machines.

As you come in to work, you may sign a time sheet, indicating the time you report. Or you may punch a time clock. The exact time of your entrance is thus recorded. After the day's work, you sign out the time you leave or punch the clock again.

In this way you have *originated* data on daily attendance at your job. By the end of the week or of some other convenient period, your time record, along with that of other employees, is checked for accuracy and edited. It perhaps is entered on some other form of record. The time records, in other words, are worked into an understandable data input.

This input is *manipulated* for output. One of the most important outputs in this case is your paycheck. Manipulation of data input here, as in many other situations, takes place in four basic steps: (1) classifying, (2) sorting, (3) calculating and recording, and (4) summarizing.

Let us look at each of these steps. When input data—the time records of employees—is *classified,* it is broken up into different groups. For example, the time records often are grouped according to the departments in the company. Each employee in a given department—such as accounting, production, or sales—usually has a number.

Note an important fact here. The data containing an employee's name, number, department, wages, and deductions exists before the current data on his working time comes in. This older data, which is gener-

Packs of disks in disk drives are the data storage bank of a large computer. Data is stored on disks when random access to the data is required.

© William Hubbell

ally considered to be permanent, is compared or combined with the new data (working time). The result is the new output (current paycheck). Reworking of older and newer records is one of the most important features of data processing.

After classification the input is *sorted*. Within each group—in this case, a company department—data is put into some kind of order, or sequence. The employees' time records may be sorted in alphabetical order (by name) or in numerical order (by the employees' numbers).

Once the data has been classified and sorted, the payroll clerks *calculate and record* wages, hours worked, time off, and deductions for taxes, pension, insurance, and social security. Another important phase of manipulation is the adding up, or *summarizing,* of all the data being processed. Much of the data connected with a payroll is very important to the managers of a firm. They use the figures to find out the labor costs of the company in a given period. Important decisions about company policies may then be made.

After the origination of data, its input and manipulation, we have an output. In our example, this includes, among other things, your paycheck. The check is usually in two attached parts. One part shows your net pay (pay after all deductions); this is the part you endorse and then cash at the bank or deposit to your own account. The other part of the check gives your gross pay, and may list the deductions—insurance, pension, taxes, and social security—that were taken out, resulting in your net pay. These deductions may vary greatly.

COMPUTER DATA PROCESSING

Basically, computerized data processing is much the same as that done by hand or by electromechanical methods. The main difference is that a computer handles all the work at one time, in one continuous operation at high speed. Certain steps, such as sorting, may be left out entirely. Very little is done by human beings.

Computer data processing takes place in three phases: *input,* the *processing* operation itself, and *output.* Data input is in such a form that the machine can "read" it for processing. Data may be read from cards or paper by optical scanning, or from magnetic tape or disks. Or it may be fed into the computer as typed or printed letters, words, numbers, or other symbols.

The computer then processes the input according to a program that has previously been put into it. In the case of a payroll operation, the computer reads all the necessary data: employee's name, number, department, pay rate, number of hours worked, time off, deductions, and so on. The computer then calculates the employee's gross pay and net pay. These figures are then released from the computer's memory, or data-storage banks, and printed on a check.

In other computers, the output is not in printed form. It might, for example, be recorded on magnetic storage media.

FILING OF DATA

A data file is a group of records. There are files on almost any subject: births, deaths, crime, and scientific research.

A familiar type of file is a collection of written and printed documents. These documents—papers, cards, pamphlets, clippings, and sometimes books—are usually kept in a definite order and stored in cabinets, in drawers, or on shelves.

Other kinds of files may be less familiar. Data may be photographed on microfilm, too small to read with the unaided eye, but taking up very little storage space. Other files may be data residing on magnetic tape or disks.

The collection of computer-stored data, be it a personal phone directory or the FBI's complete list of fingerprints, is called a *data base*. Although most data bases are proprietary, thousands are available to the public for use in large main-frame or smaller personal computers. Professional services have been organized that rent access to data bases such as business reports, journal indices, and now, even a complete reference set, *The Academic American Encyclopedia*.

Whatever its form, a file both stores records and makes them available to whomever needs them. A computer, for example, keeps records in its memory, or data-storage banks. It uses this more permanent data in processing newer data that comes in. This computer data bank is called *internal storage*. The storing of data outside the computer is called *external storage*. External data storage often exists in the form of magnetic storage media.

A file is classified in one of two ways:

1. A *functional file* is set up with a definite purpose in mind; the data is classified according to its function or use.

2. A *physically arranged file* is not set up for a specific function; its records are placed in any convenient order or location.

A functional file may be either a *master file* or a *transaction file*. A master file contains only related permanent records that need to be updated periodically. In contrast, a transaction file holds only recent records, or transactions, which represent the changes to be made in updating a master file. A transaction file is temporary.

A file classified by physical arrangement keeps its records in either of two forms: sequential or random access.

Sequential. In a sequential file, the records are stored in a serial order, or sequence. A machine must scan all previous records before it reaches the one that is desired.

The most important medium for sequential data storage is magnetic tape. Suppose we have 10 records "written" on tape in the proper numerical sequence: 01, 02, 03, 04, 05, 06, 07, 08, 09, and 10. If, for instance, record 06 is needed for processing, the machine must scan records 01 through 05 before it gets to 06.

Sequential-file storage is ideal where regular updating of all records, such as customers' accounts, is needed. Time is not lost by the computer's having to search for individual records on the tape. The accounts have simply been recorded one after another, and the computer runs through and updates them rapidly, at regular periods.

Random access. In random access files, records are usually stored on magnetic disks. The disk is similar to a long-playing

Magnetic tapes are generally used to file data in sequential order. However, it takes longer to locate specific data on tapes than on disks.

© IBM

DATA PROCESSING   457

record, except that each track on a disk is a circle. These are arranged in a pattern of smaller and smaller circles, one inside the other.

The records on a particular track, or section of track, have a specific code number. Given this code number by the person seeking a record, the machine can go directly to that track or section. Instead of going through all the records in the file, it need only "read" through those few having the same code number to find the desired record.

PROCESSING DATA FILES

Because many files are constantly tapped for information, they must be kept up to date. As new transactions take place, the data must be entered in the file. Depending on the file, this is done either by a person or by a machine.

If the records are stored on magnetic tape or disks, the data must be retrieved by computer.

Sequential files are updated by an operation called batch processing. Random-access files are updated by on-line processing.

Batch processing. Records stored in sequence on magnetic tape or punched cards call for periodic processing. Records of transactions are gathered over a period of time. Then a batch of such records are all processed at the same time.

Batch processing is ideal for payrolls, commercial accounts, and other sequential transactions. There is an obvious saving of computer time if transactions can be handled this way.

A master file may also be batch-processed. If the master file is on magnetic tape, it can be put through a computer when it needs updating. Prior to the actual computer run, the batch of new data is sorted and sequenced in the same order as the master file. When the operation begins, the computer reads each piece of data and locates the corresponding record on the master tape. This operation continues until the tape is updated. A new, revised master tape is thus produced and is used for the next updating operation.

There are several advantages to batch processing. Because a large volume of records is processed at one time, there is a saving of computer time and of money. Also, the time and money needed to prepare these records for processing are decreased. There is more effective control over processing errors. Totals of the figures can be taken both before and after each run, thus making sure that no transactions have been left out.

But batch processing also has disadvantages. It is difficult to process data at times other than those scheduled for batch processing. Updating a record right away may not be possible or may be expensive. Another disadvantage is that batch processing places a heavy load on a computer all at one time. Occasionally, serious problems may arise, such as delay or breakdown.

On-line processing. For random-access data files, an entirely different type of processing is used. On-line processing handles transactions as they come, regardless of their order. No sorting or batching is needed before the computer run.

Effective on-line processing requires the use of fast, so-called *direct-access* devices. These are magnetic disks where every data storage location and record can be reached directly.

In the on-line system a master record is updated shortly after a new transaction takes place. This system is ideal when transactions are not sequenced. It is especially useful when the situation demands more frequent runs than those made under batch processing.

Other advantages of on-line operation include faster supply of information to the user, because transactions are processed as they come. Further, a computer is not burdened with an excessive amount of work at one time, so that other data runs are not delayed. The generally faster on-line method makes the whole operation efficient from the standpoint of time saved.

But on-line processing has certain disadvantages. Large and expensive data-storage devices are needed. Processing costs can also be high, since usually a rela-

With a computer network such as the Semi-automated Business Environment system (SABRE), customer inquiries and reservations can be processed in seconds. All terminals are connected to a distant computer complex.

tively few transactions are run through at one time. Also, it is hard to trace the flow of data from input to output through an on-line system. If the system breaks down, serious delays and costly errors result.

PROCESSING OF AIRLINE RESERVATIONS

On-line processing is especially useful in the reservation of airplane passenger tickets. The actual time it takes to get a response to a request is only a matter of seconds.

American Airlines, for example, has SABRE, the Semiautomated Business Research Environment system, which allows you to make travel reservations through a central computer complex almost instantly.

When a prospective passenger calls for one or more reservations on a flight, the ticket agent punches coded information through a terminal. The terminal is connected, along with other terminals, through low-speed lines to a terminal interchange located in the same area.

The interchange holds the agent's message until the central computer complex in the Tulsa, Oklahoma, headquarters is ready to take it. When this happens, the message goes through high-speed lines to the complex. The computer then begins to search the data filed on its magnetic disks and tape for an answer to the ticket request.

When the answer is ready, it is transmitted from the computer complex to the agent who sent the original message. The agent, in turn, informs the prospective passenger whether there are seats on the desired flight or whether the reservations are confirmed. The entire operation from beginning to end takes about 3 to 5 seconds.

The program used in running SABRE consists of over 150,000 instructions. It handles, among others, the following key items of information: the passenger's name; the name of the person making the reservation; the phone number where the passenger may be reached; the name of the person picking up the tickets; special-menu requests; wheelchair requests; special shows booked at the flight destination; car-rental requests.

The complex structure of this system is shown by the fact that the entire inventory of such information can be updated, beginning as much as one year ahead of a current reservation date.

At present, SABRE terminals and other similar systems, such as the Apollo system, are operated by ticket agents all over the world. More than 80 percent of U.S. travel agencies use computerized reservation tools. About 75 percent of all airline tickets are sold through computerized systems.

DESIGN OF DATA PROCESSING

Any organization, no matter what its size, processes data. If the output is to have any meaning, the data must be processed in a systematic way.

A data system has two main jobs: creating data for files and keeping the files up to date. Thus, new output is created from time to time, such as paychecks in a payroll operation.

In order for an organization to get the desired output, a definite routine, or procedure, whether computerized or not, must be set up for the organization. This is the job of systems analysts.

Systems analysts study how equipment such as computers can improve an organization's operation. Systems analysts may have computer programming or data processing backgrounds. Often a college degree is required. Systems analysts study a company's organization and procedures, and they recommend improvements. A systems analyst may recommend computerizing the organization, or, where computers exist, integrating new and old computer systems. The analyst directs those people in charge of developing the needed procedures. The analyst must keep the managers of an organization informed about the system of data processing that best serves their interests and about the state of the system at all times.

Furthermore, the analyst explores new methods of designing more efficient data-processing systems and keeps informed about the abilities and costs of different equipment being sold on the market. The analyst must justify the need for up-to-date reports, cut out any unnecessary procedures, and switch people from dull, routine work to more creative jobs.

To suit the standards and needs discussed above, data-processing equipment must be chosen carefully. The best system is one that produces the desired output in proper form at the rate of speed required, and at the lowest possible costs in time, labor, and money.

Also important are the size of the organization, the volume of data to be processed, the complexity of the operations, and the accuracy needed. All these considerations must be weighed before a particular data-processing system is chosen.

The system chosen may be manual. This may be ideal for small organizations that handle limited amounts of simple data. The appropriate system may include conventional non-computerized equipment, or it may include computerized equipment. A large organization handling tons of data needs a computerized system. The chief operation chosen may be batch processing or on-line processing.

All in all, the combined efforts of man and machine make it possible for organizations to handle data effectively. This kind of system benefits individuals, too. Workers in large companies receive paychecks on time because of complex data-processing machines. Mathematicians and scientists process data on computers, thus completing complex calculations in minutes instead of months. Banks use data-processing systems to determine the interest on savings accounts. In these and many other instances, people who handle figures are relieved of almost endless drudgery.

selected readings

ASTRONOMY AND SPACE SCIENCE

GENERAL WORKS

Doherty, Paul. *Atlas of the Planets.* New York: McGraw-Hill, 1980; 143 pp., illus.—Handsome general guide for observing the planets and their satellites.

Ferris, Timothy. *Coming of Age in the Milky Way.* New York: Morrow, 1988; 495 pp., illus.—Lyrical account of the search for knowledge of the universe; for senior high.

Fritzsch, Harald. *The Creation of Matter: The Universe From Beginning to End.* New York: Basic Books, 1988; 273 pp., illus.—An account of the Big Bang; clearly explains the equivalence of matter and energy; for advanced readers.

Gallant, Roy A. *The Macmillan Book of Astronomy.* New York: Macmillan, 1986; 80 pp., illus.—Lucid and beautifully illustrated volume concentrating on our solar system; for grades 4–8.

Hawking, Stephen W. *A Brief History of Time: From the Big Bang to Black Holes.* New York: Bantam Books, 1988; repr. 1990; 240 pp., illus.—An advanced but highly readable account of the origin, evolution, and fate of our universe.

Kaufmann, William J., III. *Discovering the Universe.* New York: W. H. Freeman, 1987; 381 pp., illus.—Vivid and descriptive introduction to astronomy; for the general reader.

Moore, Patrick, ed. *The International Encyclopedia of Astronomy.* New York: Orion Books, 1987; 448 pp., illus.—In-depth essays and 2,500 A-to-Z articles introducing current topics in astronomy.

Moore, Patrick, and Laian Nicholson, eds. *The Universe.* New York: Macmillan, 1985; 256 pp., illus.—Handsome general guide for more-advanced students.

Sagan, Carl. *Cosmos.* New York: Random House, 1980; 365 pp., illus.—History of the universe, from the Big Bang through the evolution of human culture; a personal view.

Trefil, James. *The Dark Side of the Universe.* New York: Scribners, 1988; 256 pp., illus.—Readable account of recent discoveries by scientists exploring the mysteries of the cosmos; for older readers.

Van der Waerden, Bartel L., et al. *Science Awakening: II: The Birth of Astronomy.* New York: Oxford University Press, 1974; 347 pp., illus.—Why people began to study the stars and how their concepts of the universe developed and changed, through an examination of original sources.

CALENDARS

Krupp, E. C. *Echoes of the Ancient Skies.* New York: Harper & Row, 1983; 386 pp., illus.—A world tour of ancient temples, tombs, and observatories, illustrating how the skies were used by the people to create religion and calendars.

STUDYING THE SKY

Asimov, Isaac. *Eyes on the Universe.* Boston: Houghton Mifflin, 1975; 274 pp., illus.—A popular history of telescopes; for the nonscientist.

Berger, Melvin. *Star Gazing, Comet Tracking, and Sky Mapping.* New York: Putnam, 1985; 80 pp., illus.—Weaves mythology and current astronomy together in a discussion of the art of stargazing without a telescope.

Branley, Franklyn M. *Star Guide.* New York: Crowell, 1987; 64 pp., illus.—Information about constellations and how the night sky changes with the seasons; for grades 4–6.

Couper, Heather, and Nigel Henbest. *Telescopes and Observatories.* New York: Franklin Watts, 1987; 32 pp., illus.—Introduction for younger readers to telescopes and the observatories where they are used.

Donnelly, Marian. *A Short History of Observatories.* Eugene: University of Oregon Books, 1973; 164 pp., illus.—The story of the design and construction of observatories.

Friedman, Herbert. *The Astronomer's Universe.* New York: Norton, 1990; 359 pp., illus.—A scientific overview of modern astronomy by one of its foremost researchers.

Gallant, Roy A. *Rainbows, Mirages, and Sundogs: The Sky as a Source of Wonder.* New York: Macmillan, 1987; 112 pp., illus.—Observes and explains interesting sky phenomena and recommends activities for sky watchers.

Moore, Patrick. *The New Atlas of the Universe.* New York: Crown, 1985; 308 pp., illus.—A comprehensive tour of astronomy and the universe.

Schaaf, Fred. *The Starry Room: Naked-Eye Astronomy in the Intimate Universe.* New York: John Wiley, 1988; illus.—A fascinating guide for amateur astronomers.

Smith, Robert W. *The Space Telescope.* New York: Cambridge University Press, 1990; 478 pp., illus.—Tells the story of the space telescope as an example of how major science projects are organized and funded; for advanced readers.

THE SOLAR SYSTEM

Asimov, Isaac. *The Solar System.* Chicago: Follett, 1975; 32 pp., illus.—The basic facts about the solar system.

———. *How Did We Find Out About Neptune?* New York: Walker, 1990; 64 pp., illus.—An excellent resource for younger readers for discovering how observation and prediction can lead to new scientific knowledge.

Branley, Franklyn M. *Eclipse: Darkness in Daytime.* New York: Crowell, 1973; 33 pp., illus.—A well-illustrated account for the young reader.

Couper, Heather, and Nigel Henbest. *New Worlds: In Search of the Planets*. New York: Addison-Wesley, 1986; 144 pp., illus.—Colorful and up-to-date survey of the planets, the Moon, asteroids and comets, and the birth of the solar system; for junior high on up.

Kelch, Joseph W. *Small Worlds: Exploring the 60 Moons of Our Solar System*. New York: Julian Messner, 1990; 160 pp., illus.—Examines the origin, characteristics, and discovery of the moons in our solar system.

Lauber, Patricia. *Seeing Earth from Space*. New York: Orchard Books, 1990; 80 pp., illus.—Carefully selected and captioned NASA photos show young readers Earth as seen from the Moon.

———. *Voyagers from Space: Meteors and Meteorites*. New York: Crowell, 1989; 80 pp., illus.—A clear description of asteroids, comets, and meteors.

Moore, Patrick. *Guide to Mars*. New York: Norton, 1978; 214 pp., illus.—Carefully written description of the features of Mars and the people involved in studying the planet.

Nourse, Alan E. *The Asteroids*. New York: Franklin Watts, 1975; 59 pp.—A good history of asteroid research. Well illustrated.

Sagan, Carl. *Comet*. New York: Random House, 1985; illus. —An engaging and exhaustive tour of comet history and science.

Short, Nicholas M. *Planetary Geology*. Englewood Cliffs, N.J.: Prentice Hall, 1975; 361 pp., illus.—A well-illustrated description of how geological principles are applied to the study of other members of our solar system.

Tombaugh, Clyde W., and Patrick Moore. *Out of the Darkness: The Planet Pluto*. Harrisburg, Pa.: Stackpole Books, 1980; 160 pp., illus.—Describes astronomer Tombaugh's search for Pluto, and the discoveries of Uranus and Neptune.

Vogt, Gregory. *Halley's Comet*. New York: Franklin Watts, 1987; 96 pp., illus.—What we've learned about Halley's comet; includes its 1986 visit.

Wilford, John Noble. *Mars Beckons*. New York: Knopf, 1990; 244 pp., illus.—Outlines the history of Mars exploration and the prospects for the future; for the more advanced student.

Zim, Herbert S., and Robert Baker. *Stars*. New York: Golden Press, rev. ed., 1985; illus.—Basic information on the stars and how they are studied; for younger readers.

BEYOND THE SOLAR SYSTEM

Apfel, Necia H. *Nebulae: The Birth and Death of Stars*. New York: Lothrop, Lee & Shepard, 1988; 46 pp., illus.—A fascinating guide to the subject for younger readers.

Asimov, Isaac. *To the Ends of the Universe*. New York: Walker, rev. ed., 1976; 142 pp., illus.—Asimov discusses galaxies, nebulas, kinds of stars, and radio sources.

Bok, Bart J., and Priscilla E. Bok. *The Milky Way*. Cambridge, Mass.: Harvard University Press, 5th ed., 1981; illus.—The galaxy is studied from every aspect in this well-illustrated book; for the serious reader.

Davies, Paul. *The Edge of Infinity*. New York: Simon & Schuster, 1983; 194 pp., illus.—Mind-bending ideas about black holes and the origin of the universe.

Gallant, Roy. *Once Around the Galaxy*. New York: Franklin Watts, 1983; 128 pp., illus.—Introduction to theories about the history of our galaxy; for intermediate readers.

Kippenhahn, Rudolf. *100 Billion Suns*. New York: Basic Books, 1983; 264 pp., illus.—Nontechnical presentation of the life history of the stars.

Moore, Patrick, and Ian Nicolson. *Black Holes in Space*. New York: Norton, 1976; 128 pp.—Present theories on how stars may collapse to produce black holes.

Overbye, Dennis. *Lonely Hearts of the Cosmos*. New York: HarperCollins, 1991; 438 pp., illus.—An advanced but fascinating look at the scientific quest for the secrets of the universe.

Simon, Seymour. *Galaxies*. New York: Morrow, 1988; illus.— Clear, beautifully illustrated explanation of the origin and existence of galaxies; grades 3-6.

Sullivan, Walter. *Black Holes: The Edge of Space, the End of Time*. Garden City, N.Y.: Doubleday, 1979; 303 pp., illus.— Treats complicated topics in astrophysics in understandable terms.

Valens, Evan G. *The Attractive Universe: Gravity and the Shape of Space*. Cleveland: World, 1969; 188 pp., illus.— Well-illustrated explanation of gravitation and the way objects move in space.

SPACE EXPLORATION

Armstrong, Neil, et al. *First on the Moon: The Astronauts' Own Story*. Boston: Little, Brown, 1970; 434 pp., illus.—The story of the first manned landing on the Moon, by the astronauts who went there.

Bova, Ben. *Voyager II*. New York: Harper & Row, 1986; 352 pp., illus.—Authoritative account of the space probe.

Branley, Franklyn M. *From Sputnik to Space Shuttles: Into the New Space Age*. New York: Crowell, 1986; 64 pp., illus.— Well-organized introduction to the past, present, and future use of space technology; for grades 4-8.

Burrows, William E. *Exploring Space: Voyages in the Solar System and Beyond*. New York: Random House, 1991; 502 pp., illus.—An enthusiastic history of planetary exploration.

Embury, Barbara. *The Dream Is Alive: A Flight of Discovery Aboard the Space Shuttle*. New York: Somerville House/HarperCollins, 1990; illus.—An account of the 1984 activities of three space-shuttle crews; for younger readers.

Harris, Alan, and Paul Weismann. *The Great Voyager Adventure: A Guided Tour Through the Solar System*. New York: Julian Messner, 1990; 80 pp., illus.—Two insiders discuss the role of the Jet Propulsion Laboratory in assembling and guiding Voyagers 1 and 2 and future missions.

Jastrow, Robert. *Journey to the Stars: Space Exploration—Tomorrow and Beyond*. New York: Bantam Books, 1989— This popular account by a scientist conveys much information about the universe to nonscientists.

Maurer, Richard. *The Nova Space Explorer's Guide: Where to Go and What to See.* New York: Crown, 1985; 128 pp., illus.—Companion to the PBS television series dealing with space travel to the Moon and the near and far planets; for grades 4–8.

———. *Junk in Space.* New York: Simon & Schuster, 1989; 48 pp., illus.—Explains to young readers how technological trash and garbage got left behind during space exploration and the future problems it may cause.

Murray, Bruce. *Journey into Space: The First Thirty Years of Space Exploration.* New York: Norton, 1989; 381 pp., illus.—Engineering adventures by an impassioned space pioneer.

Murray, Charles, and Catherine Bly Cox. *Apollo: The Race to the Moon.* New York: Simon & Schuster, 1989; 506 pp., illus.—An advanced but highly readable account of Project Apollo, NASA's 1960–72 space-exploration program.

Powers, Robert M. *Shuttle: The World's First Spaceship.* Harrisburg, Pa.: Stackpole Books, 1979; 256 pp., illus.—Nontechnical, readable, and well-illustrated description of space shuttles and their uses.

Smith, Howard E. *Daring the Unknown: A History of NASA.* New York: HBJ/Gulliver, 1987; 192 pp., illus.—A history of the American space program from the challenge of Sputnik to the *Challenger* disaster; for grades 5–8.

Stoiko, Michael. *Pioneers of Rocketry.* New York: Hawthorn, 1974; 129 pp., illus.—Review of rocketry from earliest times.

Von Braun, Wernher, et al. *Space Travel: A History.* New York: Harper & Row, rev. ed., 1985; 360 pp., illus.—A classic history for readers from junior high on up.

White, Jack R. *Satellites of Today and Tomorrow.* New York: Dodd, Mead, 1985; 128 pp., illus.—Discusses the history, types, construction, and current and possible future uses of satellites.

LIFE BEYOND EARTH

Aylesworth, Thomas G. *Who's Out There? The Search For Extraterrestrial Life.* New York: McGraw-Hill, 1975; 119 pp., illus.—Compact review of arguments for and against the possibility of interstellar communications.

Knight, David C. *Those Mysterious UFOs: The Story of Unidentified Flying Objects.* New York: Parents Magazine Press, 1975; 64 pp., illus.—For elementary students; a history of UFO reports and investigations.

Krupp, E. C. *Echoes of the Ancient Skies: The Astronomy of Lost Civilizations.* New York: Harper & Row, 1983; 400 pp.—How ancient civilizations viewed the heavens.

Poynter, Margaret, and Michael J. Klein. *Cosmic Quest: Searching for Intelligent Life Among the Stars.* New York: Atheneum, 1984; 160 pp., illus.—For young readers.

Ridpath, Ian. *Messages from the Stars: Communication and Contact with Extraterrestrial Life.* New York: Harper & Row, 1978; 241 pp., illus.—The scientific background of the idea of life in other parts of the universe.

Sagan, Carl. *Cosmic Connection: An Extraterrestrial Perspective.* Garden City, N.Y.: Doubleday, 1980; 288 pp., illus.—The possibilities of life on other planets.

MATHEMATICS

Adler, Irving. *Mathematics.* Garden City, N.Y.: Doubleday, 1990; 48 pp., illus.—Introduces a variety of mathematical concepts in a simple format.

Ardley, Neil. *Making Metric Measurements.* New York: Franklin Watts, 1984; 32 pp., illus.—The metric system explained; for younger readers.

Devi, Shakuntala. *Figuring: The Joy of Numbers.* New York: Harper & Row, 1978; 157 pp.—The author's enthusiasm for numbers for their own sake communicates to the reader in descriptions of math games and shortcuts to memory.

Ecker, Michael W. *Getting Started in Problem Solving and Math Contests.* New York: Franklin Watts, 1987; 128 pp., illus.—Basic and advanced problem-solving strategies, with interesting examples; for grades 6 on up.

Hoffman, Paul. *Archimedes' Revenge: The Joys and Perils of Mathematics.* New York: Fawcett Crest, 1988; repr. 1989;—A lighthearted account of various areas where mathematics intersects with modern life.

Jacobs, Harold R. *Geometry.* San Francisco: W. H. Freeman, 2d ed., 1987; 701 pp., illus.—A well-organized approach for senior high students.

Kyle, James. *Mathematics Unraveled.* Melbourne, Fla.: Robert E. Krieger, 1984; 280 pp.—Commonsense approach.

Loomis, Lynn. *Calculus.* Reading, Mass.: Addison-Wesley, 3d ed., 1982; 1,000 pp., illus.—An intuitive approach for high school students.

Madison, Arnold, and David L. Drotar. *Pocket Calculators.* New York: Thomas Nelson, 1978; 144 pp.—Tips on selecting calculators, their history, and how to use them.

Newman, James R., ed. *The World of Mathematics: A Small Library of the Literature of Mathematics from A'h-mose the Scribe to Albert Einstein.* New York: Tempus, 1956; repr. 1989; 4 vols.—A comprehensive collection of articles about mathematics; for advanced readers.

Paulos, John Allen. *Innumeracy: Mathematical Illiteracy and Its Social Consequences.* New York: Hill & Wang, 1989; 224 pp.—Seeks to explain why so many people are numerically inept, and to inspire a mathematical way of looking at the world that does not depend much on actual mathematics; for upper-level students.

Peterson, Ivars. *The Mathematical Tourist.* New York: W. H. Freeman, 1988; repr. 1989; 240 pp., illus.—Brief pictures of virtually every branch of mathematics, including many late-breaking developments; for senior high students.

Stigler, Stephen M. *The History of Statistics.* Cambridge, Mass.: Harvard University Press, 1987; repr. 1990; 361 pp.—A well-written account of statistical analysis and its role.

Stwertka, Albert. *Recent Revolutions in Mathematics.* New York: Franklin Watts, 1987; 98 pp., illus.—Recent innovations in mathematics; focuses on answers to practical questions.

Wells, David. *Can You Solve These?* Englewood Cliffs, N.J.: Prentice Hall, 1983; 77 pp., illus.—Problems to test your thinking, with hints and answers.